Lecture Notes in Electrical Engineering

Volume 1057

The book series *Lecture Notes in Electrical Engineering* (LNEE) publishes the latest developments in Electrical Engineering—quickly, informally and in high quality. While original research reported in proceedings and monographs has traditionally formed the core of LNEE, we also encourage authors to submit books devoted to supporting student education and professional training in the various fields and applications areas of electrical engineering. The series cover classical and emerging topics concerning:

- Communication Engineering, Information Theory and Networks
- Electronics Engineering and Microelectronics
- Signal, Image and Speech Processing
- Wireless and Mobile Communication
- Circuits and Systems
- Energy Systems, Power Electronics and Electrical Machines
- Electro-optical Engineering
- Instrumentation Engineering
- Avionics Engineering
- Control Systems
- Internet-of-Things and Cybersecurity
- Biomedical Devices, MEMS and NEMS

For general information about this book series, comments or suggestions, please contact leontina.dicecco@springer.com.

To submit a proposal or request further information, please contact the Publishing Editor in your country:

China

Jasmine Dou, Editor (jasmine.dou@springer.com)

India, Japan, Rest of Asia

Swati Meherishi, Editorial Director (Swati.Meherishi@springer.com)

Southeast Asia, Australia, New Zealand

Ramesh Nath Premnath, Editor (ramesh.premnath@springernature.com)

USA, Canada

Michael Luby, Senior Editor (michael.luby@springer.com)

All other Countries

Leontina Di Cecco, Senior Editor (leontina.dicecco@springer.com)

**** This series is indexed by EI Compendex and Scopus databases. ****

Jerzy Ryszard Szymanski ·
Chandan Kumar Chanda ·
Pranab Kumar Mondal ·
Kamrul Alam Khan
Editors

Energy Systems, Drives and Automations

Proceedings of ESDA 2021

Springer

Editors
Jerzy Ryszard Szymanski
Department of Electrical Drives
and Industrial Electronics
Kazimierz Pulaski University
of Technology and Humanities
Radom, Poland

Pranab Kumar Mondal
Department of Mechanical Engineering
Indian Institute of Technology Guwahati
Guwahati, India

Chandan Kumar Chanda
Department of Electrical Engineering
Indian Institute of Engineering Science
and Technology
Howrah, West Bengal, India

Kamrul Alam Khan
Department of Physics
Jagannath University
Dhaka, Bangladesh

ISSN 1876-1100 ISSN 1876-1119 (electronic)
Lecture Notes in Electrical Engineering
ISBN 978-981-99-3690-8 ISBN 978-981-99-3691-5 (eBook)
https://doi.org/10.1007/978-981-99-3691-5

This Springer imprint is published by the registered company Springer Nature Singapore Pte Ltd.
The registered company address is: 152 Beach Road, #21-01/04 Gateway East, Singapore 189721, Singapore

Preface

This book features high quality of research papers presented at the 4th international conference on Energy Systems, Drives and Automations (ESDA2021). The book is organized in three subthemes as energy and drives, electronics and control, and computer and soft computing which includes research work of academicians and industrial experts in the field of electrical and electronics engineering, energy, mechanical, control, automations, IoT, and computers. This proceedings includes full-length papers, research in progress papers, and case studies related to all the areas of above-mentioned topics. Written by respected experts on ESDA, the book offers valuable assets for young researchers.

In this book, about 50 papers are included, and most of the papers are the outcome of study and research works of academic professors, Ph.D. students with their supervisors as co-authors and of scientists. Most of the editors have contributed chapters for this series and have given their valuable suggestions and comments to improve the quality of this book. The editors are thankful to all the authors and specially research scholars and postgraduate students who have burnt their energy to compile this series of book. We thank all the contributors, authors, experts, and reviewers. We also thank Applied Computer Technology of Kolkata as an organizer of the conference ESDA2021 for collecting, gathering, and preprocessing all documents required for publishing this book.

Howrah, India	Dr. Chandan Kumar Chanda
Radom, Poland	Dr. Jerzy Ryszard Szymanski
Dhaka, Bangladesh	Dr. Kamrul Alam Khan
Guwahati, India	Dr. Pranab Kumar Mondal

Contents

About the Editors

Jerzy Ryszard Szymanski is working as a professor in the Faculty of Transport, Electrical Engineering, and Computer Science at the University of Technology and Humanities in Radom, Poland. He graduated with Ph.D. in the Electrical Faculty at the Warsaw University of Technology. His research areas include power electronics converters in drive applications, application of power converters in renewable sources and battery electric vehicles, EMC compatibility in power converters systems, hybrid power systems in electric industry drives, energy storage for BEV, active power filters, etc. He has about 80 scientific publications and 03 patents to his credit. He has also supervised 5 doctoral students.

Chandan Kumar Chanda is working as a professor with the Department of Electrical Engineering, IIEST, Shibpur, India. He has earned a Ph.D. degree from the Department of Electrical Engineering, B.E. College (DU), Shibpur, India. Dr. C. K. Chanda has over 30 years of teaching and research experience in the diverse field of power systems engineering. His areas of interest include smart grid, resiliency, stability, and renewable energy. He is actively involved in various research projects funded by Centrally Funded Organizations like DST and UGC. He has published 135 research articles in journals and conferences of national and international repute.

Pranab Kumar Mondal is an associate professor in the Department of Mechanical Engineering at the Indian Institute of Technology Guwahati since May 2015. He received his undergraduate and postgraduate degrees from Jadavpur University, Kolkata, and completed his Ph.D. from the Indian Institute of Technology Kharagpur in 2015. His principal research interest, encompassing the broad area of microfluidics, covers various facets of micro-scale multiphase transport, electro-kinetics, and micro-scale transport of heat. He is currently working on stability analysis of flows with free surfaces, and capillary filling of bio-fluids. He has co-authored more than 100 refereed journals and conference publications.

Kamrul Alam Khan is a professor in the Department of Physics at Jagannath University, Dhaka. Prof. Khan completed his B.Sc., M.Sc., and Ph.D. from Dhaka University, Bangladesh. Prof. Khan has got over 175 papers in journals and conferences, 07 book chapters, and 04 patents published to his credit. His research interests lie in renewable energy, electricity generation from Pathor Kuchi (Bryophyllum) leaf/ biomass energy, wave and tidal energy, environmental science, and medical physics.

Energy and Drives

Co-optimization of Energy and Reserve Capacities Through Expected Load Not Served

Priyanjali Mukherjee, Sudhangshu Sarkar, Bishaljit Paul, and Chandan Kumar Chanda

Abstract Due to the ever-increasing load, generators are equipped with reserves for sudden demand, and optimization is essential to reduce the operational cost. As demand and supply vary continuously, the system's frequency changes may cause system load fluctuations and sudden power outages for any mismatch between them. This paper proposes a model for optimal dispatch of energy and reserve capacity and reveals DC load flow optimization in a three-bus system. This model has considered a day ahead electricity market which includes varying power demand bid and generates energy along with reserve capacity. The state of the art applied in this paper for minimization of the operational cost for generator production and its reserves are sequential and simultaneous methods. The energy costs at different nodes are examined by the locational marginal prices (LMPs), and how the dispatch of energy and reserve capacities are available for the market participants at the minimum price has been depicted in the paper. The reserve requirement is also calculated through a probabilistic approach and incorporated within the framework of expected load not served.

Keywords Reserve capacity · Energy and reserve · Co-optimization · Expected load not served

P. Mukherjee (✉) · S. Sarkar · B. Paul
Department of Electrical Engineering, Narula Institute of Technology, Kolkata, India
e-mail: priyanjalimukherjee.23@gmail.com

S. Sarkar
e-mail: sudhangshu.sarkar@nit.ac.in

C. K. Chanda
Department of Electrical Engineering, Narula Institute of Technology, IIEST Shibpur, Shibpur, India

J. R. Szymanski et al. (eds.), *Energy Systems, Drives and Automations*, Lecture Notes in Electrical Engineering 1057,
https://doi.org/10.1007/978-981-99-3691-5_1

1 Introduction

In the deregulated market, the concept of nodal pricing is the most important factor for trading electricity. The theory of spot pricing at various nodes was developed at different prices at various nodes or buses were due to congestion in the transmission lines. The losses in the line and congestion problem due to the limitation of thermal restrictions of the line [1] are mainly responsible for an imbalance in energy prices at various nodes. These differentiated spot prices are called locational marginal prices. Ignoring losses and in an unconstrained environment, there is a special market-clearing price [2, 3] in the system. This market-clearing price is equal to the marginal cost of the generator.

It is essential to maintain security in the power system operation. The system must be operated at all times so that the system will not be left in a dangerous condition should any credible initiating event occur. Other outages like equipment failures, transmission lines, etc., are there in the power system. Only a selected outage may lead to severe conditions in power system security [4]. The marginal pricing market approach is an efficient economic market signal for the transmission of power in the grid [5]. Restrictions of generation capacity and transmission line limits prevent supplying the least expensive energy to serve the demand and that the congestion prevails. When a transmission line does not have enough capacity to meet the demand at a certain location is a constraint in it. Locational marginal price [6] any node takes account of the energy price, congestion cost, and charge of losses if accountable. The competitive electricity market is operated with the bids submitted by market participants to the ISO. Long-term and short-term electricity trading are two different types of trading floors available depending upon the contiguity of the trading. Among two different types of trading floors, electricity trading on short-term markets is rapidly growing [7]. Often some of the electricity pools include intermediate market activities between the day ahead and the balancing markets, intended to further hedge against uncertainty and allow corrective actions in response to unexpected events and errors by market agents. These trading arrangements are generally called adjustment or intra-day markets. The day ahead market occurs on the previous day to energy delivery [8]. Before energy delivery, the balancing market takes a few minutes to balance production and consumption for un-deterministic producers, like solar or wind [9]. To minimize energy production and demand imbalances, system operators can implement reserve energy. It is the fixed amount of energy that can be planned and generated prior to demand bids for balancing the dynamic system load [10]. In the electrical power system, some ancillary services are provided by ISO to minimize imbalances between energy production and demand imbalances. Among these, reserves have a great impact on the power system to ensure reliability. Based on the bids within a very short time, it can respond to serve the load. The generator and its reserves play an essential role in the deregulated power market allocation due to line congestion.

2 Case Study

In the present work for a three-bus system, it has been shown how the optimization algorithm captures the strong bond between energy supply and provision of reserve capacity. Both sequential and simultaneous procurement have been performed. Details of the bus system are given in Table 1.

The load angle of the Bus 1, 2, and 3 are $\delta 1$, $\delta 2$, and $\delta 3$, respectively. Single line diagram of the generating unit and load information is shown in Fig. 1.

In this proposed work, the reserve and the generated energy can be co-optimized and it lies within the scope of ELD problem. In this paper, the different energy generation units have been dispatched in both simultaneous and sequential way. The reserve capacity is jointly dispatched with the generated energy. Having secured constrained economic dispatch (SCED) operation of power system, LMP at any bus or node has been calculated. Firstly, we have developed a model for the economic dispatch of four power generation units. Through linear programme, minimum of an optimization problem is specified by

$$\min\ f'^{\mathrm{T}}x \text{ such that } A \cdot x \leq b.$$

Aeq.$x \leq$ beq., where A and Aeq are matrices.

It has been considered that four power producers A, B, C, and D with 100 MW of generating capacity run a plant. Producer A proposed to sell energy at Rs. 10/MWh,

Table 1 Details of bus system

Components	Number(s)
Buses	3
Generators	4
Slack bus	1
Loads	3
Lines	3

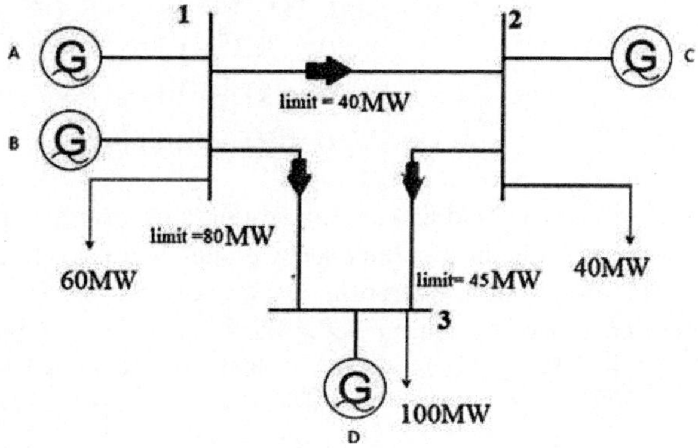

Fig. 1 Single line diagram of generating units and load information

B proposed to sell energy at Rs. 11/MWh, *C* proposed to sell energy at Rs. 13/MWh, and *D* proposed to sell energy at Rs. 12/MWh. The system operator estimates that 55 MW is required for unanticipated events. Producer *A* is willing to provide reserve at Rs. 4/MWh where *B* is offering at Rs. 6/MW, *C* is offering at Rs. 8/MW, and *D* is offering reserve at Rs. 10/MW.

3 Mathematical Formulation

To implement the proposed system, a three-bus system connected to an electricity market has been considered. Different cases have been considered for optimization for energy and reserve capacities. The objective function for cost minimization is developed or each cases in MATLAB environment.

Case IA: Furthermore, we have considered that the energy and reserve capacity are sequentially settled, and ignoring the line constraints, the energy dispatch is determined. The objective function has been formulated as

$$\text{Min } 10\text{PA} + 11\text{PB} + 13\text{PC} + 12\text{PD} \tag{1}$$

s.t: PA + PB + PC + PD = 200.

Capacities for each of the producers have been considered as 0–100 MW.

PA, PB, PC, and PD are the amounts of energy sold by producers *A*, *B*, *C*, and *D*. The clearing price results in Rs. 12/MWh.

After determination of energy dispatched, the reserve capacity market has cleared as per the following:

$$\text{Min: } 4\text{RA} + 6\text{RB} + 8\text{RC} + 10\text{RD} \tag{2}$$

s.t: RA + RB + RC + RD = 55.

Capacities are considered as

$$0 \le \text{RA} \le 100 - P * A$$
$$0 \le \text{RB} \le 100 - P * B$$
$$0 \le \text{RC} \le 100 - P * C$$
$$0 \le \text{RD} \le 100 - P * D.$$

RA, RB, RC, and RD are the amounts of reserve capacity sold by the producers. For reserve scheduling, the energy dispatched is taken as the input.

The total system operation costs, total cost (TCSEQ), include both the procurement cost of energy and reserve capacity which has been calculated from 10PA + 11PB + 13PC + 12PD + 4RA + 6RB + 8RC + 10RD.

The clearing price for the reserve capacity is Rs. 8/MW. Therefore, the profits made by the individual producers under the sequential market have been calculated.

Case IB: Now we have considered that the energy and the reserve capacity are simultaneously settled, and ignoring the line constraints, the energy dispatch is determined as follows:

$$10PA + 11PB + 13PC + 12PD + 4RA + 6RB + 8RC + 10RD \tag{3}$$

s.t: $RA + RB + RC + RD = 55$.

$PA + PB + PC + PD = 200$; capacities are considered as

$$\begin{aligned}
&PA + RA \le 100\\
&PB + RB \le 100\\
&PC + RC \le 100\\
&PD + RD \le 100\\
&PA, PB, PC, PD, RA, RB, RC, RD \ge 0.
\end{aligned}$$

The profits made by the different producers with sequential markets were also calculated.

Case IIA: Now we have considered that the energy and the reserve capacity are sequentially settled, and considering the following line constraints, the energy dispatch is determined as follows (Table 2).

For generation,

$$\text{Min } 10PA + 11PB + 13PC + 12PD \tag{4}$$

s.t: $PA + PB + PC + PD = 200$.

The clearing price results in Rs. 12/MWh by considering bus 1 as the reference bus and limit of all lines.

Capacities for each of the producers have been considered as 0–100 MW.

After determination of energy dispatched, the reserve capacity market has cleared as per the following:

$$\text{Min: } 4RA + 6RB + 8RC + 10RD \tag{5}$$

Table 2 Line parameters information

Line no	From bus no	To bus no	Reactance (℧)	Line limit (MW)
1	1	2	20	40
2	1	3	25	80
3	2	3	30	45

s.t: RA + RB + RC + RD = 55.

Capacities are considered as

$$0 \le \text{RA} \le 100 - P * A$$
$$0 \le \text{RB} \le 100 - P * B$$
$$0 \le \text{RC} \le 100 - P * C$$
$$0 \le \text{RD} \le 100 - P * D.$$

Total cost (TCSEQ) has been calculated as 10PA + 11PB + 13PC + 12PD + 4RA + 6RB + 8RC + 10RD.

Therefore, the profits made by the different producers under the sequential market have been calculated.

Case IIB: Now we have considered that the energy and reserve capacity are settled simultaneously, and considering the same line constraints of CASE II, the energy dispatch is determined as follows:

$$10\text{PA} + 11\text{PB} + 13\text{PC} + 12\text{PD} + 4\text{RA} + 6\text{RB} + 8\text{RC} + 10\text{RD} \tag{6}$$

s.t: RA + RB + RC + RD = 55;

PA + PB + PC + PD = 200.

Capacities are considered as

$$\text{PA} + \text{RA} \le 100$$
$$\text{PB} + \text{RB} \le 100$$
$$\text{PC} + \text{RC} \le 100$$
$$\text{PD} + \text{RD} \le 100$$
$$\text{PA, PB, PC, PD, RA, RB, RC, RD} \ge 0$$
$$-40 \le B12(\delta 1 - \delta 2) \le 40$$
$$-80 \le B13(\delta 1 - \delta 3) \le 80$$
$$-45 \le B23(\delta 2 - \delta 3) \le 45.$$

The simulation has shown that the simultaneous dispatch coupling between energy and reserve has been considered. Also, the profit has been calculated for this case as individually.

A probabilistic approach expected load not served has also been calculated for estimating the reserve capacities. In response to a sudden increase of load, two different balancing actions have been considered.

1. Producers increase production.
2. A part of the load, Lshed, will not be supplied. It will cause a huge economic loss of Rs. 90/MWh.

Based on 5% probability and for both with/without line constraints, it has reformulated as

$$\text{Min: } 10PA + 11PB + 13PC + 12PD + RA + 6RB + 8RC + 10RD \\ + 0.05 * (10 * rA + 11 * rB + 13 * rC + 12 * rD + 90 * Lshed)' \quad (7)$$

$$rA + rB + rC + rD + Lshed = 55.$$

rA, rB, rC and rD are the increase in production of producers.

$$\text{All variable} \geq 0$$
$$PA + RA \leq 100$$
$$PB + RB \leq 100$$
$$PC + RC \leq 100$$
$$PD + RD \leq 100$$
$$Lshed \leq 55$$
$$rA \leq RA$$
$$rB \leq RB$$
$$rC \leq RC$$
$$rD \leq RD$$

4 Result and Discussion

The cost for sequential dispatch of energy and reserve without line constraints is obtained from the simulation as given in Table 3.

Cost for simultaneous dispatch of energy and reserve without line constraints is obtained from the simulation as given in Table 4.

Table 3 Optimization problem solution of sequential dispatch (energy and reserve) without line constraints

Energy		Reserve	
$P * A$ (MW)	100	$R * A$ (MW)	0
$P * B$ (MW)	100	$R * B$ (MW)	0
$P * C$ (MW)	0	$R * C$ (MW)	55
$P * D$ (MW)	0	$R * D$ (MW)	0
Cost (Rs.)	2100	Cost (Rs.)	440
LMP (Rs./MWh)	12	LMP (Rs./MWh)	8
Total cost	Rs. 2540		

Table 4 Optimization problem solution of simultaneous dispatch (energy and reserve) without line constraints

$P*A$ (MW)	45
$P*B$ (MW)	100
$P*C$ (MW)	0
$P*D$ (MW)	55
$R*A$ (MW)	55
$R*B$ (MW)	0
$R*C$ (MW)	0
$R*D$ (MW)	0
Cost (Rs.)	2430
LMP 1 (Rs./MWh)	12
LMP 2 (Rs./MWh)	6

Table 5 Optimization problem solution of sequential dispatch (energy and reserve) with line constraints

Energy		Reserve	
$P*A$ (MW)	100	$R*A$ (MW)	0
$P*B$ (MW)	100	$R*B$ (MW)	0
$P*C$ (MW)	0	$R*C$ (MW)	55
$P*D$ (MW)	0	$R*D$ (MW)	0
Cost (Rs.)	2112	Cost (Rs.)	440
LMP (Rs./MWh)	12	LMP (Rs./MWh)	8
Total cost	Rs. 2552		

The cost for sequential dispatch of energy and reserve with line constraints is obtained from the simulation as given in Table 5.

Cost for simultaneous dispatch of energy and reserve with line constraints is obtained from the simulation as given in Table 6.

Profit for both sequential and simultaneous cases is shown in Fig. 2.

It is observed that under the solution of simultaneous optimization for energy and reserve, the profits for the producer's *A* and *B* are both Rs. 200 and Rs. 100, respectively. Moreover, under the simultaneous dispatch, the load is shredded to 36 MW, taking line constraints into account. Moreover, the energy procured from the reserve of Generator A is 19 MW without line constraints and 17 MW taking line constraints into account. Moreover, it is observed that Gen A's reserve capacity to be considered is 19 MW without line constraints, and 17.06 MW capacity for Gen A is needed taking line constraints into account. The other reserves of generators *B*, *C* and *D* need not be taken for simultaneous dispatch.

Results based on 5% probability and for both with/without line constraints when expected load not served are given in Table 7.

Table 6 Optimization problem solution of simultaneous dispatch (energy and reserve) with line constraints

$P * A$ (MW)	45
$P * B$ (MW)	100
$P * C$ (MW)	0
$P * D$ (MW)	55
$R * A$ (MW)	55
$R * B$ (MW)	0
$R * C$ (MW)	0
$R * D$ (MW)	0
Cost (Rs.)	2430
LMP 1 (Rs./MWh)	12
LMP 2 (Rs./MWh)	6

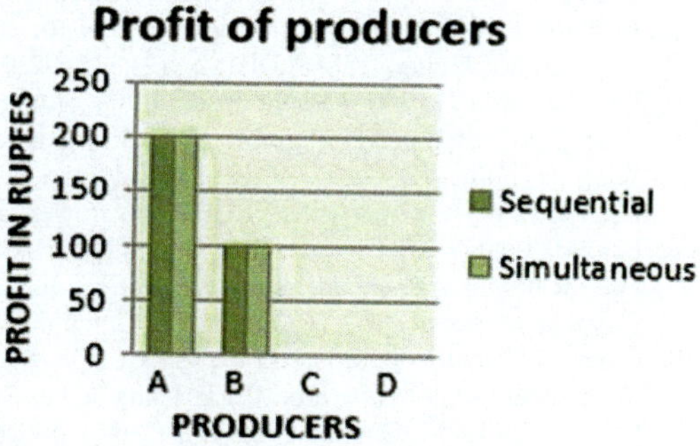

Fig. 2 Profit for different electricity producers

Table 7 Optimization problem solution of simultaneous dispatch (energy and reserve) with/without line constraints during load not served

Variables/ constraints	Without line constraints	With line constraints
Cost (Rs.)	248	225
Lshed (MW)	36	32.93
rA (MW)	19	17
RA (MW)	19	17.06

5 Conclusion

This paper exhibits its novelty in a way so that the reserve commodities are jointly dispatched with the generated energy. It has been shown that the total system operation cost in sequential energy and reserve scheduling is higher than simultaneous scheduling. Hence, while both energy and reserves are jointly dispatched using the optimization problem, it minimizes system operating costs. This paper has also shown that using expected load not served (ELNS), which is a stochastic security metric, the average amount of energy not supplied as a result of load shedding is also determined to meet economic limitations. Moreover, it is expected that due to the penetration of low and high probability of stochastic renewables such as wind and solar, the total operating cost becomes minimum concerning the energy price if renewables are not incorporated.

References

1. Wood AJ, Wollenberg BF (1996) Power generation, operation and control. Wiley, New York
2. Chao H, Peek S (1996) A market mechanism for electric power transmission. J Regul Econ 10(1):25–29
3. Hsu M (1997) An introduction to the pricing of electric power transmission. Util Policy 6(3):257–270
4. Singh MP, Arora P (2015) Comparison of different methods of contingency analysis in power system. Adv Res Electr Electron Eng 2(10):72–75
5. Litvinov E (2010) Design and operation of the locational marginal prices-based electricity markets. IET Gener Transm Distrib 4(2):315–323
6. Li Z, Daneshi H (2005) Some observations on market clearing price and locational marginal pricing. In: IEEE power engineering society general meeting, 12–16 June 2005
7. Wong S, Fuller JD (2007) Pricing energy and reserves using stochastic optimization in an alternative electricity market. IEEE Trans Power Syst 22(2):631–638
8. Paul B, Pathak M, Chanda CK, Pal J (2017) A comparison of locational marginal prices and locational load shedding marginal prices in a deregulated competitive power market. In: Calcutta conference (FALCON). IEEE, Calcutta. ISBN 978-1-5386-3744-9
9. Singh H, Papalexopoulos AD (1999) Competitive procurement of ancillary services by an independent system operator. IEEE Trans Power Syst 14(2):498–504
10. Paul B, Pathak M, Chanda CK, Pal J (2017) On transmission congestion and financial rights. IJESRT. ISSN 2277-9655
11. Papalexopoulos AD, Singh H (2001) On the various design options for ancillary services markets. Presented at the 34th international conference in system services HICSS-34, 3–5 Jan 2001
12. Gedra TW (1999) On transmission congestion and pricing. IEEE Trans Power Syst 14(1)

A Review on Power System Optimization Using PSO

Sukanya Dasgupta, Arup Halder, Prakash Chakraborty, Subhayan Saha, Sayan Banik, Sudip Ghosh, Kamalika Banerjee, Sumanta Kundu, Bishaljit Paul, Ambarnath Banerji, and Sudhangshu Sarkar

Abstract The simplicity of the PSO method inspires researchers for its uplift and modifications. Various hybridizations of PSO methods have been successfully implemented according to the optimization requirements. The development of PSO, its upgrade, modifications, comparisons, and applications are discussed here. This review paper represents a survey of sixty papers on the research progress in Particle Swarm Optimization (PSO) from 2006 to 2021. All the papers have been classified into 24 categories based on numerous features. The paper is in an integrated form that would help to learn about the various ways to implement the optimization technique of PSO all in one go. This paper shows that objectives can be different, but the path will remain the same. It also helps for futuristic works.

Keywords Particle Swarm Optimization · Particles · Stochastic algorithm

1 Introduction

Particle Swarm Optimization (PSO) is a stochastic algorithm to get an optimum solution to modified power system problems. This algorithm works on the principle of how birds or fish school together following a leading individual, just like the particles of a system which tend to follow the best particle of the group. This

S. Dasgupta · A. Halder · P. Chakraborty · S. Saha · S. Banik · S. Ghosh · K. Banerjee · S. Kundu · B. Paul · A. Banerji (✉) · S. Sarkar
Department of Electrical Engineering, Narula Institute of Technology, Kolkata, India
e-mail: ambarnathbanerji@gmail.com

K. Banerjee
e-mail: kamalika.banerjee@nit.ac.in

S. Kundu
e-mail: sumanta.kundu@nit.ac.in

S. Sarkar
e-mail: sudhangshu.sarkar@nit.ac.in

J. R. Szymanski et al. (eds.), *Energy Systems, Drives and Automations*,
Lecture Notes in Electrical Engineering 1057,
https://doi.org/10.1007/978-981-99-3691-5_2

algorithm efficiently simulates high-speed interconnects used in digital applications [1]. The algorithm can be adapted to optimize the location of devices, their types and rating [2], and security monitoring to minimize the time of fault clearing [3]. PSO is more capable of global optimization, easier to solve complex mathematical problems, depends on lesser parameters than other arithmetic algorithms [4], and is considered effective for nonlinear optimization problems [5]. This evolutionary-based technique by Kennedy and Eberhart in 1995 [6] assigns each possible solution a random velocity, and then it flows through problem hyperspace [7]. In a population, the particles' optimal solution is attained over time [8] when the neighboring particles adjust their velocity based on their momentum and track the global best particle [9]. The poor-performing particles are eliminated and scattered to arbitrary positions [10]. The PSO algorithm can also solve discrete nonlinear problems of reactive power optimization [11] to minimize active power loss in transmission networks [12]. In this genetical algorithm on swarm intelligence [13], the particles constantly update their speed to get better locations with each iteration [14]. It calculates the fitness of each particle through associated algorithms [15]. The stochastic algorithm uses a velocity vector to update every individual's current position of its swarm [16], determining the parameter values by experimentation [17]. Each swarm particle is represented as a solution to reactive [18] and functional power problems [19]. The individual particles actively participate in finding both the 'extreme particles,' the first being the 'local best' found by the individual particle themselves and the second being the 'global best' found by the whole group [20], which, when combined, serves as a powerful optimizer [21]. Here, the reserve of the power system is vital to improve transient voltage security using Dynamic Reactive Power [22], and a reactive power regulation strategy is proposed for a distribution system connected with high-penetration photovoltaic (PV) generation [23]. In 1997, another modification of the traditional PSO was given by Kennedy and Eberhart, namely the Binary Particle Swarm Optimization (BPSO), as it optimizes test functions strongly and reaches global optima quicker compared to other algorithms [24]. Meta-heuristic modifications of traditional PSO are made to avoid untimely convergence [25] to optimize overall expenses of energy production in a wind-generated/diesel-operated/PV power system [26]. In the algorithm, better results are acquired as each iteration depends upon the optimal position of each particle [27], impacting a high demand for photovoltaic systems, but being expensive, it needs some modified optimization techniques to be implemented [28, 29]. Expansion of the transmission system is complex because of its mixed-integer, nonlinear, nonconvex nature. To solve such a problem, Particle Swarm Optimization Technique is introduced. Various techniques have been proposed for PSO by eminent researchers [30, 31]. For which eminent researchers have proposed various adapted techniques of PSO. PSO is advantageous because of its rapid convergence, requirements of simple parameters [32], and minimal cost function [33]. When the logical position is initially introduced to the swarm, the objective value function results in better values and processes with each iteration. It has been conventionally used in neural network training, signal processing, etc., because of its fast convergence, the requirement of lesser parameters, simple calculations, and many more advantages [34]. While targeting optimization, another adaptation, the combined process of simulated annealing and PSO (SA-PSO), optimizes the capacity of a wind/PV system [35] as the main technical issue is the capacity and allocation of the system [36]. Here, an adaptive weighted Immune Particle Swarm Optimization Algorithm is introduced to overcome the shortcomings of the classical PSO Algorithm, including easily falling into the local optima [37]. In this article, different techniques and modifications of Particle Swarm Optimization have been reviewed as this technique is the most powerful nonlinear problem-solving algorithm of optimization [38]. A combination of Hybrid-Charged System Search and PSO (HCSS-PSO) is analyzed to minimize the loss rate in transmission lines [39]. The swarm-based optimization process [40] requiring lesser parameters and speedy search processes [41] is refined to work a little better with the help of two mutations—Cauchy mutation (Cauchy MPSO) and Gaussian mutation (Gaussian MPSO) to improve the convergence for achieving better solutions [42]. To find the value and ideal position of SVC for reduction of loss and enhancing voltage profile, PSO is proved to be more suitable than any other conventional algorithm [43]. The search technique based on predatory bird behavior [44] can also solve bad data detection [45], unit commitment issues [46], and other robust power management tasks [47]. This work investigates the reduction of active power losses through optimum dispatch of reactive power and control of voltages using the PSO technique. This work has been implemented as an engine to solve the load flow [48]. The meta-heuristic optimization algorithms, Particle Swarm Optimization, and Differential Evolution are promising candidates to be applied to OPF problems [49] and in power systems using its adapted version, the Binary PSO (BPSO) technique. Apart from bird Swarming, another optimization technique is adapted from the predatory nature of eagles (locating target, swooping down, striking, and hunting) known as the Eagle-Foraging Particle Swarm Optimization (EFPSO), which gives an inexpensive and computationally accurate solution to load flow problems in networks [50]. In Kenya, a 310 MW Wind Power Plant was commissioned, and they are clean energy technologies and relatively cheaper than fossil-fuels [51]. PSO, an arbitrary search technique based on cluster intelligence [52] modified as Binary Social Learning Particle Swarm Optimizer (BSLPSO), is proved to be beneficial for unit commitment (UC) issues in the economic dispatch of power systems [53]. When adapted as MPSO, it results in a better optimization technique to increase output power in PV systems under abnormal conditions [54]. The Multi-objective Particle Swarm Optimization (MOPSO) is proposed to resolve the constrained multi-objective OPF issue in an Electrical Power System with contradictory objectives [55]. PSS in a Multi-Machine environment is a proposed objective function and the time-domain simulation results of the PSO-based PPS under different operating conditions are demonstrated on the 16-machine 68-bus system [56].

Figure 1 shows how fishes or, as in here, particles follow a best-chosen individual to attain their optimum position [57]. Likewise, the particles in a power system follow the best individual while optimizing an intricate power system problem using PSO.

Fig. 1 Fishes schooling around the best individual of the group

2 PSO Overview

Particle Swarm Optimization (PSO) is a stochastic optimization algorithm that iterates a problem by having a population of particles and moving the particles according to the position and velocity of each particle. This algorithm is inspired by how fishes swim in shoals or bees move in swarms in a synced movement and velocity. When it is time to migrate or find food, fishes or birds do not know which path to follow, so they tend to lead the individual closest to the destination. Similarly, in PSO, the particles follow up on one particular particle and optimize its performance. A fish or a bird moves, maintaining its position and velocity regarding time t. And when it needs to change position, the velocity changes accordingly. It changes its direction and velocity from past experiences or from neighbor individuals. This concept is applied to modify an artificially simulated nonlinear problem-solving optimization technique, Particle Swarm Optimization (PSO).

3 PSO Characteristics

Each solution considered as the fish is called a particle. Each particle possesses a fitness value calculated by the objective function. All the particles are aware of their best performance and determine the group's best performance. Thus, they adjust their velocity considering their best performance and the best performance of the best particle.

4 Algorithm of PSO

The formulation of classical PSO is shown through the following equations:

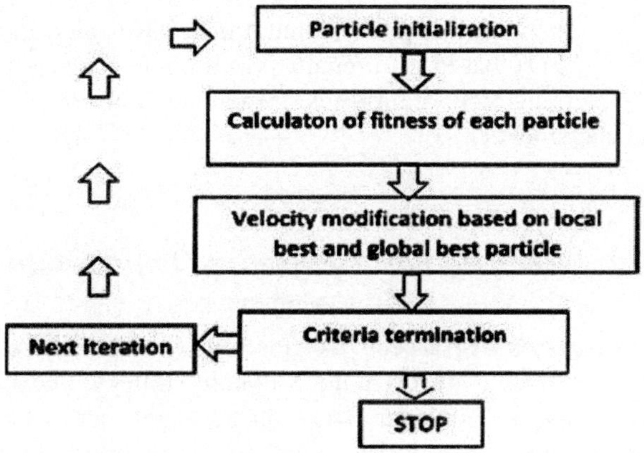

Fig. 2 Algorithm flowchart

Step (1). Updated velocity: The velocity xth particle at iteration $i + 1$.

$$D_x^{i+1} = P_\omega D_x\{K_1 * \text{arb} * (\text{Bp}_x^i - Y_x)\} + \{K_2 * \text{arb} * (\text{Bg}_x^i - Y_x)\}. \quad (1)$$

Inertia Effect in Locally Searched and Globally Searched Systems

D_x Velocity of xth particle.
D_x^{i+1} Velocity of xth particle at $i + 1$th iteration.
Y_x Position of particles.

Step (2). Updated position

K_1 and K_2 learning factors.
Bp_x^i Best personal performance.

$$Y_x^{i+1} = (Y_x^i + D_x^{i+1}) \quad (2)$$

P_ω Inertia weight factor.
Bg_x^i Best performance of the group.
arb Random numbers between 0 and 1.

Here, Fig. 2 represents the flowchart of the algorithm on which PSO works.

5 Evolution of PSO

The name, year of publishing, and methodology used by the researchers that are discussed in this paper are as in Table 1.

From Table 1, it is found that, the heuristic algorithm, Particle Swarm Optimization (PSO), based on predatory bird behavior, has been adapted in various ways by merging other algorithms or methods to use it for better energy production and management.

5.1 Modified Particle Swarm Optimization (MPSO)

It discusses five aspects—chaos-based initialization, sigmoid formulation, maximized focus distance, unique update strategy, and mutated position mechanism, respectively. It comparatively shows better results than PSO in minimizing active and reactive power losses improving the stability of the system.

5.2 Heuristic Particle Swarm Optimization (HPSO)

It utilizes a fly-back mechanism to handle the problem constraints when the particles fly in searching space and improve their performance.

5.3 Parallel Particle Swarm Optimization (PPSO)

PPSO is proposed based on a PC cluster system to overcome the limitations of heuristic algorithms.

5.4 Parallel Immune Particle Swarm Optimization (PIPSO)

It effectively gets over untimely convergence issues of PSO as it uses real and integer number encoding to execute discrete and continuous variable encoding.

5.5 Hierarchical Particle Swarm Optimization (HPSO)

HPSO gives a better solution regarding circuit designing compared to manually designed circuits or the circuits implemented by using traditional PSO or Genetic Algorithm (GA).

Table 1 Modifications of PSO through the years

S. No.	Name of the researcher	Year	Methodology applied
01	Bastola et al. [1]	2007	Particle Swarm Optimization
02	Parastar [2]	2007	Non-traditional optimization technique and Modified Particle Swarm Optimization (MPSO)
03	Hoballah [3]	2008	Particle Swarm Optimization
04	Fu [4]	2008	Heuristic Particle Swarm Optimization
05	Jeong [5]	2009	Parallel Particle Swarm Optimization
06	Al-Hinai [6]	2009	Particle Swarm Optimization
07	Wang [7]	2009	Chaos Particle Swarm Optimization (CPSO)
08	Thakker [8]	2009	Hierarchical Particle Swarm Optimization
09	Chen [9]	2009	Modified Genetic Particle Swarm Optimization
10	Lu [10]	2009	Particle Swarm Optimization
11	Xiao [11]	2010	Hybrid Particle Swarm Optimization algorithm
12	Liu [12]	2010	Particle Swarm Optimization
13	Jaini [13]	2010	Particle Swarm Optimization (PSO)
14	Wang [14]	2011	Improved Particle Swarm Optimization
15	Zhou [15]	2012	Particle Swarm Optimization
16	Mahapatra [16]	2012	Particle Swarm Optimization (PSO)
17	Shuqin [17]	2013	Adaptive Particle Swarm Optimization algorithm
18	Deepa [18]	2013	Particle Swarm Optimization and Modified Particle Swarm Optimization
19	Le Dinh [19]	2013	Self-organizing hierarchical PSO with TVAC
20	Wang [20]	2014	Parallel Immune Particle Swarm Optimization
21	Hopko [21]	2014	Particle Swarm Algorithm
22	Guangming [22]	2015	Particle Swarm Optimization
23	Yang [23]	2015	Mutation Fuzzy Adaptive Particle Swarm Optimization (MF-APSO) algorithm
24	Babu [24]	2015	Binary Particle Swarm Optimization (BPSO)
25	Mandal [25]	2015	Self-Adaptive Particle Swarm Optimization technique
26	Maleki [26]	2015	Improved Particle Swarm Optimization algorithm
27	Baharozu [27]	2015	Improved Particle Swarm Optimization Method (IPSO)
28	Sawant [28]	2016	Particle Swarm Optimization

(continued)

Table 1 (continued)

S. No.	Name of the researcher	Year	Methodology applied
29	Mandal [29]	2016	Hybrid Particle Swarm Optimization technique
30	Warudkar [30]	2017	Particle Swarm Optimization (PSO)
31	Wu [31]	2017	Improved Particle Swarm Optimization (IPSO)
32	Xia [32]	2017	Multi-objective Particle Swarm Optimization algorithm (MOPSO)
33	Zografos [33]	2017	Particle Swarm Optimization
34	Gao [34]	2017	Improved Particle Swarm Optimization (IPSO)
35	Hu [35]	2018	Simulated Annealing-Particle Swarm Optimization
36	Liqun [36]	2018	Particle Swarm Optimization
37	Lu [37]	2018	Adaptive Weighted Immune Particle Swarm Optimization
38	El-Saady [38]	2018	Particle Swarm Optimization
39	Zhang [39]	2018	Hybrid-Charged System Search and Particle Swarm Optimization (HCSS-PSO)
40	Viet [40]	2019	Particle Swarm Optimization (PSO)
41	Li [41]	2019	Adaptive discrete Binary Particle Swarm Optimization algorithm
42	Sarangi [42]	2019	Gaussian and Cauchy mutations modified Particle Swarm Optimization algorithm
43	Iqbal [43]	2019	Particle Swarm Optimization
44	Liu et al. [44]	2019	Improved PSO
45	Kumar [45]	2019	Binary Particle Swarm Optimization (BPSO)
46	Xi et al. [46]	2019	Offset Binary Particle Swarm Optimization algorithm
47	Damisa [47]	2019	Particle Swarm Optimization
48	Cabezas Soldevilla [48]	2019	Particle Swarm Optimization
49	Sennewald [49]	2020	Differential evolution and PSO
50	Ghosh [50]	2020	Eagle-Foraging Particle Swarm Optimization and PSO
51	Chepkania [51]	2020	PSO and N-R algorithm
52	He [52]	2020	Improved Particle Swarm Optimization (IPSO)
53	Zhao [53]	2020	Binary Social Learning Particle Swarm Optimization (BSLPSO)
54	Ibrahim et al. [54]	2020	Modified Particle Swarm Optimization
55	Ahmed [55]	2021	Multi-objective Particle Swarm Algorithm

(continued)

Table 1 (continued)

S. No.	Name of the researcher	Year	Methodology applied
56	Kabat [56]	2021	Binary Particle Swarm Optimization (BPSO)

5.6 *Chaos Particle Swarm Optimization (CPSO)*

CPSO minimizes the risk of voltage collapse and real power losses. This methodology is modified by adding Chaotic Local Search (CLS) to PSO.

5.7 *Modified Genetic Particle Swarm Optimization (MGPSO)*

MGPSO integrates the Genetic Algorithm, differential evolution, and PSO to give better results for economic dispatch problems rather than traditional methods.

5.8 *Adaptive Particle Swarm Optimization Algorithm (APSO)*

It is a self-adaptive strategy with conditions of boundary constraints that finds optimal global solutions to produce better solutions related to discrete reactive power problems.

5.9 *Adaptive Focusing Particle Swarm Optimization (AFPSO)*

This is proposed to optimize reactive power with better simulation output compared to other optimization algorithms.

5.10 *Mutation Fuzzy Adaptive Particle Swarm Optimization (MF-APSO)*

MF-APSO and a proposed strategy of reactive power regulation are integrated and compared with traditional ways for line loss reduction and overvoltage mitigation. As a result, MF-PSO is found to excel in solving the concerned problems regarding line losses in a PV generation system.

5.11 Binary Particle Swarm Optimization (BPSO)

This algorithm optimizes the placement of PMU inside a power system network and aids in identifying multiple interacting unwanted measurements.

5.12 Improved Particle Swarm Optimization (IPSO)

It uses probabilistic mechanisms in place of differential methods to achieve global optima and escape local optima and is easy to implement.

5.13 Multi-objective Particle Swarm Optimization (MOPSO)

It is an economical configuration of flexible power that puts forward network losses and optimizes the problem better than traditional PSO.

5.14 Simulated Annealing-Particle Swarm Optimization (SA-PSO)

SA-PSO considers economy and system reliability as its optimization goal. The optimal result is attained through MATLAB programming, and more effective and reasonable results can be seen compared to PSO.

6 Mathematical Modeling of Different Methods

Classical Particle Swarm Optimization [1]

The traditional PSO technique works according to the following equations

$$D_x^{i+1} = P_\omega D_x\{K_1 * \text{arb} * (\text{Bp}_x^i - Y_x)\} + \{K_2 * \text{arb} * (\text{Bg}_x^i - Y_x)\} \tag{3}$$

Inertia effect of Lacally Searched and Globally Searched systems

D_x^{i+1}	Velocity of particle x at iteration i.
P_ω	Inertia weight factor.
K_1, K_2	Learning factors.
arb	Random numbers between 0 and 1.
Y_x	Position of particles.

Bp_x^i Best position of particle x at iteration i.
Bp_x^i Best position of the group or neighbor particle until iteration.
I No. of iteration.
I_h The number of highest iterations.
I_r The number of present iterations.
$P\omega_{\max}$ The highest weight.
$P\omega_{\min}$ The starting weight.

$$Y_x = (Y_{x1}, \ldots, Y_{xn}), \tag{4}$$

$$D_x = (D_{x1}, \ldots, D_{xn}), \tag{5}$$

$$\mathrm{Bp}_x = \left(Y_{x1}^{\mathrm{Bp}}, \ldots, Y_{xn}^{\mathrm{Bp}}\right), \tag{6}$$

$$\mathrm{Bg}_x = \left(Y_{x1}^{\mathrm{Bg}}, \ldots, Y_{xn}^{\mathrm{Bg}}\right), \tag{7}$$

$$Y_x^{i+1} = \left(Y_x^i + D_x^{i+1}\right). \tag{8}$$

Steps of PSO Algorithm [18]

The 1st step, the 'n' number of particles are initialized arbitrarily.
In the 2nd step, all particle's fitness functions are determined.
In the 3rd step, all Parameters like Bp (Best Position Particle), Bg (Best Position of group Particle), Velocity, and Iteration Count are initialized.
In the 4th step, all iteration counts are updated.
In the 5th step, Bp and Bg are updated.
In the 6th step, the Velocity Components are updated by the main Equation.
In the 7th step, the Position Components are updated by the main Equation.
In the 8th step, Stopping Criteria are Checked in two steps, 8(a) and 8(b).
In the 8th (a) step, If the 100 iterations are completed, the process is stopped.
In the 8th (b) step, If the 100 iterations are not completed, then the update Iteration Count.

Modified Particle Swarm Optimization [18]

After initialization of Bp (best position of a particle) and Bg (Globally the best position) and also after the calculated iteration and velocity from PSO, the optimization technique has to rationalize the population by using MPSO.

The inertia weight factor:

$$P\omega_x^i = P\omega_{\min} + \frac{U_{\mathrm{Bp}}^{i-1} \times \left|U_1^{i-1} - U_{\mathrm{Bp}_x}^{i-1}\right|}{U_1^{i-1} \times \left|U_1^{i-1} - U_{\mathrm{Bg}}^{i-1}\right|} \tag{9}$$

It is to be calculated by using where U is the Scaling Factor.

The factor of modified acceleration:

$$K_{1,x}^i = \sqrt{\frac{U_x^{i-1}}{U_{\mathrm{BP}_x}^{i-1}}}, \tag{10}$$

$$K_{2,x}^i = \sqrt{\frac{U_x^{i-1}}{U_{\mathrm{Bg}}^{i-1}}}. \tag{11}$$

After that, this method will follow the rest of Particle Swarm Optimization technique.

Gaussian Modified Particle Swarm Optimization [42]

The GMPSO describes the output of an arbitrary variable by some arbitrary independent events.

The modified equation of the Gaussian function is defined as follows

$$G(z) = \frac{1}{\sqrt{2 \times \pi \times S_\sigma}} \mathrm{e}^{\frac{(z-m_\mu)^2}{2\times S_\sigma^2}}, \tag{12}$$

where

S_σ Standard Deviation.
m_μ Mean.
z Variables.

Improved Particle Swarm Optimization [31]

It depends on two things, which are Adaptive Inertia Weight and Cross-Variation, and works according to the following algorithm:

Inertia weight can be formulated as

$$D_x^{i+1} = P_\omega D_x + \left\{K_1 * \mathrm{arb}() * \left(\mathrm{Bp}_x^i - Y_x\right)\right\} + \left\{K_2 * \mathrm{arb}() * \left(\mathrm{Bg}_x^i - Y_x\right)\right\}, \tag{13}$$

$$Y_x^{i+1} = \left(Y_x^i + D_x^{i+1}\right), \tag{14}$$

$$L\omega = L\omega_{\max} - \left(\frac{L\omega_{\max} - L\omega_{\min}}{I_{\mathrm{r\,high}}}\right) \times I_{\mathrm{r}}. \tag{15}$$

Here, $L\omega$ is the Adaptive Inertia Weight.

Hybrid Particle Swarm Optimization [29]

Hybrid Particle Swarm Optimization ensures an improved method to determine different objectives. Basically, all of them depend on the classic Particle Swarm Optimization.

Black Hole Particle Swarm Optimization

This method is related to basic PSO and the creation of a black hole theory. This works with determining a current best solution and a black hole that is created near it.

So, the new position of the particle is to be determined by taking a distance between particle best and global best.

$$Y_{xn}^{i+1} = Y_{xn}^{i} + D_{xn}^{I+1} \quad \text{for} \quad M_{xn}^{i} \geq \rho, \tag{16}$$

$$Y_{xn}^{i+1} = \mathrm{Bg}_n + 2L_{xn}^{i}(l_1 - 0.5) \quad \text{for} \quad M_{xn}^{i} < \rho, \tag{17}$$

$$L_{xn}^{i} = \mu\left|\mathrm{Bg}_n - Y_{xn}^{i}\right|. \tag{18}$$

M_{in}^{i} The Probability.
L_{xn}^{i} Radius of the path, which is circular.
ρ Threshold probability.
μ Region which is specified.
l Fitness of particle.

Eagle-Foraging Particle Swarm Optimization [50]

This technique is applied by keeping the predatory nature of eagles (locating a target, swooping down, striking, and hunting) in mind. The formulation of the technique is given as follows:

$$P_q - jQ_q - y_{q1}A_1q - y_{q2}A_2q - \cdots - y_{qn}A_{qn} = 0. \tag{19}$$

$A_q = \left|A_q\right|\angle q$, i.e., voltage in polar form.
$A_q^* = \left|A_q\right|\angle \partial_q$ is its conjugate voltage.

$$z_q(t+1) = z_q(t) + D_q(t+1), \tag{20}$$

$$D_q(t) + D_q(t-1) + a_1r_1\left(\text{localbest}(t) - z_q(t-1)\right) + a_2r_2\left(\text{globalbest}(t) - z_q(t-1)\right), \tag{21}$$

where

P_q Real power produced in the bus 'q.'
Q_q Reactive power produced in the bus 'q.'

$|A_q|$ Voltage magnitude of the bus '*q*.'
y_{qk} Elements of node admittance matrix Y_{Bus}; $k = 1, 2, 3, \ldots, n$.
z_q is the position of the particle '*q*.'

Offset Binary Particle Swarm Optimization Algorithm [46]

The algorithm improving unit commitment issues of wind farms includes the following formulations:

$$\text{Min } F = \sum_{c=1}^{H}\sum_{x=1}^{T_g}\left[\left(b_{cx} f\left(L_{cx}^{g}\right)\right) + E_{cx} b_{cx}\left(1 - b_{x(c-1)}\right)\right]. \tag{22}$$

Subject to,

1. Power balance constraint

$$\sum_{x=1}^{T_g} b_{cx} L_{cx}^{g} + \sum_{x=1}^{T_u} L_{xc}^{u} = L_{\text{S}}^{c}. \tag{23}$$

2. Positive and negative rotations regulate constraints of the system after wind power is connected to the grid

$$\sum_{x=1}^{T_g} b_{cd} L_{x}^{g\,\max} - L_{\text{S}}^{c} \geq B_c + L_{f}^{c} \times f_p\%, \tag{24}$$

$$L_{\text{S}}^{c} - \sum_{x=1}^{T_g} b_{cx} L_{x}^{g\,\min} \geq L_{f}^{c} \times f_t\%. \tag{25}$$

3. Generation limit constant

$$L_{cx}^{g\,\min} \leq L_{cx}^{g} \leq L_{cx}^{g\,\max}. \tag{26}$$

4. Minimum up and downtime constraints. (27)

5. Generator climbing constant

$$\text{GM}_x \times \Delta H \leq L_{x(c+1)}^{g} - L_{cx}^{g} \leq \text{IM}_x \times \Delta H, \tag{28}$$

$$f\left(L_{cx}^{g}\right) = m_x + n_x L_{cx}^{g} + o_x L_{cx}^{g^2}, \tag{29}$$

where

F Total production cost.

$f\left(L_{cx}^{g}\right)$	Fuel cost function of the dth unit with generation output L_{cx}^{g} at cth hour.
MX, n_x, and o_x	The quadratic polynomial coefficients.
T_{g}	Number of generators.
H	Number of hours.
L_{cx}^{g}	Generation output of the dth unit at cth hour.
E_{cx}	Start-up cost of the dth unit.
b_{cx}	ON/OFF status of the dth hour and $b_{cd} = 0$ when OFF, $b_{cd} = 1$ when ON.
$f_{\mathrm{p}}\%$	Maximum rate of change when wind power reduced.
B_c	Spinning reserve at the cth hour.
L_{S}^{c}	Load demand at the cth hour (set to 10% of).
L_{f}^{c}	Wind power forecasting power at the cth hour.
$f_{\mathrm{t}}\%$	Maximum rate of change when wind power increases.
GM_x	The rate of drop in the minute of the output of unit d.
IM_x	The rate of rise in the minute of the output of unit d.
H_x^{on}	Minimum Up-Time of dth unit.
H_x^{off}	Minimum Down-Time of dth unit.
$J_{x,t-1}^{\mathrm{on}}$	Duration during which the dth unit is continuously on.
$J_{x,t-1}^{\mathrm{off}}$	Duration during which the dth unit is continuously off.
$L_{cx}^{g\ \max}$	Maximum generation limit of the dth unit.
$L_{cx}^{g\ \min}$	Minimum generation limit of the dth unit.

Gaussian and Cauchy Mutations in Modified Particle Swarm Optimization Algorithm [45]

Equations of Gaussian mutation and Cauchy mutations are introduced to PSO, and the optimization technique is modified

$$p(i) = p_{\min} + (p_{\max} - p_{\min}) \times \mathrm{e}^{-i\left(\frac{\max_i}{10}\right)}, \tag{30}$$

$p_{\max}$	The maximum weight value.
$p_{\min}$	The minimum weight value.
$\max_i$	The maximum number of iterations being used here.

Parallel PSO Algorithm [5]

In this optimization technique, one of the key factors is the population size that will affect the search performance for seeking the optimal solution. Using the PC cluster system, the Parallel PSO Algorithm is proposed and paralleled. The exchange model of evaluation information is the most important issue in parallelizing the PSO algorithm, which should be solved.

Here, each processor is involved in parallel computing to allocate each Sub-Population. For these neighboring Sub-Populations, each processor can impart as shown in Fig. 3 which represents the configuration of parallel processing.

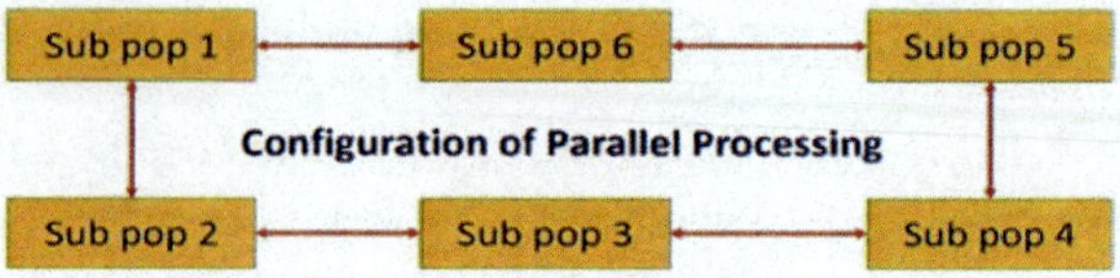

Fig. 3 Configuration of parallel processing

Initialize Population: In the Genetic Algorithm Process, the Initialize Population is the first step. Population P can also be defined as a set of chromosomes in the current generation. The initial population $P(0)$, which is the first generation, is usually unwanted.

Evaluate Fitness Function: It is also known as Evaluate Function. It is evaluated on how close a given solution is to the optimum solution of the desired problem.

Information Exchange: It is also known as Information Sharing. It means that people or other entities pass information from one to another, and it has a long history in Information Technology (IT). In the case of traditional information sharing, committed data is sent one-to-one and alternates between a sender and a receiver.

Update Position of Each Particle: A vector represents a point P on the line in relation to the origin 'O' where it is moving in a straight line. It is often thought of as a function of time, and it writes $X(t)$ for the position of the particle at the time 't.'

Multi-objective Particle Swarm Optimization Algorithm (MOPSO) [35]

This algorithm optimizes network loss problems and equates as follows:

$$D_x(n+1) = P_\omega D_x(n) + j_1 l_1\left[\mathrm{Bp}_x(n) - M_x(n)\right] + 2l_2\left[\mathrm{Bg}_d(n) - M_d(n)\right], \quad (31)$$

$$D_x(n+1) = D_x(n+1) + M_x(n), \quad (32)$$

where

n	The number of iterations.
p	The inertia weight.
j_1 and j_2	Learning factors.
l_1 and l_2	Random number (range of 0–1).
M_x	Position of the particle.

Adaptive Particle Swarm Optimization

Mathematically, the particles are manipulated according to the following equation:

$$D_x^{i+1} = P_\omega D_x + \left\{K_1 * \mathrm{arb}\left(0, \frac{\propto}{2}\right) * \left(\mathrm{Bp}_x^i - Y_x\right)\right\} + \left\{K_2 * \mathrm{arb}\left(0, \frac{\propto}{2}\right) * \left(\mathrm{Bg}^i - Y_x\right)\right\}, \quad (33)$$

$$y_{x,n}(i+1) = y_{x,n}(i) + D_{x,n}(i+1), \quad (34)$$

$$y = \frac{2}{2 - \alpha_x - \sqrt{\alpha_x^2 - 4\alpha_x}} \quad \alpha_x > 4.0. \tag{35}$$

7 Future Direction

The review study in this paper highlights several beneficial factors which are associated with improving the system performance using PSO. The future trend of the optimization process focuses on a more target-specific approach to increasing the efficiency of a system like minimum power loss, zero possibility of voltage collapse, better reactive power balance, and appropriate location detection for implementing a piece of equipment. The future version of PSO will find out optimal solutions for a complex system whose operations are associated with several problem factors.

8 Conclusion

This paper gives a brief idea about the various moderations of PSO that have been proposed by eminent researchers to solve optimization problems. While studying PSO and its modification, this paper will help readers to get an idea about the topic in an integrated form: how the methodologies work, their applications algorithm, and the mathematical modeling of every technique. A subtle comparison of the moderations of PSO and the conventional PSO is also mentioned to understand the advantages of applying the modified algorithms over the traditional one.

References

1. Bastola S, Hsu C (2007) High-speed interconnect simulation using particle swarm optimization. In: 2007 IEEE swarm intelligence symposium, Honolulu, HI, pp 100–104. https://doi.org/10.1109/SIS.2007.368032
2. Parastar A, Pirayesh A, Nikoukar J (2007) Optimal location of FACTS devices in a power system using modified particle swarm optimization. In: 2007 42nd international universities power engineering conference, Brighton, UK, pp 1122–1128. https://doi.org/10.1109/UPEC.2007.4469108
3. Hoballah A, Erlich I (2008) Generation coordination for transient stability enhancement using particle swarm optimization. In: 2008 12th international middle-east power system conference, Aswan, Egypt, pp 29–33. https://doi.org/10.1109/MEPCON.2008.4562360
4. Fu Y, Xu Z-L, Hu R, Cao J-L (2008) A heuristic particle swarm optimization method and its application in power network planning. In: 2008 first international conference on intelligent networks and intelligent systems, Wuhan, China, pp 107–110. https://doi.org/10.1109/ICINIS.2008.155

5. Jeong H, Lee H, Park J (2009) Application of parallel particle swarm optimization on power system state estimation. In: 2009 transmission & distribution conference & exposition: Asia and Pacific, Seoul, Korea (South), pp 1–4. https://doi.org/10.1109/TD-ASIA.2009.5356917
6. Al-Hinai AS, Al-Hinai SM (2009) Dynamic stability enhancement using particle swarm optimization power system stabilizer. In: 2009 2nd international conference on adaptive science & technology (ICAST), Accra, Ghana, pp 117–119. https://doi.org/10.1109/ICASTECH.2009.5409738
7. Wang C, Liu Y, Pan X (2009) A reactive power optimization solution with max power margin for a shipboard power system based on CPSO. In: 2009 2nd international conference on power electronics and intelligent transportation system (PEITS), Shenzhen, pp 162–165. https://doi.org/10.1109/PEITS.2009.5406821
8. Thakker RA, Baghini MS, Patil MB (2009) Low-power low-voltage analog circuit design using hierarchical particle swarm optimization. In: 2009 22nd international conference on VLSI design, New Delhi, India, pp 427–432. https://doi.org/10.1109/VLSI.Design.2009.14
9. Chen P, Zhao C, Li J, Liu Z (2009) Solving the economic dispatch in power system via a modified genetic particle swarm optimization. In: 2009 international joint conference on computational sciences and optimization, Sanya, China, pp 201–204. https://doi.org/10.1109/CSO.2009.475
10. Lu M, Wu H, Li X (2009) Wide-area damping control of power systems using particle swarm optimization. In: 2009 2nd international conference on power electronics and intelligent transportation system (PEITS), Shenzhen, pp 325–328. https://doi.org/10.1109/PEITS.2009.5407004
11. Xiao G, Mei J (2010) Reactive power optimization based on hybrid particle swarm optimization algorithm. In: 2010 Asia-Pacific conference on wearable computing systems, Shenzhen, China, pp 173–177. https://doi.org/10.1109/APWCS.2010.50
12. Liu S, Tang J, Li Q, Wu X, Luo Y (2010) Reactive power optimization in power system based on adaptive focusing particle swarm optimization. In: 2010 international conference on electrical and control engineering, Wuhan, China, pp 4003–4006. https://doi.org/10.1109/iCECE.2010.975
13. Jaini A, Musirin I, Aminudin N, Othman MM, Rahman TKA (2010) Particle swarm optimization (PSO) technique in economic power dispatch problems. In: 2010 4th international power engineering and optimization conference (PEOCO), Shah Alam, Malaysia, pp 308–312. https://doi.org/10.1109/PEOCO.2010.5559236
14. Wang H, Jiang H, Xu K, Li G (2011) Reactive power optimization of power system based on improved particle swarm optimization. In: 2011 4th international conference on electric utility deregulation and restructuring and power technologies (DRPT), Weihai, China, pp 606–609. https://doi.org/10.1109/DRPT.2011.5993964
15. Zhou T, Sun W (2012) Optimization of wind-PV hybrid power system based on an interactive multiobjective optimization algorithm. In: Proceedings of 2012 international conference on measurement, information and control, Harbin, China, pp 853–856. https://doi.org/10.1109/MIC.2012.6273421
16. Mahapatra S, Jha AN (2012) PSS & TCSC coordinated design using particle swarm optimization for power system stability analysis. In: 2012 2nd international conference on power, control and embedded systems, Allahabad, India, pp 1–5. https://doi.org/10.1109/ICPCES.2012.6508094
17. Shuqin S, Bingen Z, Jun W, Nan Y, Qingyun M (2013) Power system reactive power optimization based on adaptive particle swarm optimization algorithm. In: 2013 fourth international conference on digital manufacturing & automation, Shinan, China, pp 935–939. https://doi.org/10.1109/ICDMA.2013.408
18. Deepa SN, Rizwana J (2013) Power system stability by reducing power losses using optimization techniques. In: 2013 IEEE international conference on computational intelligence and computing research, Enathi, India, pp 1–4. https://doi.org/10.1109/ICCIC.2013.6724152
19. Le Dinh K, Hayashi Y (2013) Online optimal power flow based on HPSO-TVAC coordinates with centralized BESS and LRT control to stabilize voltage in a PV-supplied microgrid. In: IEEE PES ISGT Europe 2013, Lyngby, Denmark, pp 1–5. https://doi.org/10.1109/ISGTEurope.2013.6695465

20. Wang T, Yuan G, Zhu L, Yu T (2014) Reactive power optimization of electric power system incorporating wind power based on parallel immune particle swarm optimization. In: The 26th Chinese control and decision conference (2014 CCDC), Changsha, China, pp 1064–1068. https://doi.org/10.1109/CCDC.2014.6852322
21. Hropko D, Hoger M, Roch M, Altus J (2014) Reactive power optimization of generators by using particle swarm algorithm. In: 2014 ELEKTRO, Rajecke Teplice, Slovakia, pp 289–293. https://doi.org/10.1109/ELEKTRO.2014.6848904
22. Guangming Z et al (2015) A research on the control strategy of STATCOM based on particle swarm optimization algorithm. In: 2015 4th international conference on computer science and network technology (ICCSNT), Harbin, pp 745–748. https://doi.org/10.1109/ICCSNT.2015.7490850
23. Yang H, Liao J (2015) MF-APSO-based multiobjective optimization for PV system reactive power regulation. IEEE Trans Sustain Energy 6(4):1346–1355. https://doi.org/10.1109/TSTE.2015.2433957
24. Babu R, Bhattacharyya B (2015) Optimal placement of phasor measurement unit using binary particle swarm optimization in the connected power network. In: 2015 IEEE UP section conference on electrical computer and electronics (UPCON), Allahabad, India, pp 1–5. https://doi.org/10.1109/UPCON.2015.7456695
25. Mandal S, Mandal KK, Tudu B (2015) A new self-adaptive particle swarm optimization technique for the optimal design of a hybrid power system. In: 2015 IEEE power, communication and information technology conference (PCITC), Bhubaneswar, India, pp 280–285. https://doi.org/10.1109/PCITC.2015.7438175
26. Maleki A, Pourfayaz F (2015) Optimization of a grid-independent diesel-based hybrid system for power generation using improved particle swarm optimization algorithm. In: 2015 30th international power system conference (PSC), Tehran, Iran, pp 111–117. https://doi.org/10.1109/IPSC.2015.7827735
27. Baharozu E, Soykan G, Altay O, Kalenderli O (2015) An improved particle swarm optimization method to optimal reactive power flow problems. In: 2015 9th international conference on electrical and electronics engineering (ELECO), Bursa, Turkey, pp 991–995. https://doi.org/10.1109/ELECO.2015.7394600
28. Sawant PT, Bhattar CL (2016) Optimization of PV system using particle swarm algorithm under dynamic weather conditions. In: 2016 IEEE 6th international conference on advanced computing (IACC), Bhimavaram, India, pp 208–213. https://doi.org/10.1109/IACC.2016.47
29. Mandal S, Mandal KK, Tudu B (2016) A new hybrid particle swarm optimization technique for optimal capacitor placement in radial distribution systems. In: 2016 2nd international conference on control, instrumentation, energy & communication (CIEC), Kolkata, India, pp 536–540. https://doi.org/10.1109/CIEC.2016.7513744
30. Warudkar MY, Jape MV (2017) Particle swarm optimization-based transmission expansion planning. In: 2017 international conference of electronics, communication, and aerospace technology (ICECA), Coimbatore, India, pp 341–345. https://doi.org/10.1109/ICECA.2017.8212830
31. Wu X, Xu K, Wang Z, Gong Y (2017) Optimized capacity configuration of an integrated power system of wind, photovoltaic and energy storage device based on improved particle swarm optimizer. In: 2017 IEEE conference on energy internet and energy system integration (EI2), Beijing, China, pp 1–6. https://doi.org/10.1109/EI2.2017.8245465
32. Xia D, Han J, Zhang L, Zhang X, Xu J, Wang Z (2017) Research on flexible power multiobjective optimization plan method oriented to grid-connected intermittent energy. In: 2017 IEEE conference on energy internet and energy system integration (EI2), Beijing, China, pp 1–4. https://doi.org/10.1109/EI2.2017.8245511
33. Zografos D, Ghandhari M, Paridari K (2017) Estimation of power system inertia using particle swarm optimization. In: 2017 19th international conference on intelligent system application to power systems (ISAP), San Antonio, TX, pp 1–6. https://doi.org/10.1109/ISAP.2017.8071383

34. Gao S, Wang H, Wang C, Gu S, Xu H, Ma H (2017) Reactive power optimization of low voltage distribution network based on improved particle swarm optimization. In: 2017 20th international conference on electrical machines and systems (ICEMS), Sydney, NSW, Australia, pp 1–5. https://doi.org/10.1109/ICEMS.2017.8055969
35. Hu LJ, Liu K, Fu Y, Li P (2018) Capacity optimization of wind/PV/storage power system based on simulated annealing-particle swarm optimization. In: 2018 37th Chinese control conference (CCC), Wuhan, China, pp 2222–2227. https://doi.org/10.23919/ChiCC.2018.8482706
36. Liqun S, Pengwei L, Weiwei Z (2018) Capacity optimization of hybrid energy storage in wind/PV complementary power generation system based on improved particle swarm optimization. In: 2018 China international conference on electricity distribution (CICED), Tianjin, China, pp 2295–2300. https://doi.org/10.1109/CICED.2018.8592599
37. Lu M, Du W, Tian R, Li D (2018) Optimization scheduling of power system based on improved particle swarm optimization. In: 2018 international conference on power system technology (POWERCON), Guangzhou, China, pp 945–951. https://doi.org/10.1109/POWERCON.2018.8602204
38. El-Saady G, Ebrahim EA, Abdul-Ghaffar HI, Mohamed YS, El-Sayed A-H (2018) Particle-swarm optimization control of active-power filter for harmonic mitigation of hybrid electric-unbalanced traction-system. In: 2018 IEEE congress on evolutionary computation (CEC), Rio de Janeiro, pp 1–8. https://doi.org/10.1109/CEC.2018.8477653
39. Zhang Y, Chen C, Lee C (2018) Solution of the optimal reactive power dispatch for power systems by using novel charged system search algorithm. In: 2018 7th international symposium on next-generation electronics (ISNE), Taipei, Taiwan, pp 1–4. https://doi.org/10.1109/ISNE.2018.8394748
40. Viet DT, Tuan TQ, Van Phuong V (2019) Optimal placement and sizing of wind farm in Vietnamese power system based on particle swarm optimization. In: 2019 international conference on system science and engineering (ICSSE), Dong Hoi, Vietnam, pp 190–195. https://doi.org/10.1109/ICSSE.2019.8823479
41. Li J, Huang H, Lou B, Peng Y, Huang Q, Xia K (2019) Wind farm reactive power and voltage control strategy based on adaptive discrete binary particle swarm optimization algorithm. In: 2019 IEEE Asia power and energy engineering conference (APEEC), Chengdu, China, pp 99–102. https://doi.org/10.1109/APEEC.2019.8720712
42. Sarangi A, Samal S, Sarangi SK (2019) Analysis of Gaussian & Cauchy mutations in modified particle swarm optimization algorithm. In: 2019 5th international conference on advanced computing & communication systems (ICACCS), Coimbatore, India, pp 463–467. https://doi.org/10.1109/ICACCS.2019.8728452
43. Iqbal MS, Sujatha BC (2019) Multiple contingency analysis for optimal placement and estimate the value of SVC for power loss reduction employing particle swarm optimization. In: 2019 4th international conference on electrical, electronics, communication, computer technologies and optimization techniques (ICEECCOT), Mysuru, India, pp 129–133. https://doi.org/10.1109/ICEECCOT46775.2019.9114569
44. Liu H et al (2019) Reactive power optimization of power grid with photovoltaic generation based on improved particle swarm optimization. In: 2019 IEEE innovative smart grid technologies—Asia (ISGT Asia), Chengdu, China, pp 1536–1540. https://doi.org/10.1109/ISGT-Asia.2019.8880890
45. Kumar A, Gaur S (2019) Bad data processing in electrical power system using binary particle swarm optimization. In: 2019 women institute of technology conference on electrical and computer engineering (WITTON ECE), Dehradun, India, pp 1–5. https://doi.org/10.1109/WITCONECE48374.2019.9092907
46. Xi P et al (2019) Unit commitment problems in power systems with wind farms based on probability offset binary particle swarm optimization algorithm. In: 2019 IEEE international conference on energy internet (ICEI), Nanjing, China, pp 104–109. https://doi.org/10.1109/ICEI.2019.00025
47. Damisa U, Nwulu NI (2019) Particle swarm optimization for robust power management in DC prosumer microgrids with battery. In: 2019 IEEE 6th international conference on engineering

technologies and applied sciences (ICES), Kuala Lumpur, Malaysia, pp 1–6. https://doi.org/10.1109/ICETAS48360.2019.9117413
48. Cabezas Soldevilla FR, Cabezas Huerta FA (2019) Minimization of losses in power systems by reactive power dispatch using particle swarm optimization. In: 2019 54th international universities power engineering conference (UPEC), Bucharest, Romania, pp 1–5. https://doi.org/10.1109/UPEC.2019.8893527
49. Sennewald T, Linke F, Reck J, Westermann D (2020) Parameter optimization of differential evolution and particle swarm optimization in the context of optimal power flow. In: 2020 IEEE PES innovative smart grid technologies Europe (ISGT-Europe), The Hague, Netherlands, pp 1045–1049. https://doi.org/10.1109/ISGT-Europe47291.2020.9248839
50. Ghosh P, Biswas S, Purkait P (2020) Modified particle swarm optimization based technique for power flow computation. In: 2020 IEEE 17th India council international conference (INDICON), New Delhi, India, pp 1–5. https://doi.org/10.1109/INDICON49873.2020.9342566
51. Chepkania TL, Musau MP, Wekesa CW (2020) Particle swarm optimized power grid frequency stability control scheme in the presence of wind energy sources. In: 2020 IEEE PES/IAS PowerAfrica, Nairobi, Kenya, pp 1–5. https://doi.org/10.1109/PowerAfrica49420.2020.9219880
52. He S, Gong T, Chen S, Tang J (2020) Research on optimal scheduling strategy of adjustable load based on particle swarm optimization. In: 2020 management science informatization and economic innovation development conference (MSIEID), Guangzhou, China, pp 77–80. https://doi.org/10.1109/MSIEID52046.2020.00022
53. Zhao S, Yang Z, Zhu X, Wang Y (2020) A novel binary social learning particle swarm optimizer for power system unit commitment. In: 2020 IEEE 4th conference on energy internet and energy system integration (EI2), Wuhan, China, pp 1361–1365. https://doi.org/10.1109/EI250167.2020.9347191
54. Ibrahim AW et al (2020) PV maximum power-point tracking using modified particle swarm optimization under partial shading conditions. Chin J Electr Eng 6(4):106–121. https://doi.org/10.23919/CJEE.2020.000035
55. Ahmed MK, Osman MH, Shehata AA, Korovkin NV (2021) A solution of optimal power flow problem in power system based on multiobjective particle swarm algorithm. In: 2021 IEEE conference of Russian young researchers in electrical and electronic engineering (ElConRus), St. Petersburg, Moscow, Russia, pp 1349–1353. https://doi.org/10.1109/ElConRus51938.2021.9396117
56. Kabat SR, Panigrahi CK, Kumar A (2021) Computationally fast particle swarm optimization power system stabilizer design for interconnected multimachine power system. In: 2021 7th international conference on electrical energy systems (ICEES), Chennai, India, pp 496–501. https://doi.org/10.1109/ICEES51510.2021.9383712
57. Artist. Dale Sanders. School of Jacks Underwater. 700-00041243. https://www.masterfile.com/image/en/700-00041243/school-of-jacks-underwater

A Novel Architecture of Integrator and Differentiator Circuits Using VDTA

M. Anoop, Anantha Ranga Karthik Vedantham, Edulapalli Venkata Somanath Reddy, M. Jesta Kamath, and Vikash Kumar

Abstract In this paper, design of a novel integrator and differentiator circuit built using an active device like voltage difference transconductance amplifier (VDTA) has been discussed. The circuits presented in this paper illustrate the design of a differentiator using two VDTAs and a capacitor and a design of integrator using a VDTA and a capacitor. The circuits presented here work with a power supply of ± 1 V and use current sources with a bias current of 100 µA. The circuits are designed with 180 nm CMOS technology using the Virtuoso tool provided by Cadence. The integrator and differentiator have a power consumption of 400 µW and 800 µW, respectively. The voltage gain of the integrator and differentiator is 4.863 and 5.439, respectively, at 27 °C. The circuit design has been simulated. PVT analysis of the circuits has been performed, and the results are provided.

Keywords Voltage difference transconductance amplifier · Differentiator · Integrator · Active circuit

M. Anoop · A. R. K. Vedantham · E. V. S. Reddy · M. Jesta Kamath · V. Kumar (✉)
Department of Electronics and Communication Engineering, B.M.S College of Engineering, Bangalore 560019, India
e-mail: vikashkumar.ece@bmsce.ac.in

M. Anoop
e-mail: anoop.ec18@bmsce.ac.in

A. R. K. Vedantham
e-mail: vedantham.ec18@bmsce.ac.in

E. V. S. Reddy
e-mail: somanathreddy.ec18@bmsce.ac.in

M. Jesta Kamath
e-mail: mjesta.ec18@bmsce.ac.in

J. R. Szymanski et al. (eds.), *Energy Systems, Drives and Automations*,
Lecture Notes in Electrical Engineering 1057,
https://doi.org/10.1007/978-981-99-3691-5_3

1 Introduction

Over recent years, the conventional voltage-mode operation is being superseded by current-mode and transconductance mode operation due to their low power consumption, less complexity and higher slew rate. This has resulted in many active current-mode elements like current conveyor (CC) and transconductance mode elements like current difference transconductance amplifier (CDTA) and voltage difference transconductance amplifier (VDTA). In this paper, the design of an integrator and differentiator circuit using VDTA has been discussed.

Integrator is a circuit where output signal is the integral of the input signal. These are extensively used in analog to digital convertors, analog computers and wave-shaping circuits [1]. Generally, these circuits employ op-amps, but they do not provide wider dynamic range and larger bandwidth [2]. A few alternate circuits using different transconductance and current-mode elements such are VDTA, CDTA and CC are presented to overcome these drawbacks [3].

Differentiator is a circuit where output signal is the differentiation of the input signal. The operation of differentiator is the reverse of integrator. Differentiator circuits are also employed in wave-shaping circuits. One such application is to detect the high-frequency components in the input signal. Further, differentiators also play a vital role in electronic analogue computers and analogue proportional integrator differentiator (PID) controllers.

2 VDTA

VDTA is a recently introduced active building block. It was introduced by Biolek et al. [4]. It is an active transconductance mode element that provides better gain and less complexity than the operational transconductance amplifier (OTA). VDTA can operate both in current and voltage modes. It can also be used in transconductance mode. There are several applications of VDTA [5–8]. Figure 1 shows the circuit symbol of VDTA. It is a five-terminal device, i.e., two inputs, one intermediate terminal and two outputs. The five terminals in VDTA are labeled as n, p, z, x^- and x^+.

VDTA is composed of a voltage difference unit which is controlled by the current source, followed by dual output transconductance amplifier. VDTA has two

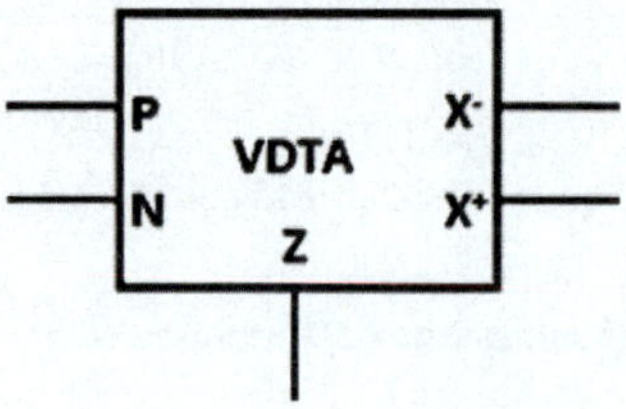

Fig. 1 Circuit symbol of the VDTA active building block

cascaded transconductances. These transconductances are called Arbel–Goldminz transconductance [9]. By varying bias currents, the transconductance values are tunable.

The transistor level depiction of the VDTA circuit is shown in Fig. 2. VDTA has two stages. The first stage has two input terminals which have very high impedance which have been labeled '*n*' and '*p*'. There is an intermediate terminal '*z*' which generates an output current (i_z). This current is proportional to the voltage difference between the input terminals '*n*' and '*p*'. This current causes a drop in voltage across terminal '*z*'. This voltage drop produces output currents at the two output terminals x^- and x^+, respectively. The output currents at x^+ and x^- terminals are identical but reverse in direction. The mathematical expressions which give the relationship between the various terminals are given.

$$\begin{bmatrix} i_p \\ i_n \\ i_z \\ i_{x^+} \\ i_{x^-} \end{bmatrix} = \begin{bmatrix} 0 & 0 & 0 \\ 0 & 0 & 0 \\ g_{mf} & -g_{mf} & 0 \\ 0 & 0 & +g_{ms} \\ 0 & 0 & -g_{ms} \end{bmatrix} \begin{bmatrix} v_p \\ v_n \\ v_z \end{bmatrix} \quad (1)$$

$$g_{mf} = \left(\frac{g_{m1} g_{m2}}{g_{m1} + g_{m2}} + \frac{g_{m3} g_{m4}}{g_{m3} + g_{m4}} \right) \quad (2)$$

$$g_{ms} = \left(\frac{g_{m5} g_{m6}}{g_{m5} + g_{m6}} + \frac{g_{m7} g_{m8}}{g_{m7} + g_{m8}} \right) \quad (3)$$

Here, g_{mk} is the transconductance of kth MOSFET which is given by

$$g_{mk} = \sqrt{I_{bias} \mu_k C_{ox} \left(\frac{W_k}{L_k} \right)}. \quad (4)$$

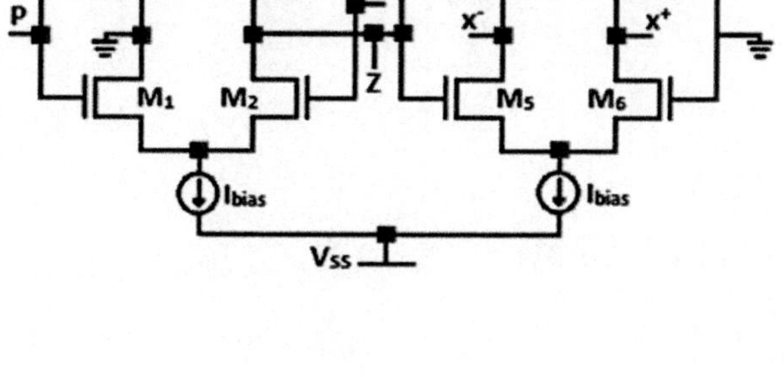

Fig. 2 Transistor-level depiction of VDTA

Here, I_{bias} signifies the bias current of VDTA; μ_{k} is the mobility of the charge carriers, C_{ox} is the oxide capacitance, and $(W_{\text{k}}/L_{\text{k}})$ is the transistor dimensions of kth MOSFET where k varies from 1 to 8.

3 Integrator and Differentiator

The presented integrator circuit is composed of a VDTA and a capacitor (C). The circuit symbol of the presented integrator is shown in Fig. 3. In the presented integrator circuit, v_{in} is the input signal that is applied at the 'p' terminal. Terminals 'x^-' and 'n' are grounded, and the output has been taken from the 'z' terminal across the capacitor (c).

$$v_{\text{z}} = \frac{1}{C}\int i_{\text{z}} \cdot \text{d}t = \frac{1}{C}\int v_{\text{in}} \cdot g_{\text{mf}} \cdot \text{d}t \tag{5}$$

$$v_{\text{out}} = R_1 \cdot g_{\text{ms}} \cdot v_{\text{z}}. \tag{6}$$

From (5) and (6),

$$v_{\text{out}} = R_1 \cdot g_{\text{ms}} \cdot \frac{1}{C} \cdot g_{\text{mf}} \int v_{\text{in}} \text{d}t. \tag{7}$$

Thus, the voltage across capacitor (v_{out}) is the integral of the applied input voltage (v_{in}).

The presented differentiator circuit is composed of a VDTA and a VDTA emulated inductor [10]. The circuit symbol of the presented differentiator circuit is presented in Fig. 4.

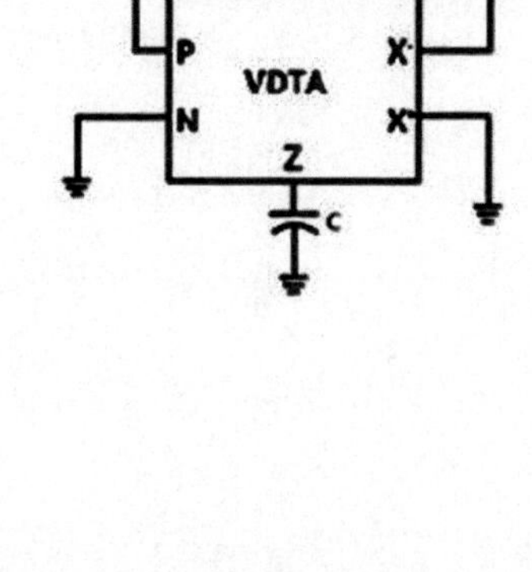

Fig. 3 Circuit symbol of proposed integrator

Fig. 4 Circuit symbol of proposed differentiator

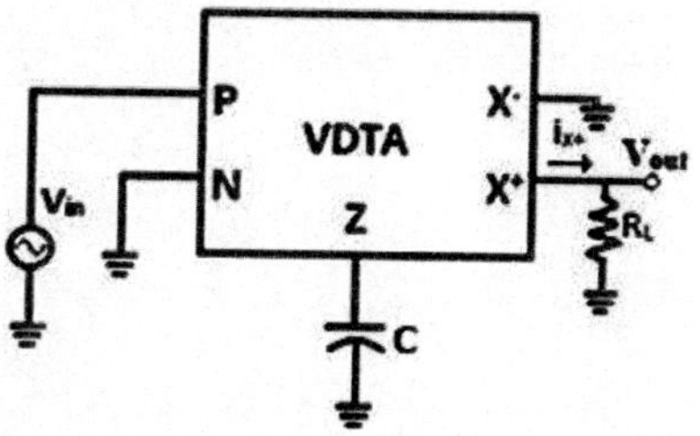

In the presented circuit, the input signal v_{in} has been applied at the '*p*' terminal, the terminals 'x^-'and '*n*' have been grounded, and the output is taken at the '*z*' terminal across the VDTA emulated inductor.

$$v_z = L \cdot \frac{di_z}{dt} = L \cdot \frac{d(v_{in} \cdot g_{mf})}{dt} \tag{8}$$

$$L = \frac{C}{g_{mf} \cdot g_{ms}} \tag{9}$$

$$v_{out} = g_{ms} \cdot v_z \times R_L \tag{10}$$

From (8)–(11),

$$v_{out} = R_L \cdot C \cdot \frac{dv_{in}}{dt}. \tag{11}$$

Thus, the voltage across inductor (v_{out}) is the differential of the input voltage (v_{in}).

4 Simulation Results

To verify the working of the presented circuits, various simulations and their corresponding observations were performed using Cadence ADE. The simulations were performed using 45-nm CMOS technology. The bias current (I_{bias}) was set at 100 μA. The supply voltages, VDD and VSS, were set at ± 1 V. The integrator and differentiator circuit dissipate the power of 600 μW and 800 μW, respectively.

4.1 Transient Response of Integrator

The transient response of presented integrator design with sine wave as input signal is displayed in Fig. 5. It plots output voltage (V_{out}) and the input voltage (V_{in}) with respect to time. The output signal is a cosine wave with 90° lag. Figure 6 shows the

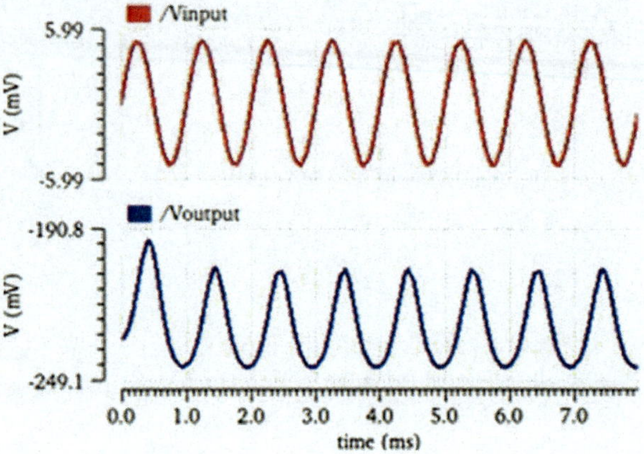

Fig. 5 Transient response of integrator with sine wave input

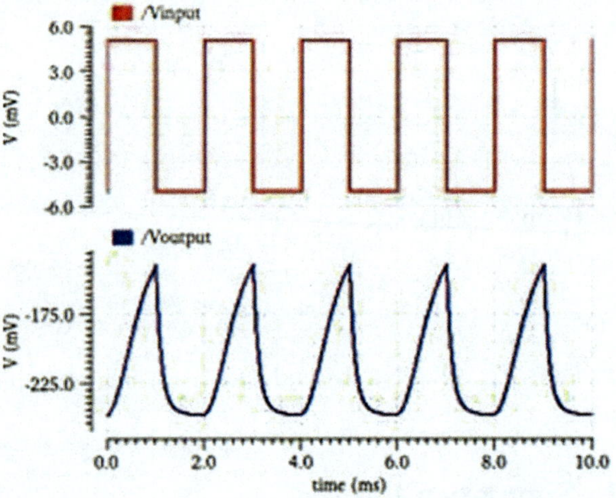

Fig. 6 Integrator transient response with square wave input

transient response when the square wave is applied as the input signal. The output obtained for square wave as input is a saw-tooth wave.

4.2 *Transient Response of Differentiator*

The transient response of presented differentiator design when the applied input signal is a sine wave is displayed in Fig. 7. The output signal is a cosine wave with a lead of 90°. Figure 8 shows the transient response where the input applied is a triangular wave signal. The output is a square wave. Figure 9 shows the transient response when the applied input signal applied is a square wave signal. The output is an impulse waveform.

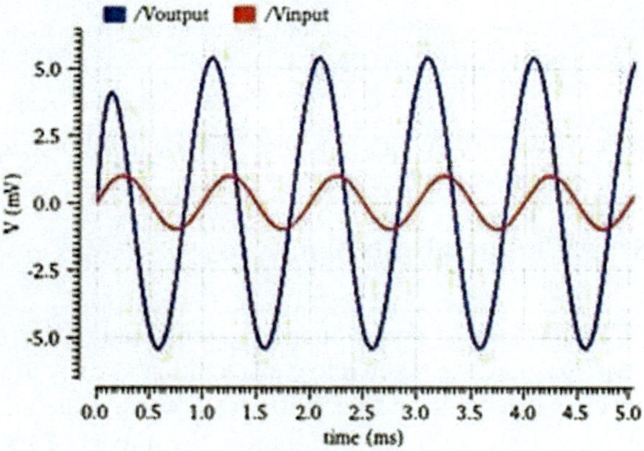

Fig. 7 Transient response of differentiator with sine wave input

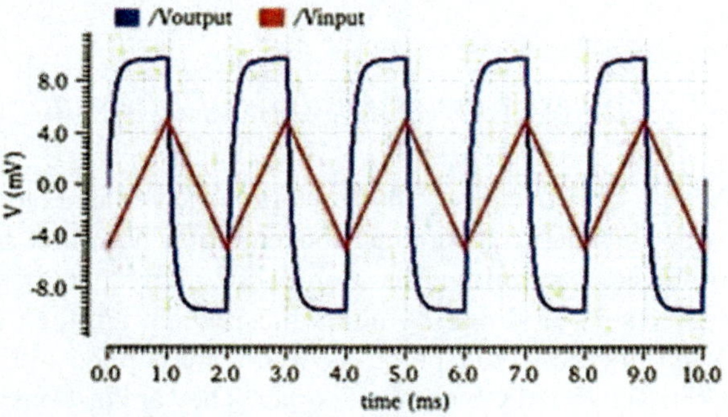

Fig. 8 Transient response of differentiator with triangular wave input

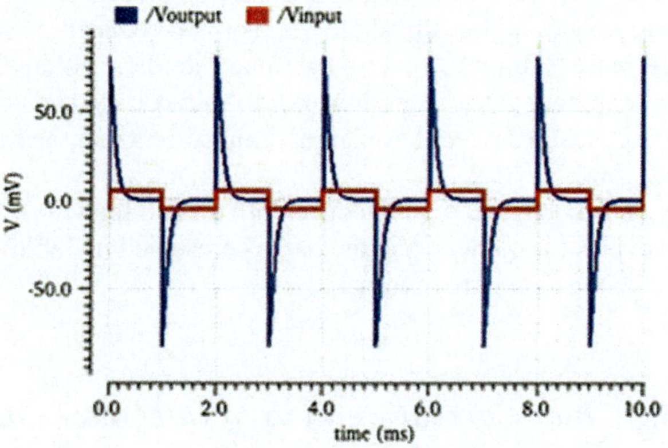

Fig. 9 Transient response of differentiator with square wave input

Table 1 Process corner analysis of integrator and differentiator

Parameter	Integrator voltage gain					Differentiator voltage gain				
	NN	SS	SF	FS	FF	NN	SS	SF	FS	FF
Voltage gain	4.863	4.94	5.133	4.611	4.789	5.439	5.524	5.36	5.435	5.45

5 PVT Variability Analysis

PVT stands for process, voltage and temperature. During CMOS designs, such PVT variations must be taken into consideration as early as possible in the design cycle. If not considered, all such variations can decrease the yield when wafers are fabricated on large scale. In following section, the impact of PVT variations on the proposed integrator and differentiator design is analyzed.

5.1 Process Corner Analysis of Integrator and Differentiator

Process-sensitive design parameters such as channel length L, channel width W, gate oxide thickness t_{ox}, doping concentration N_{CH} and others are major contributors to process variability. Variation of the above parameters at the 45-nm technology node results in different model files, which are then used to estimate the relevant design metrics. Four process corners are defined in CMOS process technology in addition to nominal–nominal corner. They are fast–fast (FF), slow–slow (SS), slow–fast (SF) and fast–slow (FS). These correspond to the combinations of fast or slow NMOS transistors and PMOS transistors, respectively. A slower device has thicker gate oxide, higher threshold voltage and reduced carrier mobility. This leads to a decrease in drain current for the same gate-drive voltage. In contrast, faster transistors have thinner gate oxides, lower threshold voltage and enhanced carrier mobility. As a result, faster devices have larger drain currents for the same gate voltage as compared to slower devices.

The voltage gain of the differentiator and integrator has been estimated for all the five process corners, and the data is presented in Table 1. The variations in the value of voltage gain are minimal.

5.2 Bias Current Analysis of Integrator and Differentiator

The impact of variations in the bias current sources on voltage gain has been studied by varying the bias current by ± 10% about the nominal value. Table 2 shows the results. Bias current analysis has been done for all the five corners. When matched VDTAs are used in differentiator (VDTAs with equal transconductances), the output voltage is independent of the transconductances (Eq. 11). From Eq. 4, it is clear that

Table 2 Bias current analysis of integrator and differentiator

Bias current (μA)	Integrator voltage gain					Differentiator voltage gain				
	NN	SS	SF	FS	FF	NN	SS	SF	FS	FF
90	3.995	4.057	4.223	3.783	3.934	5.344	5.405	5.531	5.343	5.264
94	4.335	4.403	4.58	4.107	4.269	5.394	5.452	5.412	5.358	5.309
98	4.684	4.758	4.946	4.441	4.613	5.418	5.481	5.433	5.425	5.343
100	4.863	4.940	5.133	4.611	4.789	5.439	5.524	5.450	5.435	5.360
102	5.043	5.123	5.322	4.783	4.967	5.457	5.516	5.479	5.443	5.398
106	5.410	5.496	5.705	5.134	5.329	5.487	5.543	5.510	5.483	5.420
110	5.784	5.877	6.097	5.492	5.698	5.526	5.562	5.519	5.525	5.445

transconductance of a VDTA depends on the bias current. Hence, the voltage gain of the differentiator is almost independent of bias current variations.

In case of integrator, the gain is directly proportional to transconductance of the VDTA (Eq. 7), and hence, it increases with increase in bias current.

5.3 *Temperature Analysis of Integrator and Differentiator*

Thermal instability of device threshold (Vt) and carrier mobility (μ) has a substantial impact on nanoscale integrated circuit performance and reliability. Variations of this nature are a result of temperature variations in the surroundings and/or as a result of occurrence of power dissipation in MOS transistors. In a chip, this results in both global and local temperature fluctuations. Temperature variations are induced by changes in the ambient/environmental temperature. The performance of integrated circuits varies since the chip temperature is not constant. The temperature is varied from − 100 to 100 °C, and the changes in the voltage gain are tabulated for all the five process corners (Table 3).

Table 3 Temperature analysis of integrator and differentiator

Temperature (°C)	Integrator voltage gain					Differentiator voltage gain				
	NN	SS	SF	FS	FF	NN	SS	SF	FS	FF
− 100	3.969	4.299	4.381	3.549	3.613	5.811	5.854	5.809	5.817	5.771
− 60	4.402	4.602	4.806	4.019	4.178	5.710	5.74	5.701	5.699	5.653
− 20	4.693	4.834	5.026	4.378	4.550	5.588	5.65	5.604	5.595	5.542
0	4.783	4.896	5.091	4.496	4.672	5.524	5.59	5.544	5.523	5.475
20	4.845	4.931	5.125	4.585	4.763	5.482	5.533	5.462	5.457	5.389
60	4.925	4.968	5.154	4.713	4.888	5.322	5.398	5.345	5.312	5.239
100	4.997	5.006	5.182	4.826	4.992	5.161	5.247	5.168	5.149	5.067

6 Conclusion

This paper presents a novel design of integrator and differentiator using VDTA block. The mathematical analysis showing the transfer function for the presented integrator and differentiator design is carried out. Furthermore, the performance of both the designs is verified by applying various input signals such as sine wave, triangular wave and square wave, and the results are displayed. The process corner analysis, bias current analysis and temperature analysis are done for both the integrator and differentiator circuits.

References

1. Geiger RL, Bailey GR (1982) Integrator design for high frequency active filter applications. IEEE Trans Circuits Syst 29:595–603. https://doi.org/10.1109/TCS.1982.1085200
2. Cohen DD, Zakarevivius RA (1975) Operational amplifier integrators for the measurement of the delay times of microwave transistors. IEEE J Solid-State Circuits 10(1):19–27. https://doi.org/10.1109/JSSC.1975.1050549
3. Santhoshini KM, Musala S, Avireni S (2019) An integrator circuit using voltage difference transconductance amplifier. Solid State Electron Lett 1:10–14. https://doi.org/10.1016/j.ssel.2018.08.001
4. Biolek D, Senani R, Biolkova V, Kolka Z (2008) Active elements for analog signal processing: classification review and new proposals. Radio Eng 17:15–32
5. Gupta S, Arora TS (2021) Design and experimentation of VDTA based oscillators using commercially available integrated circuits. Analog Integr Circ Sig Process 106:713–728. https://doi.org/10.1007/s10470-02001784-w
6. Srivastava J (2020) VDTA based fractional order universal filter. In: 2020 IEEE international conference for innovation in technology (INOCON). https://doi.org/10.1109/INOCON50539.2020.9298221
7. Tangsrirat W, Unhavanich S (2014) Voltage differencing transconductance amplifier-based floating simulators with a single grounded capacitor. Indian J Pure Appl Phys 52:423–428
8. Kumar V, Pal I, Aishwarya N, Nayak A, Islam A (2020) A VDTA-based robust electronically tunable memristor emulator circuit. Analog Integr Circ Sig Process 104:47–59. https://doi.org/10.1007/s10470019-01575-y
9. Arbel AF, Goldminz L (1992) Output stage for current mode feedback amplifiers, theory and applications. Analog Integr Circ Sig Process 2(3):243–255. https://doi.org/10.1007/BF00276637
10. Kumar V, Mehra R, Islam A (2017) A 2.5 GHz low power, high-Q, reliable design of active bandpass filter. IEEE Trans Device Mater Reliab 17(1):229–244. https://doi.org/10.1109/TDMR.2017.2652142

Detection of Voltage Sags and Compensation in Single Phase Power Systems

Srinivas Depally and Penta Samyuktha

Abstract Power quality is the general problem which is occurring regularly in power system network. Voltage drop is a problem which would destroy the quality of power in power system especially in distribution system. The main cause of this voltage drop is control equipment failures and use of nonlinear loads at distribution system load point. This paper represents the various voltage drop estimations. With the help of these results, this paper proposes utilization of single-phase potential sag compensators in three-phase power system networks which is to reduce the overall cost of compensator. This strategy would not eliminate voltage drops completely in the system, but for various loads there is an improvement in reduction of voltage drops. It would also reduce voltage impacts on electrical burdens. With the help of MATLAB/Simulink the proposed method has been simulated.

Keywords System reliability · Series voltage compensator · Voltage drop · Voltage source inverter (VSI)

1 Introduction

Voltage drop is the main problem of power quality, which leads the loss of quality in power and which causes the problems in modern power systems. The main cause of this problem is the control equipment which connected to the system for controlling and protection of power system network like PLC and AC contactors which were very sensitive for load variations. Some voltage drop problems may not be affected by few equipments such as high potential drives. Usually they have the ability to withstand these small variations in voltage. But these voltage sag would create the

S. Depally (✉) · P. Samyuktha
Department of Electrical and Electronics Engineering, KG Reddy College of Engineering and Technology, Hyderabad, Telangana 500075, India
e-mail: srinvasdepally@kgr.ac.in

J. R. Szymanski et al. (eds.), *Energy Systems, Drives and Automations*, Lecture Notes in Electrical Engineering 1057,
https://doi.org/10.1007/978-981-99-3691-5_4

problem for control equipment like PLC, converters and AC contactors which leads to malfunction. Because of this reason which leads to potentially shutting down the entire power system.

One of the best way to improve the power quality by reducing the voltage drop is using compensators for voltage improvement which are connected in series with line. These devices are also called as dynamic voltage restorer (DVR) [4]. DVR is a solid-state technology-based power electronic converter which used as compensator connected between load to be protected and source in series. During the voltage sag, it injects a series voltage in to line which restores the load voltage to its considered values. The most well-known strategy makes use of a series converter, supply voltage and frequency and a voltage-source inverter (VSI) connected by either a capacitor bank or a capacity structure (i.e., batteries, flywheel and so on) [5, 6]. Another method is utilization of series ac–ac converter-based AC link chopper [7]. A pulse wave voltage can be injected to the line with the help of this series converter which is a viable option [8]. This configuration has not included any sinusoidal channel and no pulse width modulation, which reduces power mismatches on VSI.

2 Previous Study

To protect the power system against three-phase voltage drops in general, three-phase compensators were recommended. This paper proposed a strategy of single-phase compensators instead of three-phase compensators to reduce the incident of three-phase voltage sags. Proposed strategy does not nullify the voltage sags which belongs to three-phase completely but converts them in to voltage sags of double phase, which can have high withstand capability for various electrical loads [9, 10].

The most effective method for coordinating voltage sags is to use series voltage compensators, also known as dynamic voltage restorers (DVR). DVR is a solid-state technology-based power electronic converter which used as compensator connected between load to be protected and source in series. It is having the capability of restoring the voltage within the specified voltage levels. The most well-known strategy is makes use of a series converter, supply voltage and frequency and a voltage-source inverter (VSI) connected by either a capacitor bank. Another thought is to use a quick strategy ac–cooling converter in conjunction with an environment control framework chopper. A strategy converter that permeates a square-wave voltage could be a best choice. This method has not included any sinusoidal channel and no pulse width control, which reduces the impact on VSI. Almost half of the shot voltage is limited by the DVR structure. This enables DVRs to provide adequate security against hangs for seasons of up to 0.1 s. Similarly, different voltage wraps a few seconds ago and gets to liberally less stood apart from half again. The excitation control voltage converters are also used to limit the destructive effects of voltage sags, unbalance in system voltage unbalance, and various other voltage drop problems.

3 Proposed System

In this proposed strategy developed based on utilization of 1-Φ voltage drop compensators in 3-Φ system, instead of 3-Φ compensators which pointing to a cost-effective game-plan. The voltage hang repeat would not be so badly harmed by this structure. A square-wave-based strategy compensator is precisely suggested in stage A for reduction in cost further more. This development stage has the most voltage records occasion and, besides the most extraordinary ones concerning extra voltage, showed up in different ways compared with stages B and C. The most sensitive loads should be introduced in this stage, with coordinates secured. This proposal replaces the use of a three-piece strategy voltage compensator with a single-orchestrate one. This proposed strategy does not completely nullify the occurrence of voltage sags but lit will reduce the consequence of occurrence of voltage sags significantly. Single-phase burdens related from stage to neutral: for this type of sensitive power loads, however long this compensator is connected before load, this strategy would completely reduce voltage hangs. Single-stage burdens linked from stage to stage: if voltage drop occurs, the proposed potential compensator will recover individual phase voltage drop, resulting in less true voltage sag.

This will deduct the effect voltage sag of sensitive loads which causes malfunction operation of control and terminal equipment of power system connected. By utilizing single-phase compensators strategy for three-stage loads will recover single-phase voltage sag problems by splitting the three-phase voltage drops in to two segment voltage drops. Regardless of the fact that it does not immediately address the difficult, the better one is less fundamental. Generally, three-phase loads are less vulnerable to 2-Φ voltage records than 3-Φ-phase ones.

4 Voltage Compensation Methods

Some relief ideas are proposed based on the examinations of the acquired estimations. The single-phase series voltage compensators can be protected in various electrical sensitive loads. It should be noted that the electrical burdens to be protected by potential compensators connected in series can be of any type.

4.1 Single-Phase Loads: Phase Substitution

A less cost supply method is to be replaced by the delicate 1-Φ loads introduced at phase A and with loads introduced at phase C. This simple transfer of load can reduce the possibility of voltage drop aggravation by 45% (Fig. 1).

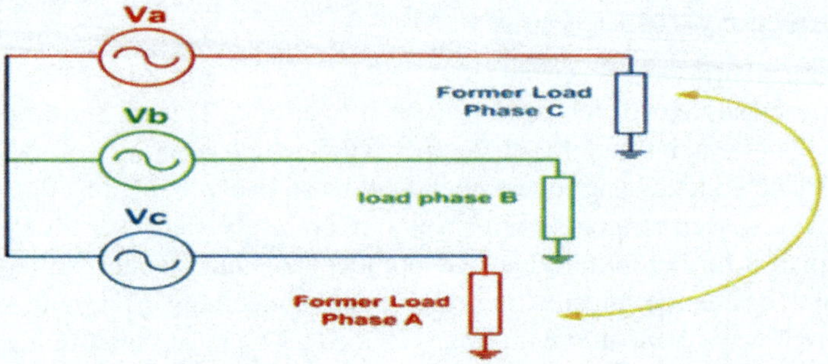

Fig. 1 Scheme A: Substitute phase A sensitive loads for phase C loads

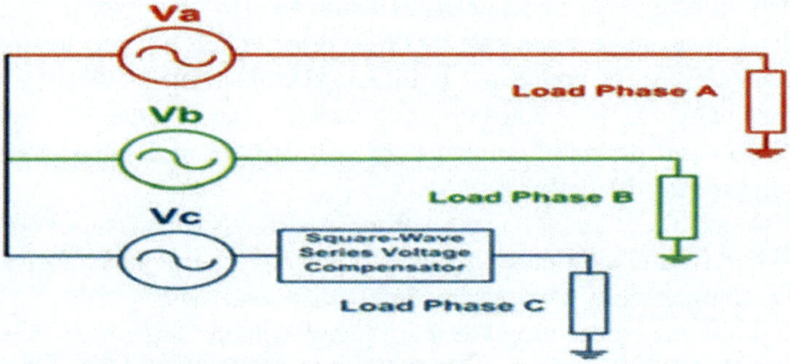

Fig. 2 Scheme B: A square-wave series voltage compensator is used to protect phase C. In this protected phase, delicate loads should be connected

4.2 At Phase C, a 1-Φ Voltage Compensator Connected in Series Is Used

Another method is utilization of AC to AC converter connected in series which is called as AC link chopper [7]. A pulse wave voltage can be injected to the line with the help of this series converter which is a viable option [8]. This configuration has not included any sinusoidal channel and no control strategy for modulation, which reduces power mismatches on VSI (Fig. 2).

4.3 Three-Phase Compensation: On Phase A, a Single-Phase Series Voltage Compensator Is Used

1-Φ loads with nonpartisan and Φ-to-Φ associations, as well as three-phase loads, would frequently be found in a real establishment, and three-phase loads would frequently be found in a modern framework. In order to have a knowledgeable

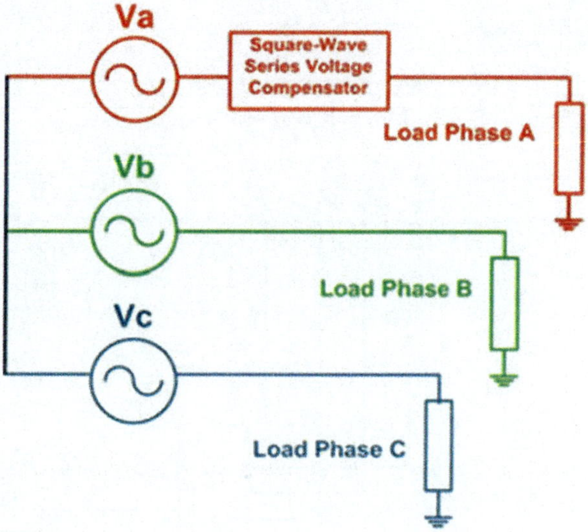

Fig. 3 Scheme C: Utilization of a voltage compensator for phase A used as square-wave potential compensator connected in series. In this protected phase, delicate loads should be connected

response for voltage lists, it is proposed to introduce a square-wave series voltage compensator only at phase A. When compared with phases B and C, this phase has the most voltage lists events as well as the most serious ones in terms of lingering voltage. After phase A has been secured, the most delicate burdens should be introduced in this phase (Fig. 3).

5 Simulation Results

Figure 4 shows the Simulink model of proposed strategy and introduced the series compensator of three-phase compensators using MATLAB/Simulink environment. The voltage sag condition has been created by connecting nonlinear load to the system (Fig. 5). The output using normal three-phase compensators has been shown in Fig. 6. The load voltage shows that there is a voltage sag in the time period of 05–07 s because of nonlinear load connected in the system without three-phase compensator.

Voltage sag which generated has been compensated using three-phase compensators has been shown. Figure 6c shows the injected voltage in the system to compensate load voltage. Figure 6 shows the introduced single-phase series compensator instead of three-phase compensators using MATLAB/Simulink environment. The voltage sag condition has been created by connecting nonlinear load to the system at single phase. The single-phase compensators have been shown in Fig. 6. The load voltage shows that there is a voltage sag in the time period of 05–07 s because of nonlinear load connected in the system without three-phase compensator. Voltage sag which generated has been compensated using single-phase compensator has been

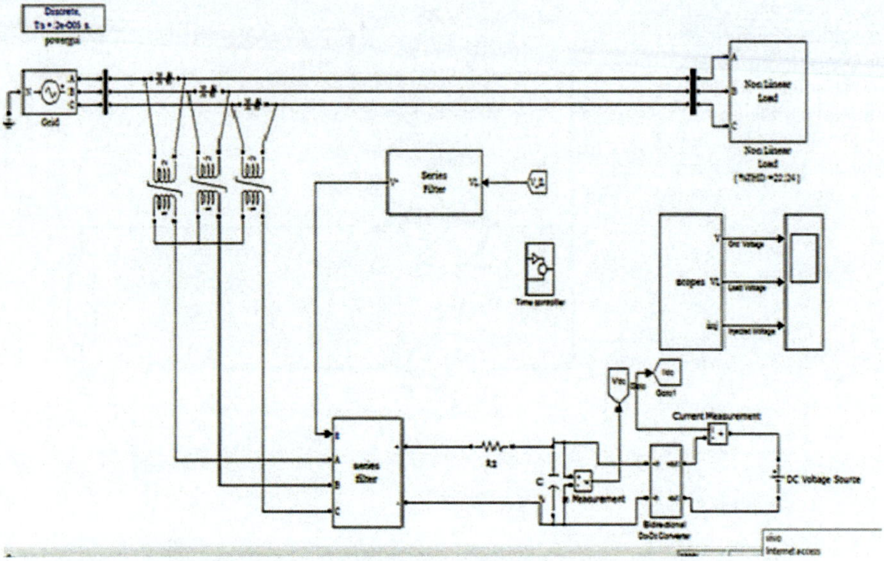

Fig. 4 Simulink model of proposed system

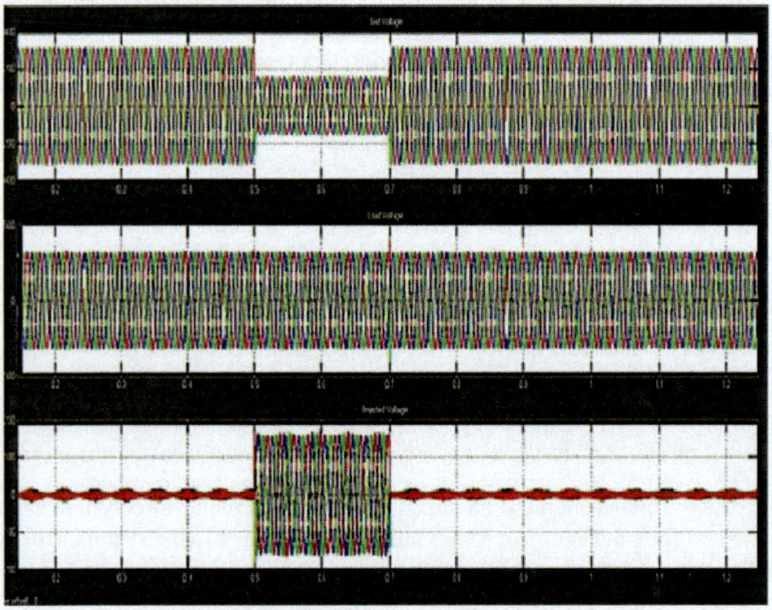

Fig. 5 Voltage sag condition using three-phase compensator

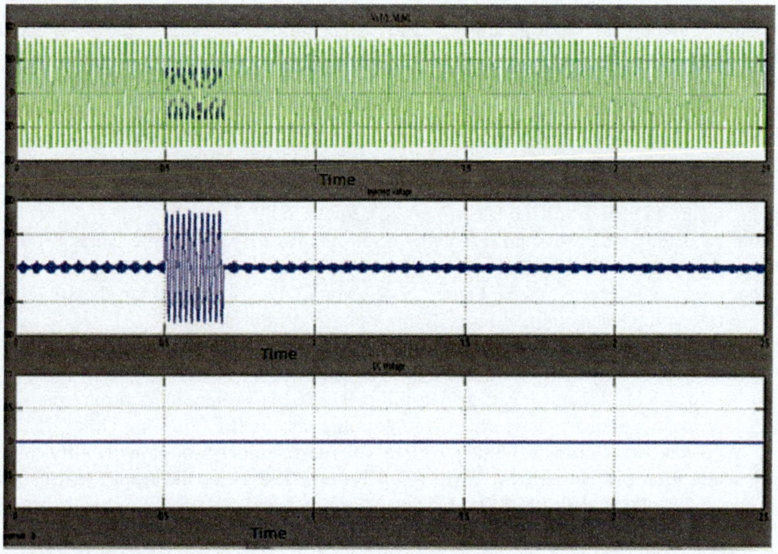

Fig. 6 Voltage sag condition using single phase

shown. Figure 6c shows the DC capacitor voltage of voltage source inverter which is used as a single-phase compensator.

6 Conclusion

Three compensation ideas were proposed based on the estimated results: reconfigure light loads introduced on phase A to phase C; utilization of a 1-Φ series voltage compensators instead of three phase compensators. The single phase at phase C, resulting in a great accessibility of the modern power system and use of a 1-Φ potential compensator connected in cascaded on phase A for three-phase supply. This proposed strategy would convert 1-Φ voltage drops into 2-Φ voltage drops, which are less than three-phase voltage drops. This method may ensure the withstand of a few three-phase loads, despite the fact that there is occurrence of some voltage sag.

References

1. IEEE (2009) Recommended practice for monitoring electric power quality. IEEE 1159/2009
2. Carlsson F, Sadarangani C, Widell B (2001) Impacts of voltage sags on a blast furnace process. In: Proceedings of the CIGRE symposium, Stockholm, Sweden

3. Carlsson F, Widell B, Sadarangani C (2000) Ride-through investigations for a hot rolling mill process. In: Proceedings of international conference on power system technology, Perth, Australia, pp 1605–1608
4. Farhoodnea M, Mohamed A, Shareef H (2014) A comparative study on the performance of custom power devices for power quality improvement. In: Proceedings of IEEE innovative smart grid technologies, Kuala Lumpur, Malaysia, 20–23 May 2014
5. Rauf AM, Khadkikar V (2015) An enhanced voltage sag compensation scheme for dynamic voltage restorer. IEEE Trans Ind Electron 62(5):2683–2692
6. Messiha M, Baraket C, Massoud A, Iqbal A, Soliman R (2015) Dynamic voltage restorer for voltage sag mitigation in oil & gas industry. In: Proceedings of the 2015 1st workshop on smart grid and renewable energy, Doha, Qatar, 22–23 Mar 2015. Pires et al. On the application of single-phase voltage sag compensators in three-phase systems 637
7. Jothibasu S, Mishra MK (2015) An improved direct AC-AC converter for voltage sag mitigation. IEEE Trans Ind Electron 62(1):21–29
8. Pires IA, Silva SM, Amaral FV, Cardoso Filho BJ (2014) Protecting control panels against voltage sags: using a square-wave series voltage compensator. IEEE Ind Appl Mag 20(5):24–33
9. Pires IA, Silva SM, Cardoso Filho BJ (2015) Increasing ride-through capability of control panels using square-wave series voltage compensator. IEEE Trans Ind Appl 51(2):1309–1316
10. Guasch L, Corcoles F, Pedra J (2004) Effects of symmetrical and unsymmetrical voltage sags on induction machines. IEEE Trans Power Delivery 19(2):774–782
11. Bollen MHJ (1998) Method of critical distances for stochastic assessment of voltage sags. IET Proc Gener Transm Distrib 145(1):70–76
12. Wang Y, Bollen MHJ, Xiao XY (2015) Calculation of the phase-angle jump for voltage dips in three-phase systems. IEEE Trans Power Delivery 30(1):480–487
13. National Electrical Manufactures Association (2006) Motors and generators. NEMA MG 1-2006
14. IEEE Standard (2004) Test procedure for polyphase induction motors and generators. IEEE Standard 112-2004

A Novel Multifunctional Active Filter Configuration

Vikash Kumar, Akash Kumar, Aryan Kumar, and Bipul Kumar Singh

Abstract This article proposes the configuration of a multifunctional active filter using voltage differencing inverting buffered amplifier (VDIBA). The filter configuration is resistorless and uses two VDIBA blocks to implement low pass, high pass and band pass filter responses. The working of the presented VDIBA-based configuration is thoroughly examined through mathematical and simulation analyses. The VDIBA-based configuration is simulated using a 45-nm CMOS technology node using ADE of Cadence at $\pm$ 0.95 V supply voltage.

Keywords VDIBA · Filter · Multifunctional · Cadence · Universal filters

V. Kumar (✉) · A. Kumar · A. Kumar · B. K. Singh
Department of Electronics and Communication Engineering, BMS College of Engineering, Bangalore, Karnataka, India
e-mail: vikashkumar.ece@bmsce.ac.in

A. Kumar
e-mail: akashk.ec19@bmsce.ac.in

A. Kumar
e-mail: aryankumar.ec19@bmsce.ac.in

B. K. Singh
e-mail: bipulkumar.ec19@bmsce.ac.in

J. R. Szymanski et al. (eds.), *Energy Systems, Drives and Automations*,
Lecture Notes in Electrical Engineering 1057,
https://doi.org/10.1007/978-981-99-3691-5_5

1 Introduction

The rapid improvements in VLSI designs have changed the dynamics of active elements. There is also a huge demand for active elements in applications such as oscillators and filters to consume as low power as possible. The operational amplifier has been the most prevalent and extensively used active block for a long time. New advances have also made it necessary for both analog and digital blocks to be implemented on the same IC.

Communication has witnessed a gradual evolution from analog-to-digital communication. However, practically signals are analog in nature. Speech signal, real time audio are all analog signals which should undergo digital communication. Analog signals are initially converted into digital signals using analog-to-digital (A/D) converters. The signals need to initially pass through an analog filter to achieve noise immunity. Analog filters, thus hold a high demand in the communication and signal processing domain.

Authors in [1] have presented the promising active elements and their future applications. In this paper, several active elements such as differential difference amplifier (DDA), four terminal floating nullor (FTFN), operational transconductance amplifier (OTA), voltage differencing inverting buffered amplifier (VDIBA), voltage differencing transconductance amplifier (VDTA), etc., were explained in brief. This paper proposes a new internal structure of the active element, VDIBA, along with its filter applications. Several operations of VDIBA, for-instance, inductor emulator [2], multifunctional filter [3], multiplier [4], oscillator [5], etc., are reported in literature. The intrinsic structure of VDIBA consists of a transconductance stage followed by an inverted buffered amplifier stage [6]. In this article, a VDIBA-based multifunctional filter configuration is proposed. The traditional transconductance phase present at the input of VDIBA is succeeded by an Arbel–Goldminz transconductance stage [7]. This Arbel–Goldminz transconductance stage eradicates the requirement for transistor matching and generates optimal outputs. Also, as in the Arbel–Goldminz transconductance stage, the transistors are placed symmetrically, this circuit is more immune to noise in contrast with earlier traditional transconductance stage. The output of the initial transconductance stage is supplied to an inverted buffer stage, which provides the necessary amplification to the circuit. A multifunctional filter can mimic the behavior of all different kind of filters. This article demonstrates the three filters namely high pass, low pass and band pass filter.

2 VDIBA Design

VDIBA is a four-port active element with block diagram depicted in Fig. 1. The functionality of VDIBA and its ports are mathematically represented by the subsequent matrix form [8]:

Fig. 1 Block diagram of the VDIBA active element

$$\begin{bmatrix} i_{\mathrm{p}} \\ i_{\mathrm{n}} \\ i_{\mathrm{z}} \\ v_{\mathrm{w}} \end{bmatrix} = \begin{bmatrix} 0 & 0 & 0 & 0 \\ 0 & 0 & 0 & 0 \\ g_{\mathrm{m}} & -g_{\mathrm{m}} & 0 & 0 \\ 0 & 0 & -1 & 0 \end{bmatrix} \begin{bmatrix} v_{\mathrm{p}} \\ v_{\mathrm{n}} \\ v_{\mathrm{z}} \\ i_{\mathrm{w}} \end{bmatrix} \tag{1}$$

As shown in block diagram, there are four terminals, i.e., *p*, *n*, *w* and *z*. The voltages v_{p}, v_{n} represent the voltages at input terminals, i_{p}, i_{n} represent the currents at input terminals. v_{z}, i_{z} and v_{w}, i_{w} represent the voltage and current at the intermediate terminal and the output terminal, respectively. The transconductance (g_{m}) of the transconductance stage is given by

$$g_{\mathrm{m}} = \left(\frac{g_{\mathrm{m1}} g_{\mathrm{m2}}}{g_{\mathrm{m1}} + g_{\mathrm{m2}}} + \frac{g_{\mathrm{m3}} g_{\mathrm{m4}}}{g_{\mathrm{m3}} + g_{\mathrm{m4}}} \right) \tag{2}$$

The internal representation of presented VDIBA configuration is shown in Fig. 2. In the representation, the first stage is realized with Arbel–Goldminz transconductance stage. This first stage contains two NMOS (M_1 and M_2) and PMOS (M_3 and M_4) transistor pairs. This stage achieves the same functionality which is achieved in a differential amplifier. The first stage is followed by an inverted voltage buffer (IVB) stage. In the second stage, a common source (CS) amplifier with a NMOS transistor load is realized. The overall configuration is provided with suitable bias current source (I_{bias}) pairs, supply voltages V_{DD} and V_{SS} to bias the transistors in saturation region.

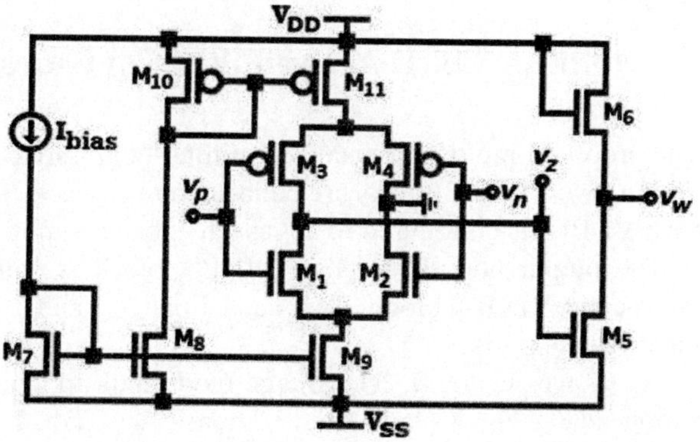

Fig. 2 Internal representation of proposed VDIBA

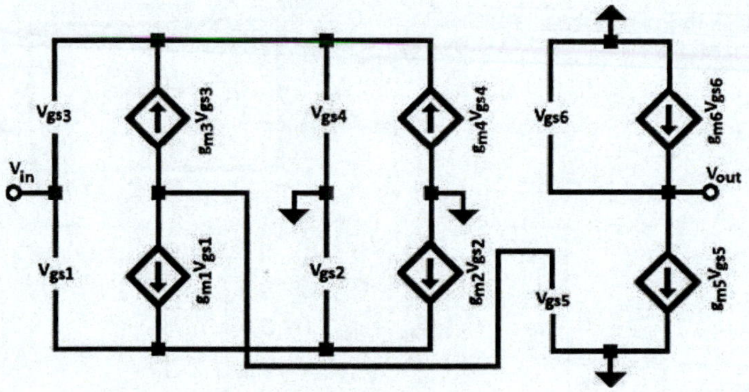

Fig. 3 Small-signal equivalent model of proposed VDIBA

The small-signal equivalent model of the presented VDIBA configuration is displayed in Fig. 3. Analysis of this equivalent model of the VDIBA configuration results in the subsequent expression of overall gain:

$$A_V = -\left(\frac{1}{g_{m1} + g_{m3}}\right)\left(\frac{g_{m1}g_{m2}}{g_{m1} + g_{m2}} + \frac{g_{m3}g_{m4}}{g_{m3} + g_{m4}}\right)\left(\frac{g_{m5}}{g_{m6}}\right) \tag{3}$$

Here, g_{mk} is the transconductance of kth NMOS/PMOS transistor. The transconductance depends on bias current I_{bias}, oxide capacitance C_{ox}, mobility μ_k and aspect ratio (W/L). The tunability of the proposed VDIBA design is provided through changing the value of I_{bias}.

$$g_{mk} = \sqrt{I_{bias}\mu_k C_{ox}\left(\frac{W}{L}\right)_k} \tag{4}$$

3 Multifunctional Active Filter Configuration

The proposed multifunctional active filter configuration is shown in Fig. 4. In this design, two VDIBA blocks are connected in cascade. The Arbel–Goldminz stage in each VDIBA is configured to act as a feedback amplifier.

The output port of the first VDIBA block is connected to the input port of the second VDIBA block. Two capacitors (C_1 and C_2) are connected at both the intermediate ports.

As shown in Fig. 4, the voltage mode multifunctional filter has multiple input voltages (V_1, V_2, V_3) and single output (V_{out}). For the purpose of implementing a

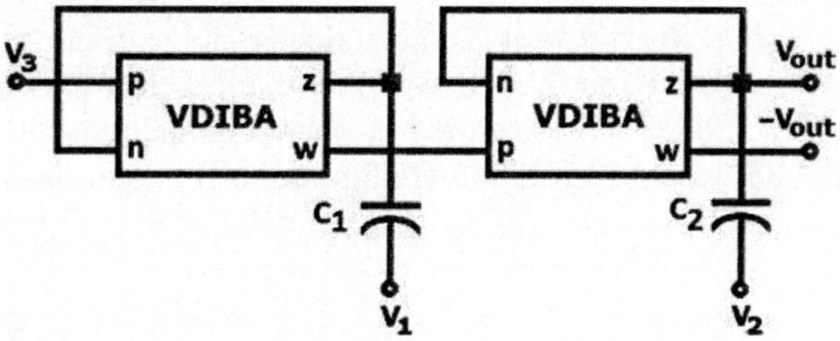

Fig. 4 Proposed multifunctional active filter configuration

particular filter configuration, i.e., lowpass, highpass and bandpass, the voltages are arranged as per the following:

(1) If voltage V_{LP} is assigned to V_3 and $V_1 = V_2 = 0$, the configuration implements second-order low pass (LP) filter. The voltage transfer function of the configuration is evaluated as

$$\frac{V_{\mathrm{out}}}{V_{\mathrm{LP}}} = -\frac{g_{\mathrm{m}}^2}{s^2 C_1 C_2 + s g_{\mathrm{m}}(C_2 + C_1) + g_{\mathrm{m}}^2} \tag{5}$$

(2) If voltage V_{HP} is assigned to $V_1 = V_2$ and $V_3 = 0$, the configuration implements second-order high pass (HP) filter. The voltage transfer function of the configuration is evaluated as

$$\frac{V_{\mathrm{out}}}{V_{\mathrm{HP}}} = \frac{s^2 C_1 C_2}{s^2 C_1 C_2 + s g_{\mathrm{m}}(C_2 + C_1) + g_{\mathrm{m}}^2} \tag{6}$$

(3) If voltage V_{BP} is assigned to V_1 and $V_2 = V_3 = 0$, the configuration implements second-order band pass (BP) filter. The voltage transfer function of the configuration is evaluated as

$$\frac{V_{\mathrm{out}}}{V_{\mathrm{BP}}} = -\frac{s g_{\mathrm{m}} C_1}{s^2 C_1 C_2 + s g_{\mathrm{m}}(C_2 + C_1) + g_{\mathrm{m}}^2} \tag{7}$$

The parameters of filter, namely cut-off frequency (ω_{c}) and bandwidth (BW) for the multifunctional filter configuration can be expressed as:

$$\omega_{\mathrm{c}} = \sqrt{\frac{g_{\mathrm{m}}^2}{C_1 C_2}}, \tag{8}$$

$$\mathrm{BW} = g_{\mathrm{m}}\left(\frac{C_1 + C_2}{C_1 C_2}\right). \tag{9}$$

From (8) and (9), it can be noticed that the cut-off frequency (ω_c) and bandwidth (BW) for the multifunctional filter can be electronically controlled. Both the filter characteristics can be varied by changing the transconductance (g_m) of the VDIBA. In addition, the proposed configuration is resistorless and employs only grounded capacitors, thereby validating its suitability for fully integrated circuit applications.

4 Simulation Results

Various analyses including AC, DC and transient were performed on the proposed design using Cadence. The performance is analyzed through simulations carried out at 45-nm CMOS. The bias supply current (I_{bias}) is fixed at 145 μA and the voltages, V_{DD} and V_{SS}, is fixed at + 950 mV and − 950 mV, respectively.

DC analysis of proposed VDIBA configuration is carried out and the result is depicted in Fig. 5. During DC analysis, the output currents, I_z and I_w, are plotted with respect to difference of two input voltages, i.e., $(V_p - V_n)$. Further, the transient analysis of proposed VDIBA configuration is performed and the result is shown in Fig. 6. Figures 7, 8 and 9, respectively, represent the frequency response of band pass, high pass and low pass obtained during AC analysis. For AC analysis, a sinusoidal signal of 10 mV amplitude and 10 kHz frequency is applied. Figures 7, 8 and 9 show the phase magnitude and gain magnitude plots which validate the working of the proposed configuration. The power dissipation of the multifunctional filter design is measured as 2.3 mW.

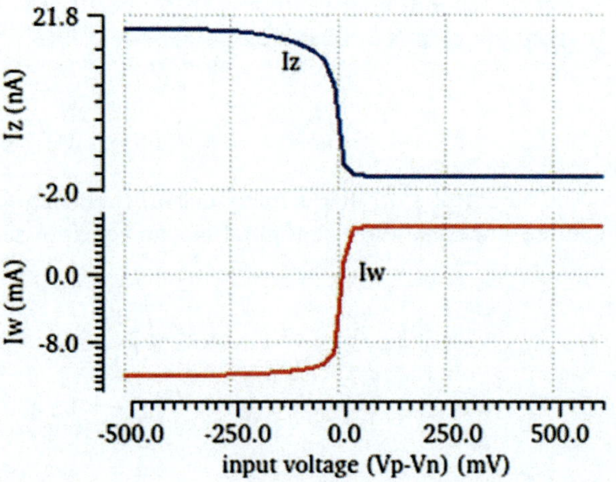

Fig. 5 DC response of proposed VDIBA configuration

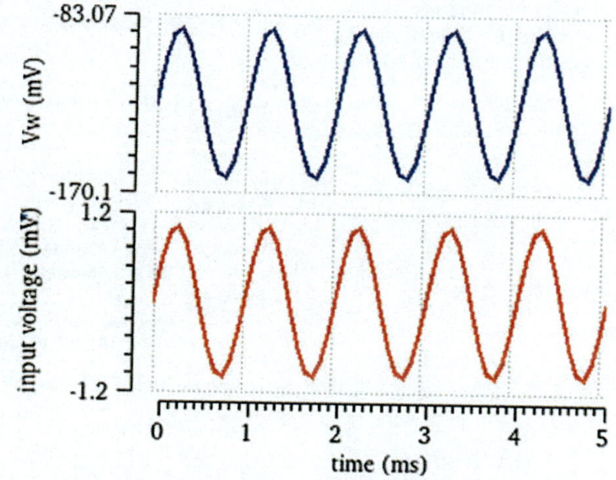

Fig. 6 Transient response of presented VDIBA design

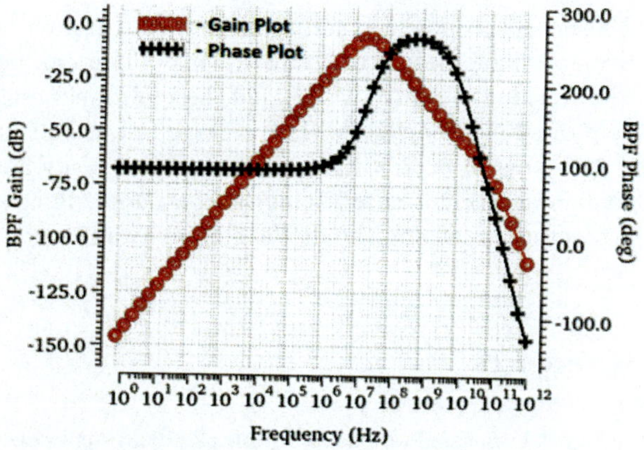

Fig. 7 AC response of band pass filter configuration with cut-off frequency of 10 MHz

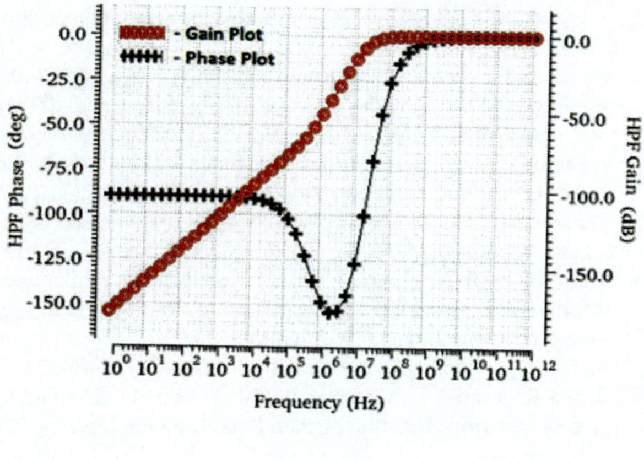

Fig. 8 AC response of high pass filter configuration with cut-off frequency of 10 MHz

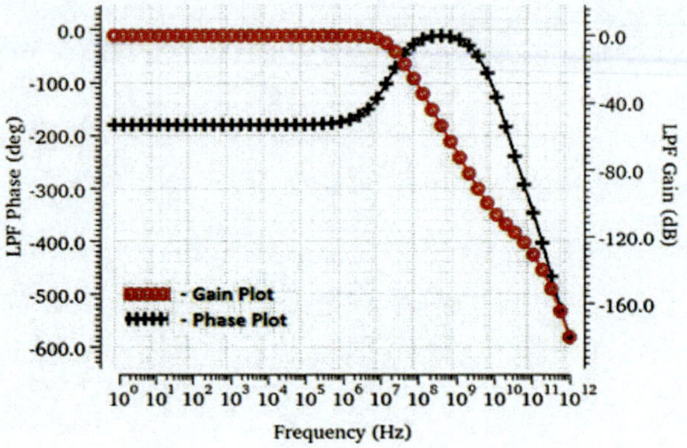

Fig. 9 AC response of low pass filter configuration with cut-off frequency of 10 MHz

5 Conclusion

This article proposes an active multifunctional filter configuration using an innovative and stable design of VDIBA. This circuit can be tuned with external current. The working of both VDIBA and the multifunctional filter are validated with the help of simulation results. The average power consumption of filter configuration is measured to be 2.3 mW with cut-off frequency around 10 MHz. This proposed configuration can be used in high performance signal processing applications which is required for analog and digital communications.

References

1. Biolek D, Senani R, Biolkova V, Kolka Z (2008) Active elements for analog signal processing: classification, review, and new proposals. Radioengineering 17(4):15–32
2. Tangsrirat W (2017) Synthetic grounded lossy inductance simulators using single VDIBA. IETE J Res 63(1):134–141
3. Kumari S, Gupta M (2018) Design and analysis of tunable voltage differencing inverting buffered amplifier (VDIBA) with enhanced performance and its application in filters. Wireless Pers Commun 100:877–894
4. Tangsrirat W (2017) Resistorless tunable capacitance multiplier using single voltage differencing inverting buffered amplifier. Rev Roum Sci Techn 62(1):72–75
5. Channumsin O, Tangsrirat W (2017) VDIBA-based sinusoidal quadrature oscillator. Prz Elektrotech 93(3):248–251
6. Kacar F, Yesil A, Noori A (2012) New CMOS realization of voltage differencing buffered amplifier and its biquad filter applications. Radioengineering 21(1):333–339
7. Arbel AF, Goldminz L (1992) Output stage for current-mode feedback amplifiers, theory and applications. Analog Integr Circ Sig Process 2(3):243–255
8. Sotner R, Jerabek J, Herencsar N (2013) Voltage differencing buffered/inverted amplifiers and their applications for signal generation. Radioengineering 22(2):490–504

PKL Electricity—A New and Innovative Idea for an Indirect Galvanic Cell (IDGC)

K. A. Khan, Md. Sayed Hossain, Salman Rahman Rasel, and Md. Akhtar-Uz-Zaman Shabuj

Abstract This work represents new approaches for the production of bio-electricity from citrus PKL extract, an indirect galvanic cell (IDGC). In this research paper, it has been designed and fabricated a Pathor Kuchi leaf (PKL) unit cell for practical utilization in Bangladesh. It has been studied the open circuit voltage, load voltage, short circuit current, load current, internal resistance, maximum power and load power. The study period was long and it was 23,000 h. It is shown that the maximum open circuit voltage was 0.99 V and the minimum open circuit voltage was 0.85 V, the maximum load voltage was 0.95 V and the minimum load voltage was 0.81 V, the maximum short circuit current was 4.14 mA and the minimum short circuit current was 2.42 mA, the maximum load current was 3.12 mA and the minimum load circuit current was 1.41 mA, the maximum internal resistance was 236.71 Ω and the minimum internal resistance was 351.24 Ω, the maximum power was 4.06 W and the minimum maximum power was 2.06 W and the load power was 2.96 W and the minimum load power was 1.14 W. The performances and different electrical parameters have been studied. It is shown that the voltage has been increased by increasing the number of cells in a series combination where the current remains constant. It is also found that the current has been increased by increasing the number of cells in a parallel combination while the voltage remains constant.

Keywords Indirect galvanic cell (IDGC) · Electrochemical cell · PKL electricity · Open circuit voltage · Short circuit current · Internal resistance · Maximum power

K. A. Khan (✉)
Department of Physics, Jagannath University, Dhaka 1100, Bangladesh
e-mail: kakhan01@yahoo.com

Md. Sayed Hossain
Center for Research Reactor, Bangladesh Atomic Energy Commission, Savar, Dhaka 1349, Bangladesh

S. R. Rasel
Local Government Engineering Department (LGED), Sherpur Sadar, Sherpur, Bangladesh

Md. Akhtar-Uz-Zaman Shabuj
Institute of Nuclear Science and Technology, Bangladesh Atomic Energy Commission, Savar, Dhaka 1349, Bangladesh

J. R. Szymanski et al. (eds.), *Energy Systems, Drives and Automations*, Lecture Notes in Electrical Engineering 1057,
https://doi.org/10.1007/978-981-99-3691-5_6

1 Introduction

Electricity is the prime need for every sector in the society. Without electricity development of all sectors cannot be imagined [1–3]. The advancement of the society fully depends on electricity. There are two types of energy—one is non-renewable energy and the other is renewable energy [4–7]. The non-renewable energy sources are oil, gas and coal, those are called fossil fuels [8–10]. The fossil fuels are not environment friendly. It pollutes the environment. It cannot be generated again and again [11–13]. It is mineral sources. It is diminishing day by day rapidly. It will be finished after certain period [14–16]. After 2100 century it will be finished from all over the world. Then the renewable energy will be needed urgently [17–19]. Renewable energy sources are pollution free and eco-friendly. Renewable energy can be generated again and again and will never run out. Fossil fuels contain particles and volatile matters. It is the source of global warming. After 2100 people will face the energy crisis. Renewable energy sources will try to meet up the energy demand [20, 21]. The population is increasing enormously day by day. People need the food and side by side they also need energy to develop the world [22]. Previously, a lot of research work has been conducted on renewable energy sources [23]. Many researches have been conducted also on vegetative and fruits [24]. Finally, it can be said that to fulfil the requirements there is no alternative way for conducting the research work on different types of vegetative and fruits for energy extraction.

2 Methodology

The methods and materials are given below.

2.1 Design and Fabrication

Figure 1 shows the experimental set-up of filled external examiner form. Here, it is needed the required apparatus: (1) battery box, (2) copper plates, (3) Zn plates, (4) different types of LED lights, (5) crocodile clips, (6) connecting wires and (7) PKL extract.

i. **Battery box**: It is made of a plastic box. It has four compartments. Each compartment makes a unit cell. Each cell has one copper and one zinc plate. The cost of the battery box is considered as a fixed cost.
ii. **Copper plates**: It acts as an anode. It is collected from the local market. The cost of the copper plate is considered as a fixed cost.
iii. **Zinc plates**: It acts as an anode. It is collected from the local market. The cost of the copper plate is considered as a variable cost.

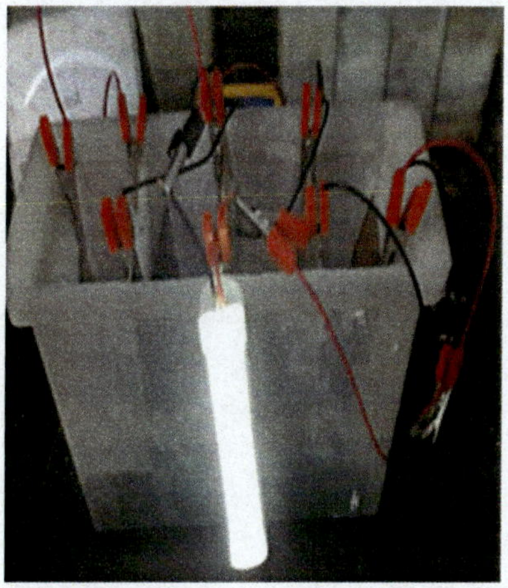

Fig. 1 Experimental set-up of LED lamp for practical utilization

iv. **LED lights**: It acts as a load. It is collected from the local market. The cost of the LED light is considered as a variable cost.
v. **Crocodile clips**: It is used as a connector. It is collected from the local market. The cost of the crocodile clips is considered as a variable cost.
vi. **Connecting wires**: It is also used as a connector. It is collected from the local market. The cost of the connecting wires is considered as a variable cost.
vii. **PKL extract**: It acts as an electrolyte. PKL is cultivated from the agricultural land. The cost of the PKL extract is considered as a fixed cost.
viii. **Internal resistance**: It is defined as: $r_{in} = (V_{oc}/I_{sc})$, r_{in} = internal resistance, V_{oc} = open circuit voltage, I_{sc} = short circuit current.
ix. **Maximum power**: It is defined as, $P_{max} = V_{oc}\, I_{sc}$, where, P_{max} = maximum power, V_{oc} = open circuit voltage, I_{sc} = short circuit current.
x. **Load power**: It is defined as, $P_L = V_L\, I_L$, where, P_L = load power, V_L = load voltage, I_L = load current.
xi. **Open circuit voltage (V_{oc})**: The voltage is taken not across any load, i.e. the voltage without any load is called open circuit voltage. It is denoted by V_{oc}.
xii. **Short circuit current (I_{sc})**: Current measured without any load is called short circuit current. Generally, it is denoted by I_{sc}.
xiii. **Load voltage (V_L)**: The voltage across the resistor is called load voltage. Generally, it is denoted by V_L.
xiv. **Load current**: The current which the appliance is drawing at that instant is known as load current, which should be always lower than the rated current of that item. Rated current is the maximum current an appliance should ever draw, or if it is a source like an outlet or a generator, then it is the maximum you should ever draw through it. Generally, it is denoted by I_L.

2.2 Chemical Reactions

The chemical reactions are given by the following:

$$Zn \rightarrow Zn^{2+} + 2e^- \text{ (Anodic reaction)} \quad (1)$$

$$2H^+ + 2e^- \rightarrow H_2 \text{ (Cathodic reaction)} \quad (2)$$

$$Zn + 2H^+ \rightarrow Zn^{2+} + H_2 \text{ (Overall reaction)} \quad (3)$$

2.3 Characteristics of PKL Leaf Comparison with Other Vegetative and Fruits

PKL leaf contains weak organic acids like citric acid, iso-citric acid and malic acid [11]. During electricity generation, the H^+ produces slowly than other vegetative and fruits [8]. As a result, the longevity of the electricity production has been increased.

2.4 Innovative Idea of the PKL Electrochemical Cell

PKL electrochemical cell has been designed and fabricated based on galvanic cell. But there is a fundamental distinction between PKL electrochemical cell and the galvanic cell. In the galvanic cell, there is a salt bridge but in the PKL electrochemical cell, there is no salt bridge. As a result internal resistance has been decreased in the PKL electrochemical cell and we got more current using the PKL electrochemical cell than the galvanic cell.

This work was based on galvanic cell. The galvanic cell was divided into two parts—firstly, the direct galvanic cell (DGC) and the indirect galvanic cell (IDGC). This research work was of the indirect galvanic cell (IDGC). It is also called modified galvanic cell. It is needed a salt bridge for the direct galvanic cell (DGC). We did not use any salt bridge in our research work that is why it is called the indirect galvanic cell (IDGC) or, the modified galvanic cell.

2.5 Theory

Every metal has a tendency to go into the solution by oxidation process and can produce a solution pressure (SP). Also, the ionic forms of the substance has tendency to be deposited on the electrode surface by reduction process which as a result creates an osmotic pressure (OP). When solution pressure is greater than osmotic pressure or SP > OP, then oxidation occurs. But, when SP < OP then reduction occurs. As the principle of oxidation, at anode, an electrochemical cell, the oxidation occurs and at cathode the reduction occurs [15].

In PKL cell, the anode used is Zn plate but Cu plate used as cathode. The electrolyte used is PKL juice with a secondary salt ($CuSO_4 \cdot 5H_2O$). Thus the reactant ionic species present in solution are H^+ and Cu^{2+} ions. The overall reaction taking place is considered to be as follows:

$$\mathrm{Zn(s) + Cu^{2+}(aq) + H^{+}(aq) \rightarrow Zn^{2+}(aq) + Cu(s) + H_2(g)} \uparrow$$

The concentration of Cu^{2+} and H^+ decreases but that of Zn^{2+} increases with time where Cu^{2+} and H^+ are the electron acceptors and Zn is the electron-donor, so the potential difference falls and current flow decreases. This is because the electron acceptors accept electrons by the reduction process. The reduction processes are explained with the help of some chemical equations:

$$\mathrm{Cu^{2+} + 2e^{-} \rightarrow Cu}$$

$$\mathrm{2H^{+} + 2e^{-} \rightarrow H_2} \uparrow$$

And the oxidation process is: $2Zn - 4e^- \rightarrow 2Zn^{2+}$.

The Nernst equations [9] for the above reactions are given below,

$$E_{\mathrm{Cu^{2+}|Cu}} = E^0_{\mathrm{Cu^{2+}|Cu}} - \frac{0.0591}{n} \log \frac{1}{[\mathrm{Cu^{2+}}]} \tag{4}$$

$$E_{\mathrm{2H^{+}|H_2}} = E^0_{\mathrm{2H^{+}|H_2}} - \frac{0.0591}{n} \log \frac{1}{[\mathrm{H^{+}}]^2} \tag{5}$$

$$E_{\mathrm{Zn^{2+}|Zn}} = E^0_{\mathrm{Zn^{2+}|Zn}} - \frac{0.0591}{n} \log \frac{1}{[\mathrm{Zn^{2+}}]^2} \tag{6}$$

2.6 *Determination of Cell Potential*

The data readings from the voltmeter give the voltage of the reaction cell or potential difference between of its two half-cells. Cell voltage, which is known as cell potential or electromotive force (emf) is shown as the symbol E_{cell}. In this experiment, the cell potential was measured by following two ways. They were open circuit potential and potential with load. The open circuit potential is generated when the circuit is open and closed with a connection of electrical instrument with some load. The potential was measured by using a multimeter in volt (V) unit.

2.7 Determination of Flow of Electron

Electrons always flow from the anode to the cathode or from the oxidation half-cell to the reduction half-cell. In terms of $E^{o}{}_{\text{cell}}$ of the half reactions, the electrons flow from the more negative half reaction side to the more positive half reaction side. In this research work, the flows of electrons were measured by following two ways. They were considered short circuit current and current with load. The short circuit current is the current when the circuit is considered open; but, when the circuit is closed with a connection of electrical instrument—is the current with load. The PKL cell current was measured by using a multimeter in Ampere (A) unit.

2.8 PKL Cell Potential

The PKL cell potential,

$$E_{\text{Cell}} = E_{\text{right (cathode)}} - E_{\text{left (anode)}} \tag{7}$$

So,

$$E_{\text{Cell}} = E_{\text{Cu}^{2+}|\text{Cu}} + E_{2\text{H}^{+}|\text{H}_2} - 2E_{\text{Zn}^{2+}|\text{Zn}} \tag{8}$$

If E_{cell} is positive the reaction is spontaneous and it is a voltaic cell. If the $E^{o}{}_{\text{cell}}$ is negative, the reaction is non-spontaneous and it is represented as an electrolytic cell.

We have,

$$E_{\text{Cu}^{2+}|\text{Cu}} = E^{0}_{\text{Cu}^{2+}|\text{Cu}} - \frac{0.0591}{n}\log\frac{1}{[\text{Cu}^{2+}]} \tag{9}$$

$$E_{2\text{H}^{+}|\text{H}_2} = E^{0}_{2\text{H}^{+}|\text{H}_2} - \frac{0.0591}{n}\log\frac{1}{[\text{H}^{+}]^{2}} \tag{10}$$

$$E_{\text{Zn}^{2+}|\text{Zn}} = E^{0}_{\text{Zn}^{2+}|\text{Zn}} - \frac{0.0591}{n}\log\frac{1}{[\text{Zn}^{2+}]^{2}} \tag{11}$$

where, n = number of moles of electrons transferred in cell reactions,

$[\text{Zn}^{2+}]$ = concentration of zinc in molality,

$[\text{Cu}^{2+}]$ = concentration of copper in molarity,

$E^{0}_{\text{Zn}^{2+}|\text{Zn}} = -0.76$ V

$E^0_{Cu^{2+}|Cu} = 0.34$ V
$E^0_{H^+|H_2} = 0.0$ V.
So,

$$\begin{aligned}
\text{PKL cell potential} &= E^0_{Cu^{2+}|Cu} - \frac{0.0591}{n}\log\frac{1}{[Cu^{2+}]} \\
&\quad - \left[E^0_{Zn^{2+}|Zn} - \frac{0.0591}{n}\log\frac{1}{[Zn^{2+}]}\right] \\
&\quad + E^0_{2H^+|H_2} - \frac{0.0591}{n}\log\frac{1}{[H^+]^2} \\
&\quad - \left[E^0_{Zn^{2+}|Zn} - \frac{0.0591}{n}\log\frac{1}{[Zn^{2+}]}\right] \\
&= E^0_{Cu^{2+}|Cu} + E^0_{2H^+|H_2} - 2E^0_{Zn^{2+}|Zn} + \frac{0.0591}{2} \\
&\left[\log\frac{1}{[Zn^{2+}]^2} - \left(\log\frac{1}{[Cu^{2+}]} + \log\frac{1}{[H^+]^2}\right)\right] \\
&= 0.34 + 0 - 2(-0.76) + \frac{0.0591}{2}\log\frac{[Cu^{2+}][H^+]^2}{[Zn^{2+}]^2} \\
&= 1.86 + \frac{0.0591}{2}\log\frac{[Cu^{2+}][H^+]}{[Zn^{2+}]}
\end{aligned} \tag{12}$$

2.9 Calculation of PKL Cell Potential

Equation (12) represents the cell potential of PKL. For any cell, the cell potential is computed below.

Here, the initial concentration of Zn^{2+} in molarity is = 0.0334 M.

The initial concentration of Cu^{2+} in molarity is = 0.3330 M.

The pH of the solution is = 3.5.

The concentration of H^+ in molarity is = antilog (− 3.5) = 3.16×10^{-4} M.

Now putting the values in Eq. (12) we get, potential of PKL cell at temperature 25 °C and 1 atm $= 1.86 + \frac{0.0591}{2}\log\frac{[0.333][3.16\times10^{-4}]}{[0.0334]} = 1.85$ V.

Thus for one unit cell, at temperature 25 °C and 1 atm, the theoretical value of PKL cell potential is 1.85. Thus the potential value for a battery with six compartments is $= 6 \times 1.85$ V = 11.1 V.

At temperature 25 °C and 1 atm, the theoretical value of one unit of PKL cell potential is 1.85 V. So, the potential of a battery with six compartments is $= 6 \times 1.85\ \text{V} = 11.1\ \text{V}$.

In experiment, the achieved potential value is only 6.4 V. Practically, it is seen that the cell potential doesn't depend on the intensity of acidity of PKL juice. But, the current flow can vary with the percent of PKL juice. Here, both the electrodes were connected in parallel but if they are connected in series then potential voltage is achieved almost 11 V.

3 Results and Discussion

The results and discussion have been discussed below.

Table 1 shows the data collection for different parameters. The different parameters were V_{oc}, V_L, I_{sc}, I_L, R_{in}, P_{max} and P_L.

Figure 2 shows the change of V_{oc} with the change of T.D. (h). It shows that the V_{oc} is same up to 8000 h and then after it has started to decrease up to 15,000 h, and then it has started to increase up to 20,000 h, and then it was almost constant up to 23,000 h.

Figure 3 shows the V_L with the change of TD (h). It shows that the V_L is almost constant up to 8000 h and then it started to decrease up to 15,000 h, and then it started to increase up to 20,000 h and finally it was almost constant up to 23,000 h.

Figure 4 shows the change of I_{sc} with the change of TD (h). It shows that it started to decrease almost exponentially up to 20,000 h and finally it started to decrease linearly up to 23,000 h.

Figure 5 shows the change of I_L with the change of TD (h). It shows that it started to decrease almost exponentially up to 20,000 h and finally it started to decrease linearly up to 23,000 h.

Figure 6 shows the change of R_{in} with the change of TD (h). It shows that it started to decrease almost exponentially up to 15,000 h and then started to increase rapidly and finally it started to increase exponentially up to 23,000 h.

Figure 7 shows the change of P_{max} with the variation of time duration (h). It shows that it started to decrease almost exponentially up to 7000 h and then it started to decrease almost linearly, and then it started to increase exponentially up to 20,000 h and finally it started to decrease linearly up to 23,000 h.

Figure 8 shows the change of P_L with the change of TD (h). It show that it started to decrease almost exponentially up to 14,000 h and then started to it increase exponentially up to 20,000 h, and finally it started to decrease linearly up to 23,000 h.

Table 1 Table for data collection

Time duration (min)	Open circuit voltage, V_{oc} (V)	Load voltage, V_L (V)	Short circuit current, I_{sc} (mA)	Load current, I_L (mA)	Internal resistance, r_{in} (Ω)	Maximum power, P_{max} (W)	Load power, P_L (W)
00	0.98	0.95	4.14	3.12	236.71	4.06	2.96
140	0.99	0.94	3.98	2.97	248.74	3.94	2.79
180	0.99	0.94	3.89	2.87	254.50	3.85	2.70
260	0.99	0.94	3.83	2.82	258.49	3.79	2.65
360	0.99	0.93	3.76	2.75	263.30	3.72	2.56
1440	0.98	0.93	3.53	2.52	277.62	3.46	2.34
1500	0.98	0.92	3.52	2.50	278.40	3.45	2.30
2940	0.98	0.92	3.44	2.43	284.88	3.37	2.24
3240	0.98	0.92	3.41	2.40	287.39	3.34	2.21
4380	0.97	0.91	3.37	2.36	296.73	3.27	2.15
5820	0.97	0.90	3.29	2.28	294.83	3.19	2.05
6960	0.97	0.90	3.26	2.25	297.55	3.16	2.03
7060	0.95	0.89	3.24	2.23	293.21	3.08	1.98
7540	0.94	0.89	3.23	2.22	291.02	3.04	1.98
8260	0.88	0.86	3.19	2.18	275.86	2.81	1.87
8990	0.88	0.84	3.22	2.21	273.29	2.83	1.86
9230	0.87	0.83	3.23	2.21	269.35	2.81	1.83
9840	0.85	0.82	3.25	2.23	261.54	2.76	1.83
10,560	0.81	0.80	3.24	2.22	250.00	2.62	1.78
11,280	0.82	0.80	3.23	2.21	253.87	2.65	1.77
12,000	0.81	0.79	3.20	2.19	253.13	2.59	1.73
13,440	0.80	0.80	3.17	2.16	252.37	2.54	1.73
13,800	0.89	0.83	3.16	2.14	316.46	2.81	1.78
15,240	0.93	0.90	3.14	2.13	296.18	2.92	1.92
16,000	0.93	0.90	3.13	2.11	297.12	2.91	1.90
17,400	0.93	0.90	3.10	2.08	300.00	2.88	1.87
18,900	0.92	0.90	3.01	2.00	305.65	2.77	1.80
20,000	0.87	0.85	2.73	1.70	318.60	2.38	1.45
23,000	0.85	0.81	2.42	1.41	351.24	2.06	1.14

4 Conclusions

This work is a research work on indirect galvanic cell (IDGC) used with PKL gel for studying of green energy generation. It is found that the experimental and theoretical results were almost same. This cell was fabricated-based galvanic cell. It

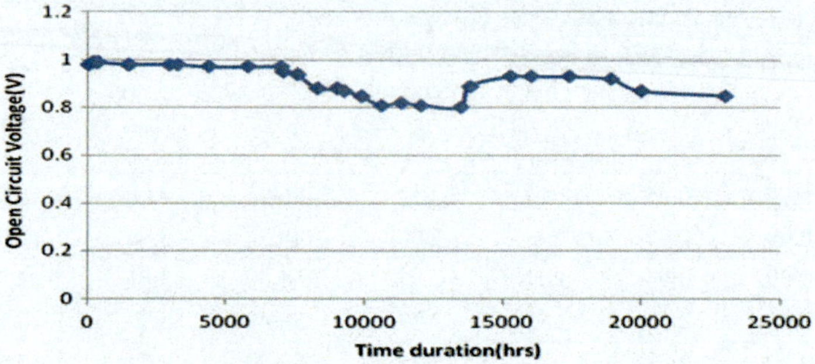

Fig. 2 Change of V_{oc} with T.D. (h)

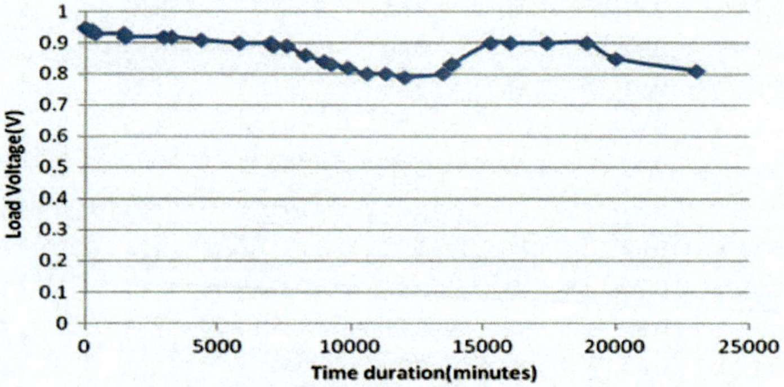

Fig. 3 Change of V_L with T.D. (h)

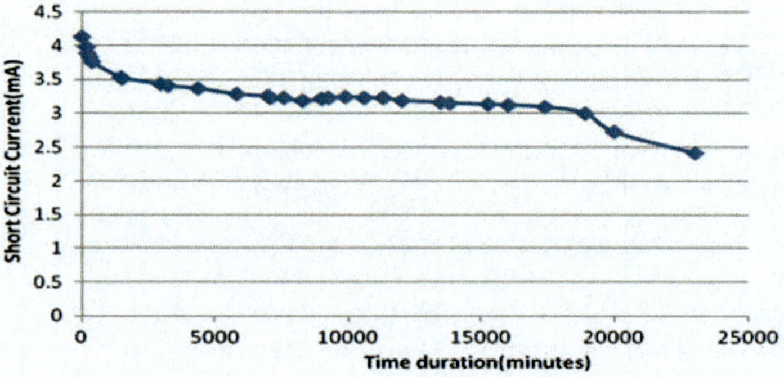

Fig. 4 Change of I_{sc} with T.D. (h)

is also called bio-voltaic cell. In this experiment, it is shown that the current flow depends on the percentage of PKL extract. It is found that when the electrodes were connected in parallel connections the current was increased and when the electrodes were connected in series connection the voltage was increased.

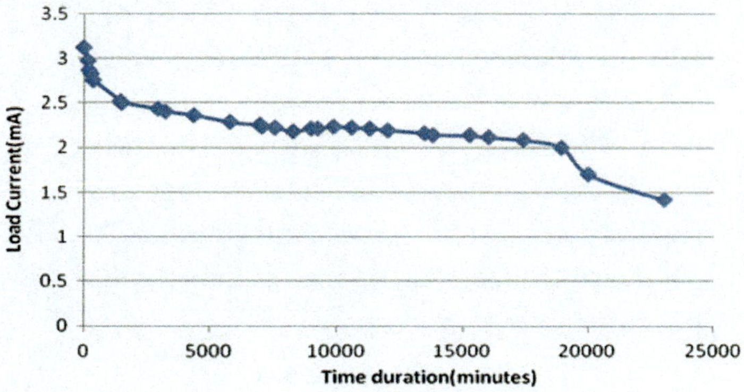

Fig. 5 Change of I_L with T.D. (h)

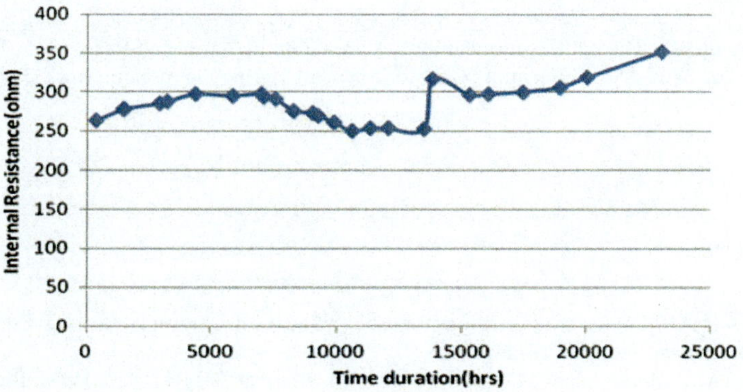

Fig. 6 Change of R_{in} with T.D. (h)

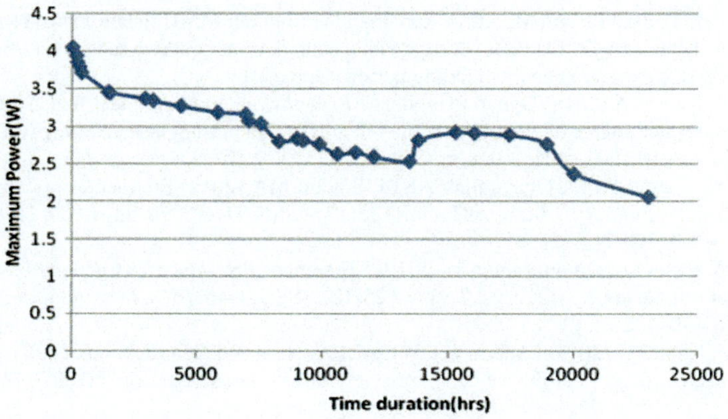

Fig. 7 Change of P_{max} with T.D. (h)

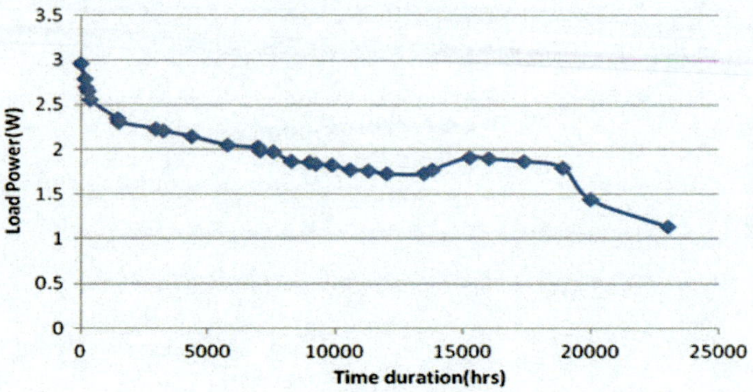

Fig. 8 Change of P_L with T.D. (h)

Acknowledgements The authors are thankful to the Grant of Advanced Research in Education (GARE) project, Ministry of Education, GoB for financing during the research work (Project/User ID: PS2019949).

References

1. Khan KA (1999) Copper oxide coatings for use in a linear solar Fresnel reflecting concentrating collector. Renew Energy 17(4):603–608
2. Ruhane TA, Tauhidul Islam M, Saifur Rahman M, Bhuiyah MMH, Islam JMM, Bhuiyah TI, Khan KA, Khan MA (2017) Impact of photo electrode thickness annealing temperature on natural dye sensitized solar cell. Sustain Energy Technol Assess. https://doi.org/10.1016/j.seta.2017.01.012
3. Ruhane TA, Tauhidul Islam M, Saifur Rahaman M, Bhuiyan MMH, Islam JMM, Newaz MK, Khan KA, Khan MA (2017) Photo current enhancement of natural dye sensitized solar cell by optimizing dye extraction and its loading period. Optik
4. Hasan M, Khan KA (2018) Dynamic model of *Bryophyllum pinnatum* leaf fueled BPL cell: a possible alternate source of electricity at the off-grid region in Bangladesh. Microsyst Technol. https://doi.org/10.1007/s00542-018-4149-y
5. Khan KA, Hazrat Ali M, Obaydullah AKM, Wadud MA (2019) Production of candle using solar thermal technology. Microsyst Technol 25(12). ISSN 0946-7076. https://doi.org/10.1007/s00542-019-04390-7
6. Khan KA, Rasel SR, Ohiduzzaman M (2019) Homemade PKL electricity generation for use in DC fan at remote areas. Microsyst Technol 25(12). ISSN 0946-7076. https://doi.org/10.1007/s00542-019-04422-2
7. Hasan M, Khan KA (2019) Experimental characterization and identification of cell parameters in a BPL electrochemical device. SN Appl Sci 1:1008. https://doi.org/10.1007/s42452-019-1045-8
8. Hassan L, Khan KA (2019) A study on harvesting of PKL electricity. Microsyst Technol 26(3):1031–1041. https://doi.org/10.1007/s00542-019-04625-7
9. Khan KA, Mamun MA, Ibrahim M, Hasan M, Ohiduzzaman M, Obaydullah AKM, Wadud MA, Shajahan M (2019) PKL electrochemical cell: physics and chemistry. SN Appl Sci 1:1335. https://doi.org/10.1007/s42452-019-1363-x

10. Hazrat Ali M, Chakma U, Howlader D, Tawhidul Islam M, Khan KA (2019) Studies on performance parameters of a practical transformer for various utilizations. Microsyst Technol. https://doi.org/10.1007/s00542-019-04711-w
11. Khan KA, Hassan L, Obaydullah AKM et al (2018) Bioelectricity: a new approach to provide the electrical power from vegetative and fruits at off-grid region. Microsyst Technol. https://doi.org/10.1007/s00542-018-3808-3
12. Khan KA, Bhuyan MS, Mamun MA, Ibrahim M, Hasan L, Wadud MA (2018) Organic electricity from Zn/Cu-PKL electrochemical cell, chap 9. In: Mandal JK et al (eds) Contemporary advances in innovative and applicable information technology. Advances in intelligent systems and computing, vol 812. Springer Nature Singapore Pvt. Ltd., pp 75–90
13. Atique Ullah AKM, Mahbubul Haque M, Akter M, Hossain A, Tamanna AN, Mottaleb Hosen M, Fazlbria AKM, Khan MNI, Khan MKA (2020) Green synthesis of *Bryophyllum pinnatum* aqueous leaf extract mediated biomolecule capped dilute ferromagnetic α-MnO_2 nanoparticles. Mater Res Express 7(1):015088
14. Badwal SPS, Giddey S, Kulkarni A, Goel J, Basu S (2015) Direct ethanol fuel cells for transport and stationary applications—a comprehensive review. Appl Energy 145:80–103. https://doi.org/10.1016/j.apenergy.2015.02.002
15. Bennetto HP (1990) Electricity generation by microorganism. Biotechnol Educ 1(4):163–168
16. Carvalho J, Ribeiro A, Castro J, Vilarinho C, Castro F (2011) Biodiesel production by microalgae and macroalgae from north littoral Portuguese coast. In: 1st international conference held by centre for waste valorisation (CVR)—wastes: solutions, treatments and opportunities, Guimarpes, Portugal, CVR, 12–14 Sept 2011
17. Christi Y (2007) Biodiesel from microalgae. Biotechnol Adv 25:294–296
18. Dincer I (2000) Renewable energy and sustainable development. Renew Sustain Energy Rev 4:157–175
19. Franco D (2005) Volta and the 'pile'. In: Electrochemistry encyclopaedia. Case Western Reserve University
20. Goff M, Nicholas SB, Shailesh L, William RS, Galen JS (2004) Acid-catalysed alcoholysis of soybean oil. J Am Oil Chem Soc 81:415–420
21. Goodisman J (2001) Observations on lemon cells. J Chem Educ 75(4):516–518
22. Ha JH, Shah N, Islam M, Park J (2010) Production of bioethanol by simultaneous saccharification and fermentation process using waste from beer fermentation broth. J Biotechnol 150(1):147–148
23. Hossain ABMS, Mekhled MA (2010) Biodiesel fuel production from waste canola cooking oil as sustainable energy and environmental recycling process. Aust J Crop Sci 4(7):543–549
24. Stirling JL, Bennetto HP, Delaney GM, Mason JR, Roller SD, Tanaka K, Thurston CF (1984) Microbial fuel cells. Biochem Soc Trans 11(4):451–453. https://doi.org/10.1042/bst0110451

A Study on Pulsating Nature of Pathor Kuchi Leaf (PKL) Electricity

Md. Hazrat Ali, K. A. Khan, and Md. Obaidur Rahman

Abstract Direct current (DC) is divided into two parts. One is pure DC, and another is pulsating DC. Direct current follows ohm's law. To keep it in mind that it has been studied the three different characteristics of the Pathor Kuchi Leaf (PKL) electricity like ohmic characteristic, pure DC characteristic and pulsating DC characteristic. Since DC follows the ohm's law, so that PKL electricity follows the ohm's law. Nine different parameters like open circuit voltage, short circuit current, maximum power, energy density, specific energy density, power density, specific power density, energy and capacity of the PKL cell have been studied. Results have been tabulated and graphically discussed. This work will help to generate and storage for practical utilization of PKL electricity as an alternative source of power generation like solar energy.

Keywords Open circuit voltage · Short circuit current · Maximum power · Energy density · Specific energy density · Power density · Specific power density · Energy · Capacity

1 Introduction

The number of people is increasing day by day in the world. So that the energy demand is also increasing day by day in the world [1, 2]. The new and innovative energy sources are necessary for the human being after a few years. The traditional energy sources like oil, gas and coal are diminishing rapidly [3, 4]. To keep it in

Md. Hazrat Ali
Department of Electrical and Electronic Engineering, European University of Bangladesh (EUB), Gabtoli, Mirpur, Dhaka 1216, Bangladesh

K. A. Khan (✉)
Department of Physics, Jagannath University, Dhaka 1100, Bangladesh
e-mail: kakhan01@yahoo.com

Md. Obaidur Rahman
Department of Physics, Jahangirnagar University, Savar, Dhaka 1342, Bangladesh

J. R. Szymanski et al. (eds.), *Energy Systems, Drives and Automations*,
Lecture Notes in Electrical Engineering 1057,
https://doi.org/10.1007/978-981-99-3691-5_7

mind, this work has been designed and fabricated. As a solution, research work has been conducted on PKL electricity [5, 6]. Basically, PKL electricity is DC electricity. By using an inverter, it is possible to get alternating current (AC) [7, 8]. The PKL electricity has several advantages. It works both day and night time equally, whereas solar photovoltaic electricity works only day time. Even in rainy season, PKL electricity woks properly. But solar photovoltaic electricity does not work in the rainy season properly. Energy is needed in all sectors in our country including educational institutions, agriculture, industry, domestic purposes, commercial locations and transportation [9, 10]. Without energy, it is not possible to survive human civilizations and to develop modern technology. Most of the power comes from non-renewable energy sources like oil, gas, and coal. It is mineral sources. These are also known as non-renewable energy sources. It will be finished within 2100 centuries [11, 12]. Then this type of renewable energy sources will be needed. That is why we have no time to stay sit only with conventional energies. The non-renewable energy sources are polluted, whereas renewable energy sources are pollution free and environment friendly. PKL is the example of biomass energy [13, 14]. It contains citric acid, iso-citric acid, malic acid, sugar and other ingredients with huge chemical energies which are possible to convert into electrical energy [15, 16]. The reaction which occurs here is called redox reaction.

2 Methodology

The methods and materials are given below.

2.1 Materials for the Study

It is shown in Fig. 1 about the PKL extract production from the PKL tree. The PKL extract was prepared by a blender and then filtered by a Whatman Paper 41.

Figure 2 shows an experimental design of PKL electricity generation system. The electrodes zinc and copper are immersed in to the PKL extract in a falcon tube. There are four falcon tubes here. One falcon tube is for one unit cell. Thus, there are four unit cells that are in series connected with each other. The zinc plate is used as an anode, and the copper plate is used as a cathode.

Figure 3 shows an experimental setup of PKL electricity with load LED light. The extract of the PKL was an electrolyte. The copper and zinc plate were cathode and anode, respectively.

Fig. 1 PKL extract production from the Pathor Kuchi Leaf (PKL) tree

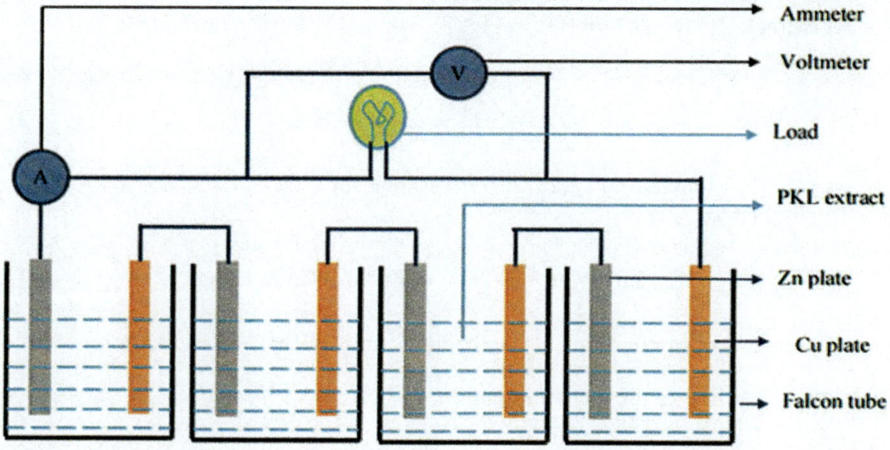

Fig. 2 Experimental design of PKL electricity generation system

Fig. 3 Experimental setup

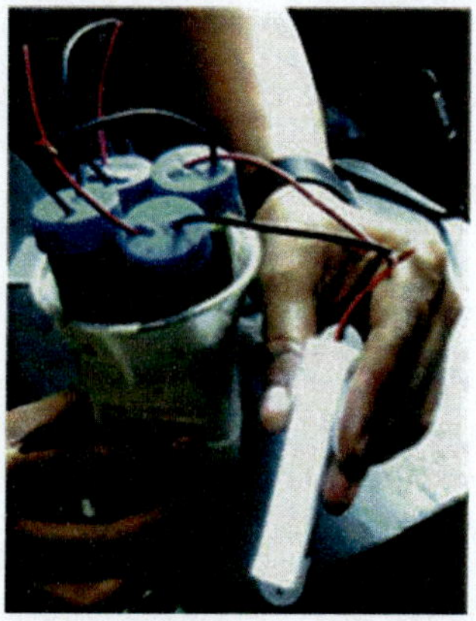

2.2 *Chemical Reactions of the PKL Cell*

(1) At zinc: $Zn = Zn^{2+} + 2e^-$, where Zn^{2+} = product ion. Here, Zn is called sacrificial element.
At copper: $Cu^{++} + 2e^- = Cu$, where Cu^{2+} = reactant ion. Here, Cu is called cathode.
Adding:

$$Zn + Cu^{2+} = Zn^{2+} + Cu, \tag{1}$$

where Cu^{2+} = reactant ion and Zn^{2+} = product ion.

(2) At zinc: $Zn = Zn^{2+} + 2e^-$, where Zn^{2+} = product ion.
At extract: $2H^+ + 2e^- = H_2$ (in Fig. 3), where H^+ = reactant ion.
Adding:

$$Zn + 2H^+ = Zn^{2+} + H_2, \tag{2}$$

where H^+ = reactant ion and Zn^{2+} = product ion.
Total reaction,

$$Zn + Cu^{2+} + 2H^+ = Zn^{2+} + Cu + H_2, \tag{3}$$

where Cu^{2+} = reactant ion, H^+ = reactant ion and Zn^{2+} = product ion.

Hence, according to the Nernst equation,

$$E_{cell} = E^o_{cell} - (RT/nF)\ln Q, \tag{4}$$

where

$$Q = [\text{Product Ion}]/[\text{Reactant Ion}] = \left[Zn^{2+}\right]/\left[Cu^{2+}\right]\left[H^+\right]. \tag{5}$$

Now, combing Eqs. (4) and (5), we have

$$E_{cell} = E^o_{cell} - (RT/nF)\ln\left[Zn^{2+}\right]/\left[Cu^{2+}\right]\left[H^+\right], \tag{6}$$

where $[Zn^{2+}]$ = zinc ion concentration = product ion concentration, $[Cu^{2+}]$ = copper ion concentration = reactant ion concentration, $[H^+]$ = hydrogen ion concentration = reactant ion concentration.

3 Results and Discussion

The results and discussions are given in Table 1.

Table 1 Table for data collection

Time duration (h)	V_{oc} (V)	I_{sc} (mA)	P_{max} (mW)	Solute quantity (kg)	Solvent quantity (L)	Energy (Wh)	Power density (W/kg)	Specific power density (W/L)	Energy density (Wh/kg)	Specific energy density (Wh/L)	Electrode power density (W/kg)	Charge capacity, C (mAh)
0.0	5.18	2.87	14.87	0.80	3.20	0.00	0.02	0.005	0.00	0.000	0.005	0.00
168	5.05	2.94	14.85	0.80	3.20	2.50	0.02	0.005	3.13	0.782	0.005	495.39
336	5.16	2.85	14.71	0.80	3.20	4.95	0.02	0.005	6.19	1.55	0.005	959.03
504	5.05	2.94	14.85	0.80	3.20	7.49	0.02	0.005	9.36	2.34	0.005	1483.23
672	5.17	2.86	14.79	0.80	3.20	9.94	0.02	0.005	12.43	3.12	0.005	1923.35

Table 1 shows the data collection for different parameters. The data was collected by a calibrated multimeter. The time duration was measured by a clock.

Figure 4 shows the variation of open circuit voltage (V) with the variation of time duration (h). It is shown that when switch was on, then the open circuit voltage has been decreased up to 168 h, and when the switch was off, then the open circuit voltage has been increased up to 336 h. Again when the switch is on, then the open circuit voltage has been decreased up to 504 h. Finally, when the switch is off, then the open circuit voltage has been increased up to 672 h. This study was for 672 h. The switching system shows that the PKL electricity has pulsating nature.

Figure 5 shows the variation of short circuit current (I_{sc}) with the variation of time duration (h). It is shown that when switch was on, then the short circuit current (I_{sc}) has been increased up to 168 h, and when the switch was off, then the open circuit voltage has been decreased up to 336 h. Again when the switch is on, then the short circuit current (I_{sc}) has been increased up to 504 h. Finally, when the switch is off, then the short circuit current (I_{sc}) has been decreased up to 672 h. This study was for 672 h. The switching system shows that the PKL electricity has pulsating nature. Furthermore, it is shown that the pulsating nature of the short circuit current (I_{sc}) is totally opposite to the pulsating nature of the open circuit voltage (V_{oc}).

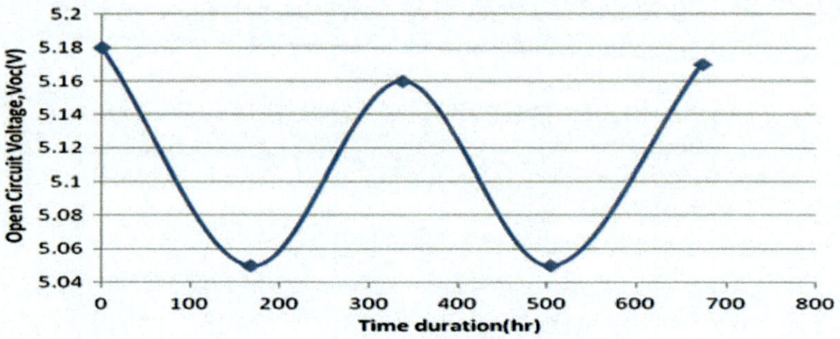

Fig. 4 V_{oc} versus T.D. (h) curve

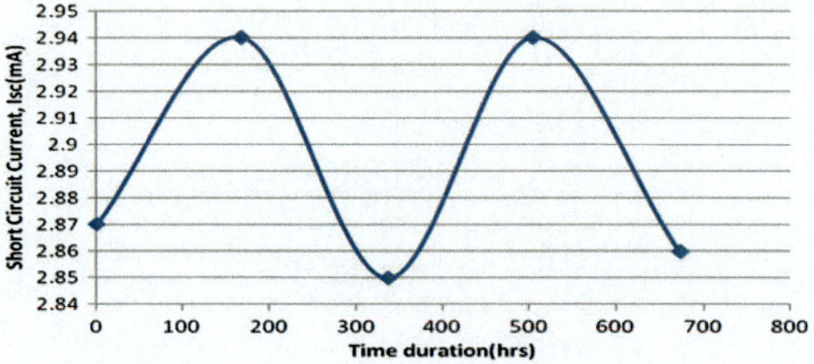

Fig. 5 I_{sc} versus T.D. (h) curve

Figure 6 shows the variation of maximum power (P_{max}) with the variation of time duration (h). It is shown that when switch was on, then the maximum power (P_{max}) has been decreased up to 168 h, and when the switch was off, then the maximum power (P_{max}) has been increased up to 336 h. Again when the switch is on, then the maximum power (P_{max}) has been decreased up to 504 h. Finally, when the switch is off, then the maximum power (P_{max}) has been increased up to 672 h. This study was for 672 h. The switching system shows that the PKL electricity has pulsating nature.

Figure 7 shows the variation of energy (Wh) with the variation of time duration (h). It is shown that when switch was on, then the energy (Wh) has been increased linearly up to 168 h, and when the switch was off, then the energy (Wh) has been increased also linearly up to 336 h. Again when the switch is on, then the energy (Wh) has been increased up to 504 h. Finally, when the switch is off, then the energy (Wh) has been increased up to 672 h. This study was for 672 h. The switching system shows that the PKL electricity has ohmic nature.

Figure 8 shows the variation of power density (W/kg) with the variation of time duration (h). It is shown that when switch was on, then the power density (W/kg) was constant up to 168 h, and when the switch was off, then the power density (W/kg) was also constant up to 336 h. Again when the switch is on, then the power

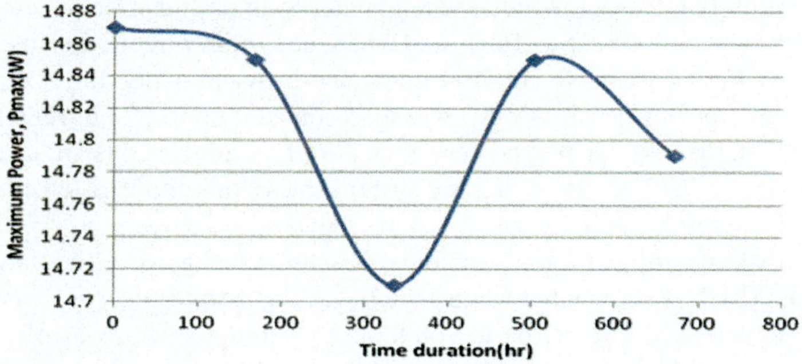

Fig. 6 P_{max} versus T.D. (h) curve

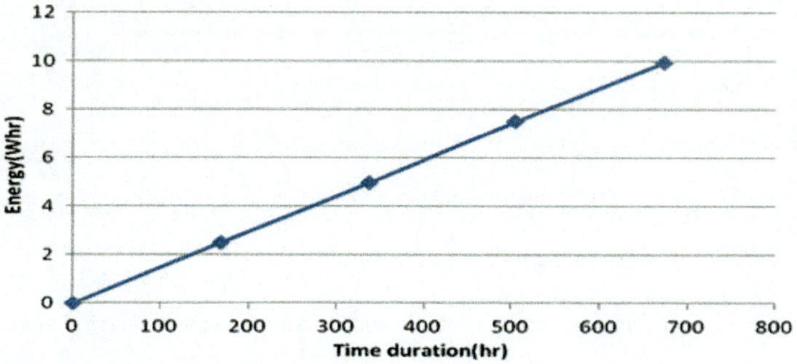

Fig. 7 Energy (Wh) versus T.D. (h) graph

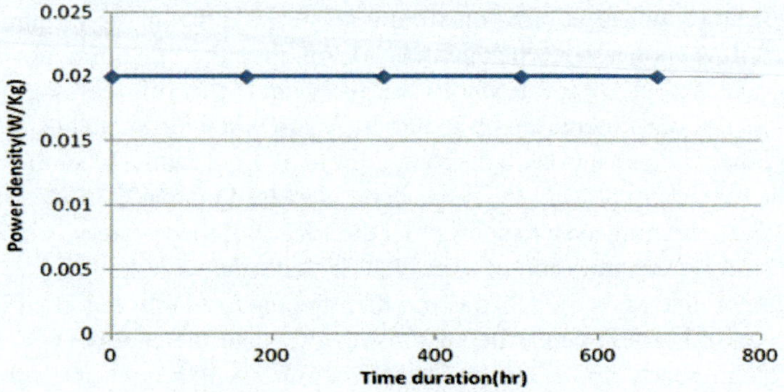

Fig. 8 Power density (W/kg) versus T.D. (h) curve

density (W/kg) was also constant up to 504 h. Finally, when the switch is off, then the power density (W/kg) was also constant up to 672 h. This study was for 672 h. The switching system shows that the PKL electricity has the pure direct current (DC) nature. That means PKL electricity has constant amplitude with changing time.

Figure 9 shows the variation of specific power density (W/L) with the variation of time duration (h). It is shown that when switch was on, then the specific power density (W/L) was constant up to 168 h, and when the switch was off, then the specific power density (W/L) was also constant up to 336 h. Again when the switch is on, then the specific power density (W/L) was also constant up to 504 h. Finally, when the switch is off, then the specific power density (W/L) was also constant up to 672 h. This study was for 672 h. The switching system shows that the PKL electricity has the pure direct current (DC) nature.

Figure 10 shows the variation of energy density (Wh/kg) with the variation of time duration (h). It is shown that when switch was on, then the energy density (Wh/kg) has been increased linearly up to 168 h, and when the switch was off, then the energy density (Wh/kg) has been increased also linearly up to 336 h. Again when

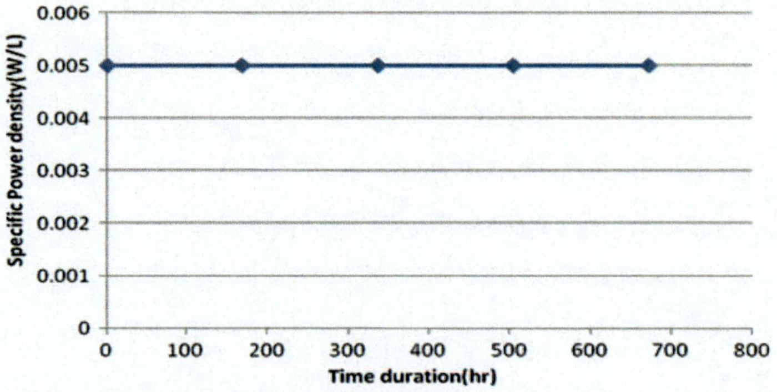

Fig. 9 Specific power density (W/L) versus T.D. (h) curve

the switch is on, then the energy density (Wh/kg) has been increased up to 504 h. Finally, when the switch is off, then the energy density (Wh/kg) has been increased up to 672 h. This study was for 672 h. The switching system shows that the PKL electricity has ohmic nature.

Figure 11 shows the variation of specific energy density (Wh/L) with the variation of time duration (h). It is shown that when switch was on, then the specific energy density (Wh/L) has been increased linearly up to 168 h, and when the switch was off, then the specific energy density (Wh/L) has been increased also linearly up to 336 h. Again when the switch is on, then the specific energy density (Wh/L) has been increased up to 504 h. Finally, when the switch is off, then the specific energy density (Wh/L) has been increased up to 672 h. This study was for 672 h. The switching system shows that the PKL electricity has ohmic nature.

Figure 12 shows the variation of charge capacity (mAh) with the variation of time duration (h). It is shown that when switch was on, then the charge capacity (mAh) has been increased linearly up to 168 h, and when the switch was off, then the charge capacity (mAh) has been increased also linearly up to 336 h. Again when the switch

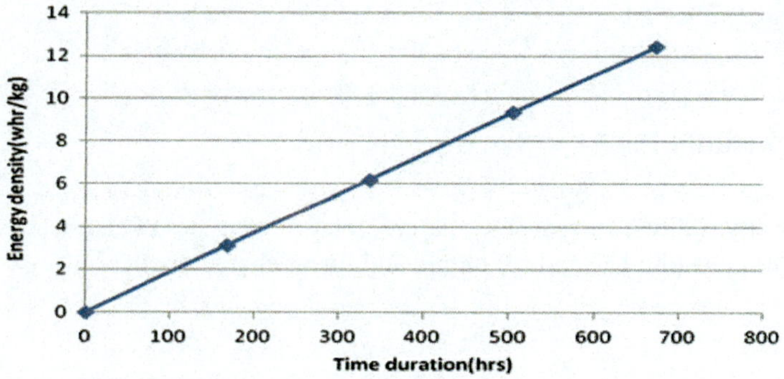

Fig. 10 Energy density (Wh/kg) versus T.D. (h) curve

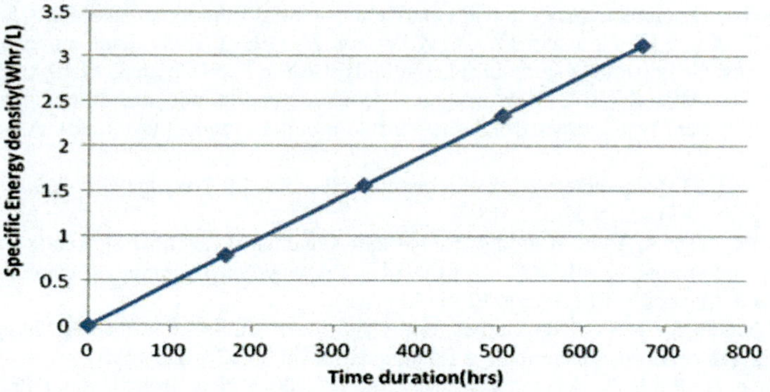

Fig. 11 Specific energy density (Wh/L) versus T.D. (h) curve

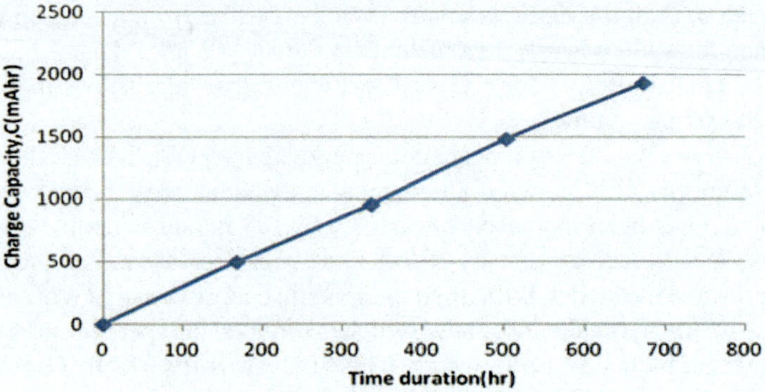

Fig. 12 Charge capacity (mAh) versus T.D. (h) curve

is on, then the charge capacity (mAh) has been increased up to 504 h. Finally, when the switch is off, then the charge capacity (mAh) has been increased up to 672 h. This study was for 672 h. The switching system shows that the PKL electricity has ohmic nature.

4 Conclusions

PKL electricity is direct current (DC) nature. It has three types of characteristics: (a) pure DC, (b) pulsating DC and (c) ohmic and linear characteristics.

References

1. Saad B, Abdelmoneim I, Adam G, Elghazali L (2006) Traditional Arab herbal medicine, evidence-based complementary and alternative medicine. J Ethnopharmacol 90:625–627
2. Larrey DA, Chan JE, Ngogang JY, Cai X, Vinson JA (1994) Antioxidant capacity of some herbs/spices: a comparative study of two methods. J Agric Food Chem 53:6819–6824
3. Jagessar RC, Allen R (2012) Phytochemical screening and atomic absorption spectroscopic studies of solvent type extract from leaves of *Terminalia catappa*, (Almond). Acad Res Int 3:17–26
4. Hamid MR (2013) Characterization of a battery cell fueled by *Bryophyllum pinnatum* sap. Int J Sci Eng Res 9(3). ISSN 2229-5518
5. Hamid MR, Yusuf A, Abdul Wadud AM, Mosfiqur Rahaman M (2016) Design and performance test of a prototype of a 12 volt DC battery fueled by *Bryophyllum pinnatum* sap and improvement of its characteristics. Int J Electron Electr Eng 4(5)
6. Pal S, Chaudhuri AKN (1991) Studies on the anti-ulcer activity of a *Bryophyllum pinnatum* leaf extract in experimental animals. J Ethanopharmacol 33:97–102
7. Supratman U, Fujita T, Akiyama K, Hayashi H (2000) New insecticidal bufadienolide, bryophyllin C, from *Kalanchoe pinnata*. Biosci Biotechnol Biochem 64:1310–1312
8. Ojewole JAO (2005) Antinociceptive, anti-inflammatory and antidiabetic effects of *Bryophyllum pinnatum* (Crassulaceae) leaf aqueous extract. J Ethnopharmacol 99:13–19

9. Asiedu-Gyekye IJ, Antwi DA, Bugyei KA, Awortwe C (2012) Comparative study of two kalanchoe species: total flavonoid, phenolic contents and antioxidant properties. Afr J Pure Appl Chem 6:65–73
10. Khan KA (1999) Copper oxide coatings for use in a linear solar Fresnel reflecting concentrating collector. Renew Energy 17(4):603–608
11. Ruhane TA, Tauhidul Islam M, Saifur Rahman M, Bhuiyah MMH, Islam JMM, Bhuiyah TI, Khan KA, Khan MA (2017) Impact of photo electrode thickness annealing temperature on natural dye sensitized solar cell. Sustain Energy Technol Assess. https://doi.org/10.1016/j.seta.2017.01.012
12. Ruhane TA, Tauhidul Islam M, Saifur Rahaman M, Bhuiyan MMH, Islam JMM, Newaz MK, Khan KA, Khan MA (2017) Photo current enhancement of natural dye sensitized solar cell by optimizing dye extraction and its loading period. Optik
13. Hasan M, Khan KA (2018) Dynamic model of *Bryophyllum pinnatum* leaf fueled BPL cell: a possible alternate source of electricity at the off-grid region in Bangladesh. Microsyst Technol. ISSN 0946-7076. https://doi.org/10.1007/s00542-018-4149-y
14. Khan KA, Hazrat Ali M, Obaydullah AKM, Wadud MA (2019) Production of candle using solar thermal technology. Microsyst Technol 25(12). ISSN 0946-7076. https://doi.org/10.1007/s00542-019-04390-7
15. Khan KA, Rasel SR, Ohiduzzaman M (2019) Homemade PKL electricity generation for use in DC fan at remote areas. Microsyst Technol 25(12). ISSN 0946-7076. https://doi.org/10.1007/s00542-019-04422-2
16. Hasan M, Khan KA (2019) Experimental characterization and identification of cell parameters in a BPL electrochemical device. SN Appl Sci 1:1008. https://doi.org/10.1007/s42452-019-1045-8

Analysis of a Power System Under Contingency Using Load Flow Studies

Arup Halder, Prakash Chakrabarty, Sayan Banik, Subhayan Saha, Sukanya Dasgupta, Somenath Guha, Bishaljit Pal, Sudhanshu Sarkar, and Ambarnath Banerji

Abstract As per the perceived generation and load, a power system requires conducting load flow studies. Further analysis of the performance of the power system under contingencies and for future expansion is an important matter. This paper analyzes the power system using various standard IEEE buses. The load flow studies on the MATLAB. The platform uses the MATHSWORK library programs. This analysis forecasts the maximum loading of any line under contingencies. It allows for coordination of the various protection relays and modification of the power system to cater to the contingencies.

Keywords Load flow studies · Contingencies · IEEE buses · Power flow · Line overloading

1 Introduction

Designing a proposed power system with a proposed set of loads and a proposed set of generation capacity requires some power calculation to understand the power system's electric response. It is not desirable to build the power system and subject it to the generation and load and then see the response because, in that case, the power system may be subjected to dangerous/damaging conditions. A load flow study estimates the voltage and currents, and power flow in various power system lines [1, 2]. IEEE Std 3002.2 lays down the methods used for conducting the load flow studies. According to this standard—load flow are also known as power flow. It is the name of a network solution that predicts the steady state, voltage and active and reactive power flowing through each branch and bus of a system. Power flow studies can

---

A. Halder · P. Chakrabarty · S. Banik · S. Saha · S. Dasgupta · S. Guha · B. Pal · S. Sarkar · A. Banerji (✉)
Department of Electrical Engineering, Narula Institute of Technology, Kolkata, West Bengal, India
e-mail: ambarnathbanerji@gmail.com

S. Sarkar
e-mail: sudhangshu.sarkar@nit.ac.in

J. R. Szymanski et al. (eds.), *Energy Systems, Drives and Automations*, Lecture Notes in Electrical Engineering 1057,
https://doi.org/10.1007/978-981-99-3691-5_8

simulate operating conditions whose exposure can be dangerous to the actual physical systems because the system has not been built to operate under such operating conditions and it is not wise to expose the real physical system to such conditions. The ultimate goal of destructive power flow studies is not always to obtain rigorous numerical performance parameters. Often, the goal is to gain insight into how the system behaves under a range of operating conditions. Power flow is an important part of power system operation and planning. Since transmission and distribution lines, wires and the parameters of components such as transformers are constant, so the impedance of the power system network is mostly constant. However, power flow problems typically involve constant kVA loads, generators and tap changers, meaning that the relationship between voltage and current varies depending on the type of load. Busbar active and the same applies to the relationship between the reactive power consumption or the active power produced by the generator busbar and the rated voltage level. Therefore, load flow calculations solve equations involving constant impedance, constant power and sometimes constant current loads [3]. A large section of the modern-day load is sensitive loads. They are sensitive to voltage variation and require highly regulated voltage. Indulkar and Ramalingam present a load flow analysis of power systems having sensitive loads [4, 5]. Electric Vehicles (E.V.s) and their charging stations increase the load on the present-day power systems. A load flow analysis is carried out by incorporating an E.V. model in the power system model designed on P.S.A.T. to study the impact of ever-increasing E.V. loads [6]. The recent years have seen more and more generation of electrical power using renewable energy resources, which are connected to the distribution part of the power system [7]. Gianto has proposed a load flow study incorporating the steady model of the wind power plant [8]. With both generation and load, there is a certain number of uncertainties. The load on any bus is estimated. It is difficult to assign a value with certainty. Saraiva et al. [9] proposes fuzzy-based load flow studies and compares its performance with Monte Carlo-based simulation.

Various conditions are incorporated, and the load flow studies are carried out for steady-state analysis of the power system. Also, various devices are connected to the power system to improve the load flows. The distribution system has feeders coming out of a single substation and a tree-like structure of radial feeders. The normal load flow studies for transmission systems cannot be directly applied to a distribution system. Bahmanyar et al. [10] proposes a modified load flow study for distribution systems using the non-synchronized measurements collated from the distribution substation and the terminals of the distributed generators and that of the microgrid. The communication system of the smart grid is used for collecting these data. The synchronization angle is considered an unknown variable and is determined as a part of the load flow study. Several computational tools are available for conducting load flow studies. Some of them are N.E.P.L.A.N., Power World, P.S.A.T. and MANPOWER. Kumar et al. [11] provides a comparison of these tools for conducting power system analysis based on load flow studies.

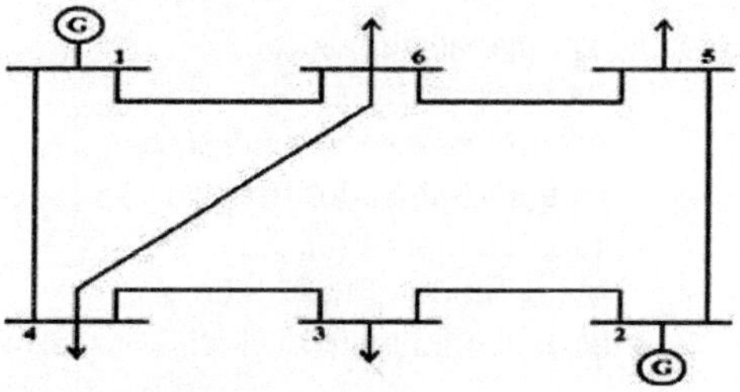

Fig. 1 Schematic diagram of IEEE 6 bus system

The present paper delves into the load flow analysis of a proposed power system. The power system is subjected to contingencies like tripping a line and adding a line. The load flow studies are carried out to estimate the loading of the various lines under the contingencies. This study estimates power flow in the various lines that exceed the relay tripping settings and cause further or cascaded tripping of the lines.

2 The Architecture of the Power System

To carry out the load flow studies in the laboratory, IEEE bus systems are used like the IEEE 6 Bus, 9 Bus, etc. The circuit diagram of the IEEE 6 Bus is given in Fig. 1.

Present work has included IEEE 6 Bus system which is taken as a standard value for comparison purposes as shown in as Table 2.

3 Algorithm for Load Flow Studies

Newton–Raphson method is a repetitive process to solve several nonlinear equations which contain equal number unknowns. Newton–Raphson method has two kind of solutions. First one uses rectangular coordinates for variables and the second one uses the form of polar coordinate. In between them, the polar coordinate form is popular to be used. Basically, to solve algebraic nonlinear equations, N-R method is used, that gives a quicker answer and proper convergence than the Gauss–Seidal method.

Power flow equations:

$$P_i(\text{Real Power}) = |V_i| \sum_{j=1}^{m} \left(|V_j||Y_{ij}| \cos(\phi_{ij} + \delta_j - \delta_i) \right) \tag{1}$$

$$Q_i(\text{Reactive Power}) = -|V_i| \sum_{j=1}^{m} \left(|V_j||Y_{ij}| \cos(\phi_{ij} + \delta_j - \delta_i) \right) \tag{2}$$

V_i = voltage at ith bus
V_j = voltage at jth bus
Y_{ij} = admittance of ith and jth bus
ϕ_{ij} = angle of the admittance
δ_j = phase angle of the jth bus
δ_i = phase angle of the ith bus.
J is the Jacobian matrix which is used for solving the N.R. method.

$$J = \begin{bmatrix} \frac{\mathrm{d}p}{\mathrm{d}\delta} & \frac{\mathrm{d}p}{|V|} \\ \frac{\mathrm{d}Q}{\mathrm{d}\delta} & \frac{\mathrm{d}p}{|V|} \end{bmatrix} \tag{3}$$

$$J = \begin{bmatrix} Y_{11} & \ldots\ldots & Y_{ij} \\ \ldots\ldots & \ldots\ldots & \ldots\ldots \\ Y_{ji} & \ldots\ldots & Y_{jj} \end{bmatrix}, \tag{4}$$

where Y_{ij}, is known as the bus matrix.

Step 1: Consider one bus as slack bus in a system whose voltage and phase angle are $1 \angle 0$ and assume all all-other buses as P.Q. and P.V. buses.

Step 2: In the rth iteration,

$$P_i^r = |V_i|^r \sum_{j=1}^{m} \left(|V_j||Y_{ij}| \cos(\phi_{ij} + \delta_j - \delta_i) \right) \tag{5}$$

$$Q_i^r = -|V_i|^r \sum_{j=1}^{m} \left(|V_j||Y_{ij}| \cos(\phi_{ij} + \delta_j - \delta_i) \right) \tag{6}$$

Let,

$$e_i^r = |V_i|^r \cos \delta_i^r \text{ and } f_i^r = |V_i|^r \sin \delta_i^r \tag{7}$$

$$G_{ij} = |Y_{ij}| \cos \phi_{ij} \tag{8}$$

$$B_{ij} = |Y_{ij}| \sin \phi_{ij} \tag{9}$$

Then Calculate,

$$\Delta P_i^r = P_i(\text{Scheduled}) - P_i^r \text{ for PV and PQ buses} \tag{10}$$

$$\Delta Q_i^r = Q_i(\text{Scheduled}) - Q_i^r \text{ for PQ buses} \tag{11}$$

Here if all the values of ΔP_i^r and ΔQ_i^r are less than the tolerance, iterations are stopped, calculate $P1$ and $Q1$ and the solution is obtained.

Step 3: Calculate the Jacobian matrix elements if the convergence criteria are not obtained.

Step 4: Correction of voltage magnitude and phase angles.

Step 5: Next update the voltage magnitude and phase angles.

$$e_i^r = |V_i|^r \cos \delta_i^r \text{ and } f_i^r = |V_i|^r \sin \delta_i^r \tag{12}$$

$$G_{ij} = |Y_{ij}| \cos \phi_{ij} \tag{13}$$

$$B_{ij} = |Y_{ij}| \sin \phi_{ij} \tag{14}$$

The algorithm for the load flow studies is taken from the MATHLAB library. The algorithm has provisions for selecting the type of IEEE bus from several listed IEEE buses like 6 Bus, 9 Bus, 14 Bus, 26 Bus, 30 Bus and 57 Bus. The algorithm also provides options to select from the method to conduct the iterations of the load flow studies. The method options to select from are N.R., G.S. and Fast Decoupled.

4 Results of Load Flow Calculation with Contingencies

Displayed equations the power system to be studied is represented by the IEEE six-bus–seven-line power system shown in Fig. 1. The proposed bus voltages, the scheduled generation, and loads are presented in Table 1. The transmission line admittance and the entire line charging admittance at each bus is also presented in Table 1. The load flow algorithm is run and a 6 Bus system is selected from the option for buses. Then against the options for selection of the method. Newton–Raphson method is selected. This method uses the bus admittance matrix to obtain a load flow solution. The results are presented below. Table 2 gives bus data under normal steady-state.

Table 1 Some parameters of IEEE 6 bus system

Line characteristics				
Line No.	Starting busbar	Ending busbar	Line impedance's	
			R (Ω)	X (Ω)
1	6	3	4.88187	20.55942
2	6	4	3.17520	14.68530
3	4	3	3.84993	16.15383

Table 2 Bus data under normal steady state

Bus No.	Voltage mag.	Angle degree	Load		Generation	
			MW	Mvar	MW	Mvar
1	1.050	0.000	0.000	0.000	0.117	– 73.549
2	1.100	– 1.012	0.500	0.000	0.000	13.234
3	1.110	– 1.127	0.600	0.000	0.000	– 1.016
4	1.091	– 0.669	0.000	0.000	0.700	0.700
5	1.103	– 0.934	0.000	0.000	0.700	0.700
6	1.110	– 1.118	0.000	0.000	0.700	0.700
Total			1.100	0.000	2.217	– 59.232

Power Flow Solution by Newton–Raphson Method

Elapsed time is 0.133038 s

Maximum Power Mismatch = 0.000282118

No. of Iterations = 9.

The load flow solution using Newton–Raphson method is presented in Table 3. It presents the power (MW and MVAR) at the different buses and the power flow in different lines line connecting the buses. Table 3 also presents the power losses incurred in the various lines.

A contingency is created by removal of a line between Bus 2 and Bus 4. Again a load flow study was carried out. The results are presented below.

Elapsed time is 0.036595 s

Power Flow Solution by Newton–Raphson Method

Maximum Power Mismatch = 0.000249826

No. of Iterations = 9.

The forced tripping of the line between the buses 2 and 4 results in a contingency which changes the bus data namely the voltage, load and generation on the different buses. Table 4 presents these new Bus data. This contingency also results in changes in the power (MW and MVAR) at different buses and the power flow in the various lines. Table 5 presents these powers at the buses, the power flow in the lines and the losses in the various lines.

Table 3 Line flow and losses

Line		Power at bus and line flow			Line loss	
From	To	M.W.	Mar	M.V.A.	MW	Mar
1		0.117	73.549	73.549		
	2	– 2.308	27.211	27.309	0.572	– 3.481
	3	1.218	24.098	24.129	0.218	– 3.714
	5	1.249	22.238	22.273	0.261	– 5.980
2		– 0.500	13.234	13.243		
	1	2.880	23.730	23.904	0.572	– 3.481
	3	0.104	– 8.050	8.050	0.008	– 7.286
	4	– 1.885	9.424	9.605	0.048	– 2.304
	5	– 0.851	– 3.339	3.446	0.001	– 4.851
	6	– 0.778	– 8.490	8.525	0.018	– 6.057
3		– 0.600	– 1.016	1.180		
	2	– 0.096	0.764	0.770	0.008	– 7.286
	5	– 0.216	– 0.110	0.242	0.009	– 6.105
	6	– 0.288	– 1.655	1.680	0.000	– 6.057
4		0.700	0.700	0.990		
	1	– 1.000	20.384	20.408	0.218	– 3.714
	2	1.903	– 11.728	11.882	0.048	– 2.304
	5	– 0.203	– 7.995	7.958	0.017	– 9.597
5		0.700	0.700	0.990		
	1	– 0.988	16.258	16.288	0.261	– 5.980
	2	0.853	– 1.511	1.735	0.001	– 4.851
	3	0.225	– 5.995	5.999	0.009	– 6.105
	4	0.220	– 1.642	1.657	0.017	– 9.597
	6	0.390	– 6.409	6.421	0.006	– 7.332
6		0.700	0.700	0.900		
	2	0.795	2.432	2.559	0.018	– 6.057
	3	0.288	– 0.810	0.860	0.000	– 2.465
	5	– 0.383	– 0.922	0.999	0.006	– 7.332

Table 4 Bus data after contingency

Bus No.	Volt mag.	Angle degree	Load		Generation	
			MW	Mvar	MW	Mvar
1	1.050	0.000	0.000	0.000	0.117	– 73.549
2	1.100	– 1.012	0.500	0.000	0.000	13.234
3	1.110	– 1.127	0.600	0.000	0.000	– 1.016

(continued)

Table 4 (continued)

Bus No.	Volt mag.	Angle degree	Load		Generation	
			MW	Mvar	MW	Mvar
4	1.091	− 0.669	0.000	0.000	0.700	0.700
5	1.103	− 0.934	0.000	0.000	0.700	0.700
6	1.110	− 1.118	0.000	0.000	0.700	0.700
Total			1.100	0.000	2.217	− 59.232

Table 5 Line flow and losses after contingency

Line		Power at bus and line flow			Line loss	
From	To	M.W.	Mar	M.V.A.	MW	Mar
1		0.223	− 75.141	75.141		
	2	− 2.316	− 27.207	27.306	0.572	− 3.841
	4	1.222	− 24.376	24.407	0.224	− 3.694
	5	1.354	− 23.555	23.594	0.299	− 5.863
2		− 0.500	12.329	12.339		
	1	2.888	23.726	23.901	0.572	− 3.481
	3	− 0.000	− 3.631	3.631	0.000	− 7.392
	4	− 1.958	8.891	9.104	0.044	− 2.314
	5	− 0.844	− 4.700	4.775	0.005	− 4.856
	6	− 0.586	− 11.920	11.934	0.046	− 6.010
3		− 0.600	1.425	1.546		
	2	0.000	− 3.762	3.762	0.000	− 7.393
	5	− 0.002	2.478	2.478	0.030	− 6.134
	6	− 0.599	2.724	2.789	0.003	− 2.488
4		0.700	0.700	0.990		
	1	− 0.998	20.682	20.706	0.224	− 3.694
	2	2.001	− 11.206	11.383	0.044	− 2.314
	5	− 0.303	− 8.776	8.781	0.027	− 9.615
5		0.700	0.700	0.990		
	1	− 1.055	17.692	17.723	0.299	− 5.863
	2	0.849	− 0.156	0.863	0.005	− 4.856
	3	0.032	− 8.612	8.612	0.030	− 6.134
	4	0.331	− 0.839	0.902	0.027	− 9.615
	6	0.544	− 7.385	7.405	0.011	− 7.382
6		0.700	0.700	0.900		
	2	0.632	5.910	5.943	0.046	− 6.010
	3	0.601	− 5.213	5.247	0.003	− 2.488
	5	− 0.533	0.003	0.533	0.011	− 7.382

5 Conclusion

Here in this paper, results of a neat load flow study using Newton–Raphson method is presented. Also, a clear difference between normal condition and fault condition has been shown. By this study, the effect of any expansion or deletion of transmission lines in the power system can be evaluated. The effect on the relay coordination can be estimated in advance and incorporated in the design of the power system. It can be used on Transformer, SVC, Wind Power, Sensitive load utilization instruments, Electric Vehicle, etc. Here, in this paper, the transformer's characteristics are avoided which can be used for the future work.

References

1. Shipley RB (1976) Introduction to matrices and power systems. Wiley, New York
2. Gomez-Exposito A, Conejo AJ, Canizares C (2009) Electric energy systems: analysis and operation. C.R.C. Press
3. IEEE STD.3002.2 (2018) Recommended practice for conducting load-flow studies and analysis of industrial and commercial power systems. Technical Books Coordinating Committee of the IEEE Industry Applications Society
4. Banerji A, Biswas SK, Singh B (2016) Enhancing quality of power to sensitive loads with microgrids. IEEE Trans Ind Appl 52(1):360–368
5. Indulkar CS, Ramalingam K (2008) Load flow analysis with voltage-sensitive loads. In: Proceedings of 2008 joint international conference on power system technology and IEEE power India conference
6. Kongjeen Y, Bhumkittipich K (2016) Modeling of electric vehicle loads for power flow analysis based on P.S.A.T. In: Proceedings of 2016 13th international conference on electrical engineering/electronics, computer, telecommunications and information technology (ECTI-CON)
7. Banerji A, Bhadra R, Banerji A, Banerji A, Biswas SK (2020) A renewable energy based hybrid autonomous grid. In: IEEE Kolkata conference FALCON 2020, 28–29 Feb 2020
8. Gianto R (2020) Steady-state model of wind power plant for load flow study. In: Proceedings of international seminar on intelligent technology and its applications (S.I.T.I.A.)
9. Saraiva JT, Miranda V, Matos MACC (1991) Generation and load uncertainties incorporated in load flow studies. In: Proceedings of IEEE 6th Mediterranean electrotechnical conference
10. Bahmanyar A, Estebsari A, Bahmanyar A, Bompard E (2017) Nancy load flow: smart grid load flow using non-synchronized measurements. In: 2017 IEEE international conference on environment and electrical engineering and 2017 IEEE industrial and commercial power systems Europe (EEEIC/I&CPS Europe)
11. Kumar Y, Devabhaktuni VK, Vemuru S (2015) Comparison of power system simulation tools with load flow study cases. In: Proceedings of 2015 IEEE international conference on electro/information technology (E.I.T.)

Analysis of Different Pulse Width Modulation Methods for Cascaded Multilevel Inverter Used as a Shunt Active Power Filter

Srinivas Depally, N. Bhoopal, and K. Naga Sujatha

Abstract In the Present era, the utilization of non-linear loads and power electronic equipment has enhanced, which causes degeneration voltage and shows an impact on voltage for consumers at a particular point, i.e. various consumers of common coupling point. In a power distribution system, reduce or compensate the voltage and current disturbances with the help of two configurations stand-alone APFs, either active or passive; these consist of series and shunt filters. Compared to series active filter, shunt filter is a cost-effective method for harmonics compensation in low and medium power systems. Based on the harmonics, the control scheme is selected to reduce the harmonics, and the 3-phase shunt filter is used for the Series *H*-Bridge Multilevel inverter. This paper introduces a unipolar multicarrier pulse width modulation technique for the multilevel inverter. The total Harmonic Distortion was investigated with the prospective influence of the multicarrier pulse width. The proposed sinusoidal pulse width modulation method called carrier-based pulse width modulation was simulated using the MATLAB/SIMULINK.

Keywords Power quality · Voltage source inverter · Active power filter · Non-linear load · Multilevel inverter · Modulation techniques · Total harmonic distortion

S. Depally (✉) · K. N. Sujatha
Ph.D Scholar, Department of Electrical and Electronics Engineering, JNTU, Hyderabad, Telangana, India
e-mail: shrinudepally@gmail.com

N. Bhoopal
Department of Electrical and Electronics Engineering, BV Raju Institute of Technology, Medak, Telangana, India
e-mail: bhoopal.neerudi@bvrit.ac.in

K. N. Sujatha
Professor & HoD, EEE Department, Senior IEEE member, JNTU, Hyderabad, Telangana, India

J. R. Szymanski et al. (eds.), *Energy Systems, Drives and Automations*,
Lecture Notes in Electrical Engineering 1057,
https://doi.org/10.1007/978-981-99-3691-5_9

1 Introduction

The term PQ means power quality, which is a significant concern in industries for better performance of motors to operate a drive with the product quality. The PQ is one of the most important things. The power is the combination of voltage and current. The shape of the voltage and current waveforms have to be maintained in the form of pure sinusoidal. Extensive use of static power converter leads to the harmonics in the voltage and current (this will show the effect on the quality of electric power). Due to the voltage sag, voltage swell, and voltage interruption, zero and negative sequence components originated using single-phase and unbalanced loads, etc. [1].

Some methods are used to increase the power quality, like passive filters and active power filters. Still, passive filters have more disadvantages such as fixed compensation, volume, electromagnetic interference, possible resonances. The active power filter is the best approach; the design of stand-alone APFs, [2] can be made either active or passive, including the series filter and shunt filter. Compared to series filter, shunt filter is a cost-effective method for compensation of harmonic in low and medium power system, the range of the power system depends on the length of the transmission lines and range of the voltages [3].

For different voltages ratings, they are named as different levels like low voltage level, medium voltage level, and high voltage level. For the voltage at a low level, traditional two-level inverters are used. To gain the required level of voltage for medium voltage levels, the active devices are needed, which has been to connect in series with the system, otherwise plug-in the transformers between the traditional inverter and source terminals. The resource is the one it is helpful to attain the necessary voltage with the help of low voltage rating devices that is multilevel inverter [4]. Therefore by this work, the shunt APF is acknowledged utilizing the cascaded *H*-Bridge Multilevel inverter; the multilevel inverter is not designed for over modulation. In shunt active power filters, the demanded output is sufficiently large. Whenever the modulation exceeds beyond the limits, i.e. (over modulation) extinguish Ma than the modulation index waveform that is in the shape of a triangular accordingly realistic fundamental triangular component does not follow but linearly, and the control is saturated as the over modulation increases, the modulator loses control over the waveform in the period of saturation intervals, and then the output initiates additional harmonics of the fundamental component [5, 6]. To achieve lossless control over the fundamental voltage waveform in the over modulation region, various pulse width modulation strategies are possible.

There are various types of PWM strategies available. It is started with single pulse width modulation and multiple pulse width modulation. The sinusoidal pulse width modulation technique was one of the well-known methods of multiple pulse width modulation. We need to achieve lossless control and operate the cascaded multilevel inverter in the over modulation region. This can be achieved using various sinusoidal pulse width modulation methods incorporating multicarrier methodologies.

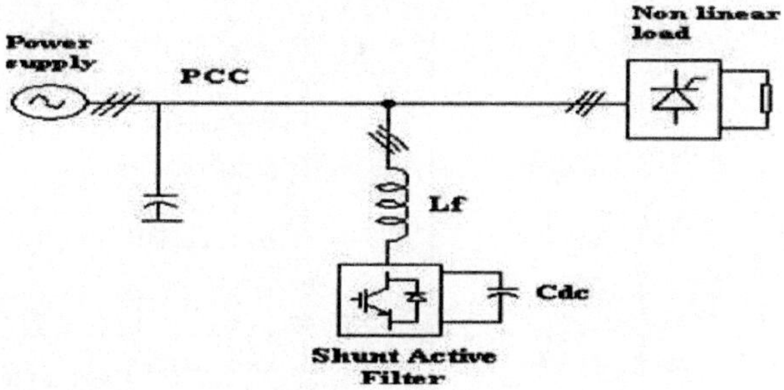

Fig. 1 Shunt active power filter block diagram

2 Shunt Active Power Filter

To address the power quality issues and get the possible solutions are the main reasons behind the design of Active Power Filter [7].

The functions of Active Power Filter:

1. Injecting the Compensating currents by withdrawing harmonic currents.
2. Eliminating the reactive power from fundamental frequency components
3. Preserve a constant dc voltage at the capacitor. The connection diagram of Shunt Active Power Filter is shown in Fig. 1.

The Multilevel Voltage Source is connected to the inductance along with the Capacitor these are connected to the line [8].

3 Cascaded *H*-Bridge Multilevel Inverter

The circuit diagram of 3phase *H*-Bridge Multilevel inverter is shown in Fig. 2. It uses the less Components as compare to other multilevel inventers [9], so then harmonics also less than the losses will be less like this in the cascaded multilevel inverter having many advantages because of this reasons which is more preferable in the applications of power systems and drives [10].

4 Different Multicarrier PWM Methods

A lot of PWM schemes are presented in this paper we are discussing a proposed control scheme and different PWM control techniques.

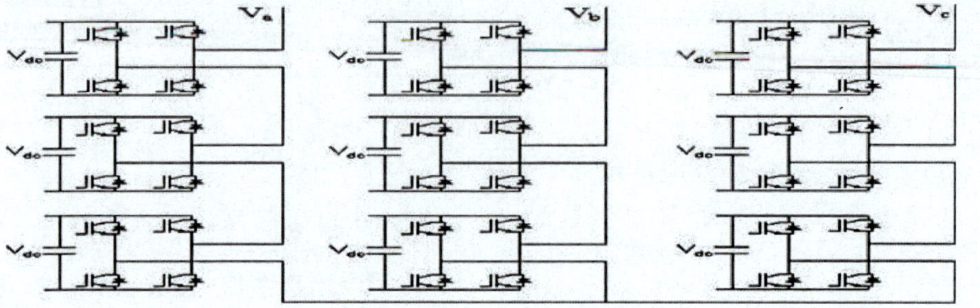

Fig. 2 Seven level cascaded H-Bridge inverter

4.1 Phase Shifted Pulse Width Modulation

To attain the harmonic sideband cancellation, additionally phase shifting is required, which shifts the carrier of each bridge. It is the principle of phase-shifted PWM. It can also occur in the multiple groups around the even carrier [11, 12]. The waveform of the seven level cascaded H-Bridge inverter with the phase shift PWM technique is shown in Fig. 3. We can reach the optimum harmonic cancellation by shifting the phase in each carrier wave displaced by $(K - 1)/n$, k indicates the kth inverter. The number of single-phase inverters connected in the series is represented with the letter "n". The series connection of three H-Bridges having the phase shift of 60° and attained the side bands of harmonics cancellation around the multiples of 6fc. The cancellation of harmonics is independent of the carrier/ Fundamental frequency ratio [13].

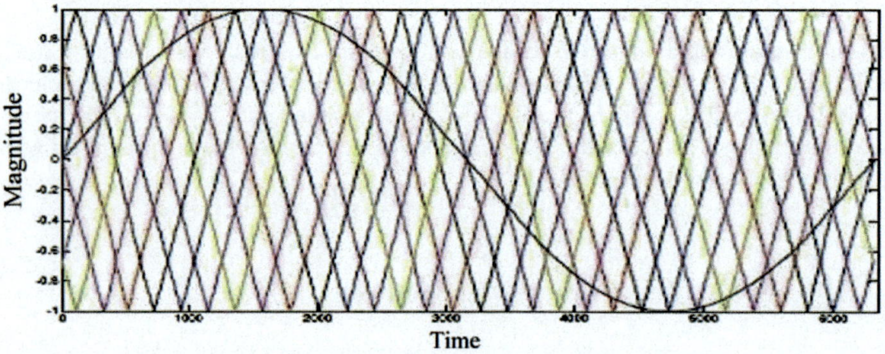

Fig. 3 Reference and carrier waveforms of PSPWM technique for three level cascaded H-Bridge inverter

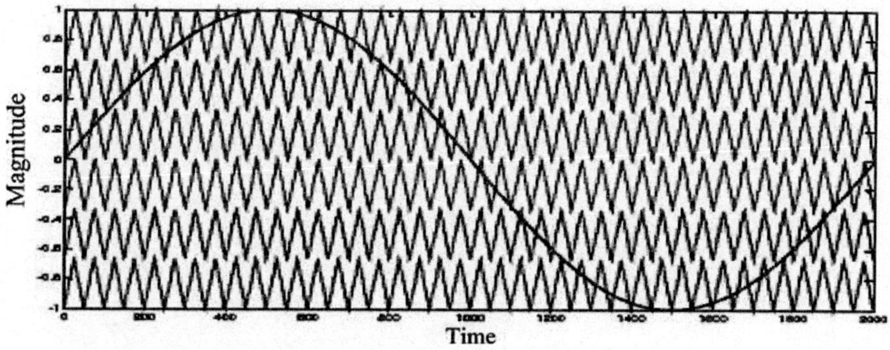

Fig. 4 Phase disposition PWM method for three level bridge

4.2 Disposition of Carrier Wave PWM

In the CDPWM method, to achieve the modulation, $L - 1$ triangular carriers are required. Here "L" indicates the levels of the voltage. To occupy the continuous bands in the order, the carries have to be arranged; the order is $(L - 1)$Vdc/2 to $(L - 1)$Vdc/2 (for L odd) [12]. To find out the level of switched voltage, the reference of a single sinusoidal has to be compared with the carriers. For the CDPWM technology level of freedom is given as:

$$M_a = \frac{2 \times A_r}{(L - 1) \times A_c} \tag{1}$$

where A_r, A_c, and L indicates the Amplitude of Reference wave, amplitude of the carrier waveform and number of levels. This Carrier Disposition PWM strategy includes PDPWM, PODPWM, and APODPWM full forms. The total Number of carriers in the PDPWM technology are in phase across the total number of bands described in Fig. 4. The explanation of PDPWM is given in the Fig. 5. In the PODPWM technique the total number of carriers are divided into two categories.

1. The carriers above the zero reference value
2. The carriers below the zero reference value.

Carriers above zero reference carriers are also in phase but these are 180° out of phase with the reference above zero. The carriers of the adjacent APODPWM bands are 180° out of phase (Fig. 6).

4.3 Unipolar Multicarrier PWM Schemes

The pulse width modulation of the unipolar multicarrier scheme is obtained by differentiating the reference of Rectified sinusoidal waveform with above the zero level

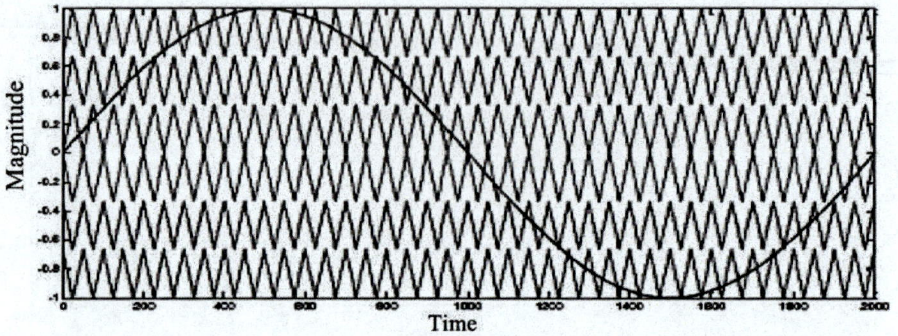

Fig. 5 Phase opposition and disposition PWM technique for three level *H*-Bridge inverter

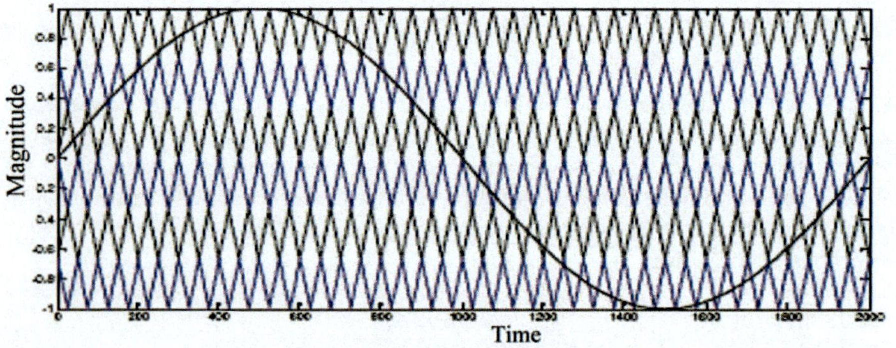

Fig. 6 Alternate phase opposition and disposition PWM technique for three level *H*-Bridge inverter

of multicarriers shown in Fig. 7. In unipolar carrier disposition, the carriers are portioned above the zero reference [1]. In this type of scheme, the number of carriers are depends on the "*n*" number of cells of cascaded *H*-Bridge inverter, distinct from the above schemes. In other techniques, the number of output voltages depends on the number of carriers. Ex: CDPWM in this technique, only three carrier waves are required to generate the seven levels. The grade of choice for this scheme is given by:

$$M_a = \frac{A_r}{n \times A_c} \tag{2}$$

where A_r is the amplitude of the reference and A_c is the amplitude of carrier and "*n*" is the number of *H*-Bridge cells The carriers of the unipolar and sinusoidal reference PSPWM are shifted all the carriers are arrange above the zero level. To obtain the seven level output voltage to the *H*-Bridge inverter the carried are phase shifted by 60°, to obtain seven levels. The unipolar-phase shifted PWM for seven level inverter is shown in the Fig. 8.

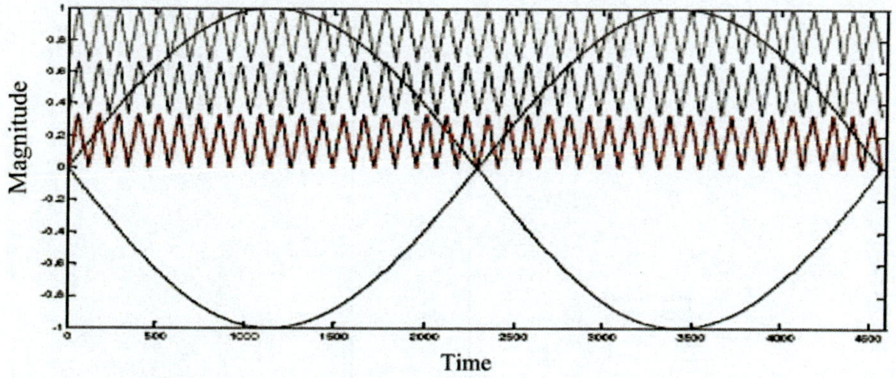

Fig. 7 Unipolar-phase disposition PWM method

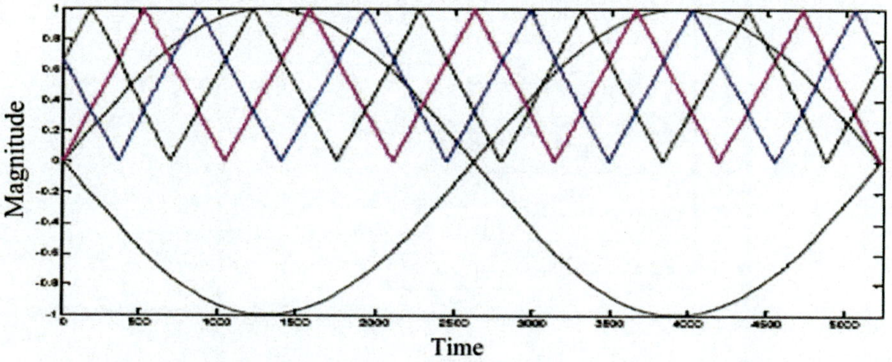

Fig. 8 Unipolar-phase shifted PWM method for 7 level inverter

5 Closed Loop Control Scheme

By utilizing synchronous reference frame extraction of reference current with closed loop control is shown in Fig. 9. The ripple identifying module operation is given in Fig. 9 separately. The Park's transformation method is base for synchronous reference frame method in which three phase line current quantities are converted in to two phase quantities [14].

Constant dc voltage across the capacitor is most important for the effective operation of the shunt active filter. In a shunt active filter, the active power transfer directly depends upon the dc capacitor voltage [14]. There is no active power transfer when we keep the voltage constant. However, some small amount of active power is required because of losses in switching devices and other components. Then actual value of dc link input voltage is compared with base voltage, as shown in Fig. 10. By using a PI controller, regulation of capacitor voltage can be achieved. Then, by comparison of detected ripple current from synchronous reference frame method (*I* hd) and the

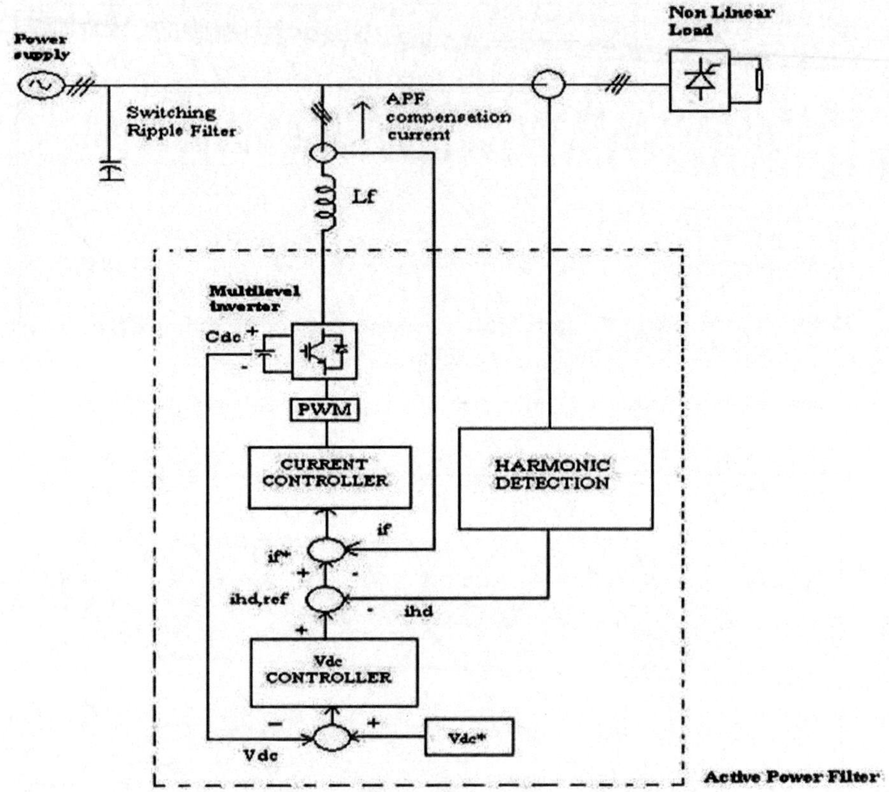

Fig. 9 Shunt active power filter closed loop control

current of voltage controller (*I* hd, ref), we can calculate the actual current of shunt active filter, and it is now compared with filter reference current. Reference current will obtained by regulating the output current of the current controller. By utilizing this reference current there will be a generation of required harmonic compensation current which will reduce the effect of dominate ripple components which will present in the line currents.

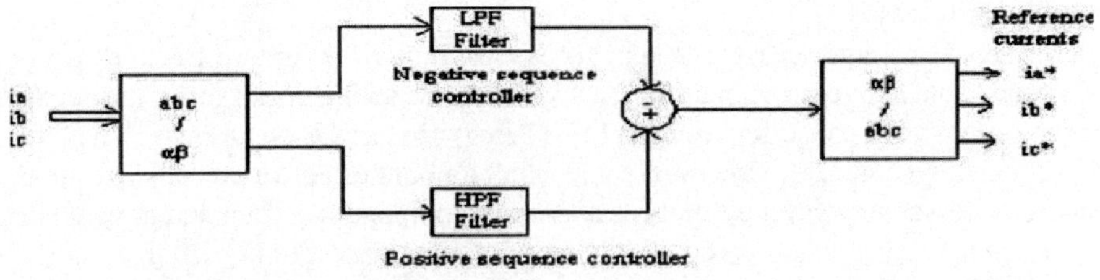

Fig. 10 Function of ripple identification module

6 MATLAB Design

See Figs. 11 and 12.

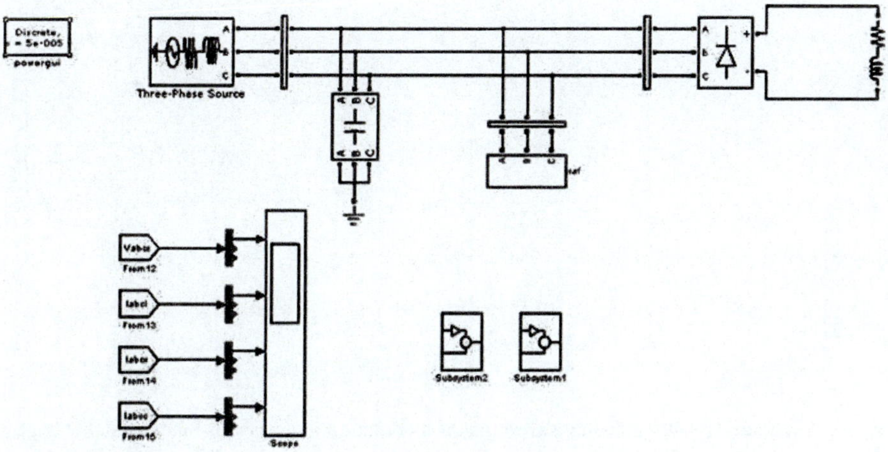

Fig. 11 Simulation circuit of the test system

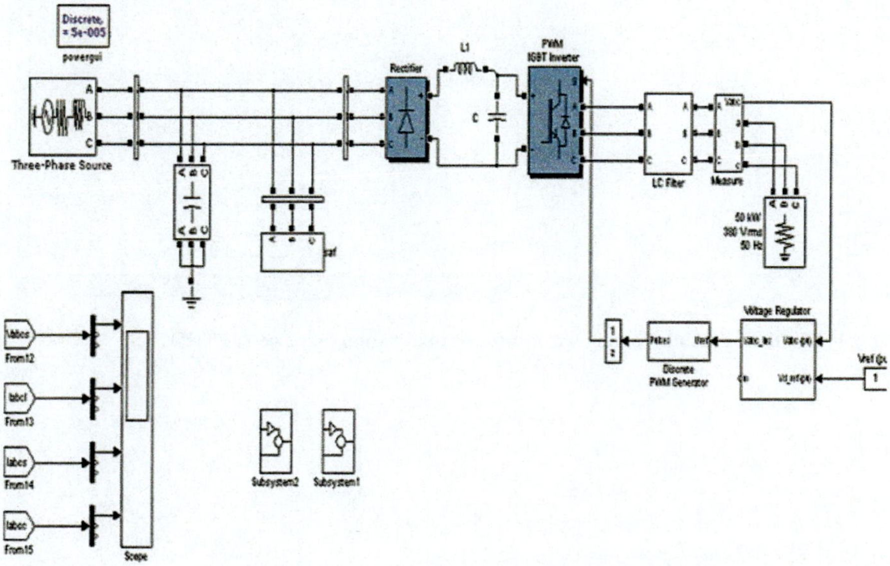

Fig. 12 Simulation model of the test system when AC-DC-AC converter is used as load

7 Simulation Results

See Figs. 13, 14, 15 and 16.

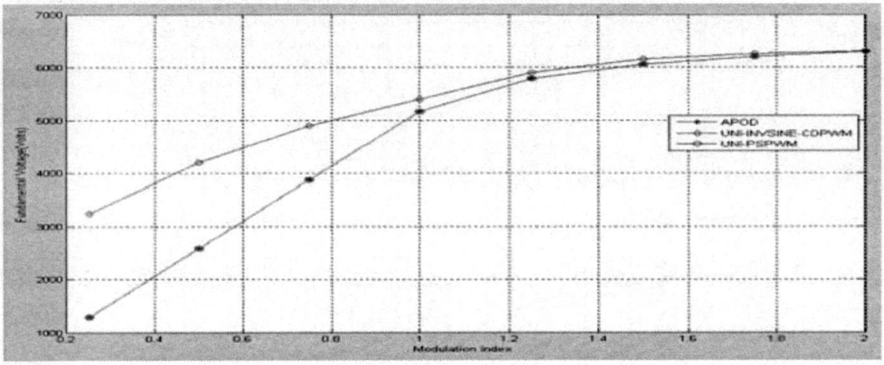

Fig. 13 Simulation results showing waveforms of modulation index versus fundamental voltage

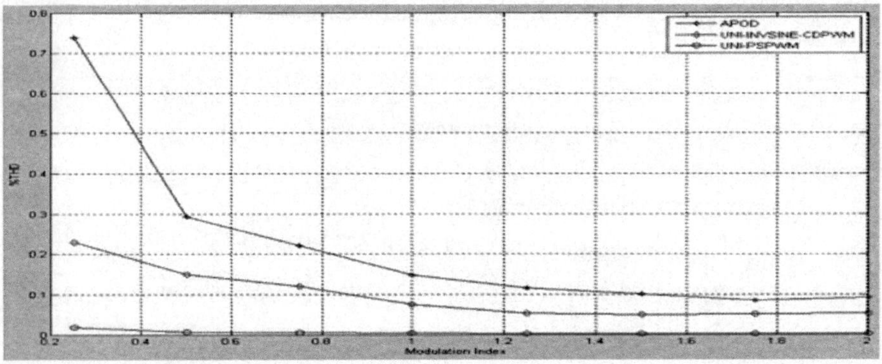

Fig. 14 Simulation result showing waveforms of modulation index versus THD

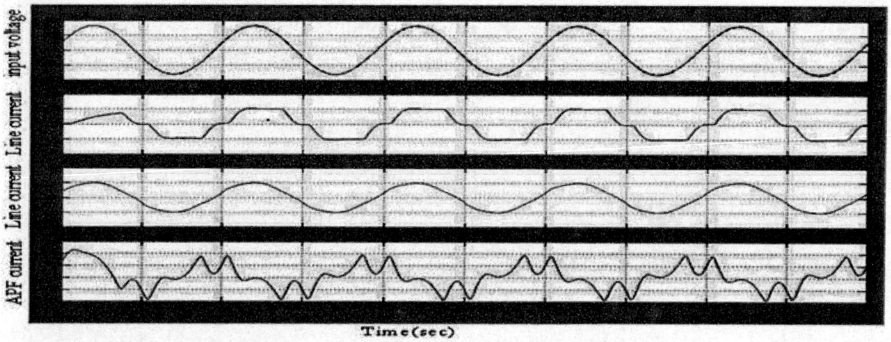

Fig. 15 Simulation result when uncontrolled rectifier is used as load

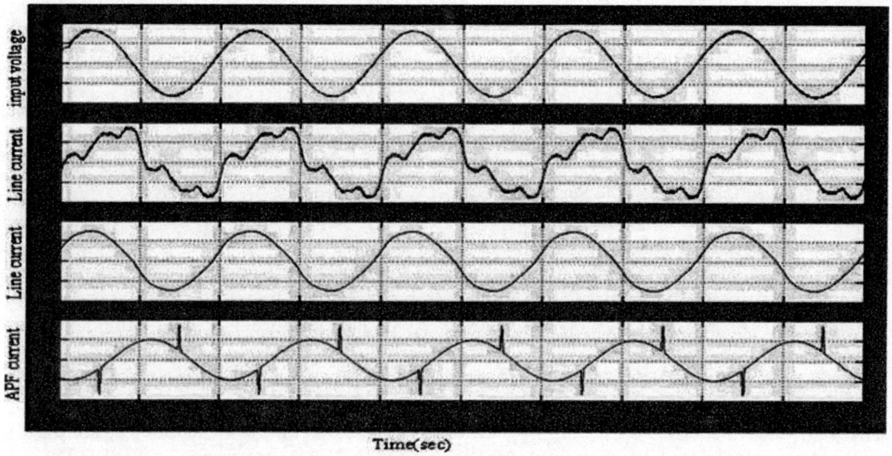

Fig. 16 Simulation result when AC-DC converter is used as load

8 Conclusion

A novel modulation approach is introduced for a three phase cascaded *H*-Bridge Multilevel inverter which will reduce the harmonic content in over modulation when compared with regular modulation strategies. In addition, there is a techniques reduction in complexity of pulse generation when compared with existing PWM strategies like PSPWM and CDPWM because required carrier signals have become half. Also, there will be a wide range of variations in the modulation index. With the reduction of lower order harmonics due to the doubling effect, a reduction in total harmonic distortion occurs with an increase in modulation index more significant than one. The seven level cascaded *H*-Bridge with a unipolar pulse width modulation scheme is used as a shunt active power filter. The decrease in the quality of the power supply has been demonstrated as an increase in the harmonic content due to non-linear loads, particularly with power electronic components. Then, the compensation of the ripple currents was obtained using a shunt active power filter. Using the results of the simulation, it can be concluded that there is a significant improvement in the quality of the power supply by reducing the harmonic content in the line currents and the total harmonic distortion.

References

1. Kundu M An introduction to power quality concerns. ICPC (I). George M, Basu KP (2008) Three phase shunt active power filter. Am J Appl Sci 5(8):909–916
2. Lascu C, Asiminoaei L, Boldea I, Blaabjerg F (2007) High performance current controller for selective harmonic compensation in active power filters. IEEE Trans Power Electron 22(5):1826–1834. Postan HM, Beig AR (2008) A three phase active filter based on three level diode clamp inverter. IEEE Trans Ind Appl
3. McGrath BP, Holmes DG (2002) Sinusoidal PWM of multilevel inverters in the overmodulation region. In: Proceedings of IEEE conference, pp 520–525
4. McGrath BP, Holmes DG, Meynard T (2006) Reduced PWM harmonic distortion for multilevel inverters operating over a wide modulation range. IEEE Trans Power Electron 21(4)
5. Moran SA, Brennen MB (1995) Active power line conditioner with fundamental negative sequence compensation. U.S. Patent 5384696
6. Mattavelli P (2001) A closed-loop selective harmonic compensation for active filters. IEEE Trans Ind Appl 37(1):81– 89
7. Xiao P, Venayagamoorthy GK, Corzine KA (2009) Seven-level shunt active power filter for high-power drive systems. IEEE Trans Power Electron 24(1):6–13
8. Fang-Zen P, Zhao-Ming Q (2003) Applications of cascade multilevel inverters. J Zhejiang Univ SCI A 4(6):658–665
9. McGrath BP, Holmes DG (2002) Multicarrier PWM strategies for multilevel inverters. IEEE Trans Ind Electron 49(4)
10. Wu H, Deng Y, Liu Y, He X (2002) A new clew for research on PWM methods of multilevel inverters: principles and applications. In: Proceedings of IEEE conference, pp 1251–1256
11. Feng C, Agelidis VG (2002) On the comparison of fundamental and high frequency carrier-based PWM techniques for multilevel NPC inverters. In: Proceedings of IEEE conference, pp 520–525
12. Aiello M, Cataliotti A, Cosentino V, Nuccio S (2007) Synchronization techniques for power quality instruments. IEEE Trans Instrum Meas 56:1511–1519
13. Bashi SA, Mailah NF, Kadir MZ, Leong KH (2008) Generation of triggering signals for multilevel converter. Eur J Sci Res 24(4):548–555
14. Vardar K, Akpinar E, Surgevil T (2009) Evaluation of reference current extraction methods for DSP implementation in active power filters. Electr Power Syst Res

A Technological Scrutinize on Energy Harvested Wireless Sensor for IoMT Healthcare System

Piyush Samant and Bhanu Priyanka Valluri

Abstract The emerging technology is providing a solution to transforming our lives in a way that no one could think of before. Be it security, commercial centers, banking industry, or any industry, without any hesitation, the technology is proving a smart and feasible solution for every question. One of such domains is medical or healthcare services; technology is helping to simplify our lives by reducing or limiting the need for clinic/hospital visits. We are utilizing the genuine force of the Internet of Things (IoT) to eliminate the burden on medical / healthcare services outlines by permitting patients to associate with their medical advisors by moving the clinical information in a protected climate. The Internet of medical things (IoMT) is an assortment of various clinical appliances as per the applications that can associate with all the data networks of health services systems through different approaches. Medical equipment will gather, examine, and send data across the web using this technology. IoMT is making its place in the world at a fast pace with 65% of global healthcare organizations already making use of it, and it's likely to increase by 27% by the end of 2021. To power up different applications of IoT systems and to make the emerging technology sustainable, there is a high demand for energy harvesting schemes. In this paper, we will be discussing the introduction, areas of applications, types of IoMT devices, advantages, challenges, and energy-efficient methods.

Keywords Internet of medical things (IoMT) · Challenges · Opportunities · Applications · Energy-efficient system

---

P. Samant (✉)
AIT-CSE, Chandigarh University, Gharuan, Punjab 140413, India
e-mail: piyushsamantpth@gmail.com

B. P. Valluri
Department of Electronics and Communication Engineering, Chandigarh University, Gharuan, Punjab 140413, India

J. R. Szymanski et al. (eds.), *Energy Systems, Drives and Automations*,
Lecture Notes in Electrical Engineering 1057,
https://doi.org/10.1007/978-981-99-3691-5_10

1 Introduction

Internet of medical things (IoMT) empowers machine-to-machine interface and attempts to provide innovative solutions for real-time problems which will profoundly renovate the healthcare provision by thriving toward its development, affordability, and dependability shortly. Additionally, the increasing patient engagement in decision-making will also enhance the healthcare services' amenabilities [1]. As a result, the technology takes on the rate of growth in the market to meet the sophisticated demand.

This technological development is also leading toward an innovative approach by providing the individual data drive treatment, augmented equipment, and devices to meet the demand of physiological requirements. IoMT is on the path to encourage personalized care and provide a high standard of living. Besides, recent research in developing fields such as big data realms, cloud services, accessible sensor networks, security, artificial intelligence, machine learning, and argument reality [2] are helping to lead affordable medical devices and allied health networks. This report showcases thorough detail analysis, methods, and discusses future sides of Internet of medical things (IoMT) technology and applications. On the path to understanding the concept of IoT in the field of medical technology (MedTech), many queries were raised. This report also attempts to answer a few of them.

The sphere of health care often involves the interactions between patients, equipment, and devices starting from bandages to surgical instruments, diabetic testing kits to monitoring devices, pacemakers to sophisticated scanners, etc. The MedTech industry is leading toward the designing and manufacturing of a series of equipment to analyze, diagnose, observe, monitor, etc., to achieve reliable patient outcomes, minimize healthcare expenditures, improve efficiency, and innovate ways of engaging and empowering patients. The flexibility in this field is possible due to the advancement in collaborative technologies such as wireless communication and computing techniques [3] leading toward the advancement of an increasing number of connected medical devices that can create, gather, evaluate, and transmit data. In combination with all the stages of data along with devices themselves as helped in evolving the Internet of medical things (IoMT).

Internet of things is a reliant system, which empowers real-time data acquisition, device connectivity, data transmissions, and analytics to control end user applications as the IoT ecosystem has a very intricate architecture, in which numerous components relate with each other to enable various solutions for the end user. It also provides a reliable environment, depending on the physical systems, which adapts social interference with computerized systems and accommodations data-driven decision processes. In recent times, IoT comprehends technologies such as smart cities, smart towns, smart homes, intellectual logistics, augmented through various devices such as sensors, actuators, tags, and many more along with communication network protocols [2]. IoT proposes various concurrent solutions through the assimilation of data analysis and sensors embedded on machines.

To meet the requirement of the healthcare sector, consequently many attempts are in the field of MedTech. This approach led to designing some set of new regulations by using the possibilities such as digitization, data methodologies, and collaborative technologies for the development of a value-based healthcare system. This also showcased to change product-based models to value-based systems that are driven by software-based service and solutions to produce a new value-based paradigm.

In simple words, IoMT can be understood as a connected infrastructure of medical devices and equipment, software applications, and healthcare services. IoMT is rapidly renovating MedTech's role and interactions within health services. Todays, interconnecting devices, are being intended to increase efficiencies, lower care costs, and drive better outcomes in health care. So, along with the collaborative technologies with the other industry segments such as energy and power distribution, construction, management, manufacturing helped the healthcare system to be poised for a transformation through IoT into IoMT and its applications by connecting devices that digitalized and transmit sensors vital data in real-time accordingly [4].

IoMT tools are hastily changing healthcare delivery by extending their ability to collect, gather, examine, understand, and transmit data. These applications are contributing to tracking and preventing chronic illness and they're poised to evolve the future of care for patients and clinicians. But the question comes down to how exactly does this connected ecosystem works?

The IoMT ecosystem spreading out is paving the approach for other new technologies too. These new approaches will further enable clinicians to monitor and treat patients remotely for ever-increasing need for patients even in rural communities as they brawl to recruit and retain medical professionals. Irrespective of a patient's environs, this evolution of the IoMT ecosystem will provide impactful benefits for better access care as connected medical devices continue to find their way [5]. In the domain of MedTech companies manufacture more than 576,280 different types of medical equipment, including wearable external medical devices, implanted medical devices, and stationary medical devices. Most patient interactions with healthcare systems involve the use of medical equipment and devices.

IoMT is also known as healthcare IoT, because of its applications and infrastructure. As we know it is made up of large volumes of sensor-based tools, which include standalone devices and wearable's that can be used for remote monitoring of patients. The ecosystem streamlines clinical workflow and leads to improvements in patient care. Figure 1 explains the basic components that are involved in it.

2 Methodology

The healthcare industry was very much pleased with the growth of IoT as it is given impeccable results. With the same inspiration, the study for this paper has started with the influence of IoT in the medical industry [6]. During the process of the search, IoMT was introduced and hence the study begin by report observations of the online surveys conducted by various healthcare organizations, research experts, etc. Many

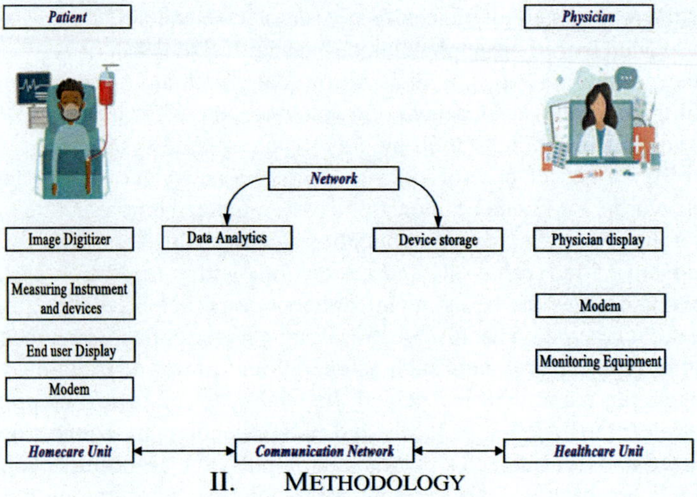

Fig. 1 Components of IoMT

companies have also discussed the benefits, trends, challenges, applications, and implementations of healthcare IOT on various platforms.

This paper mainly showcases the overview of IoMT along with its applications.

3 IoMT Ecosystem

With the increase in demand for IoMT, the attention toward the growth in the number of connected medical devices is also increasing. During this process, the data goes through a set of steps that involves the generation of data by collecting, evaluating, or transmitting health data or images and connecting them to health care provider networks, which transmits data to either a cloud depository or core servers [7]. Figure 2 showcases the main stakeholders in this ecosystem. The healthcare IoT links both the physical and digital realm and can observe patient performance in real time to manage different medical conditions.

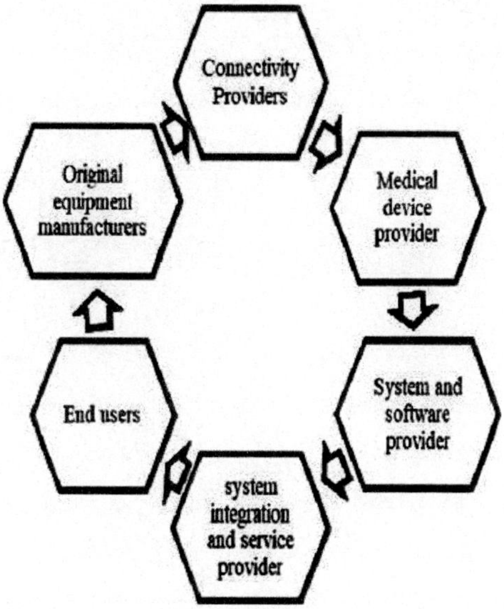

Fig. 2 IoMT designed ecosystem

4 Benefits and Limitations of IoMT

As we know the healthcare IoT describes almost any type of Internet-connected medical wired and wireless or wellness devices such as smart stethoscopes to wearable's lead the domain to the most refined approach. Along with the refinement, any imminent technologies will have their benefits and limitations.

A. *Benefits*

To transform IoT to IoMT, three factors are responsible for that.

1. The need for high-speed technology such as 5G (wireless technologies).
2. The need for high-end data compression techniques by using advanced computing power.
3. The need of minimizing complex and huge electronic medical devices.

Along with these factors, many others are also responsible for a smooth approach such as infrastructure, portability of devices, cost of the equipment, maintenance, etc. The benefits can be categorized into the patient, equipment manufactories, and healthcare providers (Fig. 3).

a. Patient advantages

   - Cost reduction
   - Real-time intervention
   - Reducing the financial burden
   - Reduces human interaction in surgery

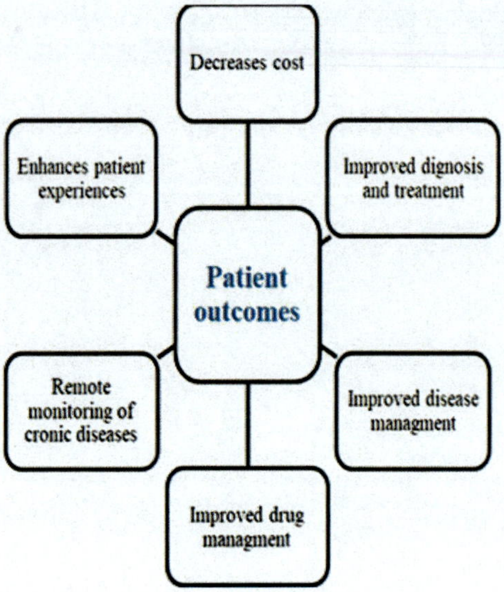

Fig. 3 Impact of IoMT on health care

b. Equipment manufacturers

- Uniformity of data availability
- Standardization of data analysis
- Communicating information to remote locations
- The capability of using high-end technologies

c. Healthcare providers

- Optimization of resources
- Minimizing the response time in an emergency.

The possibility of getting greater benefits by connecting devices or things in every industry is turning out an advantageous way. The same has been proved to be true with IoMT as well [8]. Not only the above-mentioned benefits but there are also other advantages related to the performance of the system such as accessibility, improvement in efficiency, faster implementation, and low cost.

B. *Limitations*

The limitations can be categorized into two factors as technical challenges and market challenges.

a. Technical challenges

- Data security threats
- Lack of communication protocol and standards
- Errors in data handling
- Integration of data
- Lack of medical expertise

- Diversity
- Interoperability
- Performance

b. Market challenges

- Compliance with security policy
- Mobile hesitation
- Healthcare faculty overloading data
- Storing patient information in devices
- Physical compliance

Despite the limitations that are specified above there are some other challenges are addressed below leading toward the development in research:

- Design issues
- Privacy and security
- Analysis of big data

5 Technologies Enduing IoMT

About the micro-level architecture of IoT, IoMT also comprises three basic layers, namely local system and control layer, device and data connectivity layer, application and analytic solution layer (Fig. 4).

A. *Local system and control layer*

Among the different elements of IoT decentralized intelligence is an important one. The main object of this element is to build a medical device with intelligent control capabilities and interoperability techniques [9]. Generally, smart devices are equipped with things like sensors and actuators to measure parameters, digitizing the inputs, process and analyze data to make the real-time decision reliable, and also share the information with other devices and servers with efficient network interface such as diagnostic devices, wearable monitors, and encrypting devices.

Along with the decentralized intelligence, compatibility, and integration of advanced devices are also driving factors to get effective solutions for real-time problems. For the transmission of data and to track the performance an impactful ecosystem is a need that has been designed based on IoT architecture. Protocols such as NFC, Bluetooth low (BLE), Zigbee, etc. for communication, technologies such as wireless cellular, satellite technologies, and on-air programming for analysis are implemented for better performance.

B. *Device connectivity and data layer*

This layer focuses on data aggregation and implementation. The device collects data from the network devices and stores it in predefined data stores in the form of a database for further analysis. There are no specific technologies for this layer though

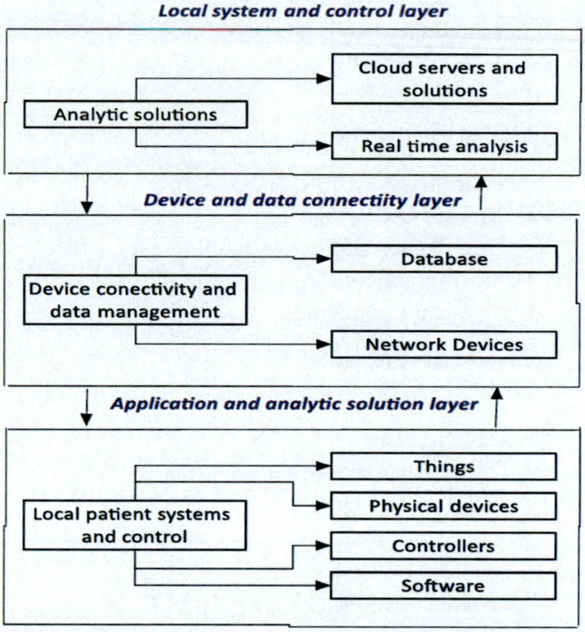

Fig. 4 IoMT basic architecture

patient monitoring data is unique for a particular disease [6]. The most important aspect that needs to be concentrated on is security. To process and analyze secured medical data transfer, technologies have also been improved, basically to manage large volumes of data and ensure the quality during processing at both the user and system end based on the requirement.

C. *Application and analytic solution layer*

In the field of health care, many techniques are involved to be efficient results. The data collets either a central server or remote server from multiple devices as per requirement over a network by considering their key components. The server is designed with inbuilt algorithms to analyze real-time data for conclusions. The methods will vary with change in devices utility that is the method of analysis for smart devices is different from methods used in remote monitoring and chronic disease management.

6 Connected Medical Devices

The evolution of artificial intelligence (AI) and machine learning (ML1) technologies is leading to reducing the cost by improving the efficiency of health care in collaboration with IoT by proving solutions using IoMT. The number of connected

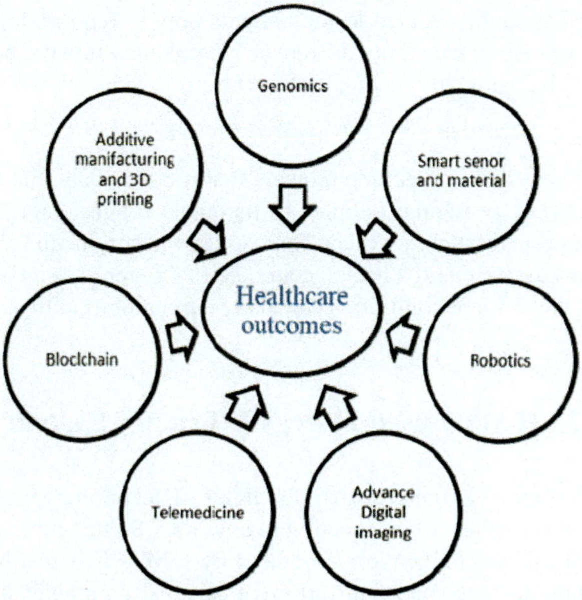

Fig. 5 Descriptive technologies involved in the improvement of IoMT

devices is increasing day by day so to make it more friendly even in the medical industry the adoption of smartphones is also considered [10].

This growth includes the development of organizations, clinicians, physician, and patients that are ready to accept IoMT solutions (Fig. 5).

So, based on researchers study the connected medical devices can be categorized into three such as

1. Stationary medical devices
2. Implanted medical devices
3. Wearable external medical devices.

A. *Stationary medical devices*

To measure the physiological parameters, in general devices such as mammography devices, CT and MRI scanners, ultrasound machines, X-ray, and imaging devices are used physically; this process involves time consumption, physicians' presence, cost, etc. But instead of using manual input, using high-end devices that can transmit images remotely to the physicians at any location improves the need for such devices [10]. Hospitals, clinics, and diagnostics centers can be incorporated with sophisticated equipment to observe and give real-time results eventually. Stationary medical devices are critical to diagnosis and include high-end applications to overlap patient data handling for precise decision-making.

B. *Implanted medical devices*

These devices include various biosensors which process different signals such as a pacemaker, hip replacement, cardiac conditions, diaphragm, and nerve stimulators.

That simply relates to the patients how to require constant monitoring. In another way as per the name, the device is implanted into the body either by surgical method or by using any medical intervention.

C. *Wearable external medical devices*

These devices include trackers that produce data that can be monitored by medical experts or regular people. As the name suggests, they are in contact with the body externally such as skin patches, smartwatches, insulin pumps for diabetes monitoring, and oxy meter to check the amount of oxygen present along with BP ratings. Patients with chronic conditions can also be monitored with such devices.

7 IoMT-Based Energy Efficiency System

The exponential growth of IoT helps the technology to achieve the concept of anytime and anyplace by connecting to anything. So this motive leads to the development of healthcare IoT which is nothing by IoMT. This involves various devices, methods that are designed with different sensors, data analysis tools, display devices, and many more which help to build a user-friendly environment [7]. With the demand for increasing smart devices and applications, there is a possibility of an increase in power consumption as well. On considering different technologies, along with the benefits there would consent as well. Among the different consents, energy efficiency and power management are one.

During the process of connecting things, there should be conservation between one machine and another machine, i.e., this type of conversation is called machine-to-machine communication (M2M). Figure 6 illustrates the M2M communication using IoT without human intervention. With the help of Internet or radio signals via a network, the application devices are connected M2M gateway using RFID, WSN technologies for further communication. The link can be a wired medium or wireless medium based on availability and device configuration.

When talking about M2M communication in health care, the chip can be placed either insider the body or a wearable circuit externally. IoT in collaboration with WSN technologies has created a new platform called wireless body area network (WBAN) which offers mobility and flexibility in MedTech [4]. As we already understood that using IoMT as its implantations in the field of chronic diseases, remote healthcare monitoring, physiological signal monitoring, fitness, etc., almost all the WBAN are designed with small and lightweight batteries which lead to the energy-driven system. Thus, the need for energy-efficient IoMT healthcare systems is critically needed to maintain the network for longer.

The organizations such as CobeBlue and MASN are focusing on designing a system that produces low-power sensors to monitor significant body signals as per the requirement. The communication procedure includes the reception and transmission of data between sensor nodes, gateways, and databases. Mostly during the data

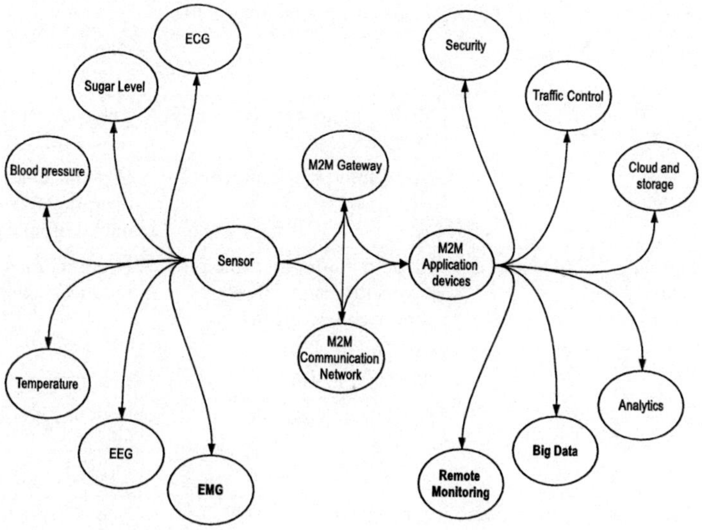

Fig. 6 IoMT-based M2M communication model

sensing, data processing, and data communication, the utilization and consumption of energy are observed. Generally, data transmission happens in two phases.

- Phase-1: Communicating between sensor node to system central node
- Phase-2: Communicating the gathered data to a database that is a remote medical station.

Data communication involves different stages, so during this multi-tasking process, there are high chances of energy wastage. The factors that contribute toward energy consumption are packet collision, overheating, ideal listing, over-emitting, etc.

A. *Energy Conservation*

A continuous source of power is required for the competent operation of any IoT-based system. During the process of data communication, different technologies are involved to get efficient results. With the help of wireless sensor systems, they are trying to provide secondary storage devices battery. The requirement of battery regular charging and has a regulated life span which may cause intrudes in network operations. So, to overcome this problem there is a need for a self-sustainable energy solution [6]. There are different types of energy sources present in the surroundings such as solar energy, thermal energy, electrostatic energy, vibrational energy, acoustic energy, wind energy, tidal energy, RF energy, and many more as such. The energy sources depend on various factors such as climatic conditions, environment, difficulty to harvest, storage, and infrastructure. All these sources have their own merits and demerits (Table 1).

Table 1 Merits and demerits of energy harvesting systems

S. No.	Energy type	Source	Merits	Demerits
1	Solar energy	Sunlight	Most commonly used as it is easily accessible	The source of energy is limited (day time only)
			Non-depletable resources	High installation and operational cost, low output efficiency
2	Thermal energy	Heat	Based on different principles the energy can be directly converted into electrical energy	Needs a larger area, but produces low power
			Available all time	Low output voltage
			Scalable and durable	Very low conservation efficiency
				Less power density
3	Vibrational energy	Vibration	In most cases, the used systems are light in weight	Needs larger area, power out is not fixed
			High voltage output	Not appropriate for large scale harvesting
			The exploitation of waste mechanical energy	
4	Electrostatic energy	Potential energy	High output voltage	Effects with environment conditions (insulators)
			Simple integration	Need of polarization source
			Gathers low frequency	Complex power circuit mechanism
				Devices have less life spam
5	Wind energy	Wind	Efficient use of landscape	Intermittent
			Low operating cost	Noise pollution
			Clean source of energy	Depends on climatic condition
			High power received	
6	Tidal energy	Waves	Easy to operate and maintain	Depends on waves
			Short start uptime	Can only be used in the area where sustainable waves are produced
7	RF energy	EM waves	Long-distance energy transfer	Ultra-low output power
			Freely available throughout, no limit	Low efficiency
				Distance is limited

B. *Factors for Energy wastage and communication*

As discussed earlier, in the entire process the wastage of energy mostly happens sensors node and mostly during the processing of data communication. The data can be transmitted and received from the sensor to the devices, but to achieve precise sensed data and a well-functioning healthcare monitoring system the entire process should be minimized. Sensors are battery-operated, if the sensor device is placed inside the human body, it would be very tough to replace or change the batter without employing a medical procedure. For the same reason, it needs to be in working for more than fifteen years. If the battery drains, that means the entire sensor network stop functioning and this is an unacceptable condition. So to overcome this issue and provide reliability, cost-effective, flexible, effective, long-term health monitoring system, new methods and techniques need to be developed, and this was made possible by technologies as WBAN and IoMT.

Energy optimization methods were advanced to combine physical and data link layer operational features of wireless devices which helped in increasing battery spam. Data security and analysis happens depends on a communication protocol such as MAC, IEEE 802.15.4, and IEEE 802.15.1. Such protocols help in maintaining data security and battery life.

8 Applications of IoMT

The existing healthcare systems can be replaced by IoMT systems to sense real-time data for patient monitoring through sophisticated equipment such as sensors, signal converters, communication devices. These devices have been transformed in numerous ways built on the requirement and usage such as smart wearable, domestic usage medical devices, and kits, professional usage devices and kits, remote or mobile applications which can be communicated with medical experts at any location as required [8]. Not only for the performance analysis but these devices can also be used for disease prevention, fitness, remote intervention in emergencies (Fig. 7). There is much application of IoMT but some of those applications areas are discussed as follows:

A. *Tele Heath*

Gathering and processing requirement-specific data aids healthcare automation, which further analyzes data against previous records to make decisions in a future course. Data from network devices is maintained at central locations mostly at the local office. This technology-enabled intelligence helps in providing service transfer for monitoring, administration to IoMT machines, which leads to reducing cost experienced from implementing resources and utilization of infrastructure efficiently. Furthermore, remote monitoring has diminished in rates, rise in healthcare productivity [5]. The need for a guardian is also reduced. Systems like cardiac monitoring required continuous observation and personal assistance the development of remote

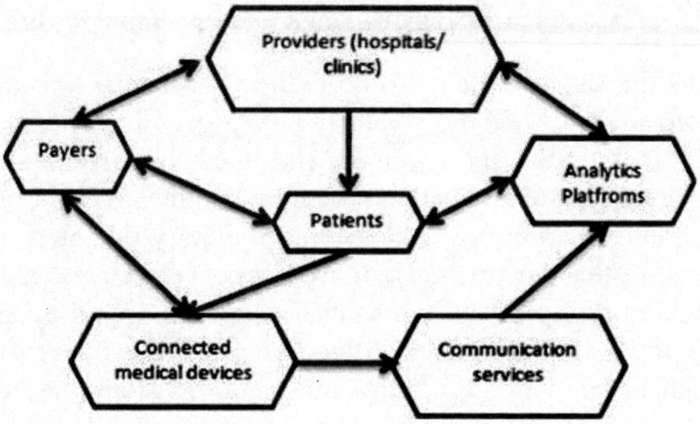

Fig. 7 Connected medical device collaboration model

mentoring systems has shown a solution by gathering the patient's information and observation data by security to provide reliable solutions.

B. *Improved drug management*

Things such as tags are also considered as a part of IoMT devices; they are mostly used to manage drug availability problems. The tags with RFID technology are mostly used in such cases. As per the suggested guidelines of RFID by FDA and drug supply management, they are also used on medication packaging, which enables companies to maintain supply chain quality. By adding this technology to medication few organizations are developing comestible IOT smart pills to track the drug doses given to the patient at times.

C. *Chronic disease management*

To manage chronic disease conditions such as diabetes, cardiac failure, and BP, IoMT embedded devices offer promising alternatives. These devices are used to monitor various parameters inside the body based on the requirement and treatment method [6]. To predict the disease they help in processing high-level analysis of gathered data by real-time sources to provide future alternative treatments and dose changes eventually. This data can be collected and stored for further research and can be useful in studying the epidemiological trend of a particular disease.

There are many other application domains such as

- Providing training paramedic staff
- Providing assistance in rehabilitation centers
- Providing access to health information electron records, etc.

9 Conclusion and Prospects of IoMT

Though IoT-based MedTech applications are still in a nascent stage of development, the application of connecting devices will improve the healthcare industry significantly. Transparent data flow will be maintained in all levels of execution that is from the device level to cloud storage level. With the help of interoperability techniques and remote monitoring, the technique will reduce the burden of cost, compliance, etc., and also leverages the use of smart devices to provide instantaneous health status.

Automation in health care would increase the demand for operational efficiency, but it may also raise the risk factor during implementation including security. Because of the benefits and challenges over IoMT seems to be a favorable approach to improve the healthcare sector.

This technology depicts the evaluation of a new era in personalized health care for better living standards throughout the world by providing distinct data-driven treatment methods. The recent research and development in various sectors such as cloud storage and security, big data analytics, artificial intelligence, augmented reality networks, computing, and wireless communication are also trying to provide their impact in this field to create an affordable smart medical system and connected healthcare ecosystem.

References

1. Vishnu S, Ramson SJ, Jegan R (2020) Internet of medical things (IoMT)—an overview. In: 2020 5th International conference on devices, circuits, and systems (ICDCS). IEEE, pp 101–104
2. Al-Turjman F, Nawaz MH, Ulusar UD (2020) Intelligence in the internet of medical things era: a systematic review of current and future trends. Comput Commun 150:644–660
3. Magsi H, Sodhro AH, Chachar FA, Abro SAK, Sodhro GH, Pirbhulal S (2018) Evolution of 5G in internet of medical things. In: 2018 International conference on computing, mathematics and engineering technologies (iCoMET). IEEE, pp 1–7
4. Irfan M, Ahmad N (2018) Internet of medical things: architectural model, motivational factors and impediments. In: 2018 15th Learning and technology conference (L&T). IEEE, pp 6–13
5. Rubí JNS, Gondim PRDL (2020) Interoperable internet of medical things platform for e-Health applications. Int J Distrib Sens Netw 16(1):1550147719889591
6. Abdullah WANW, Yaakob N, Elobaid ME, Warip MNM, Yah SA (2016) Energy-efficient remote healthcare monitoring using IoT: a review of trends and challenges. In: Proceedings of the international conference on internet of things and cloud computing, pp 1–8
7. Kotronis C, Routis I, Politi E, Nikolaidou M, Dimitrakopoulos G, Anagnostopoulos D, Amira A, Bensaali F, Djelouat H (2019) Evaluating internet of medical things (IoMT)-based systems from a human-centric perspective. Internet Things 8:100125
8. Singh RP, Javaid M, Haleem A, Vaishya R, Ali S (2020) Internet of medical things (IoMT) for orthopaedic in COVID-19 pandemic: roles, challenges, and applications. J Clin Orthop Trauma 11(4):713–717
9. Limaye A, Adegbija T (2017) A workload characterization for the internet of medical things (IoMT). In: 2017 IEEE computer society annual symposium on VLSI (ISVLSI). IEEE, pp 302–307
10. Joyia GJ, Liaqat RM, Farooq A, Rehman S (2017) Internet of medical things (IoMT): applications, benefits and future challenges in healthcare domain. J Commun 12(4):240–247
11. Qureshi F, Krishnan S (2018) Wearable hardware design for the internet of medical things (IoMT). Sensors 18(11):3812
12. Manogaran G, Chilamkurti N, Hsu CH (2018) Emerging trends, issues, and challenges in internet of medical things and wireless networks. Pers Ubiquit Comput 22(5):879–882

Analysis of Electric Vehicle-To-Grid (V2G) Simulation Model and Optimal Peak Shaving in Smart Grids

Souvik Banerjee, Arup Halder, Prakash Chakraborty, Sukanya Dasgupta, Subhayan Saha, Pallav Dutta, Arkendu Mitra, Ambarnath Banerji, Bishaljit Paul, Sumanta Kundu, and Sudhangshu Sarkar

Abstract Vehicle-to-grid (V2G) model is very potential and it has been developed for Hybrid Electric Vehicle (HEV) implementation largely. A modified V2G control model has established and different kinds of renewable energy sources, idle time of vehicles, and generation of power have been taken into consideration simultaneously as per the day schedule of the user of vehicle. Vehicle-to-grid technology is controlled by keeping the relation with the desired state of charge of battery and plug-in material detection. In this paper, the proposed model is implemented and represented by MATLAB. In this paper, we have researched that how to shave the house load with the help of vehicle-to-grid technology. Our main objective is to use the vehicle-to-grid concept to optimize transportation and supply electric power by converting electric cars into power plants virtually. Our aim is also to erect the V2G technology as a scientific method to reduce the peak load and this reduction of peak load is very much essential for smart grids.

Keywords Vehicle-to-grid · Electrical Vehicle · Smart grids · Peak shaving

---

S. Banerjee · A. Halder · P. Chakraborty · S. Dasgupta · S. Saha · P. Dutta · A. Mitra · A. Banerji (✉) · B. Paul · S. Kundu · S. Sarkar
Department of Electrical Engineering, Narula Institute of Technology, Kolkata, India
e-mail: ambarnathbanerji@gmail.com

P. Dutta
e-mail: pallav.dutta@nit.ac.in

A. Mitra
e-mail: arkendu.mitra@nit.ac.in

S. Kundu
e-mail: sumanta.kundu@nit.ac.in

S. Sarkar
e-mail: sudhangshu.sarkar@nit.ac.in

J. R. Szymanski et al. (eds.), *Energy Systems, Drives and Automations*,
Lecture Notes in Electrical Engineering 1057,
https://doi.org/10.1007/978-981-99-3691-5_11

1 Introduction

Nowadays, Electrical Vehicle has proven its importance as a replacement of conventional petrol–diesel run vehicles. The demand of EVs is increasing day by day, and gradually, the most used vehicles will be Electrical Vehicles. Now many car companies are manufacturing EVs. In the upcoming days, there will be the rise of demand of Electrical Vehicles with bidirectional chargers in the Electrical Vehicles' market. This tendency will materialize the vehicle-to-grid (V2G) technology concept. Several research works and development activities are performed nowadays [1–26] Therefore, significant studies are found in the journals and articles regarding this vehicles-to-grid (V2G) technology. Vehicle-to-grid (V2G) is a technology which has the power for transformation of the energy system. V2G is a technology which enables energy which is to be pushed back to the power grid from the battery of Electric Vehicle. By Electrical Vehicle-to-grid technology—which is also known as vehicle-to-grid—a vehicle battery can be charged and discharged depending upon different signals—such as production of energy or consumption, and V2G technology helps us to make the best use of the existing population of Electrical Vehicles. V2G is a solution of energy storage. Peak shaving generally refers to point out the peak use of Electricity by industrial and commercial power users [1]. The biggest advantage of peak shaving is to save the Electricity bill of consumers. In this research paper, we have shown V2G technology as an method of energy storage to reduce the peak load which is essential criteria for smart grids [2]. Vehicle-to-grid network is seen from the following model and the frequency of a microgrid by this system during a cycle of 24 h of events occurring. There are four important components in this proposed V2G model—(1) the diesel generator—it is the major component of generation of power. (2) The photovoltaic farm—it is combined with a wind energy farm, and it is the main sources of generation of renewable energy. (3) The V2G system—it carries the pivotal role for the transfer of two-way energy. (4) The household load system—it is the requirements of overhead energy and it is fulfilled by various sources of energy. The thousand household community is represented by this microgrid model during a working day of summer and it is windy day also. An amount of hundred Electrical Vehicles are there in this model of V2G, and it gives a quantitative ratio of 1:10 among the Electrical Vehicles and the household [3] (Fig. 1).

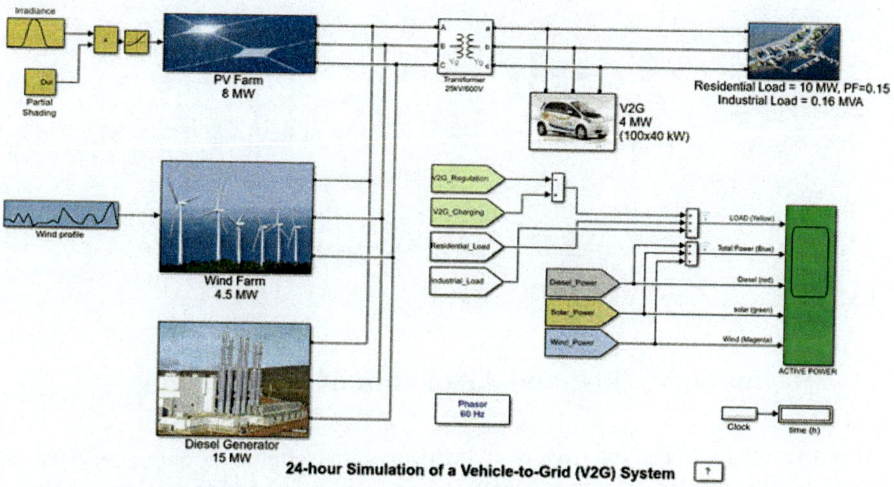

Fig. 1 MATLAB model of V2G system

2 Methods and Regulations

A lot of modifications are performed in the MATLAB R2014a Simulink model and it is mentioned in this research paper. This model has studied in a very well manner and the changes are done in some sectors.

A. **Change of Power Factor for Residential Load** The residential load used the power factor of 0.7 in the original model. This power factor is changed and increased up to 0.95 for energy consumption in a better manner. In this way, the losses are also reduced.

B. **Electrical Vehicle-to-Grid (V2G) Simulation Model** The state of charge (SOC) for all types of Electrical Vehicles is changed for bringing gradual variations. When the car is traveling, then the plug state is zero. Then, the SOCs are reduced at 0.1 in per hour units. When the car is not used, then if the plug state is zero [4], then the SOCs are unchanged. If the plug state is 1, then the SOCs are changed and increased up to 0.95. It gives the charge in the grid during massive consumption of energy or in case of random closeness of any farm of energy. The block which controls the SOC calculation is changed to get a changeable SOC when a PHEV's plug state is zero and it is responsible for the reduction of SOC of used PHEV. Here, Fig. 2 represents the strategy for Modified Control [5].

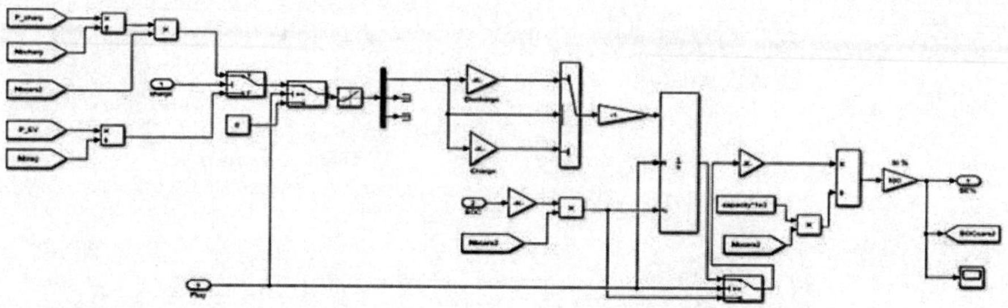

Fig. 2 Strategy for modified control

3 Systematic Method and Algorithm of Peak Shaving

This research regulates intensively at the level of household without battery, PV cell, etc. The development of optimal peak shaving algorithm and the Electrical Vehicle storage system has forwarded this research. Our main aim is to shave the load of house and to charge up the vehicle's battery [6]. The daily load profile of the house is found by the optimal peak shaving algorithm. Here, the offline strategy is chosen as it helps to calculate the optimum results theoretically. The detailed study of several cases is the best practice to assume the principle of operation, but there is difficulty to make the conclusions generally [7]. Therefore, this optimal peak shaving algorithm is applied on 1000 practical cases to make the real elements for countering this situation. The EV charging cost optimization algorithm has pushed our research to another level [2]. This algorithm is applied on 1000 practical cases for making the conclusions generally which is in relation to the EV charging cost reduction capacity. In this research, we have to realize the charging cost which is achieved by the above two optimal algorithms. By this research, we will be able to know the charging cost optimization of peak shaving and vehicle-to-grid (V2G) [5]. By this research, the average shapes of daily load profile will be also achieved. Figures 3 and 4 represent the synopsis of algorithm of optimal peak shaving and the energy of the system.

4 Results of Simulation

There are three parts in this result and simulation section. The results of simulation of the peak shaving algorithm are in Sect. 4.1. The results of algorithm of optimization for cost of charging of EV for V2G model are in Sect. 4.2. Realization of results of cost of charging and the daily load profile shapes achieved by the above algorithms are in Sect. 4.3 [3].

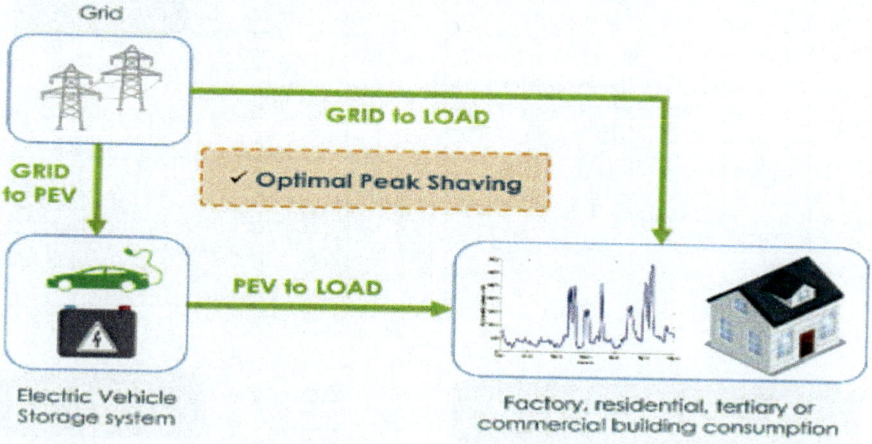

Fig. 3 Synopsis of algorithm of optimal peak shaving

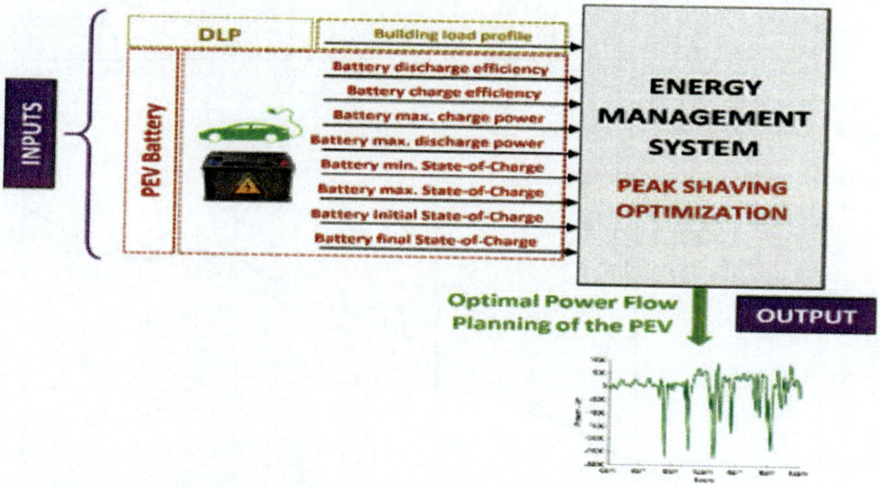

Fig. 4 Flow of energy of the system

4.1 Results of Simulation of Peak Shaving

For realization of algorithm of peak shaving, four cases are erected here. A case of EV charging is presented in the first case. The initial and final states of charge of battery of EV are described in the second case. There are same data in the third and fourth cases, and it is assumed that the Electrical Vehicle charging and discharging capacity at maximum level are not strong enough [3] (Figs. 5 and 6).

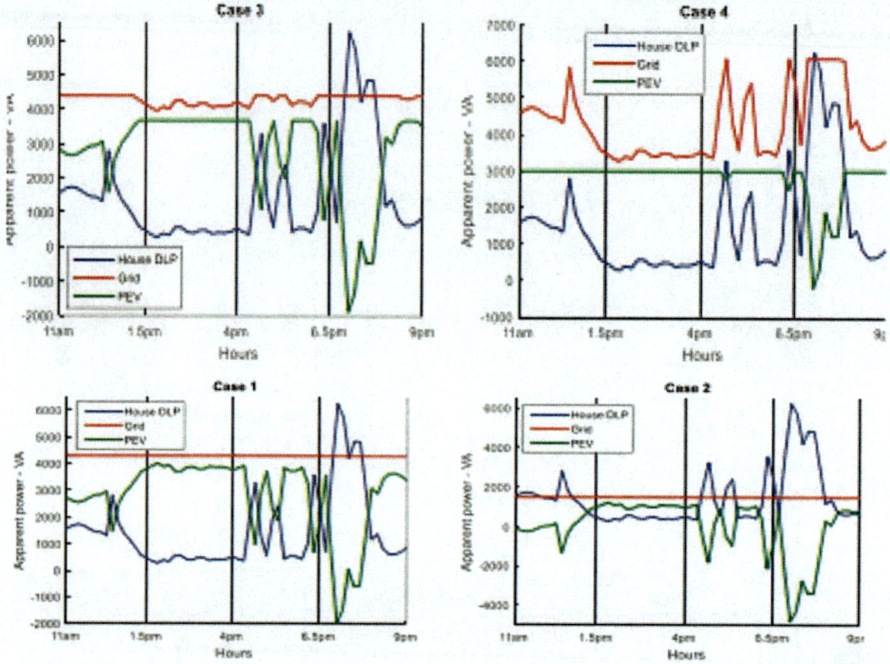

Fig. 5 Optimization of peak shaving—case (1), case (2), case (3), case (4)

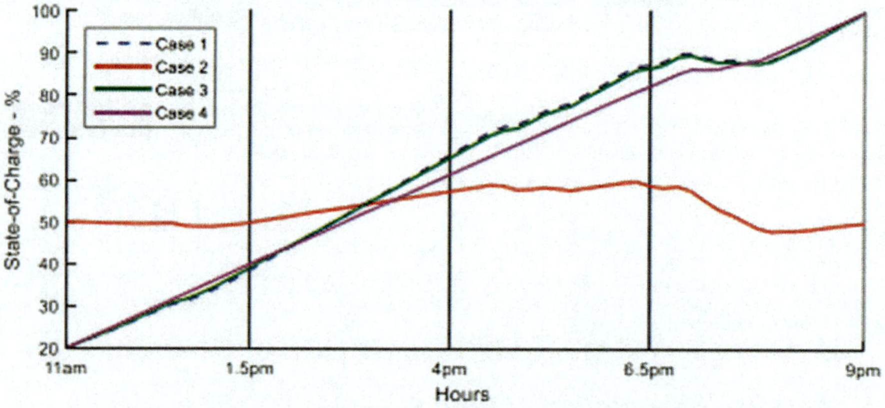

Fig. 6 Optimization of power shaving—evolution of state of charge of battery

4.2 Algorithm of Optimization of Charging Cost of V2G

The algorithm of optimization of charging cost of V2G is described by this section. The main aim of optimization of charging cost is the reduction of cost of charging

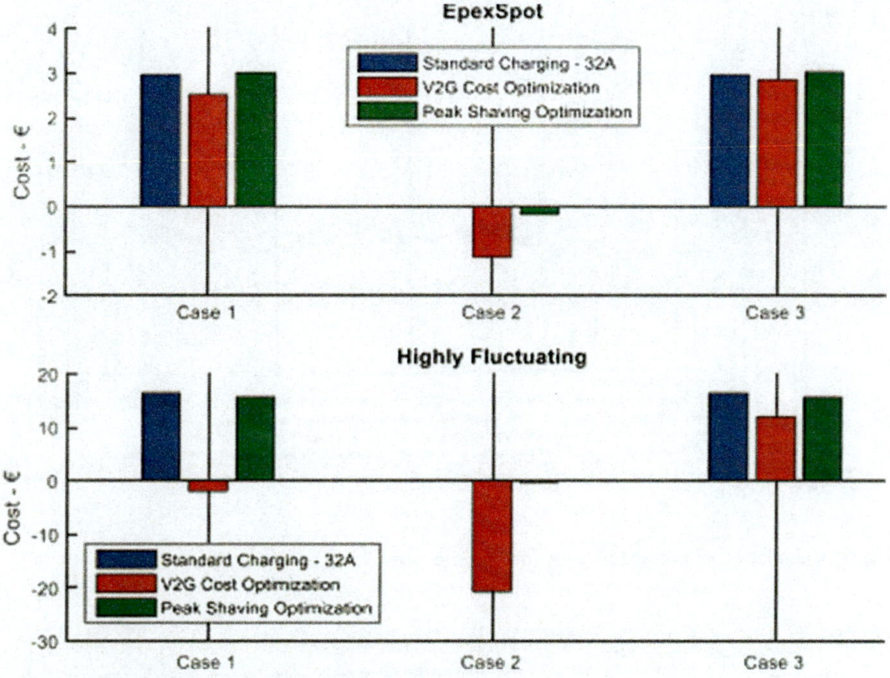

Fig. 7 Cost of charging of EV results of simulation of practical cases

and there are two main targets, and they are: (1) grid planning optimally up to the flow of Electrical Vehicles' power by the use of suitable daily energy price for the reduction of cost of charging. (2) The maximization of the process of selling of power from vehicles-to-grid [1] (Fig. 7).

4.3 Comparative Analysis of the Above Two Cases

The cost of charging achieved by applying on one thousand practical cases by above algorithms, to get the shaving of peak load, and optimized cost of charging of vehicles-to-grid are described in this section for using two profiles of price of Daily Energy. The cost of charging of Electrical Vehicles are achieved by the above cases in averagely. The algorithm of shaving the peak load optimally decreases the cost of charging by the profile of Daily Energy Price [2]. But in opposite, the profile of Daily Energy Price in hugely variable case has no decrease. The algorithm of optimized cost of charging of vehicles-to-grid has shown its effectiveness [3] in the above cases. Figures 8 and 9 describe the possibilities of Electrical Vehicles for the decrease of cost of charging and gaining profit and money.

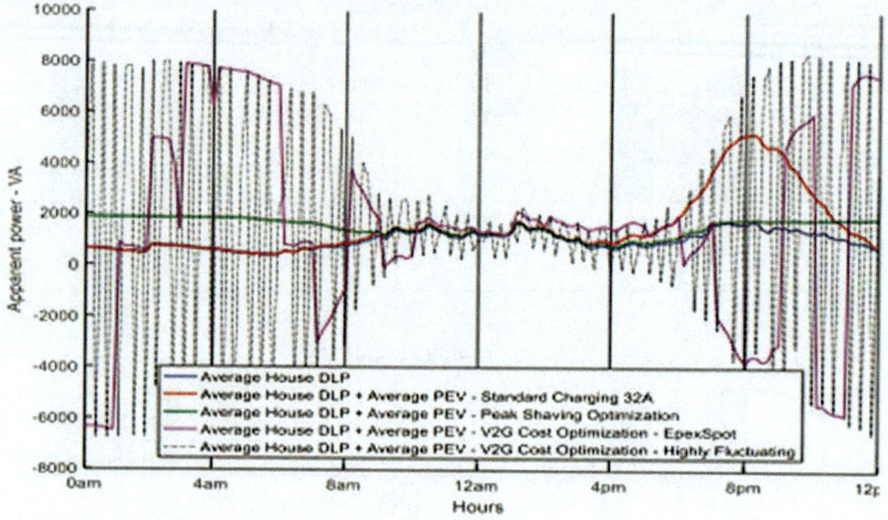

Fig. 8 Calculation of average of daily load profile—results as per database

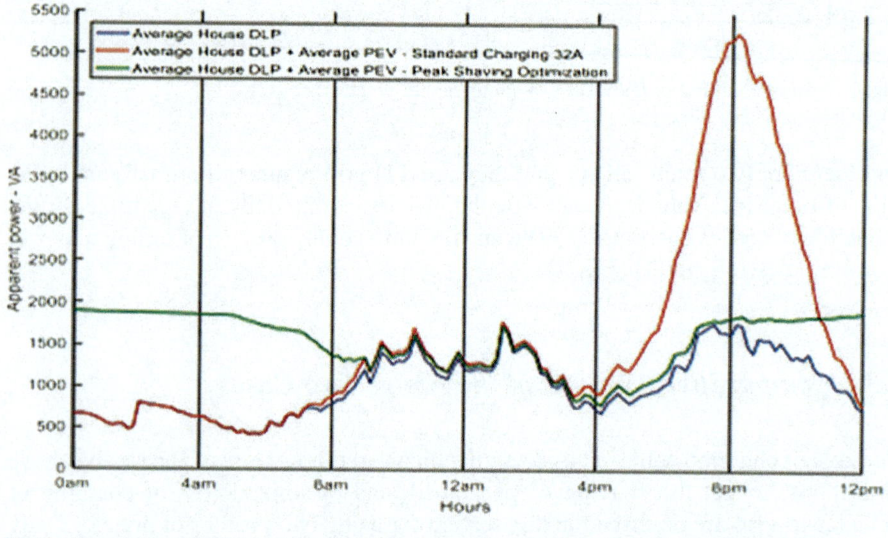

Fig. 9 Calculation of average of daily load profile—by algorithm of optimal peak shaving

5 Conclusion

The MATLAB results describes the capability of algorithm of peak shaving, and it shows the path of decrease of cost of charging of Electrical Vehicles [6]. Now, it is clear that the reduction of cost of charging of Electrical Vehicles will make a new level of consumption and the algorithm of peak shaving will not be able for the decrease of cost of charging of Electrical Vehicles competitively. Two sides will be erected there: (1) optimization of cost of charging of Electrical Vehicles and the shaving of load together by an algorithm which is multifunctional. (2) The processes shown in this paper are not online, and therefore, it is necessary to establish an algorithm which will be online in which the daily load of home is not known to the developer. Therefore, the research paper will bring some new sides of solutions to sort out the above problems in later [8].

References

1. Turker H, Colak I (2020) Optimal peak shaving with vehicles-to-grid capability of electrical vehicles in smart grids. In: 7th International conference on renewable energy research and applications, Paris-France. https://doi.org/10.1109/ICRERA.2018.8566962
2. Vuille LF, Kuhn D (2017) A review of the state of research on vehicle-to-grid (V2G) progress and barriers to deployment. In: European battery, hybrid and fuel cell electrical vehicle congress. Geneva. https://doi.org/10.17226/21725
3. Gupta A, Bansal HO, Jaiswal P, Kumar R (2020) Modelling and analysis of V2G scheme: a concept on smart grid. In: 2020 International conference on emerging trends in communication, control and computing (ICONC3), MODY University of Science and Technology, Laksmangarh. https://doi.org/10.1109/ICONC345789.2020.9117408
4. Krajcar S, Mescarj P (2015) Home demand-side management of integrated with electric vehicles and renewable energy sources. Energy Build 108:1–9. https://doi.org/10.1016/j.enbuild.2015.09.001
5. Pani P, Athreya AR, Bansal HO (2019) Integration of vehicle-to-grid (V2G) technology. In: National conference on emerging trends of technology. California. https://doi.org/10.1109/EnergyEconomics.2015.7235108
6. Guo Y, Xiong J, Xu S, Su W (2016) Two-stage economic operation of microgrid-like electric vehicle parking deck. In: 2016 IEEE/PES Transmission and distribution conference and exposition (T&D), Dallas, TX. https://doi.org/10.1109/TSG.2015.2424912
7. Xue Y, Yu XE, Sirouspour S (2012) Microgrid and transportation electrification. In: IEEE transportation electrification conference and expo (ITEC). https://doi.org/10.1109/ITEC.2012.6243464
8. Shi L, Lv T, Wang Y (2019) Vehicle-to-grid (V2G) service management logic and formulation. J Mod Power Syst Clean Energy 7(4):935–947. https://doi.org/10.1007/s40565-018-0464-7
9. Su W, Wang J, Roh J (2014) Stochastic energy scheduling in microgrids with intermittent renewable energy resources. IEEE Trans Smart Grid 5(4):1876–1883. https://doi.org/10.1109/TSG.2013.2280645
10. Tushar W, Yuen C, Huang S, Smith DB, Poor HV (2016) Cost minimization of charging stations with Photovoltaics: an approach with EV classification. IEEE Trans Intell Transp Syst 17(1):156–169. https://doi.org/10.1109/TTIS.2015.2462824
11. Wi YM, Lee JU, Joo SK (2013) Electric vehicle charging method for smart homes/buildings with a photovoltaic system. IEEE Trans Consum Electron Field 59(2):323–328. https://doi.org/10.1109/TCE.2013.6531113
12. Chen C, Duan S, Cai T, Liu B, Hu G (2011) Smart energy management system for optimal microgrid economic operation. IET Renew Power Gener 5(3):258–267
13. Montoya M, Sherick R, Haralson P, Neal R, Yinger R (2013) Islands in the storm: integrating microgrids into the larger grid. IEEE Power Energy Mag 11(4):33–39
14. Washom B et al. (2013) Ivory tower of power: microgrid implementation at the University of California, San Diego. IEEE Power Energy Mag 11(4):28–32

15. Bakre A, Kumar P, Pandita S (2019) Technical study of electric vehicles and charging infrastructure. E&Y Publications, p 24
16. Seethapathy R, Ghatikar R, Ahuja A (2017) Implementation plan for electrification of public transport in Kolkata, p 32
17. Chen TD, Kockelman KM, Khan M (2013) The electric vehicle charging station location problem: a parking-based assignment method for Seattle. In: 92nd Annual meeting transportation research board, vol 340, pp 13–1254. Supervisory control, vol 3, Dec 2019
18. He X, Ding CH, Simon HD (2005) On the equivalence of nonnegative matrix factorization and spectral clustering, in SDM. In: Proceedings of the 2005 SIAM international conference on data mining, vol 5, pp 606–610
19. Chen WY, Song Y, Bai H, Lin CJ, Chang EY (2011) Parallel spectral clustering in distributed systems. IEEE Trans Pattern Anal Mach Intell 33(3):568–586
20. Ng AY, Jordan MI, Weiss Y (2002) On spectral clustering: analysis and an algorithm. Adv Neural Inf Process Syst 2:849–856
21. Martin CH (2012) Spectral clustering: a quick overview. Retrieved from https://charlesmartin14.wordpress.com/2012/10/09/spectralclustering/. Accessed on 05 Jan 2017
22. Press WH (2007) Numerical recipes 3rd edition: the art of scientific computing. Cambridge University Press
23. Huang ZK, Chau KW (2008) A new image thresholding method based on Gaussian mixture model. Appl Math Comput 205(2):899–907
24. Kim SC, Kang TJ (2007) Texture classification and segmentation using wavelet packet frame and Gaussian mixture model. Pattern Recogn 40(4):1207–1221
25. Reynolds DA, Quatiere TF, Dunn RB (2000) Speaker verification using adapted Gaussian mixture models. Digit Signal Process 10(1):19–41
26. Catalbus MC, Yildirim M, Issever D, Gulten A, Kurum H (2016) An approach to adaptive structuring element size selection for feature extraction. Int J Comput Sci Inf Secur

Short Circuit Analysis in Unbalanced Radial Distribution System Considering Mutual Impedances

A. Sriker and G. Yesuratna

Abstract In this paper unsymmetrical short circuit analysis algorithm for unbalanced radial three-phase distribution networks, based on two matrices is presented. Two matrices, the bus-injection to branch-current (BIBC) matrix and the branch-current to bus-voltage (BCBV) matrix for each phase are used to represent the special topological characteristics of distribution networks. These two matrices for each phase are built by simple building algorithms and are easily implemented. Most of the short circuit computation methods in literature neglect mutual impedances between the phases. This paper computes the short circuit currents considering the mutual impedances between phases by building two new BIBC and BCBV matrices per phase. This improves the accuracy of the short circuit computation and also simulates the practical distribution system (DS). The proposed short circuit analysis method is applied to 55 bus system and the results are summarized.

Keywords Unbalanced radial distribution · BIBC · BCBV · Distribution system · Fifty-five bus system

1 Introduction

Electric systems occasionally experience short circuits resulting in abnormally high currents. The faults should be isolated with no damage to the equipment under the constraint of minimal disruption to the operation of the plant using protective devices, such as circuit breakers and fuses.

The computation of short circuit currents for unbalanced faults in a normally balanced three-phase system has traditionally been accomplished by the application of symmetrical components. However, this method is not well suited to a distribution feeder that is inherently unbalanced. The unequal mutual coupling between phases leads to mutual coupling between sequence networks. From this aspect, several

A. Sriker (✉) · G. Yesuratna
Department of Electrical Engineering, Osmania University, Hyderabad, Telangana, India
e-mail: sannangi@yahoo.com

J. R. Szymanski et al. (eds.), *Energy Systems, Drives and Automations*,
Lecture Notes in Electrical Engineering 1057,
https://doi.org/10.1007/978-981-99-3691-5_12

short circuit analysis methods based on phase co-ordinates, hybrid compensation method and three-phase impedance matrix have been proposed [1–3]. To the best of our knowledge systematic short circuit analysis method for unbalanced distribution systems [4] method is the most popular method of computing short circuit currents. However this method neglects mutual impedances between phases which contributes significantly while computing short circuit currents.

In this paper short circuit analysis algorithm for radial three-phase distribution networks is presented considering the per phase self and mutual impedances. Two new matrices BIBC(m) and BCBV(m) per phase representing mutual impedances are built from the already existing per phase self-impedance BIBC(s) and BCBV(s) matrices. The proposed short circuit analysis method is developed from these 4 matrices BIBC(s), BCBV(s), BIBC(m) and BCBV(m) for each phase.

2 Proposed Method

In the literature [5] BIBC and BCBV matrices are constructed considering only the balanced radial distribution systems explained in Sect. 2.1. The proposed method builds BIBC and BCBV matrices for unbalanced radial distribution system including mutual impedances between phases which will be explained in Sects. 2.2 and 2.3, respectively.

2.1 BIBC and BCBV Matrices for Three-Phase Balanced System

Practically most of the loads of feeders in DS are unbalanced loads. Figure 1 shows an unbalanced three-phase distribution system model with mutual impedances.

$$\begin{bmatrix} Va \\ Vb \\ Vc \end{bmatrix} = \begin{bmatrix} VA \\ VB \\ VC \end{bmatrix} - \begin{bmatrix} Zaa-n & Zab-n & Zac-n \\ Zba-n & Zbb-n & Zbc-n \\ Zca-n & Zcb-n & Zcc-n \end{bmatrix} * \begin{bmatrix} Ia \\ Ib \\ Ic \end{bmatrix} \quad (1)$$

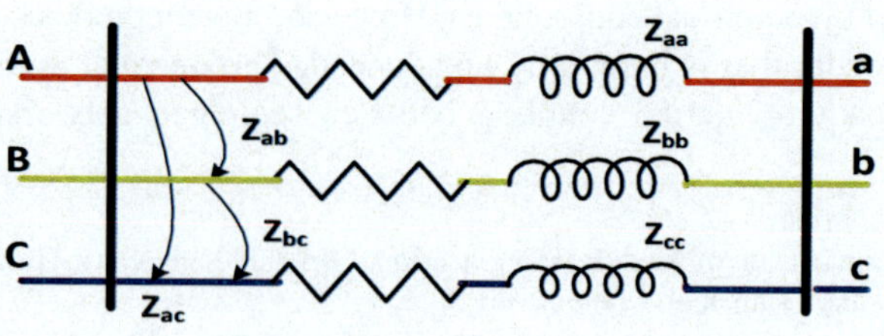

Fig. 1 Unbalanced three-phase distribution system model

In above equation, terminating end voltages (*Va*, *Vb* and *Vc*) are expressed in terms of sending end voltages (*VA*, *VB* and *VC*) and branch currents flows (*Ia*, *Ib*, and *Ic*) in three-phase matrix form.

Branch-currents to bus-voltages (BCBV) and bus-injections to branch-currents (BIBC) matrices together are used to get power-flow solution of DS in case of balanced network. The detailed formation of BIBC and BCBV matrices are given in [5].

$$[B] = [\text{BIBC}] * [I] \tag{2}$$

The constant bus-injection to branch-current (BIBC) matrix is an upper triangular matrix and contains values of 0 and + 1 only.

$$[\Delta V] = [\text{BCBV}] * [B] \tag{3}$$

where BCBV is the branch-current to bus-voltage matrix.

For a simple 6 bus network shown in Fig. 2, BIBC matrix is given in Eq. 2 and BCBV matrix is given in Eq. 3 [5].

$$\begin{bmatrix} B1 \\ B2 \\ B3 \\ B4 \\ B5 \end{bmatrix} = \begin{bmatrix} 1\,1\,1\,1\,1 \\ 0\,1\,1\,1\,1 \\ 0\,0\,1\,1\,0 \\ 0\,0\,0\,1\,0 \\ 0\,0\,0\,0\,1 \end{bmatrix} \begin{bmatrix} I2 \\ I3 \\ I4 \\ I5 \\ I6 \end{bmatrix} \tag{4}$$

$$\begin{bmatrix} V1 \\ V1 \\ V1 \\ V1 \\ V1 \end{bmatrix} - \begin{bmatrix} V2 \\ V3 \\ V4 \\ V5 \\ V6 \end{bmatrix} = \begin{bmatrix} Z12 & 0 & 0 & 0 & 0 \\ Z12 & Z23 & 0 & 0 & 0 \\ Z12 & Z23 & Z34 & 0 & 0 \\ Z12 & Z23 & Z34 & Z45 & 0 \\ Z12 & Z23 & 0 & 0 & Z36 \end{bmatrix} \begin{bmatrix} B1 \\ B2 \\ B3 \\ B4 \\ B5 \end{bmatrix} \tag{5}$$

BIBC and BCBV matrices in above equations are for self-impedances and for balanced three-phase system.

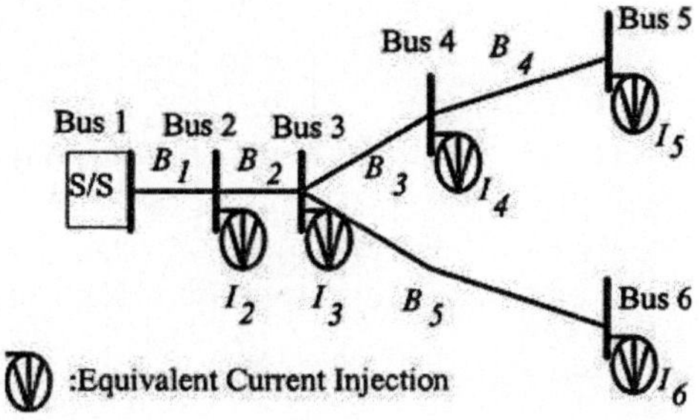

Fig. 2 Simple distribution system

The above approach has two limitations:

(i) Applicable to three-phase balanced system.
(ii) Mutual impedances are not considered.

With the above two limitations, the short circuit computations in the practical unbalanced distribution system with mutual impedances are erroneous. To overcome these limitations the following methodology is proposed in this paper.

2.2 *BIBC and BCBV Matrices for Three-Phase Unbalanced System*

The above matrices (Eqs. 4 and 5) are derived for each phase, i.e. BIBC will now have three matrices BIBC*a*, BIBC*b* and BIBC*c* representing *A*, *B* and *C* phases, respectively. Similarly BCBV will also have three matrices BCBV*a*, BCBV*b* and BCBV*c* representing *A*, *B* and *C* phases, respectively.

For the same 6 bus network shown in Fig. 2, assuming each branch is three-phase, BIBC matrix per phase is given in Eqs. 6a–6b and BCBV matrix per phase is given in Eqs. 7a–7f.

$$\begin{bmatrix} B1a \\ B2a \\ B3a \\ B4a \\ B5a \end{bmatrix} = \begin{bmatrix} 1\,1\,1\,1\,1 \\ 0\,1\,1\,1\,1 \\ 0\,0\,1\,1\,0 \\ 0\,0\,0\,1\,0 \\ 0\,0\,0\,0\,1 \end{bmatrix} \begin{bmatrix} I2a \\ I3a \\ I4a \\ I5a \\ I6a \end{bmatrix} \tag{6a}$$

$$\begin{bmatrix} B1b \\ B2b \\ B3b \\ B4b \\ B5b \end{bmatrix} = \begin{bmatrix} 1\,1\,1\,1\,1 \\ 0\,1\,1\,1\,1 \\ 0\,0\,1\,1\,0 \\ 0\,0\,0\,1\,0 \\ 0\,0\,0\,0\,1 \end{bmatrix} \begin{bmatrix} I2b \\ I3b \\ I4b \\ I5b \\ I6b \end{bmatrix} \tag{6b}$$

$$\begin{bmatrix} B1c \\ B2c \\ B3c \\ B4c \\ B5c \end{bmatrix} = \begin{bmatrix} 1\,1\,1\,1\,1 \\ 0\,1\,1\,1\,1 \\ 0\,0\,1\,1\,0 \\ 0\,0\,0\,1\,0 \\ 0\,0\,0\,0\,1 \end{bmatrix} \begin{bmatrix} I2c \\ I3c \\ I4c \\ I5c \\ I6c \end{bmatrix} \tag{6c}$$

$$\begin{bmatrix} V1a \\ V1a \\ V1a \\ V1a \\ V1a \end{bmatrix} - \begin{bmatrix} V2a \\ V3a \\ V4a \\ V5a \\ V6a \end{bmatrix} = \begin{bmatrix} Z12aa & 0 & 0 & 0 & 0 \\ Z12aa & Z23aa & 0 & 0 & 0 \\ Z12aa & Z23aa & Z34aa & 0 & 0 \\ Z12aa & Z23aa & Z34aa & Z45aa & 0 \\ Z12aa & Z23aa & 0 & 0 & Z36aa \end{bmatrix} \begin{bmatrix} B1a \\ B2a \\ B3a \\ B4a \\ B5a \end{bmatrix} \tag{7a}$$

$$\text{Where } [\text{BCBV}]a = \begin{bmatrix} Z12aa & 0 & 0 & 0 & 0 \\ Z12aa & Z23aa & 0 & 0 & 0 \\ Z12aa & Z23aa & Z34aa & 0 & 0 \\ Z12aa & Z23aa & Z34aa & Z45aa & 0 \\ Z12aa & Z23aa & 0 & 0 & Z36aa \end{bmatrix} \tag{7b}$$

$$\begin{bmatrix} V1b \\ V1b \\ V1b \\ V1b \\ V1b \end{bmatrix} - \begin{bmatrix} V2b \\ V3b \\ V4b \\ V5b \\ V6b \end{bmatrix} = \begin{bmatrix} Z12bb & 0 & 0 & 0 & 0 \\ Z12bb & Z23bb & 0 & 0 & 0 \\ Z12bb & Z23bb & Z34bb & 0 & 0 \\ Z12bb & Z23bb & Z34bb & Z45bb & 0 \\ Z12bb & Z23bb & 0 & 0 & Z36bb \end{bmatrix} \begin{bmatrix} B1b \\ B2b \\ B3b \\ B4b \\ B5b \end{bmatrix} \tag{7c}$$

$$\text{Where } [\text{BCBV}]b = \begin{bmatrix} Z12bb & 0 & 0 & 0 & 0 \\ Z12bb & Z23bb & 0 & 0 & 0 \\ Z12bb & Z23bb & Z34bb & 0 & 0 \\ Z12bb & Z23bb & Z34bb & Z45bb & 0 \\ Z12bb & Z23bb & 0 & 0 & Z36bb \end{bmatrix} \tag{7d}$$

$$\begin{bmatrix} V1c \\ V1c \\ V1c \\ V1c \\ V1c \end{bmatrix} - \begin{bmatrix} V2c \\ V3c \\ V4c \\ V5c \\ V6c \end{bmatrix} = \begin{bmatrix} Z12cc & 0 & 0 & 0 & 0 \\ Z12cc & Z23cc & 0 & 0 & 0 \\ Z12cc & Z23cc & Z34cc & 0 & 0 \\ Z12cc & Z23cc & Z34cc & Z45cc & 0 \\ Z12cc & Z23cc & 0 & 0 & Z36cc \end{bmatrix} \begin{bmatrix} B1c \\ B2c \\ B3c \\ B4c \\ B5c \end{bmatrix} \tag{7e}$$

$$\text{Where} [\text{BCBV}]c = \begin{bmatrix} Z12cc & 0 & 0 & 0 & 0 \\ Z12cc & Z23cc & 0 & 0 & 0 \\ Z12cc & Z23cc & Z34cc & 0 & 0 \\ Z12cc & Z23cc & Z34cc & Z45cc & 0 \\ Z12cc & Z23cc & 0 & 0 & Z36cc \end{bmatrix} \tag{7f}$$

Here *Zaa*, *Zbb* and *Zcc* are per phase self-impedances of individual loads.

2.3 BIBC and BCBV Matrices for Three-Phase Unbalanced System Considering Mutual Impedances

Algorithm for constructing BIBC matrix with mutual impedances:

Step 1 Take [BIBC]a matrix from Eq. (6a).
Step 2 Add each element of first row in [BIBC]a and store the sum in a variable say sum 1.
Step 3 Take [BIBC]b matrix from Eq. (6b)
Step 4 Add each element of first row in [BIBC]b and store the sum in a variable say sum 2.
Step 5 If sum 1 and sum 2 are greater than zero ((sum 1) & (sum 2) > 0), take first row of [BIBC]b matrix and store in a new [BIBC]$a - b$ matrix as first row. If the sum1 and sum 2 is equal to 0 ((sum 1) & (sum 2) = 0), then replace the first row of [BIBC]$a - b$ matrix with all zeros. Repeat step 5 for all remaining rows.
Step 6 Follow step 1 to step 5 with Eqs. (6b) and (6c) for constructing [BIBC]$b - c$ and follow step 1 to step 5 with Eqs. (6c) and (6a) for constructing [BIBC]$c - a$ matrices.

The resultant [B] matrix considering mutual impedances is shown as below:

$$[B]a - b = [\text{BIBC}]a - b * [I]b \tag{8a}$$

$$[B]b - c = [\text{BIBC}]b - c * [I]c \tag{8b}$$

$$[B]c - a = [\text{BIBC}]c - a * [I]a \tag{8c}$$

Algorithm for constructing BCBV matrix with mutual impedances:

Step 1: Take [BCBV]a and [BCBV]b matrices from Eqs. (7b) and (7d).
Step 2: Take element by element of first row in both the matrices and perform the logical "AND" operation. Store the result in [BCBV]$a - b$ matrix as first row.
Step 3: Repeat step 2 for remaining rows and construct [BCBV]$a - b$ matrix.
Step 4: Replace all the non-zero impedances of [BCBV]$a - b$ matrix with the $Za - b$ mutual impedances.
Step 5: Follow step 1 to step 4 with Eqs. (7d) and (7f) for constructing [BCBV]$b - c$ and follow step 1 to step 4 with Eqs. (7f) and (7b) for constructing [BCBV]$c - a$ matrices.

The resultant [ΔV] matrix considering mutual impedances is shown as below:

$$[\Delta V]a = [\text{BCBV}]a * [B]a + [\text{BCBV}]a - b * [B]a - b \tag{9a}$$

$$[\Delta V]b = [\text{BCBV}]b * [B]b + [\text{BCBV}]b - c * [B]b - c \tag{9b}$$

$$[\Delta V]c = [\text{BCBV}]c * [B]c + [\text{BCBV}]c - a * [B]c - a \tag{9c}$$

3 Short Circuit Computation

Most of faults that occur on distribution system are unsymmetrical faults. Unsymmetrical faults include single line-to-ground (SLG), double line-to-ground (DLG), line-to line (L-L) and three line-to-ground (LLLG). For demonstration purposes let us assume that all the unsymmetrical faults (SLG, DLG, L-L and LLLG) are at Bus 4 (Fig. 2).

SLG Fault

This fault is the most common fault in the distribution system. This fault occurs between a phase and the ground. The fault currents flows at a very high magnitude and due to this it can be safely assumed the load currents are zero while computing the short circuit analysis. Let us assume phase A at Bus 4 is short circuited with a fault impedance Z_f and the fault current is I_f.

Matrix formulation

The BIBC matrices (6a–6c) are modified to arrive at the post fault BIBC per phase matrices. These matrices are represented below.

$$\begin{bmatrix} B1a \\ B2a \\ B3a \\ B4a \\ B5a \end{bmatrix} = \begin{bmatrix} 1\,1\,1\,1\,1 \\ 0\,1\,1\,1\,1 \\ 0\,0\,1\,1\,0 \\ 0\,0\,0\,1\,0 \\ 0\,0\,0\,0\,1 \end{bmatrix} \begin{bmatrix} 0 \\ 0 \\ I_f \\ 0 \\ 0 \end{bmatrix} \tag{10a}$$

$$\begin{bmatrix} B1b \\ B2b \\ B3b \\ B4b \\ B5b \end{bmatrix} = \begin{bmatrix} 1\,1\,1\,1\,1 \\ 0\,1\,1\,1\,1 \\ 0\,0\,1\,1\,0 \\ 0\,0\,0\,1\,0 \\ 0\,0\,0\,0\,1 \end{bmatrix} \begin{bmatrix} 0 \\ 0 \\ 0 \\ 0 \\ 0 \end{bmatrix} \tag{10b}$$

$$\begin{bmatrix} B1c \\ B2c \\ B3c \\ B4c \\ B5c \end{bmatrix} = \begin{bmatrix} 1\,1\,1\,1\,1 \\ 0\,1\,1\,1\,1 \\ 0\,0\,1\,1\,0 \\ 0\,0\,0\,1\,0 \\ 0\,0\,0\,0\,1 \end{bmatrix} \begin{bmatrix} 0 \\ 0 \\ 0 \\ 0 \\ 0 \end{bmatrix} \tag{10c}$$

Simplifying the above will result in the following matrix

$$\begin{bmatrix} B1a \\ B2a \\ B3a \\ B4a \\ B5a \end{bmatrix} = \begin{bmatrix} 1 \\ 1 \\ 1 \\ 0 \\ 0 \end{bmatrix} |I_f| \tag{11}$$

From the above it is clear that the column of the BIBC matrix where the fault has occurred should only be considered. All the other columns need not be considered for the calculations.

Also

$[\mathrm{BIBC}]b = 0$, $[\mathrm{BIBC}]c = 0$ $[\mathrm{BIBC}]a - b = 0$, $[\mathrm{BIBC}]b - c = 0$ and $[\mathrm{BIBC}]c - a = 0$ since all the load currents in phase B and phase C are very low compared to the fault current in phase A at Bus 4.

The BCBV matrices (Eqs. 7a–7f) are modified to arrive at the post fault BCBV per phase matrix. This matrix is represented below

$$\begin{bmatrix} V1a \\ V1a \\ V1a \\ V1a \\ V1a \end{bmatrix} - \begin{bmatrix} V2a \\ V3a \\ I_f Z_f \\ V5a \\ V6a \end{bmatrix} = \begin{bmatrix} Z12aa & 0 & 0 & 0 & 0 \\ Z12aa & Z23aa & 0 & 0 & 0 \\ Z12aa & Z23aa & Z34aa & 0 & 0 \\ Z12aa & Z23aa & Z34aa & Z45aa & 0 \\ Z12aa & Z23aa & 0 & 0 & Z36aa \end{bmatrix} \begin{bmatrix} B1a \\ B2a \\ B3a \\ B4a \\ B5a \end{bmatrix} \tag{12}$$

Since the fault is at Bus 4 with impedance Z_f the voltage at Bus 4 post fault would be $I_f * Z_f$.

Calculating I_f

The fault current I_f is calculated by picking up the row corresponding to 4th bus from Eq. 12

$$\left[V1a - I_f Z_f\right] = \left[Z12aa\ Z23aa\ Z34aa\ 0\ 0\right] \begin{bmatrix} 1 \\ 1 \\ 1 \\ 0 \\ 0 \end{bmatrix} |I_f| \tag{13}$$

Solving Eq. 13 would result in

$$\mathrm{I_f} = V1a/\left(Z12aa + Z23aa + Z34aa + Z_f\right) \tag{14}$$

With known pre fault voltage at Bus 1 for phase a ($V1a$) and known self and fault impedances the fault current can be computed from Eq. 14.

The post fault voltages at other buses for phase A can be calculated after computing fault current from Eq. 27 and using matrix Eq. 12.

Similar calculations can be performed for the fault at other buses and for other phases.

In this case the effect of other phases mutual impedances can be ignored since the load currents in other phases are assumed to be zero as compared to the fault current I_f.

DLG Fault

This fault occurs between any 2 phases and the ground. The fault currents flows at a very high magnitude and due to this it can be safely assumed the load currents are zero while computing the short circuit analysis. Let us assume phase A and B at Bus 4 are short circuited to ground with a fault impedance Z_f and the fault currents are Ia_f and Ib_f for phase A and phase B, respectively.

Matrix formulation

The BIBC matrices (6a–6c and 8a–8c) are modified to arrive at the post fault BIBC per phase matrices. These matrices are represented below

$$\begin{bmatrix} B1a \\ B2a \\ B3a \\ B4a \\ B5a \end{bmatrix} = \begin{bmatrix} 1\,1\,1\,1\,1 \\ 0\,1\,1\,1\,1 \\ 0\,0\,1\,1\,0 \\ 0\,0\,0\,1\,0 \\ 0\,0\,0\,0\,1 \end{bmatrix} \begin{bmatrix} 0 \\ 0 \\ Ia_f \\ 0 \\ 0 \end{bmatrix} \tag{15a}$$

$$\begin{bmatrix} B1b \\ B2b \\ B3b \\ B4b \\ B5b \end{bmatrix} = \begin{bmatrix} 1\,1\,1\,1\,1 \\ 0\,1\,1\,1\,1 \\ 0\,0\,1\,1\,0 \\ 0\,0\,0\,1\,0 \\ 0\,0\,0\,0\,1 \end{bmatrix} \begin{bmatrix} 0 \\ 0 \\ Ib_f \\ 0 \\ 0 \end{bmatrix} \tag{15b}$$

$$\begin{bmatrix} B1c \\ B2c \\ B3c \\ B4c \\ B5c \end{bmatrix} = \begin{bmatrix} 1\,1\,1\,1\,1 \\ 0\,1\,1\,1\,1 \\ 0\,0\,1\,1\,0 \\ 0\,0\,0\,1\,0 \\ 0\,0\,0\,0\,1 \end{bmatrix} \begin{bmatrix} 0 \\ 0 \\ 0 \\ 0 \\ 0 \end{bmatrix} \tag{15c}$$

Simplifying the above matrices and considering the mutual impedance matrix will result in the below matrices

$$\begin{bmatrix} B1a \\ B2a \\ B3a \\ B4a \\ B5a \end{bmatrix} = \begin{bmatrix} 1 \\ 1 \\ 1 \\ 0 \\ 0 \end{bmatrix} |Ia_f| \quad (16a)$$

$$\begin{bmatrix} B1b \\ B2b \\ B3b \\ B4b \\ B5b \end{bmatrix} = \begin{bmatrix} 1 \\ 1 \\ 1 \\ 0 \\ 0 \end{bmatrix} |Ib_f| \quad (16b)$$

$$\begin{bmatrix} B1a-b \\ B2a-b \\ B3a-b \\ B4a-b \\ B5a-b \end{bmatrix} = \begin{bmatrix} 1 \\ 1 \\ 1 \\ 0 \\ 0 \end{bmatrix} |Ib_f| \quad (16c)$$

From the above it is clear that the column of the BIBC matrix where the fault has occurred should only be considered. All the other columns need not be considered for the calculations.

Also

$[\text{BIBC}]c = 0$, $[\text{BIBC}]b - c = 0$ and $[\text{BIBC}]c - a = 0$ since all the load currents in phase C are very low compared to the fault current in phase A and phase B at Bus 4.

Since the fault is at Bus 4 with impedance Z_f the voltage at Bus 4 post fault would be $(Ia_f + Ib_f) * Z_f$ for both phase A and phase B.

The BCBV matrices (Eqs. 7a–7f and 9a–9c) are modified to arrive at the post fault BCBV per phase matrix. This matrix is represented below. Here both the self-impedance matrix and mutual impedance matrix have to be considered for computing the fault currents.

$$\begin{bmatrix} V1a \\ V1a \\ V1a \\ V1a \\ V1a \end{bmatrix} - \begin{bmatrix} V2a \\ V3a \\ (Ia_f + Ib_f)Z_f \\ V5a \\ V6a \end{bmatrix}$$
$$= \begin{bmatrix} Z12aa & 0 & 0 & 0 & 0 \\ Z12aa & Z23aa & 0 & 0 & 0 \\ Z12aa & Z23aa & Z34aa & 0 & 0 \\ Z12aa & Z23aa & Z34aa & Z45aa & 0 \\ Z12aa & Z23aa & 0 & 0 & Z36aa \end{bmatrix} \begin{bmatrix} B1a \\ B2a \\ B3a \\ B4a \\ B5a \end{bmatrix}$$

$$+\begin{bmatrix} Z12ab & 0 & 0 & 0 & 0 \\ Z12ab & Z23ab & 0 & 0 & 0 \\ Z12ab & Z23ab & Z34ab & 0 & 0 \\ Z12ab & Z23ab & Z34ab & Z45ab & 0 \\ Z12ab & Z23ab & 0 & 0 & Z36ab \end{bmatrix}\begin{bmatrix} B1a-b \\ B2a-b \\ B3a-b \\ B4a-b \\ B5a-b \end{bmatrix} \quad (17)$$

Following is the post fault voltage difference phase B matrix with fault at Bus 4

$$\begin{bmatrix} V1b \\ V1b \\ V1b \\ V1b \\ V1b \end{bmatrix} - \begin{bmatrix} V2b \\ V3b \\ (Ia_f + Ib_f)Z_f \\ V5b \\ V6b \end{bmatrix} = \begin{bmatrix} Z12bb & 0 & 0 & 0 & 0 \\ Z12bb & Z23bb & 0 & 0 & 0 \\ Z12bb & Z23bb & Z34bb & 0 & 0 \\ Z12bb & Z23bb & Z34bb & Z45bb & 0 \\ Z12bb & Z23bb & 0 & 0 & Z36bb \end{bmatrix}\begin{bmatrix} B1b \\ B2b \\ B3b \\ B4b \\ B5b \end{bmatrix} \quad (18)$$

Calculating Ia_f and Ib_f

Fault currents Ia_f and Ib_f are calculated by picking up the row corresponding to 4th bus from Eqs. 17 and 18

$$\left[V1a - (Ia_f + Ib_f)Z_f\right] = \left[Z12aa\ Z23aa\ Z34aa\ 0\ 0\right]\begin{bmatrix} 1 \\ 1 \\ 1 \\ 0 \\ 0 \end{bmatrix}\left|Ia_f\right| + \left[Z12ab\ Z23ab\ Z34ab\ 0\ 0\right]\begin{bmatrix} 1 \\ 1 \\ 1 \\ 0 \\ 0 \end{bmatrix}\left|Ib_f\right| \quad (19)$$

$$\left[V1b - (Ia_f + Ib_f)Z_f\right] = \left[Z12bb\ Z23bb\ Z34bb\ 0\ 0\right]\begin{bmatrix} 1 \\ 1 \\ 1 \\ 0 \\ 0 \end{bmatrix}\left|Ib_f\right| \quad (20)$$

With known pre fault voltage at Bus 1 for phase a ($V1a$) and phase b ($V1b$) and known self, mutual and fault impedances the fault currents Ia_f and Ib_f are computed by solving Eqs. 19 and 20.

The post fault voltages at other buses for phase A and phase B can be calculated after computing fault currents and using matrix Eqs. 17 and 18.

Similar calculations can be performed for the fault at other buses and for other phases.

In this case the effect of other mutual impedances, i.e. $b - c$ and $c - a$ can be ignored since the load currents in phase C are assumed to be zero as compared to the fault currents Ia_f and Ib_f in phase A and phase B, respectively.

LL Fault

This fault occurs between any 2 phases. The fault currents flows at a very high magnitude and due to this it can be safely assumed the load currents are zero while computing the short circuit analysis. Let us assume phase A and B at Bus 4 are short circuited with a fault impedance Z_f and the fault currents are Ia_f and Ib_f for phase A and phase B, respectively.

Matrix formulation

The BIBC matrices (6a–6c and 8a–8c) are modified to arrive at the post fault BIBC per phase matrices. These matrices are represented below

$$\begin{bmatrix} B1a \\ B2a \\ B3a \\ B4a \\ B5a \end{bmatrix} = \begin{bmatrix} 1\,1\,1\,1\,1 \\ 0\,1\,1\,1\,1 \\ 0\,0\,1\,1\,0 \\ 0\,0\,0\,1\,0 \\ 0\,0\,0\,0\,1 \end{bmatrix} \begin{bmatrix} 0 \\ 0 \\ Ia_f \\ 0 \\ 0 \end{bmatrix} \tag{21a}$$

$$\begin{bmatrix} B1b \\ B2b \\ B3b \\ B4b \\ B5b \end{bmatrix} = \begin{bmatrix} 1\,1\,1\,1\,1 \\ 0\,1\,1\,1\,1 \\ 0\,0\,1\,1\,0 \\ 0\,0\,0\,1\,0 \\ 0\,0\,0\,0\,1 \end{bmatrix} \begin{bmatrix} 0 \\ 0 \\ Ib_f \\ 0 \\ 0 \end{bmatrix} \tag{21b}$$

$$\begin{bmatrix} B1c \\ B2c \\ B3c \\ B4c \\ B5c \end{bmatrix} = \begin{bmatrix} 1\,1\,1\,1\,1 \\ 0\,1\,1\,1\,1 \\ 0\,0\,1\,1\,0 \\ 0\,0\,0\,1\,0 \\ 0\,0\,0\,0\,1 \end{bmatrix} \begin{bmatrix} 0 \\ 0 \\ 0 \\ 0 \\ 0 \end{bmatrix} \tag{21c}$$

Simplifying the above matrices and considering the mutual impedance matrix will result in the below matrices

$$\begin{bmatrix} B1a \\ B2a \\ B3a \\ B4a \\ B5a \end{bmatrix} = \begin{bmatrix} 1 \\ 1 \\ 1 \\ 0 \\ 0 \end{bmatrix} |Ia_f| \tag{22a}$$

$$\begin{bmatrix} B1b \\ B2b \\ B3b \\ B4b \\ B5b \end{bmatrix} = \begin{bmatrix} 1 \\ 1 \\ 1 \\ 0 \\ 0 \end{bmatrix} |Ib_f| \tag{22b}$$

$$\begin{bmatrix} B1a-b \\ B2a-b \\ B3a-b \\ B4a-b \\ B5a-b \end{bmatrix} = \begin{bmatrix} 1 \\ 1 \\ 1 \\ 0 \\ 0 \end{bmatrix} |Ib_f| \tag{22c}$$

From the above it is clear that the column of the BIBC matrix where the fault has occurred should only be considered. All the other columns need not be considered for the calculations.

Also

[BIBC]$c = 0$, [BIBC]$b - \text{c} = 0$ and [BIBC]$c - a = 0$ since all the load currents in phase C are very low compared to the fault current in phase A and phase B at Bus 4.

Since the fault is at Bus 4 with impedance Z_f between phase A and phase B the current Ia_f and Ib_f are opposite to each other. Mathematically it can expressed as

$$Ib_f = -\left(Ia_f\right) \tag{23}$$

The voltage equation at Bus 4 post fault can be mathematically expressed as

$$V4^f a - V4^f b = Ia_f * Z_f \tag{24}$$

where $V4^f a$ and $V4^f b$ are post fault bus voltages at Bus 4 for phase A and phase B, respectively.

The BCBV matrices (7a–7f and 9a–9c) are modified to arrive at the post fault BCBV per phase matrix. This matrix is represented below. Here both the self-impedance matrix and mutual impedance matrix have to be considered for computing the fault currents

$$\begin{bmatrix} V1a \\ V1a \\ V1a \\ V1a \\ V1a \end{bmatrix} - \begin{bmatrix} V2a \\ V3a \\ (V4fa \\ V5a \\ V6a \end{bmatrix} = \begin{bmatrix} Z12aa & 0 & 0 & 0 & 0 \\ Z12aa & Z23aa & 0 & 0 & 0 \\ Z12aa & Z23aa & Z34aa & 0 & 0 \\ Z12aa & Z23aa & Z34aa & Z45aa & 0 \\ Z12aa & Z23aa & 0 & 0 & Z36aa \end{bmatrix} \begin{bmatrix} B1a \\ B2a \\ B3a \\ B4a \\ B5a \end{bmatrix}$$

$$+\begin{bmatrix} Z12ab\ 0 & 0 & 0 & 0 \\ Z12ab\ Z23ab\ 0 & 0 & 0 \\ Z12ab\ Z23ab\ Z34ab\ 0 & & 0 \\ Z12ab\ Z23ab\ Z34ab\ Z45ab\ 0 & & \\ Z12ab\ Z23ab\ 0 & 0 & Z36ab \end{bmatrix} \begin{bmatrix} B1a-b \\ B2a-b \\ B3a-b \\ B4a-b \\ B5a-b \end{bmatrix} \tag{25}$$

Following is the post fault voltage difference phase *B* matrix with fault at Bus 4

$$\begin{bmatrix} V1b \\ V1b \\ V1b \\ V1b \\ V1b \end{bmatrix} - \begin{bmatrix} V2b \\ V3b \\ (V4fb \\ V5b \\ V6b \end{bmatrix} = \begin{bmatrix} Z12bb\ 0 & 0 & 0 & 0 \\ Z12bb\ Z23bb\ 0 & 0 & 0 \\ Z12bb\ Z23bb\ Z34bb\ 0 & & 0 \\ Z12bb\ Z23bb\ Z34bb\ Z45bb\ 0 & & \\ Z12bb\ Z23bb\ 0 & 0 & Z36bb \end{bmatrix} \begin{bmatrix} B1b \\ B2b \\ B3b \\ B4b \\ B5b \end{bmatrix} \tag{26}$$

Calculating Ia_f and Ib_f

Fault currents Ia_f and Ib_f are calculated by picking up the row corresponding to 4th bus from Eqs. 25 and 26 and also using Eqs. 23 and 24.

$$[V1a - V4fa] = \begin{bmatrix} Z12aa\ Z23aa\ Z34aa\ 0\ 0 \end{bmatrix} \begin{bmatrix} 1 \\ 1 \\ 1 \\ 0 \\ 0 \end{bmatrix} |-Ia_f| + \begin{bmatrix} Z12ab\ Z23ab\ Z34ab\ 0\ 0 \end{bmatrix} \begin{bmatrix} 1 \\ 1 \\ 1 \\ 0 \\ 0 \end{bmatrix} |-Ia_f| \tag{27}$$

where Ib_f is substituted as $-\ Ia_f$ as per Eq. 23

$$[V1b - V4fb] = \begin{bmatrix} Z12bb\ Z23bb\ Z34bb\ 0\ 0 \end{bmatrix} \begin{bmatrix} 1 \\ 1 \\ 1 \\ 0 \\ 0 \end{bmatrix} |-Ia_f| \tag{28}$$

where Ib_f is substituted as $-\ Ia_f$ as per Eq. 23

Also as per Eq. 24

$$V4^f a - V4^f b = Ia_f * Z_f$$

With known pre fault voltage at Bus 1 for phase a ($V1a$) and phase b ($V1b$) and known self, mutual and fault impedances the fault currents the fault current Ia_f, post fault voltages $V4^f a$ and $V4^f b$ is computed using the above Eqs. 27, 28 and 24.

The post fault voltages at other buses for phase A and phase B can be calculated after computing fault currents and using matrix Eqs. 25 and 26.

Similar calculations can be performed for the fault at other buses and for other phases.

In this case the effect of other mutual impedances, i.e. $b - c$ and $c - a$ can be ignored since the load currents in phase C are assumed to be zero as compared to the fault currents Ia_f and Ib_f in phase A and phase B, respectively.

LLLG Fault

This fault occurs between any three-phases and the ground. This fault is the most severe fault. The fault currents flows at a very high magnitude and due to this it can be safely assumed the load currents are zero while computing the short circuit analysis. Let us assume phase A, B and C at Bus 4 are short circuited to ground with a fault impedance Z_f and the fault currents are Ia_f, Ib_f and Ic_f for phase A, phase B and phase C, respectively.

Matrix formulation

The BIBC matrices (6a–6c and 8a–8c) are modified to arrive at the post fault BIBC per phase matrices. These matrices are represented below

$$\begin{bmatrix} B1a \\ B2a \\ B3a \\ B4a \\ B5a \end{bmatrix} = \begin{bmatrix} 1\,1\,1\,1\,1 \\ 0\,1\,1\,1\,1 \\ 0\,0\,1\,1\,0 \\ 0\,0\,0\,1\,0 \\ 0\,0\,0\,0\,1 \end{bmatrix} \begin{bmatrix} 0 \\ 0 \\ Ia_f \\ 0 \\ 0 \end{bmatrix} \tag{29a}$$

$$\begin{bmatrix} B1b \\ B2b \\ B3b \\ B4b \\ B5b \end{bmatrix} = \begin{bmatrix} 1\,1\,1\,1\,1 \\ 0\,1\,1\,1\,1 \\ 0\,0\,1\,1\,0 \\ 0\,0\,0\,1\,0 \\ 0\,0\,0\,0\,1 \end{bmatrix} \begin{bmatrix} 0 \\ 0 \\ Ib_f \\ 0 \\ 0 \end{bmatrix} \tag{29b}$$

$$\begin{bmatrix} B1c \\ B2c \\ B3c \\ B4c \\ B5c \end{bmatrix} = \begin{bmatrix} 1\,1\,1\,1\,1 \\ 0\,1\,1\,1\,1 \\ 0\,0\,1\,1\,0 \\ 0\,0\,0\,1\,0 \\ 0\,0\,0\,0\,1 \end{bmatrix} \begin{bmatrix} 0 \\ 0 \\ Ic_f \\ 0 \\ 0 \end{bmatrix} \tag{29c}$$

Simplifying the above matrices and considering the Mutual impedance matrix will result in the below matrices

$$\begin{bmatrix} B1a \\ B2a \\ B3a \\ B4a \\ B5a \end{bmatrix} = \begin{bmatrix} 1 \\ 1 \\ 1 \\ 0 \\ 0 \end{bmatrix} |Ia_f| \quad (30a)$$

$$\begin{bmatrix} B1b \\ B2b \\ B3b \\ B4b \\ B5b \end{bmatrix} = \begin{bmatrix} 1 \\ 1 \\ 1 \\ 0 \\ 0 \end{bmatrix} |Ib_f| \quad (30b)$$

$$\begin{bmatrix} B1b \\ B2b \\ B3b \\ B4b \\ B5b \end{bmatrix} = \begin{bmatrix} 1 \\ 1 \\ 1 \\ 0 \\ 0 \end{bmatrix} |Ic_f| \quad (30c)$$

$$\begin{bmatrix} B1a-b \\ B2a-b \\ B3a-b \\ B4a-b \\ B5a-b \end{bmatrix} = \begin{bmatrix} 1 \\ 1 \\ 1 \\ 0 \\ 0 \end{bmatrix} |Ib_f| \quad (30d)$$

$$\begin{bmatrix} B1b-c \\ B2b-c \\ B3b-c \\ B4b-c \\ B5b-c \end{bmatrix} = \begin{bmatrix} 1 \\ 1 \\ 1 \\ 0 \\ 0 \end{bmatrix} |Ic_f| \quad (30e)$$

$$\begin{bmatrix} B1c-a \\ B2c-a \\ B3c-a \\ B4c-a \\ B5c-a \end{bmatrix} = \begin{bmatrix} 1 \\ 1 \\ 1 \\ 0 \\ 0 \end{bmatrix} |Ic_f| \quad (30f)$$

Since the fault is at Bus 4 with impedance Z_f the voltage at Bus 4 post fault would be $(Ia_f + Ib_f + Ic_f) * Z_f$ for phase *A*, phase *B* and phase *C*, respectively.

The BCBV matrices (7a–7c and 9a–9c) are modified to arrive at the post fault BCBV per phase matrix. This matrix is represented below. Here both the self-impedance matrix and mutual impedance matrix have to be considered for computing the fault currents

$$\begin{bmatrix} V1a \\ V1a \\ V1a \\ V1a \\ V1a \end{bmatrix} - \begin{bmatrix} V2a \\ V3a \\ (Ia_f + Ib_f + Ic_f)Z_f \\ V5a \\ V6a \end{bmatrix} = \begin{bmatrix} Z12aa & 0 & 0 & 0 & 0 \\ Z12aa & Z23aa & 0 & 0 & 0 \\ Z12aa & Z23aa & Z34aa & 0 & 0 \\ Z12aa & Z23aa & Z34aa & Z45aa & 0 \\ Z12aa & Z23aa & 0 & 0 & Z36aa \end{bmatrix} \begin{bmatrix} B1a \\ B2a \\ B3a \\ B4a \\ B5a \end{bmatrix} + \begin{bmatrix} Z12ab & 0 & 0 & 0 & 0 \\ Z12ab & Z23ab & 0 & 0 & 0 \\ Z12ab & Z23ab & Z34ab & 0 & 0 \\ Z12ab & Z23ab & Z34ab & Z45ab & 0 \\ Z12ab & Z23ab & 0 & 0 & Z36ab \end{bmatrix} \begin{bmatrix} B1a-b \\ B2a-b \\ B3a-b \\ B4a-b \\ B5a-b \end{bmatrix} \quad (31)$$

Following is the post fault voltage difference phase *B* matrix with fault at Bus 4

$$\begin{bmatrix} V1b \\ V1b \\ V1b \\ V1b \\ V1b \end{bmatrix} - \begin{bmatrix} V2b \\ V3b \\ (Ia_f + Ib_f + Ic_f)Z_f \\ V5b \\ V6b \end{bmatrix} = \begin{bmatrix} Z12bb & 0 & 0 & 0 & 0 \\ Z12bb & Z23bb & 0 & 0 & 0 \\ Z12bb & Z23bb & Z34bb & 0 & 0 \\ Z12bb & Z23bb & Z34bb & Z45bb & 0 \\ Z12bb & Z23bb & 0 & 0 & Z36bb \end{bmatrix} \begin{bmatrix} B1b \\ B2b \\ B3b \\ B4b \\ B5b \end{bmatrix} + \begin{bmatrix} Z12bc & 0 & 0 & 0 & 0 \\ Z12bc & Z23bc & 0 & 0 & 0 \\ Z12bc & Z23bc & Z34bc & 0 & 0 \\ Z12bc & Z23bc & Z34bc & Z45bc & 0 \\ Z12bc & Z23bc & 0 & 0 & Z36bc \end{bmatrix} \begin{bmatrix} B1b-c \\ B2b-c \\ B3b-c \\ B4b-c \\ B5b-c \end{bmatrix} \quad (32)$$

Following is the post fault voltage difference phase *C* matrix with fault at Bus 4

$$\begin{bmatrix} V1c \\ V1c \\ V1c \\ V1c \\ V1c \end{bmatrix} - \begin{bmatrix} V2c \\ V3c \\ (Ia_f + Ib_f + Ic_f)Z_f \\ V5c \\ V6c \end{bmatrix}$$

$$= \begin{bmatrix} Z12cc & 0 & 0 & 0 & 0 \\ Z12cc & Z23cc & 0 & 0 & 0 \\ Z12cc & Z23cc & Z34cc & 0 & 0 \\ Z12cc & Z23cc & Z34cc & Z45cc & 0 \\ Z12cc & Z23cc & 0 & 0 & Z36cc \end{bmatrix} \begin{bmatrix} B1c \\ B2c \\ B3c \\ B4c \\ B5c \end{bmatrix}$$

$$+ \begin{bmatrix} Z12ca & 0 & 0 & 0 & 0 \\ Z12ca & Z23ca & 0 & 0 & 0 \\ Z12ca & Z23ca & Z34ca & 0 & 0 \\ Z12ca & Z23ca & Z34ca & Z45ca & 0 \\ Z12ca & Z23ca & 0 & 0 & Z36ca \end{bmatrix} \begin{bmatrix} B1c - a \\ B2c - a \\ B3c - a \\ B4c - a \\ B5c - a \end{bmatrix} \quad (33)$$

Calculating Ia_f, Ib_f and Ic_f

Fault currents Ia_f, Ib_f and Ic_f are calculated by picking up the row corresponding to 4th bus from Eqs. 31, 32 and 33

$$\left[V1a - (Ia_f + Ib_f + Ic_f)Z_f\right] = \left[Z12aa\ Z23aa\ Z34aa\ 0\ 0\right] \begin{bmatrix} 1 \\ 1 \\ 1 \\ 0 \\ 0 \end{bmatrix} |Ia_f|$$

$$+ \left[Z12ab\ Z23ab\ Z34ab\ 0\ 0\right] \begin{bmatrix} 1 \\ 1 \\ 1 \\ 0 \\ 0 \end{bmatrix} |Ib_f| \quad (34)$$

$$\left[V1b - (Ia_f + Ib_f + Ic_f)Z_f\right] = \left[Z12bb\ Z23bb\ Z34bb\ 0\ 0\right] \begin{bmatrix} 1 \\ 1 \\ 1 \\ 0 \\ 0 \end{bmatrix} |Ib_f|$$

$$+\begin{bmatrix} Z12bc & Z23bc & Z34bc & 0 & 0 \end{bmatrix}\begin{bmatrix} 1 \\ 1 \\ 1 \\ 0 \\ 0 \end{bmatrix}|Ic_f| \quad (35)$$

$$\begin{bmatrix} V1c - (Ia_f + Ib_f + Ic_f)Z_f \end{bmatrix} = \begin{bmatrix} Z12cc & Z23cc & Z34cc & 0 & 0 \end{bmatrix}\begin{bmatrix} 1 \\ 1 \\ 1 \\ 0 \\ 0 \end{bmatrix}|Ic_f|$$
$$+\begin{bmatrix} Z12ca & Z23cb & Z34cb & 0 & 0 \end{bmatrix}\begin{bmatrix} 1 \\ 1 \\ 1 \\ 0 \\ 0 \end{bmatrix}|Ia_f| \quad (36)$$

With known pre fault voltage at Bus 1 for phase a ($V1a$), phase b ($V1b$) and phase c ($V1c$) and known self, mutual and fault impedances the fault currents Ia_f, Ib_f and Ic_f are computed by solving Eqs. 34, 35 and 36.

The post fault voltages at other buses for phase A and phase B can be calculated after computing fault currents and using matrix Eqs. 31, 32 and 33.

Similar calculations can be performed for the fault at other buses.

In this case the effect of mutual impedances for all the phases are considered since the fault has occurred between all the phases and ground for a particular bus.

4 Simulation Results

Table 1 shows the results of fault currents without mutual impedances. The fault is simulated at Bus 11 for the feeder data in Appendix A. SLG fault is simulated on phase A and DLG, LL fault is simulated on phase A and phase B. The fault currents are Ia_f, Ib_f and Ic_f for phase A, phase B and phase C in pu, respectively, with fault impedance $(Z_f) = 0$.

Table 2 shows the results of fault currents with mutual impedances. The fault is simulated at Bus 11 for the feeder data in Appendix A. SLG fault is simulated on phase A and DLG, LL fault is simulated on phase A and phase B. The fault currents are Ia_f, Ib_f and Ic_f for phase A, phase B and phase C in pu, respectively, with fault impedance $(Z_f) = 0$.

Table 1 Results of fault currents without mutual impedances

Fault type	Ia_f	Ib_f	Ic_f
SLG	4.0691	–	–
LL	3.5239	3.5239	
DLG	4.0691	4.0691	–
LLLG	4.0691	4.0691	4.0691

Table 2 Results of fault currents with mutual impedances

Fault type	Ia_f	Ib_f	Ic_f
SLG	4.0691	–	–
LL	4.1147	4.1147	
DLG	2.0557	4.0691	–
LLLG	4.5143	4.6447	4.5607

5 Conclusions

In this paper, the short circuit analysis for unbalanced radial distribution system with mutual impedances using the modified direct approach method is introduced. The proposed method was applied on 55 bus three-phase unbalanced network with mutual impedances in matlab simulation environment and short circuit currents for SLG, LL, DLG and LLLG are calculated. The inclusion of mutual impedances to the short circuit calculations represents the practical distribution system and the results are shown in Tables 1 and 2. The results obtained from the proposed method have been compared with the results obtained from the method given in reference [4]. From these results it is obvious that there are significant differences in the estimated fault currents.

References

1. Grainger JJ, Stevenson WD (1994) Power system analysis. McGraw-Hill International Editions
2. Brandwajn V, Tinney WF (1985) Generalized method of fault analysis. IEEE Trans PAS-104(6):1301–1306
3. Kersting WH, Phillips WH (1990) Distribution system short circuit analysis. In: Proceedings of 25th intersociety conference, IECEC-90, vol 1, pp 310–315
4. Teng JH (2005) Systematic short-circuit-analysis method for unbalanced distribution systems. IET Proc Gener Transm Distrib 152(4):549–555
5. Teng JH (2003) A direct approach for distribution system load flow solutions. IEEE Trans PWRD-18(3):882–887
6. Brown HE (1985) Solution of large networks by matrix methods. Wiley, New York
7. Glover JD, Sarma M (1994) Power system analysis and design. PWS Publishing Co.
8. Chen T-H, Chen M-S, Lee W-J, Kotas P, Van Olinda P (1992) Distribution system short circuit analysis-A rigid approach. IEEE Trans PWRS-7(1):444–450
9. Chen TH, Chen MS, Inoue T, Chebli EA (1991) Three-phase cogenerator and transformer models for distribution system analysis. IEEE Trans PWRD-6(4):1671–1681

10. Berman A, Xu W (1998) Analysis of faulted power systems by phase coordinates. IEEE Trans PWRD-13(2):587–595
11. Zhang X, Soudi F, Shirmohammadi D, Cheng CS (1995) A distribution short circuit analysis approach using hybrid compensation method. IEEE Trans PWRS-10(4):2053–2059

Analysis of a Solar Photovoltaic Cell Array Characteristics Using MATLAB Simulink Tool

Divya Bharti, Girdhar Gopal, Arun Kishor Johar, Yogendra Kumar Upadhyaya, Ashutosh Tripathi, and Tarun Varma

Abstract In this research article, model of a solar triple layer solar cell is developed on MATLAB Simulink software. Triple layer solar cell has high efficiency system. The photovoltaic characteristic curves of triple layer solar cell are produced using a mathematical model using a MATLAB Simulink tool. The performance characteristics of power voltage or current voltage of photovoltaic modules are investigated by variation of the solar irradiance and temperatures environments. The various maximum power-point system approaches are presented in this study paper. The properties of photovoltaic cells are generally determined by a nonlinear mathematical equation that is difficult to solve analytically. Therefore, in this article the peak power monitoring of photovoltaic modules has been studied that is very important for solar missile systems.

Keywords Triple layer solar cell · Photovoltaic · Nonlinear mathematical equation

1 Introduction

Nowadays, one of the major crises that the world is experiencing is an energy crisis, specifically an electric energy crisis. The ever-increasing energy demand places a stress on alternative energy sources. As a result, we need alternative energy sources that can supply us with energy in a long-term way. There are a variety of alternative energy sources available, but the sun is the most consistent, as most of the globe receives more than 250 days of sufficient sunlight each year. In a single year, the quantity of renewable radiation reaching the earth's surface is nearly twice that of all

D. Bharti · G. Gopal · A. K. Johar (✉) · T. Varma
Department of Electronics and Communication Engineering, Malaviya National Institute of Technology Jaipur, Jaipur, Rajasthan, India
e-mail: arun.e12006@cumail.in

A. K. Johar · Y. K. Upadhyaya · A. Tripathi
Department of Electronics and Communication Engineering, Chandigarh University, Gharuan, Mohali, Punjab, India

J. R. Szymanski et al. (eds.), *Energy Systems, Drives and Automations*,
Lecture Notes in Electrical Engineering 1057,
https://doi.org/10.1007/978-981-99-3691-5_13

non-renewable resources on the planet, including coal, oil, natural gas, and mined uranium. As a result, it is critical to make advantage of this huge energy source [1, 2]. Multijunction solar cells, which can be used in concentrated photovoltaic system, vary from silicon photovoltaic cell in that they can transform a significant solar radiation into electricity with a maximum efficiency. The maximum conversion effectiveness of solar cells with many interfaces is greater than 40% [3]. This work develops a photovoltaic cell simulation model based on the photovoltaic cell mathematical model. MATLAB Simulink is used to generate photovoltaic cell characteristic curves. Changes in sun irradiance and external temperature are used to compare photovoltaic characteristic curves. The ground test of the photovoltaic cell and simulation analysis yield the corresponding conclusions [4, 5]. We found a linear rise in the I_{sc} short circuit current under solar radiation ranging from 160 to 1000 W/m^2. The values for the irradiation are 0.8232A for the W/m^2 and 5.1465A for the 1000 W/m^2 at varied levels of current I_{sc} with solar radiation. The ideality factor n linearly rises to more than 350 W/m^2 with solar radiation. When irradiation ranges from 160 to 1000 W/m^2, the efficiency rises from 3.14 to 15.59% [6, 7]. The article is utilized to build a highest possible detection power point from solar system through artificial neural network. A peak value of power tracking system foundation on artificial neural network is also compared to conventional maximum power transfer (MPPT) approaches [8]. A variant adaptive perceptive particle swarm optimization was employed in this paper (APPSO). The solar array features are multiple maximums in the power voltage feature under partially shaded conditions. The technique proposed is for controlling several solar arrays with only one pair of sensors. In contrast to the earlier 96.41% with the use of particle swarm optimization, this result results in lower costs and a higher level of precision of 97.7%. For tracking and monitoring purposes, the proposed tracking technology was mapped to the MSP430FG4618 micro-controller [9]. The number of equations and unknown factors in a double diode model rises, making computations a bit more difficult but producing better exact response compare to a single diode model. Because the equations are simpler, the single diode design has fewer mistakes but also is quicker than the two-diode model in terms of mathematical computations and iterations [10–12]. The basic combination of photovoltaic array is discussed, as well as varied output characteristics of photovoltaic system under multiple situations also including solar radiation variation, temperature fluctuations. Inner variable resistance is demonstrating how series parallel solar panels have various effects [13, 14].

2 Methodology

2.1 *Proposed Modal of Photovoltaic Cell*

The most basic type of photovoltaic system is *p*–*n* junction diode. Electron and hole pairs are often generated in the depletion zone, where the inherent voltage and electric field drive electrons to n area and holes to *p*-region. Extra electrons travel through to the loading and interact with the massive amounts of holes when an outside stress is placed. Photovoltaic cells are a unique technology that transforms solar radiation energy into electrical. As a result, the working properties of these specific photovoltaic modules are investigated, and the analysis must be simulated using an analogous circuit and an external load. Figure shows the photovoltaic cell's corresponding schematic diagram.

2.2 *The Ideal Circuit of Photovoltaic Cell*

This model described in the following equation consists of one photovoltaic cell equivalent circuit linked in parallel with a photocurrent source

$$I = I_{\text{ph}} - I_0\left(\exp^{\left(\frac{qV_{\text{d}}}{akT}\right)} - 1\right). \tag{1}$$

2.3 *Photovoltaic Cell of One Diode*

A single diode variant (as shown in Fig. 1), which can be utilized for the output current, has a series resistance R_{s} and a sharp resistance R_{p}.

$$I = I_{\text{ph}} - I_0\left(\exp^{\left(\frac{V+IR_{\text{s}}}{aV_T}\right)} - 1\right) - \frac{V + IR_{\text{s}}}{R_{\text{p}}} \tag{2}$$

With

$$V_{\text{T}} = \frac{N_{\text{s}}KT}{q}. \tag{3}$$

The foundation of the photo-current is

$$I_{\text{ph}} = \left(I_{\text{ph,STC}} + K_1\Delta T\right)\frac{G}{G_{\text{STC}}}. \tag{4}$$

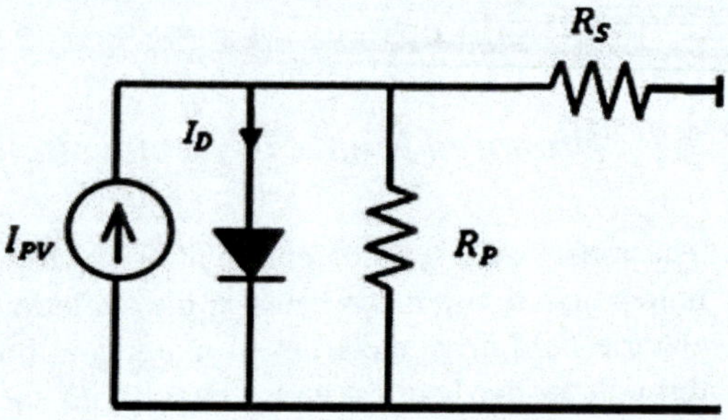

Fig. 1 Singular diode photovoltaic similar circuit model

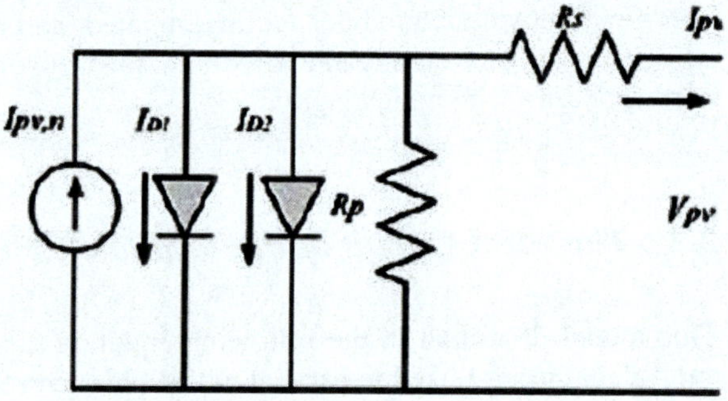

Fig. 2 Solar cell dual diode equivalent circuit model

2.4 Photovoltaic Cell of Two Diode

The two-diode models (as shown in Fig. 2) contain an extra diode in the analogue circuit of one diode, and they are linked to the first diode in parallel. The following expression specifies the output current in this example.

$$I = I_{\text{ph}} + I_{01}\left[\exp\left(\frac{V + IR_{\text{s}}}{\alpha_1 V_{T2}}\right) - 1\right] - I_{02}\left[\exp\left(\frac{V + IR_{\text{s}}}{\alpha_2 V_{T2}}\right) - 1\right] - \frac{V + IR_{\text{s}}}{R_{\text{p}}}, \tag{5}$$

$$V_{\text{T1}} = V_{\text{T2}} = \frac{N_{\text{s}}KT}{q}. \tag{6}$$

2.5 Modeling Solar Cells with Single and Two-Diode Types

The use of preceding equations becomes difficult for vast arrays consisting of N_{ss} * N_{pp} modules

$$I = I_{\text{ph}} N_{\text{pp}} + I_0 N_{\text{pp}} \left(\exp\left(\frac{V + I R_s \frac{N_{ss}}{N_{pp}}}{\alpha V_T N_{ss}} \right) - 1 \right) - \frac{V + I R_s \frac{N_{ss}}{N_{pp}}}{R_s \left(\frac{N_{ss}}{N_{pp}} \right)} \tag{7}$$

$$\begin{aligned} I = & I_{\text{ph}} N_{\text{pp}} + I_{01} N_{\text{pp}} \left(\exp\left(\frac{V + I R_s \frac{N_{ss}}{N_{pp}}}{\alpha_1 V_{T1} N_{ss}} \right) - 1 \right) \\ & - I_{02} N_{\text{pp}} \left(\exp\left(\frac{V + I R_s \frac{N_{ss}}{N_{pp}}}{\alpha_2 V_{T2} N_{ss}} \right) - 1 \right) - \frac{V + I R_s \frac{N_{ss}}{N_{pp}}}{R_s \left(\frac{N_{ss}}{N_{pp}} \right)}. \end{aligned} \tag{8}$$

2.6 *InGaP, InGaAs, and Ge Triple Photovoltaic Cell Junction Circuit*

The solar cell is made up of three layers as shown in Fig. 3.

(1) From the top to the bottom, the sub-cells are built with smaller energy gaps.
(2) This arrangement reduces losses from hot carrier thermalization and low-energy photon transmission, allowing solar energy to be turned into electricity more effectively than single junction cells.
(3) Multiple intersections solar cells made of InGaP/InGaAs/Ge are known for their high efficiency and used in space applications.

The current in a solar cell may be expressed as

$$I_c = I_{\text{Li}} - I_{\text{Di}} - I_{\text{shunt1}}. \tag{9}$$

If $i = 1$ indicates the upper layer, $i = 2$ indicates the middle layer, and $i = 3$ indicates the below layer. The triple layer photovoltaic cell's light produced current is provided by

$$I_{\text{Li}} = R K_c \left[I_{\text{sci}} + \alpha \left(T_c - T_{c,\text{ref}} \right) \right]. \tag{10}$$

When critical reference temperature T_c, ref in deg C, The temperature coefficient is a of the short circuit current in A deg C, The concentration ratio is represent KC, and the irradiation "*R*" in kWm^2 The current flow in diode is presented by

$$\begin{aligned} I_{\text{Di}} &= i_{0i} \left(\exp\left(\frac{q V_{\text{Di}}}{n_i K_B T} \right) - 1 \right) \\ V_{\text{Di}} &= V_i + I_c R_{\text{si}} \\ I_{0i} &= K_i * T^{(3+Y_i/2)} \left[\exp\left(\frac{-E_{\text{gi}}}{n_i K_B T} \right) \right]. \end{aligned} \tag{11}$$

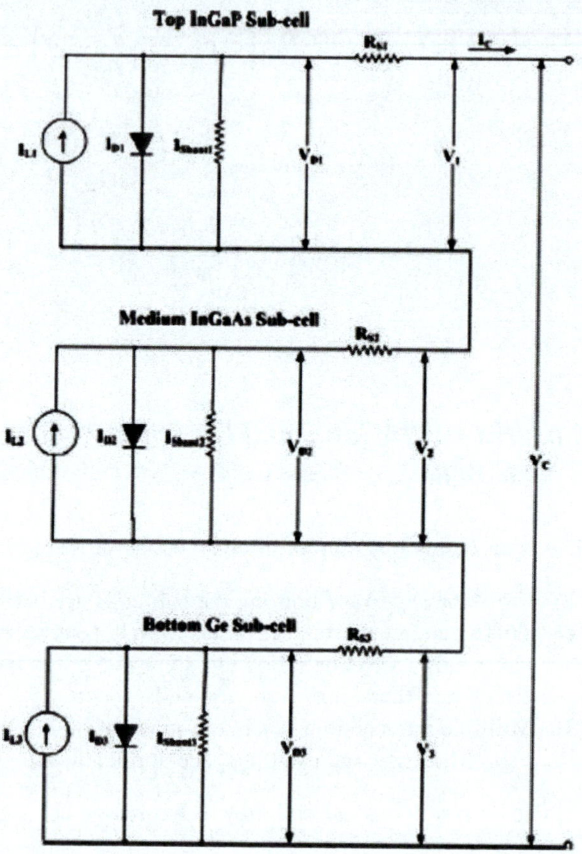

Fig. 3 Three junction photovoltaic cell equivalent circuit model

If charge of electron is q, the ideality factor of diode is n, Boltzmann constant is K_B, band gap energy is E_g, constants are K and c, the absolute temperature is T, and the resistance of the photovoltaic system is R_S. Temperature has an impact on the band gap energy E_g. The band gap energy fluctuates, depends on the relative of temperature as shown below graph depends on the relative of temperature. Shunt current can also be disregarded when shunt resistance is high enough.

$$E_g(T) = E_g(0) + \frac{\alpha T^2}{T + \beta} \tag{12}$$

Individual voltages V_1, V_2, and V_3 of three distinct layers can be represented as follows, ignoring the shunt current:

$$V_1 = \frac{n_1 K_B T}{q} \ln \left[\frac{I_{L1} - I_c}{I_{01}} + 1 \right] - I_c * R_{s1}$$

$$V_2 = \frac{n_2 K_B T}{q} \ln\left[\frac{I_{L2} - I_c}{I_{02}} + 1\right] - I_c * R_{s2}$$

$$V_3 = \frac{n_3 K_B T}{q} \ln\left[\frac{I_{L3} - I_c}{I_{03}} + 1\right] - I_c * R_{s3}$$

$$V_c = \frac{n_1 K_B T}{q} \ln\left[\frac{I_{L1} - I_c}{I_{01}} + 1\right] + \frac{n_2 K_B T}{q} \ln\left[\frac{I_{L2} - I_c}{I_{02}} + 1\right] + \frac{n_3 K_B T}{q} \ln\left[\frac{I_{L3} - I_c}{I_{03}} + 1\right] - I_c * R_s, \quad (13)$$

where $R_s = R_{s1} + R_{s2} + R_{s3}$.

3 Designing of a Triple Junction Photovoltaic Cell in MATLAB Simulink Tool

3.1 Photovoltaic Parameter Specification and Data

Figure 4 presents the photovoltaic parameter specification used for simulating the three-layer solar cell in MATLAB Simulink environment.

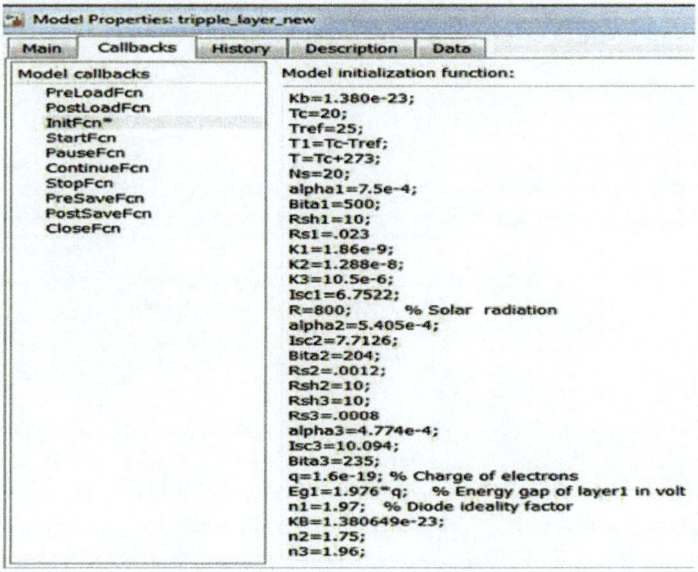

Fig. 4 Description and values of photovoltaic parameters mention

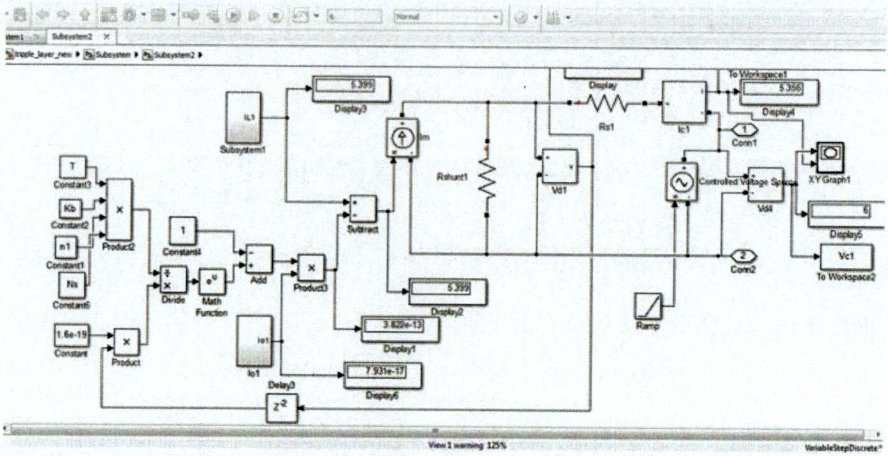

Fig. 5 Triple solar system first layer

3.2 Triple Solar System First Layer

See Fig. 5.

3.3 Triple Solar System Second Layer

See Fig. 6.

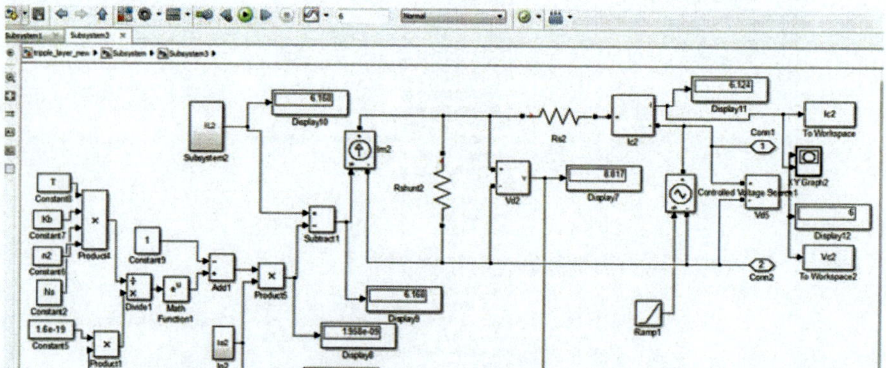

Fig. 6 Triple solar system second layer

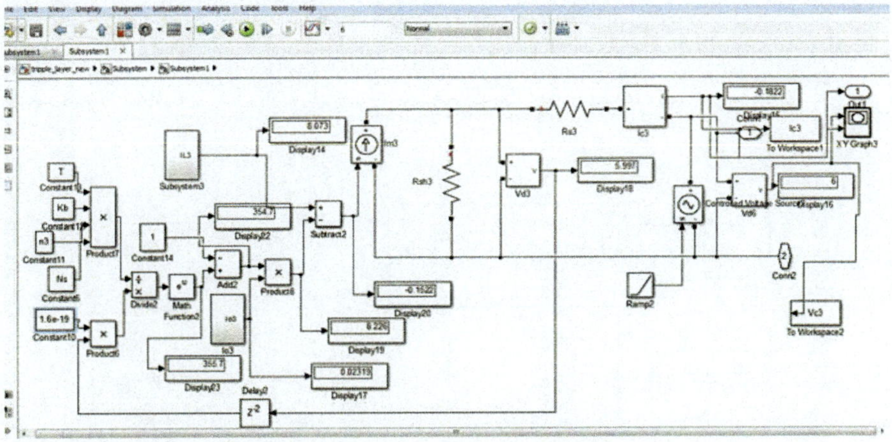

Fig. 7 Triple solar system third layer

3.4 Triple Solar System Third Layer

See Fig. 7.

3.5 Completed Modal of Solar System

The design of various layers of triple layer solar photovoltaic systems is shown in Figs. 5, 6, and 7, whereas the complete triple layer solar photovoltaic system is presented in Fig. 8.

4 Result and Discussion

The current voltage characteristics curve and the power voltage characteristics curve of the proposed photovoltaic system are shown in Figs. 9 and 10, respectively. In Fig. 9, *X*- and *Y*-axis represent voltage and current of the proposed photovoltaic system, respectively. Similarly, in Fig. 10, *X*- and *Y*-axis represent voltage and power of the proposed photovoltaic system, respectively.

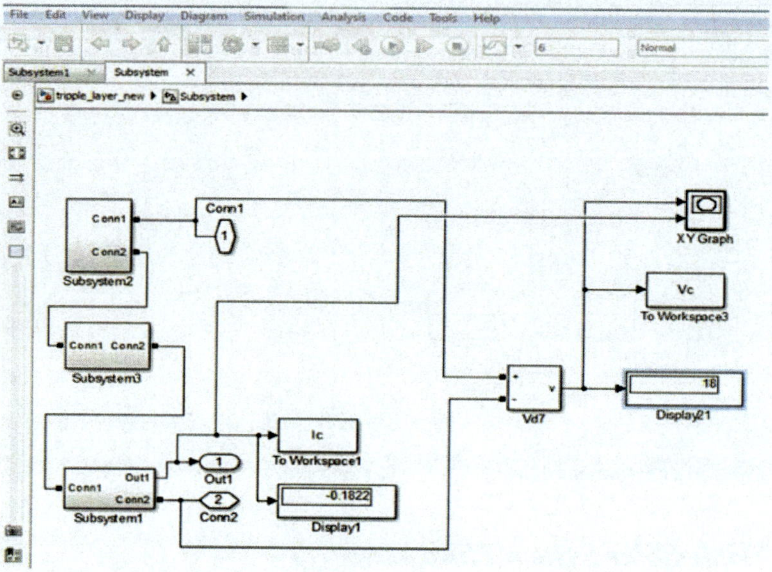

Fig. 8 Triple layer of photovoltaic system

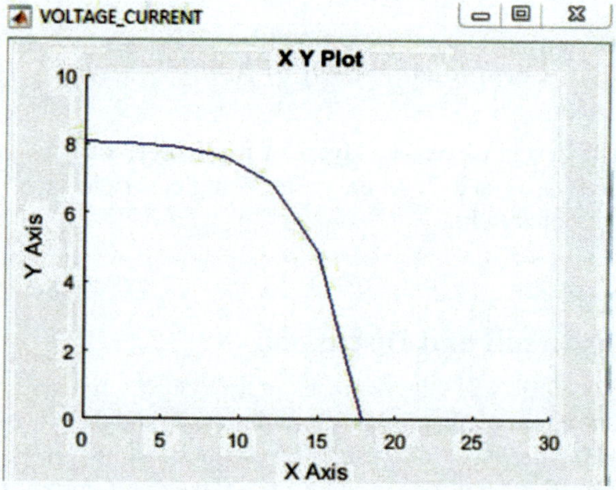

Fig. 9 Current voltage characteristics curve of photovoltaic system

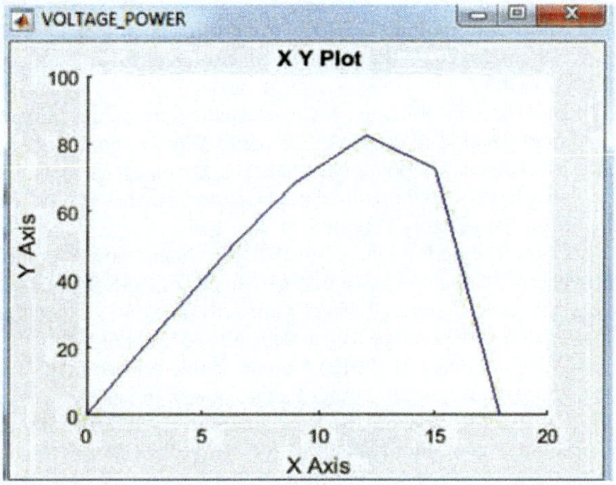

Fig. 10 Power voltage characteristics curve of photovoltaic system

5 Conclusion

MATLAB Simulink model of the triple layer InGaP, InGaAs, Ge high-performance photovoltaic modules is proposed. The model shown combines the photovoltaic cells, module, and array on the simulation platform with simple usage. The Simulink model of photovoltaic cell accepts parameter values and produces current voltage and power voltage characteristics under many circumstances, consisting the effects of the temperature and solar radiation. The solar cell's characteristic curve is achieved through the use of the mathematical operating model of photovoltaic cell.

References

1. Sims R (2000) Energy for tomorrow's world. A renewable energy perspective. Renew Energy World 3(4):24–30
2. Rezk H, Abou Hashema M (2013) Sizing of a stand alone concentrated photovoltaic system in Egyptian site. Int J Electr Power Energy Syst 45(1):325–330
3. Kinsey GS, Hebert P, Barbour KE, Krut DD, Cotal HL, Sherif RA (2008) Concentrator multi-junction solar cell characteristics under variable intensity and temperature. Prog Photovoltaics Res Appl 16(6):503–508
4. Chenni R, Makhlouf M, Kerbache T, Bouzid A (2007) A detailed modeling method for photovoltaic cells. Energy 32(9):1724–1730
5. Wang H, Shen J (2018) Analysis of the characteristics of solar cell array based on MATLAB/Simulink in solar unmanned aerial vehicle. IEEE Access 6:21195–21201
6. Rezk H, Hasaneen ES (2015) A new MATLAB/Simulink model of triple-junction solar cell and MPPT based on artificial neural networks for photovoltaic energy systems. Ain Shams Eng J 6(3):873–881
7. Chegaar M, Hamzaoui A, Namoda A, Petit P, Aillerie M, Herguth A (2013) Effect of illumination intensity on solar cells parameters. Energy Procedia 36:722–729

8. Jyothy LP, Sindhu MR (2018) An artificial neural network based MPPT algorithm for solar PV system. In: 2018 4th International conference on electrical energy systems (ICEES). IEEE, pp 375–380
9. Bana S, Saini RP (2016) A mathematical modeling framework to evaluate the performance of single diode and double diode based SPV systems. Energy Rep 2:171–187
10. Bouraiou A, Hamouda M, Chaker A, Sadok M, Mostefaoui M, Lachtar S (2015) Modeling and simulation of photovoltaic module and array based on one and two diode model using Matlab/ Simulink. Energy Procedia 74:864–877
11. Chowdhury SR, Saha H (2010) Maximum power point tracking of partially shaded solar photovoltaic arrays. Sol Energy Mater Sol Cells 94(9):1441–1447
12. Ishaque K, Salam Z, Taheri H (2011) Simple, fast and accurate two-diode model for photovoltaic modules. Sol Energy Mater Sol Cells 95(2):586–594
13. Wang X, Liang H (2012) Output characteristics of PV array under different insolation and temperature. In: 2012 Asia-Pacific power and energy engineering conference. IEEE, pp 1–4
14. Villalva MG, Gazoli JR, Ruppert Filho E (2009) Comprehensive approach to modeling and simulation of photovoltaic arrays. IEEE Trans Power Electron 24(5):1198–1208

A Comparative Study of Sliding Mode Controller for Non-Isolated DC–DC Converters

M. Manikandan and Afzal Sikander

Abstract Appropriate DC–DC converter selection and control is a critical aspect that has a considerable impact on the overall performance of power systems. Therefore, motivated by various types of the converter and their control, a comparative analysis of sliding mode controller (SMC) for non-isolated DC–DC converter is presented in this study. To analyze the performance of SMC, three different types of converter are considered, namely buck, boost, and single-ended primary-inductor converter (SEPIC). The concept of pulse width modulation (PWM) is used to design an SMC controller for these converters. It is observed that the performance of the SEPIC converter with SMC is better in terms of the efficiency of the converter.

Keywords Sliding mode controller · DC–DC converter · Single-ended primary-inductor · PWM

1 Introduction

Switched-mode power supplies (SMPSs) are usually referred to as converters since they convert one voltage level to another. Using an SMPS, a voltage can be stepped up, down, or inverted. The DC/DC SMPS, also known as a DC/DC converter, will be assessed in this research. Prior to the introduction of SMPS, the market for power conversion was dominated by linear regulators [1]. Although linear regulators are cost-effective, they have the disadvantage of producing a lot of heat, which necessitates the addition of an unwelcome large heat sink to the circuits. SMPS provides a reasonable answer, but they come with their own set of disadvantages. SMPSs are beneficial in systems where distinct sub-circuits or components demand varying

M. Manikandan · A. Sikander (✉)
Department of Instrumentation and Control Engineering, Dr. B. R. Ambedkar National Institute of Technology Jalandhar, Jalandhar, India
e-mail: afzals@nitj.ac.in

M. Manikandan
e-mail: manikandanmp.ic.20@nitj.ac.in

J. R. Szymanski et al. (eds.), *Energy Systems, Drives and Automations*,
Lecture Notes in Electrical Engineering 1057,
https://doi.org/10.1007/978-981-99-3691-5_14

voltage levels. Because of their higher efficiency, smaller and lighter size, lower cost, and ability to work over wide input voltage ranges, SMPSs are in high demand in today's market. Most practical converter applications may have noted several forms of non-idealities such as switching nonlinearity, parametric uncertainty, load disturbances, and so on. As a result, traditional controllers may occasionally fail to keep the switching converters' regulations at the desired levels [2, 3].

The design and implementation of PWM-based SMC on the buck, boost, and SEPIC converters are performed in this paper. Results were obtained, and a comparative analysis is made.

1.1 Sliding Mode Control (SMC)

Sliding mode control (SMC), which is based on variable structure system theory, has gotten a lot of interest recently because of its resiliency to parameter changes and external disturbances under matching conditions. The system states are transferred from their initial states to a specified manifold in the state space, referred to as the sliding surface, in sliding mode control (SMC). The system becomes completely insensitive to parametric uncertainty and external disturbances once it reaches the sliding manifold. The reaching phase is the motion of the trajectory from the initial condition to the sliding surface until it hits the sliding surface. The system is not resilient during the reaching phase, and even minimal uncertainty can have an impact on system performance. The integral sliding mode (ISM) [4] concept was presented in to overcome this problem. The advantages of SMC are that it is a robust control algorithm, and the systemic design is well known. The basic first-order sliding mode controller consists of two types of controllers which are switching control and equivalent control. The main objective of the switching control is to enforce the system to the sliding surface and to maintain the state trajectories on this sliding surface. The Lyapunov technique is used to determine the stability of the system.

2 Modeling of Converter Using State-Space Analysis

Figure 1 depicts the fundamental structure of a DC–DC buck converter. The circuit makes up of a switch, diode, inductor, filter capacitor, and output load. Kirchhoff's voltage law and Kirchhoff's current law are used in the system to produce the dynamic equations for the converter.

There are two possible states in buck converter which are switch on and switch off state as shown in Fig. 2.

$$L\frac{\mathrm{d}i_L}{\mathrm{d}t} = uV_i - V \tag{1}$$

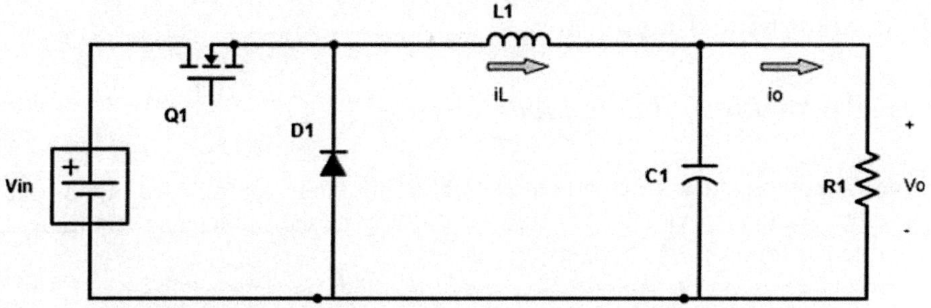

Fig. 1 DC–DC buck converter circuit [5]

$$C\frac{dV_o}{dt} = i_L - i \tag{2}$$

where i_L represents the inductor current, V_o represents the voltage at the output also its equal to the voltage of the capacitor, V_i is the input voltage, and u represents the position of the switch which takes the value 1 when switch is ON and 0 switch is OFF.

$$i_0 = \frac{V_o}{r_L} \tag{3}$$

state equations by taking i_L and V_o as the system states [5]

$$\frac{di_L}{dt} = \frac{V_i}{L}u - \frac{V_o}{L} \tag{4}$$

$$\frac{dV_o}{dt} = \frac{i_L}{C} - \frac{V_o}{r_L C} \tag{5}$$

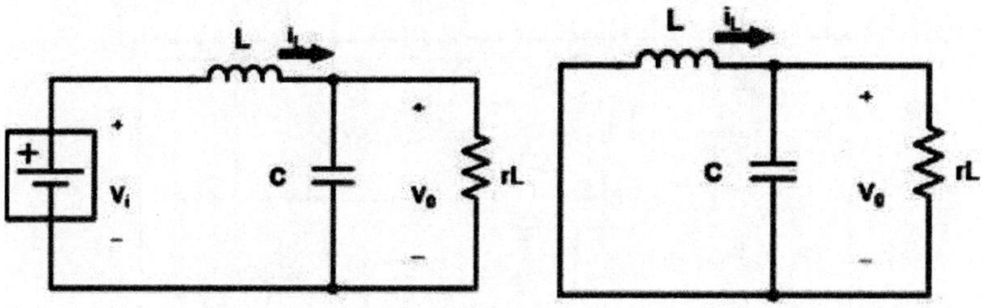

Fig. 2 Turn ON and turn OFF state of buck converter

3 System Modeling

3.1 PWM-Based Sliding Mode Control

Figure 3 represents the schematic model of sliding mode control used in the buck converter. The control variable x in the case of PWM-based sliding mode control can be written as follows:

$$x_1 = V_{\text{ref}} - \beta V_o \tag{6}$$

$$x_2 = \dot{x}_1 = \frac{\beta V_o}{r_L C} + \int \frac{\beta(V_o - V_i u)}{\text{LC}} \text{d}t \tag{7}$$

$$x_3 = \int x_1 \text{d}t = \int (V_{\text{ref}} - \beta V_o) \text{d}t \tag{8}$$

where x_1, x_2, and x_3 denote voltage error, rate of change of voltage error, and integral of voltage error, respectively. Where β is the voltage divider ratio.

The state-space equations for the design of the controller are shown in equation number 8 where x_1, x_2, and x_3 are the state variables.

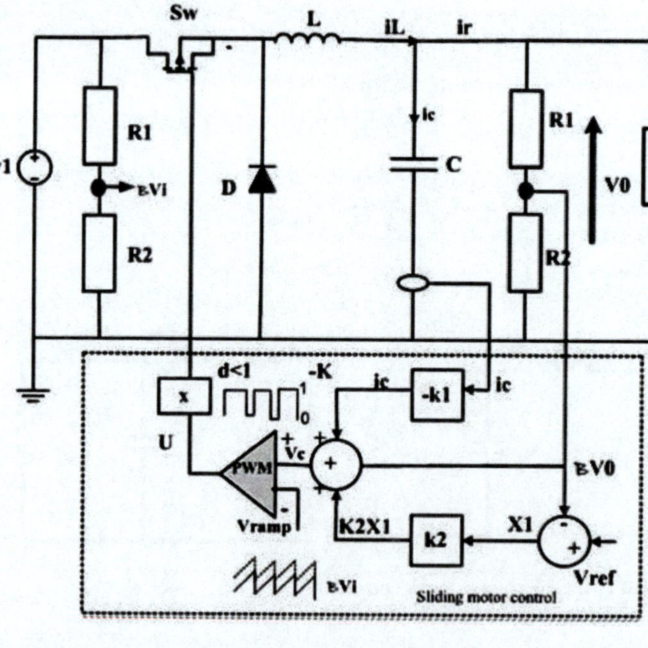

Fig. 3 Schematic of a buck converter using sliding mode [2]

$$\begin{bmatrix} \dot{x}_1 \\ \dot{x}_2 \\ \dot{x}_3 \end{bmatrix} = \begin{bmatrix} 0 & 1 & 0 \\ 0 & \frac{-1}{r_L C} & 0 \\ 1 & 0 & 0 \end{bmatrix} \begin{bmatrix} x_1 \\ x_2 \\ x_3 \end{bmatrix} + \begin{bmatrix} 0 \\ \frac{-\beta V_i}{\mathrm{LC}} \\ 0 \end{bmatrix} u + \begin{bmatrix} 0 \\ \frac{\beta V_o}{\mathrm{LC}} \\ 0 \end{bmatrix} \tag{9}$$

Now defining a switching function (S) for the system like $u = 0$ when $S = 0$ and u=1 when S is greater than 0, here S represents the current state variables. The equation for the switching function is shown in Eq. 10.

$$S = \alpha_1 x_1 + \alpha_2 x_2 + \alpha_3 x_3 = J^T x \tag{10}$$

$$J^T = \begin{bmatrix} \alpha_1 & \alpha_2 & \alpha_3 \end{bmatrix} \tag{11}$$

The sliding coefficients are represented by α_1, α_2, and α_3. After the steady state is reached, it is considered to be a fixed point, and the integral control is eliminated. After that, the control signal (v_c) was transformed to PWM proportional derivative linear controller form, providing the formula

$$v_c = -K_1 i_c + K_2(V_{\text{ref}} - \beta V_o) + \beta \tag{12}$$

K1 and K2 can be calculated by the following:

$$K_1 = \beta L\left(\frac{\alpha_1}{\alpha_2} - \frac{1}{r_L C}\right) \tag{13}$$

$$K_2 = \mathrm{LC}\left(\frac{\alpha_3}{\alpha_2}\right) \tag{14}$$

K_1 and K_2 are the feedback signal gain constants, and their values can be calculated using converter component L, C, and r_L, as well as the values of sliding parameters α_1, α_2, and α_3. By setting $S = 0$ and comparing this equation to the normal second-order system form, the sliding coefficients can be determined.

$$\alpha_1 x_1 + \alpha_2 \frac{\mathrm{d}x_2}{\mathrm{d}t} + \alpha_3 \int x_1 = 0 \tag{15}$$

where

$w_n = \sqrt{\frac{\alpha_3}{\alpha_2}}$ is the undamped natural frequency

$\varepsilon = \frac{\alpha_1}{2\sqrt{\alpha_2 \alpha_3}}$ is the damping ratio

$f_{\text{Bw}} = \frac{w_n}{2\pi}\sqrt{\frac{\alpha_3}{\alpha_2}}$ Bandwidth of controller response

$$\frac{\alpha_1}{\alpha_2} = 4\pi f_{\text{Bw}} \tag{16}$$

Table 1 Buck converter specification [2]

Description	Parameter	Nominal values
Input voltage	V_s	20 V
Load Resistance	R	2.5 Ω
Inductance	L	2 mH
Capacitance	C	47 μF
Switching frequency	f	20 kHz
Output voltage	V_o	12 V

$$\frac{\alpha_3}{\alpha_2} = 4\pi^2 f_{\mathrm{Bw}}^2 \tag{17}$$

The CCM refers to a continuous current mode in which the i_L never hits zero for any amount of time. As a result, the values for the filters L and C are as follows:

$$L_{\min} = \frac{(1 - D) r_L}{2f} \tag{18}$$

$$C_{\min} = \frac{(1 - D) V_o}{8 V_r L f^2} \tag{19}$$

4 Simulation Results and Analysis

4.1 Simulation of a Buck Converter Using PWM-Based SMC

The simulation was performed using the MATLAB/Simulink software, and the buck converter parameters are listed in Table 1. The Simulink model of a buck converter with SM control is shown in Fig. 4. The output voltage V_o and current I_o for an SMC-controlled buck converter system with input voltage $V_s = 20$ V and output load resistance $R_l = 6\ \Omega$, resulting in output voltage $V_o = 12.23$ V and output current $I_o = 5.05$ A and efficiency of the designed buck converter using SMC is 92.45% as shown in Fig. 5.

4.2 Simulation of Boost Converter Using PWM-Based SMC [6]

The Simulink model of a DC/DC boost converter based on SMC is shown in Fig. 6. The output voltage V_o and current I_o for an SMC-controlled boost converter system with input voltage $V_s = 12$ V and output load resistance $R_l = 60\ \Omega$, resulting in output

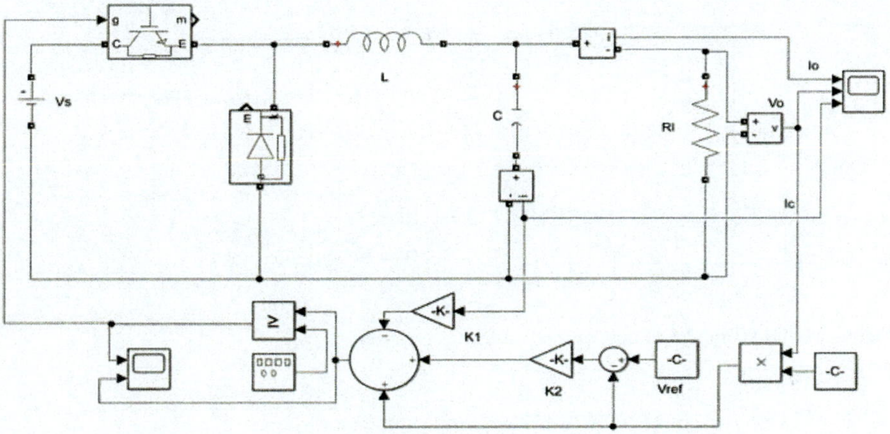

Fig. 4 MATLAB model of buck converter based on SMC

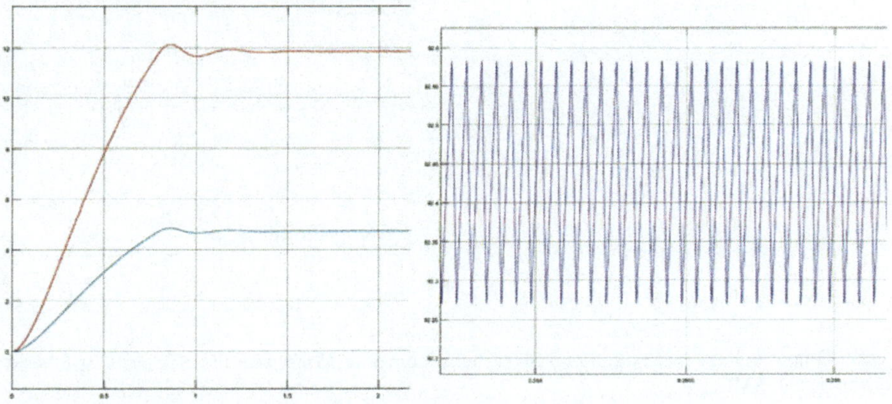

Fig. 5 Output voltage and output current of buck converter along with the efficiency of a buck converter with SMC

voltage $V_o = 23.74$ V and output current $I_o = 0.38$ A, efficiency of the designed boost converter using SMC is 93.5% as seen in Fig. 7.

The boost converter parameters are listed in Table 2.

4.3 Simulation of SEPIC Converter Using PWM-Based SMC [7]

The SEPIC converter parameters are listed in Table 3.

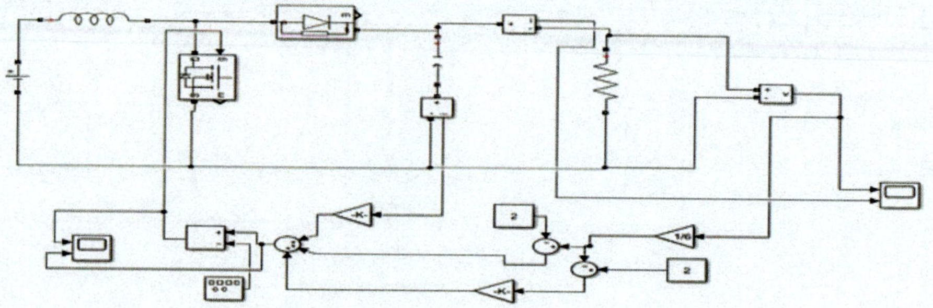

Fig. 6 MATLAB model of boost based on SMC converter

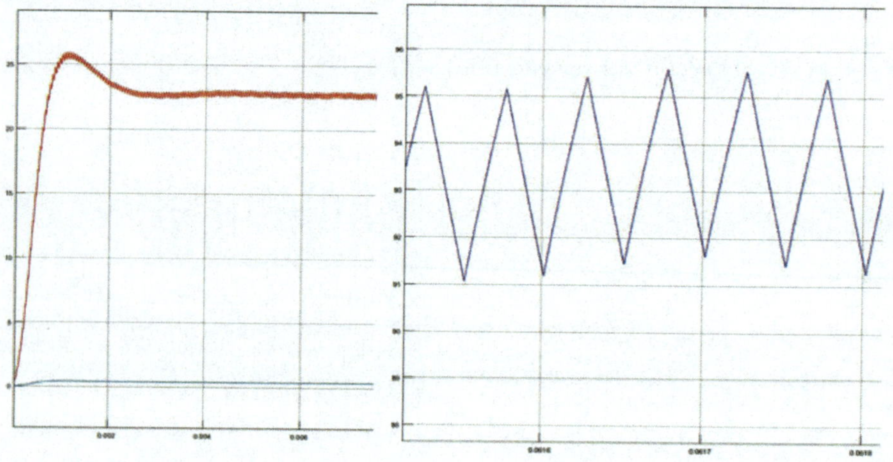

Fig. 7 Output voltage and output current of boost converter along with the efficiency of a boost converter with SMC

Table 2 Boost converter specification [6]

Description	Parameter	Nominal values
Input voltage	V_s	12 V
Load resistance	R	60 Ω
Inductance	L	2.5 mH
Capacitance	C	20 μF
Switching frequency	F	20 kHz
Output voltage	V_o	24 V

The Simulink model of a DC/DC SEPIC converter based on SMC is shown in Fig. 8. The output voltage V_o and current I_o for an SMC-controlled SEPIC converter system with input voltage $V_s = 12$ V and output load resistance $R_l = 5\ \Omega$, resulting

Table 3 SEPIC converter specification

Description	Parameter	Nominal values
Input voltage	V_s	20 V
Load resistance	R	1.5 Ω
Inductance	L_1	0.2946 mH
Inductance	L_2	0.2946 mH
Capacitance	C_1	10 μF
Capacitance	C_2	50 μF
Switching frequency	F	20 kHz
Output voltage	V_o	12 V

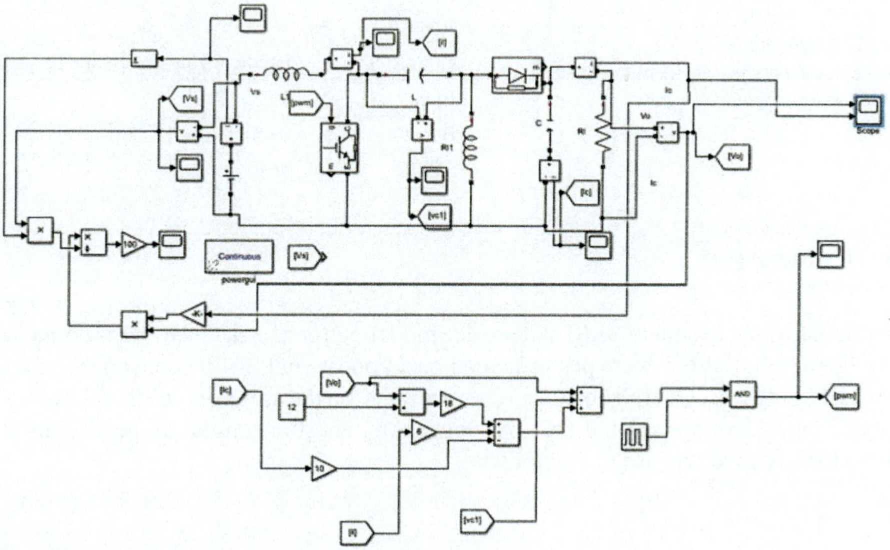

Fig. 8 MATLAB model of SEPIC converter based on SMC

in output voltage $V_o = 12$ V and output current $I_o = 7.85$ A and the efficiency of the designed SEPIC converter using SMC is 96.4% as seen in Fig. 9.

Table 4 shows the comparison of efficiencies of sliding mode control for the considered types of topologies.

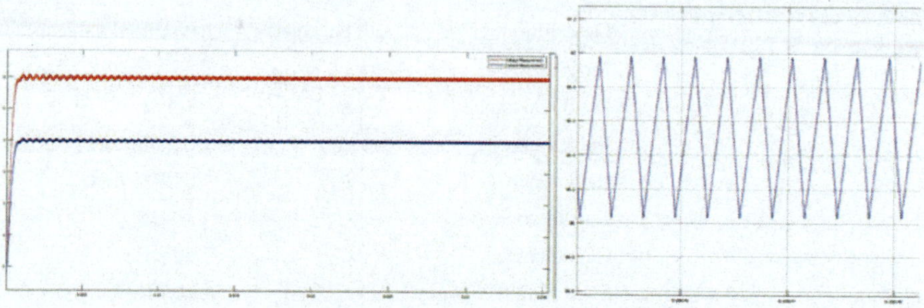

Fig. 9 Output voltage and output current of SEPIC converter and efficiency of SEPIC converter with SMC

Table 4 Efficiency comparison for the considered types of topologies

Sr. No.	Converter	Efficiency in percentage
1	Buck	92.45
2	Boost	93.5
2	SEPIC	96.4

5 Conclusion

In this study, the design of SMC state-space modeling for a buck converter is derived. MATLAB simulations were performed on buck, boost, and SEPIC converters using PWM-based SMC. Overall observations were made on the usage of traditional sliding mode control in non-isolated DC–DC converters, and the simulation results show that SEPIC converters are more efficient.

References

1. Mumtaz F, Zaihar Yahaya N, Tanzim Meraj S, Singh B, Kannan R, Ibrahim O (2021) Review on non-isolated DC-DC converters and their control techniques for renewable energy applications. Ain Shams Eng J 12(4):3747–3763. https://doi.org/10.1016/j.asej.2021.03.022
2. Fadel M, Shehab M, Al-Gizi A, Al-Saadi M, Al-Qaisi M (2019) High performance DC/DC buck converter using sliding mode controller. Int J Power Electron Drive Syst 10:1806–1814. https://doi.org/10.11591/ijpeds.v10.i4.1806-1814
3. Safari A, Ardi H (2018) Sliding mode control of a bidirectional Buck/Boost DC-DC converter with constant switching frequency. Iran J Electr Electron Eng 14:69–84. https://doi.org/10.22068/IJEEE.14.1.69
4. Komurcugil H, Biricik S, Guler N (2020) Indirect sliding mode control for DC–DC SEPIC converters. IEEE Trans Ind Inform 16(6):4099–4108. https://doi.org/10.1109/TII.2019.2960067
5. Drakunov SV, Reyhanoglu M, Singh B (2009) Sliding mode control of DC-DC power converters. IFAC Proc Vol 42(19):237–242. https://doi.org/10.3182/20090921-3-TR-3005.00043

6. Podey AP, Chandle JO (2020) Stability analysis of DC-DC boost converter using sliding mode controller. In: 2020 IEEE PES/IAS PowerAfrica, pp 1–5. https://doi.org/10.1109/PowerAfrica49420.2020.9219954
7. Gireesh G, Seema PN (2015) High frequency SEPIC converter with PWM integral sliding mode control. In: 2015 International conference on technological advancements in power and energy (TAP energy), pp 393–397. https://doi.org/10.1109/TAPENERGY.2015.7229651

Electronics and Control

Measurement of Beam Profile of a Linear Accelerator (LINAC) for 6 and 10 MV Photon Energies

Rejoana Najnin Sruti, Muhammad Masud Rana, M. Moinul Islam, K. A. Khan, and M. M. H. Bhuiyan

Abstract Beam Profile is one of the essential parameters which must be obtained before clinical use. This essential parameter is obtained during the commissioning of a linear accelerator (LINAC) machine. To measure beam Profile curves for 6 and 10 MV Photon beams, Varian LINAC with different field sizes were used. The measured Profile curves compared with TPS calculated Profile using Beam Analysis Tool of TPS (Varian Eclipse). The measurements were performed for 12 field sizes (e.g., 3 × 3 (smallest field size), 4 × 4, 6 × 6, 8 × 8, 10 × 10, 12 × 12, 15 × 15, 20 × 20, 25 × 25, 30 × 30, 35 × 35 and 40 × 40 (largest field size)) cm^2 for 6 and 10 MV beams of Photon, maintaining the conditions (e.g., voltage and polarity of ion chamber, geometry, incremental direction and step) same. The MEPHYSTO MC^2 software and two Semiflex ionization chambers, electrometer DOSE 1 from PTW for the Varian Linear Accelerator Model D-2300CD were used in this study. The flatness for reference field size (FS) 10 × 10 cm^2 at 10 cm depth is 2.28% and 2.38%, and the symmetry is 100.75% and 101.00% for 6 and 10 MV energies of Photon, respectively. The obtained result for flatness and symmetry is within the limit. The Profile curves which is measured for 6 and 10 MV Photon energies show good agreement with the TPS calculated Profile using Beam Analysis Tool of TPS (Varian Eclipse).

Keywords LINAC · Photon · Energy · Measurement · Profile · TPS · Field size · Beam analysis tool

R. N. Sruti · K. A. Khan (✉)
Department of Physics, Jagannath University, Dhaka 1100, Bangladesh
e-mail: kakhan01@yahoo.com

M. M. Rana
Department of Oncology, BSMMU, Dhaka, Bangladesh

M. M. Islam · M. M. H. Bhuiyan
INST, Bangladesh Atomic Energy Commission, Savar, Bangladesh

R. N. Sruti
Princeton University, Princeton, NJ 08544, USA

J. R. Szymanski et al. (eds.), *Energy Systems, Drives and Automations*,
Lecture Notes in Electrical Engineering 1057,
https://doi.org/10.1007/978-981-99-3691-5_15

1 Introduction

A linear accelerator (LINAC) basically accelerates charged particles by increasing the kinetic energy of charged particles or ions. The charged particles are subjected to a series of oscillating electric potentials along a linear beamline. It is one of the most common devices which is used to treat tumors or cancer with external beam radiation [1–4]. Several built-in safety measures are featured on it to ensure that it will not deliver a lower or higher dose than the prescribed dose and is checked regularly by the medical physicist to ensure that it is properly working [5, 6]. For the production of electrons at relativistic speeds, high power LINACs are being developed [7]. Measurements of some dosimetric parameters such as Percentage Depth Dose (PDD), Profile, and Output characterization are essential to validate the treatment planning systems (TPS) [8]. The measurement of these essential parameters is the part of commissioning of a LINAC system [9–11]. After completion of the process of pre-commissioning testing, the LINAC is used to treat patients for radiotherapy treatment [12, 13]. One basic dosimetric parameter Profile with different FSs for 6 and 10 MV Photon beams was measured in this study. These measurements have been performed at the Radiation Oncology department of the National Institute of Cancer Research and Hospital, Mohakhali, Dhaka for the Varian Linear Accelerator Model D-2300CD machine [14]. The scientific methods which are used for the commissioning of modern Linear Accelerator need dedication in work and the procedure is time-consuming [15]. The measurement is done in a three-dimensional (3D) computer-controlled water phantom (PTW MP3) using DOSE 1 (S/N 10680), Semiflex ionization chamber (S/N 1217), and Semiflex ionization chamber (S/N 1222) [16, 17]. This data set which is taken during the pre-commissioning of LINAC can be used as the standard data for clinical issue [18]. This work aims to measure beam Profile which is one of the essential dosimetric parameters for the commissioning of a LINAC. Measured dosimetric parameters are used as reference data to make sure that the experimentally obtained parameters have not changed during the regular operation of the LINAC system [19].

2 Methodology

2.1 Instrumentation

In this study, the model of the linear accelerator was Varian D-2300 CD. It has dual-energy mode with both Electron beam and Photon beam facility. It has five electron energies (e.g., 4, 6, 9, 12, and 15 meV) and two Photon energies, e.g., 6 and 10 MV. A 3D water phantom, PTW 31010 Semiflex chamber (S/N 1222) as a reference chamber which remains stationary, PTW 31010 Semiflex chamber (S/N 1217) as a field chamber which moved in the water phantom as programmed, DOSE 1 (S/

N 10680) electrometer, and MEPHYSTO mc^2 software version 1.6 were used to measure beam Profile for 6 and 10 MV Photon energies.

2.2 *Transverse Beam Profile Measurement*

In the case of a Linear Accelerator machine, dose measurements for longitudinal and transverse beam Profiles are done by a radiation detector to characterize the radiation beams from medical linear accelerators. To create these radiation dose Profiles, a water phantom and an ionization chamber were used. The water phantom was set up under the linear accelerator machine. Water phantom is used because water is considered to be close to tissue equivalent. Transverse dose measurements, that is, dose Profile measurements were performed in the x (cross-plane) or y (in plane) directions perpendicular to the radiation beam, and at a given depth (z) in the phantom [1]. For treatment planning, the distribution of dose at any point within a radiation beam is required to be known. To characterize the dose at points of the central axis transverse beam Profiles are measured. The transverse beam Profiles were measured at four different depths, and the required measurement fields were smallest field 3 × 3, 4 × 4, 6 × 6, 8 × 8, 10 × 10, 12 × 12, 15 × 15, 20 × 20, 25 × 25, 30 × 30, 35 × 35 and largest field size (40 × 40) cm^2.The first beam Profile was measured at d_{max} of the respective field size, and the other five beam Profiles were measured at 5 cm, 10 cm, 20 cm, and 30 cm, respectively.

2.3 *Measurement Setup for Transverse Beam Profiles*

To measure beam Profile, water phantom and ionization chambers were placed at the isocenter alignment of a linear accelerator machine system having 6 and 10 MV Photon energies. The MP3 water tank was rotated to match crosshair projection on the floor of the tank and leveled with a clinometer. The source to water surface distance (SSD) was 100 cm.

A reference chamber, i.e., PTW 31010 Semiflex chamber (S/N 1222) was set in the corner of the measuring field just above the water surface which remain constant during the procedure. Figure 1 shows the Phantom setup under the linear accelerator machine for Profile measurements.

The Profiles were measured in five depths d_{max}, 50, 100, 200, and 300 mm having step size 2.5 mm in the plateau region and outside of the field, and 1 mm in the penumbra region. Profiles were measured to at least 50% outside of the respective fields. PTW MEPHYSTO mc^2 software version 1.6.2 is used to control the scanning.

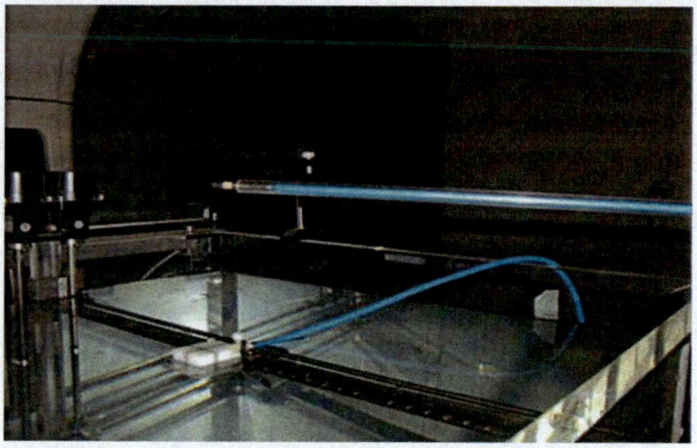

Fig. 1 Water phantom setup for Profile measurements

2.4 Transverse Half Beam Profiles

The transverse half beam Profiles were measured at five different depths. The first beam Profile was measured at d_{max} of the respective field size, and other beam Profiles were measured at 50, 100, 200, and 300 mm, respectively. To measure half beam Profile phantom center shifted by 15 cm to the right for right half beam Profile measurement since it is not possible to measure full beam Profiles for the largest field size (40×40 cm^2) due to the limitation in phantom size.

2.5 Diagonal Profiles

The diagonal Profiles were measured only for the largest field size (40×40 cm^2) at the same depths as Profiles. To calculate the point dose it is essential to measure the diagonal Profile at different depths of the phantom so that the measurement at any point of the cubic object can be determined. For this study, the diagonal Profile of the largest field size (40×40) cm^2 was measured with the same SSD (100 cm) and at the same four depths (d_{max}, 50, 100, 200, and 300) mm as the Open field *X*-Profiles. Due to the large field size (40×40) cm^2, it is not possible to measure full beam Profile due to the limitation in phantom size. Therefore, 3D water phantom needs to be set for measuring diagonal Profiles phantom center should shift by 15 cm to the right from the isocenter and the collimator should rotate at 45°. In the present study to measure the diagonal Profiles, the 3D water phantom was set to 15 cm right from the isocenter and the collimator was rotated 45° angle of the LINAC. The nominal field size entered into MEPHYSTO mc^2 version 1.6 software was 40×40 cm. The finer 1 mm step size of the penumbra region had to be moved to the correct position by other means. This was done by defining a separate "steps and speeds" setting for the diagonal Profiles, with 40×40 cm as the reference field size.

3 Results and Discussion

The measured values of beam Profile for 6 and 10MV Photon energies for 12 field sizes, e.g., smallest 3 × 3, 4 × 4, 6 × 6, 8 × 8, 10 × 10, 12 × 12, 15 × 15, 20 × 20, 25 × 25, 30 × 30, 35 × 35, largest 40 × 40 cm^2 were plotted to obtain Profile curves. The curves of the measured Profile for reference field size 10 × 10 cm^2 at reference depth 10 cm at 100 cm SSD of 6 and 10 MV Photon energies are shown in Fig. 2a, b.

The flatness and symmetry of 6 and 10 MV Photon beams for 10 × 10 cm^2 FS at 10 cm depth obtained from Fig. 2a, b are given in Table 1.

The flatness for FS 10 × 10 cm^2 for 6 and 10 MV Photon energies at 10 cm depth are 2.28% and 2.38% and the symmetry are 100.75% and 101.00%, respectively. According to the IEC 60,976 scale, the flatness is considerable up to 3%, and symmetry is up to 103% [15]. From the result, the measured flatness and symmetry of 6 and 10 MV Photon beam Profiles were found within the permissible limit.

Comparison of 6 and 10 MV Profiles in TPS using Beam Analysis Tool

The measured values of Profile for 6 MV Photon energy were compared with the TPS calculated 6 MV Profile values by using the Beam Analysis Tool and similarly measured Profile values for 10 MV Photon energy also compared with TPS calculated values. The compared curves between TPS calculated Profile values and measured Profile values for 6 and 10 MV Photon beams for reference field size 10 × 10 cm^2 is given in Fig. 3a, b.

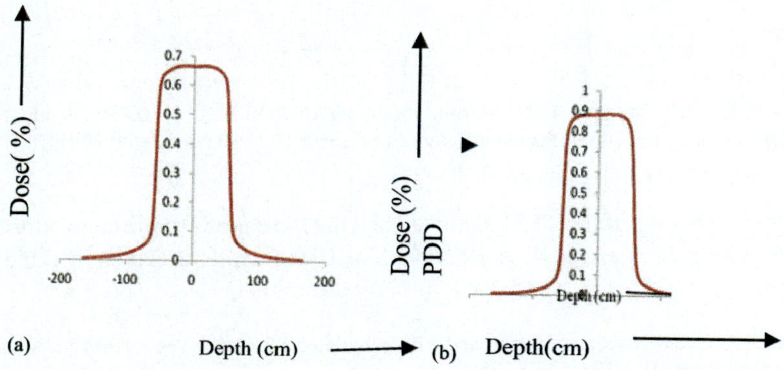

Fig. 2 **a** Measured Profile for 6MV energy and **b** measured Profile for 10MV energy

Table 1 Flatness and symmetry for FS 10 × 10 cm^2 of 6MV and 10MV Photon energies

Energy (MV)	Field size (cm^2)	Depth (cm)	Flatness (%)	Symmetry (%)
6	10 × 10	10	2.28	100.75
10	10 × 10	10	2.38	101.00

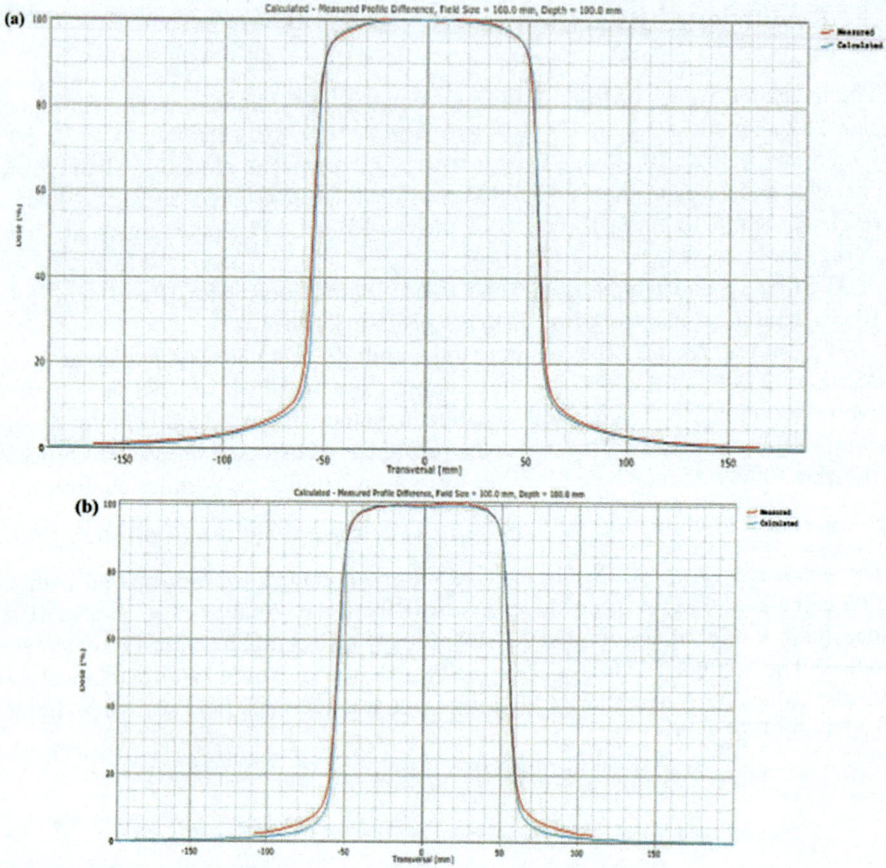

Fig. 3 **a** Calculated—measured 6MV Profile difference [field size 10 × 10 cm^2, depth 10 cm], **b** calculated—measured 10MV Profile difference [field size 10 × 10 cm^2, depth 10 cm]

Compared results of the TPS calculated and the measured Profile curves for 6 and 10 MV Photon energies for 10 × 10 cm^2 FS at 10 cm depth are given in Table 2.

Table 2 Difference between measured and TPS calculated Profile curves for 6MV and 10 MV Photon energies for 10 × 10 cm^2 FS

Energy (MV)	Left side off-axis dist. Diff. (mm) at 80% dose	Left side off-axis dist. Diff. (mm) at 50% dose	Left side off-axis dist. Diff. (mm) at 20% dose	Right side off-axis dist. Diff. (mm) at 80% dose	Right side off-axis dist. Diff. (mm) at 50% dose	Right side off-axis dist. Diff. (mm) at 20% dose
6	– 0.132	0.665	1.842	0.481	– 0.258	– 1.483
10	0.219	0.332	0.722	– 0.149	– 0.365	– 0.788

According to AAPM TG-40 protocol, using Beam Analysis Tool, the difference between the calculated Profile by TPS and the measured Profile for 6 and 10 MV energies of Photon were found within ± 2 mm [2].

4 Conclusions

The commissioning process of a LINAC machine includes comprehensive measurements of essential dosimetric parameters such as Percentage Depth Dose (PDD), Profiles, and Output characterization for a series of field sizes (FSs). Since the outcome of radiation treatment is directly related to the precession in the delivered dose to the patient that is dependent on the accuracy of beam data used in the treatment planning process. The dosimetric parameters are needed to measure correctly to operate a LINAC system, that's why the beam Profile is needed to measure carefully. From Table 2, the differences between measured Profiles and TPS calculated Profiles for 6 and 10 MV Photon energies are ± 2 mm. The International Electrotechnical Commission (IEC 60976) protocol was followed to verify the measured results. The American Association of Physicists in Medicine-Task Group-40 (AAPM TG 40) was followed for the Quality Assurance (QA) purpose. Flatness and Symmetry for 6 and 10 MV beams of Photon were found within the limit mentioned in the IEC 60976 scale [2]. In this work, the obtained Profile data for 6 and 10 MV Photon energies will be useful for the initial commissioning of LINAC system. However, to verify the measured data for clinical use further work can be done.

References

1. https://en.wikipedia.org/wiki/Dose_profile
2. AAPM Task Group 40 (1994) American Association of Physicists in Medicine. Comprehensive QA for Radiation Oncology AAPM TG-40
3. Khan KA (1999) Copper oxide coatings for use in a linear solar Fresnel reflecting concentrating collector. Renew Energy Int J 17(4):603–608. Pergamon. WREN (World Renewable Energy Network), UK
4. Ruhane TA, Tauhidul Islam M, Rahman MS, Bhuiyah MMH, Islam JMM, Bhuiyah TI, Khan KA, Khan MA (2017) Impact of photo electrode thickness annealing temperature on natural dye sensitized solar cell. Sustain Energy Technol Assess. Elsevier. https://doi.org/10.1016/j.seta.2017.01.012
5. Ruhane TA, Tauhidul Islam M, Rahman MS, Bhuiyah MMH, Islam JMM, Newaz MK, Khan KA, Khan MA (2017) Photo current enhancement of natural dye sensitized solar cell by optimizing dye extraction and its loading period. Optik Int J Light Electron Opt. Elsevier
6. Hasan M, Khan KA (2018) Dynamic model of *Bryophyllum pinnatum* leaf fueled BPL cell: a possible alternate source of electricity at the off-grid region in Bangladesh. Microsyst Technol. ISSN: 0946-7076. Microsystem Technologies Micro- and Nanosystems Information Storage and Processing Systems, Springer. https://doi.org/10.1007/s00542-018-4149-y
7. Mittal VK, Verma RC, Gupta SC (2003) Introduction to nuclear and particle physics, 2nd edn, p 247

8. Sahoo SK, Rath AK, Mukharjee RN, Mallick B (2012) Commissioning of a modern LINAC for clinical treatment and material research, vol 3, p 19
9. Khan KA, Hazrat Ali M, Obaydullah AKM, Wadud MA (2019) Production of candle using solar thermal technology. Microsyst Technol 25(12). ISSN: 0946-7076. Microsystem Technologies Micro- and Nanosystems Information Storage and Processing Systems, Springer. https://doi.org/10.1007/s00542-019-04390-7
10. Khan KA, Rasel SR, Ohiduzzaman M (2019) Homemade PKL electricity generation for use in DC fan at remote areas. Microsyst Technol 25(12). ISSN: 0946-7076. Microsystem Technologies Micro- and Nanosystems Information Storage and Processing Systems. https://doi.org/10.1007/s00542-019-04422-2
11. Hasan M, Khan KA (2019) Experimental characterization and identification of cell parameters in a BPL electrochemical device. SN Appl Sci 1:1008. Springer. https://doi.org/10.1007/s42452-019-1045-8
12. Hassan L, Khan KA (2019) A study on harvesting of PKL electricity. Microsyst Technol 26:1031–1041. Springer. https://doi.org/10.1007/s00542-019-04625-7
13. Khan KA, Mamun MA, Ibrahim M, Hasan M, Ohiduzzaman M, Obaydullah AKM, Wadud MA, Shajahan M (2019) PKL electrochemical cell: physics and chemistry. SN Appl Sci 1:1335. Springer. https://doi.org/10.1007/s42452-019-1363-x
14. Hazrat Ali M, Chakma U, Howlader D, Tawhidul Islam M, Khan KA (2019) Studies on performance parameters of a practical transformer for various utilizations. Microsyst Technol. Springer. https://doi.org/10.1007/s00542-019-04711-w
15. Ibbott GS (2006) IEC 60976. A physicist's guide to the international electrotechnical commission
16. Khan KA, Hassan L, Obaydullah AKM et al (2018) Bioelectricity: a new approach to provide the electrical power from vegetative and fruits at off-grid region. Microsyst Technol. https://doi.org/10.1007/s00542-018-3808-3
17. Khan KA, Bhuyan MS, Mamun MA, Ibrahim M, Hasan L, Wadud MA (2018) Organic electricity from Zn/Cu-PKL electrochemical cell (chapter 9). In: Mandal JK et al. (eds) Contemporary advances in innovative and applicable information technology. Advances in intelligent systems and computing, vol 812. Springer Nature Singapore Pvt. Ltd., pp 75–90
18. Atique Ullah AKM, Haque MM, Akter M, Hossain A, Tamanna AN, Hosen MM, Fazle Kibria AKM, Khan MNI, Khan MKA (2020) Green synthesis of *Bryophyllum pinnatum* aqueous leaf extract mediated bio-molecule capped dilute ferromagnetic α-MnO2 nanoparticles. Mater Res Express 7(1):015088. IOP publishing Ltd.
19. Khan KA, Hazrat Ali M, Mamun MA, Mahbubul Haque M, Atique Ullah AKM, Islam Khan MN, Hassan L, Obaydullah AKM, Wadud MA (2020) Bioelectrical characterization and production of nanoparticles (NPs) using PKL extract for electricity generation. Microsyst Technol. Springer. https://doi.org/10.1007/s00542-020-04774-0

A Low-Cost Foot Step Power Generation Using Piezoelectric Sensors

L. Shiva Kumar, Vijaya Bhasker Reddy, G. Ramya, K. Ajay Reddy, and Y. Pavani

Abstract Electrical energy is important and had been demand improved in the course of the world. Conventional approach of producing energy has a direct impact on environment; a lot of electricity is wasted and exhausted with the aid of humans. So, that energy is utilized and converting kinetic energy to electrical electricity as the stress exerted by way of foot step and by using the usage of transducers is essentially referred to as foot step power generation system. There exists range of power harvesting strategies through piezoelectric powered sensors. This mission would be mounted in crowded locations so when a human walks on a certain arrangement that unused energy which is in the form of stress or force is transformed to electrical energy. The stress is exerted by using human weight on piezoelectric powered sensors device and successfully generates energy to run the AC or DC loads. So, using prosed system we can generate power and give charge for lightening systems.

Keywords Electricity generation · Piezo-sensor · Foot step strain

1 Introduction

Electricity is one of the many times used power, and it is growing alongside the population. Conventional electricity has been demanded a critical technique of producing energy and has a direct effect on environment. The alternative method of producing electrical energy with availability of different quantity of techniques by using which electrical energy can be produced[1– 8], out of such approaches generating foot step power can be an wonderful approach of generate electrical Electricity. The object in this invention is to use the growing of human population and give high impact

L. S. Kumar · V. B. Reddy (✉) · G. Ramya · K. A. Reddy · Y. Pavani
Department of Electronics and Communication Engineering, KG Reddy College of Engineering & Technology, Hyderabad, India
e-mail: vijayabhaskar458@gmail.com

J. R. Szymanski et al. (eds.), *Energy Systems, Drives and Automations*, Lecture Notes in Electrical Engineering 1057,
https://doi.org/10.1007/978-981-99-3691-5_16

on in growing the electricity while lowering the poor impact on environment. This makes use of electricity additionally that do no longer rely on matter on the local weather condition. Human electricity transducers have been in existence since the time immemorial in the form of walking, going for walks, etc.

Electricity is most used resource in many instances. Now-a-days demand for utilization of power is increasing and which is lifeline for most of the people in their daily life. Due to this, variety of electricity assets are generated and wasted. Generation of electricity can be done from assets like wind, water, biomass, waves, solar, etc., to generate the electrical energy from those set of sources improvement of electric energy massive flora is wanted, having excessive preservation charges. Some different power sources are additionally high-priced and purpose cause pollution. They are now not cheap to frequent people. Current scenario electricity becomes to be a necessary asset for all the human being; hence, it is required that wasted strength have to be utilized.Taking walks is the most frequent undertaking completed through human being. While taking walks through pedestrian, electricity is wasted in the form of charge motion to the surface. And this kind of power can be transformed into electrical energy; the same can be used for various applications like street lights, operating small hand-held electrical devices, etc.

Using the precept is referred to as piezoelectric effect. Piezoelectric effect is the impact in which mechanical vibrations are converted in to electric energy. Pressure or pressure utilized to piezoelectric fabric is transformed into electrical form.

Modern technological know-how has led to improvement of machines to enhance use of human energy in extra environment friendly manner. The conversion of strength that exhausted and wasted while on foot or running is done using sensors. The power is transformed to electrical energy. This is the present-day fashion in electrical strength producing, and it is attaining from people foot step. When man or woman utilized strain on the platform related with sensors their physique weight compresses the setup of device and cutting-edge is produced and saved in battery. The voltage will increase as a consequence alongside the weight will increase and strain utilized on the piezoelectric powered sensors (Fig. 1).

Fig. 1 Foot step power generation using steps [9]

2 Literature Survey

R. Meier, N. Kelly et al. proposed a mechanism of power generation through piezoelectric technique in shoe, which is a limited work and did not mention applications for the same [10].

R. K. Datta and S. Rahman et al. discussed about foot step electricity generation using micro-generator system, even though it's a low cost but not up to the mark as require power generation which is designed based on a simple op-amp [11].

T. R. Deshmukh et al. discussed about a paper deals with plan and modeling of components of the model of a system those foot step energy technology system the usage of 3D modeling software program. This procedure consists quantity of simple setup that is mounted below the strolling or standing platform.

A. D. Triono et al., "Utilization of Pedestrian Movement-based power generation using piezo-sensor for side walk" design becomes little expensive where a special attraction is required [12].

A. Bhaumik et al., Task machine chips away at the standard of changing over the direct movement in light of the fact that to tension of strides into pivoting movement via rack and pinion game plan. This component comes up short in case there is any frequency of variable burden prompts adjusting kind issues power isn't created for the term of return development of rack. Sasank Shekhar Pandas paper is principally founded on driving rod, fly wheel, and stuff plan. This sort of strides power age machine is qualified to be introduced in jam-packed spots and country regions [13]. Along these lines, this is an awesome innovation to outfit pleasant response to control related issues to minimal expense degree. This will be the most legitimate abilities of giving energy to the spots that incorporates hardships of transmission.

T. Sarala et al. express those piezoelectric substances having glasslike structure. They can change over mechanical energy in the electrical energy as well as the other way around. The delivered electrical energy from piezoelectric precious stone is extremely low in the request for 2–3 V and is saved in battery to charge regulator, taking into account that it is presently not plausible to charge 12v battery through gem yield [14].

To intensify the voltage, the lift converter circuit is utilized. Examination between different piezoelectric controlled texture proposes that PZT is unrivaled in qualities. Likewise, by examination it used to be found that series-equal blend association is more reasonable. The weight applied on the tile and comparing voltage produced is considered, and they are detectable to have direct connection. It is predominantly alluring for execution in jam-packed regions. Jose Ananth Vino express the project that utilization of simple power instrument which incorporate rack and assessment meeting and chain drive component. The change of the tension or strain energy into electrical energy is discussed [1–3]. The power age is exceptionally high yet. The underlying expense of this framework is high. There is no need of energy from the mains, and this framework is eco amicable. It is exceptionally advantageous at the jam-packed spots and on all streets and as appropriately as all assortment of stride which is utilized to produce the power. Upkeep and oil is expected opportunity to time as a by product of the experiment. Power is not generally created during return movement of rack.

3 Proposed System

First of all, we need to maintain piezoelectric sensors on crowed areas or places. Piezoelectric tile converts stress into electrical energy. The strain can be both from weight of moving motors or from the weight of human beings taking walks on it. The produced output is in the variable form. So bridge rectifier circuit is used to convert variable voltage into linear voltage. An AC filter is used to filter out this output voltage, and it is saved in rechargeable battery. We are using Arduino Uno. Arduino is open supply digital prototyping platform based totally on flexible, handy to use hardware and software. When gadget is on it shows a message on LCD and then it is a primarily based safety device lets in solely approved humans to use this system.

The energy harvesting circuit consists of piezo-sensor transducers positioned on a tile. At the point when a unique pressing factor is applied on them [15], voltage across the piezo-transducer is created. The field of piezoelectric sensor is not steady so we use bridge rectifier to change over the variable into a direct voltage. Around 8–12v from this piezo-transducer bank is created. Bridge rectifier is given to capacitor of 2000uF followed by SPDT switch. The voltage is produced when pressure is applied on piezoelectric sensors. That is voltage is conveyed to battery, and here the battery utilized is lead acid battery, 12v. The voltage divider setup brings the voltage down to the microcontroller as we can't took care of 12v straightforwardly to the capacitor so it used to partition the voltage. Diode impedes the flow moving from capacitor to piezoelectric sensor. Pointer segment incorporates LED to show the force age. Inverter changes 12v DC over to 230v AC. After perusing the battery, voltage and the voltage created from the piezoelectric transducer are shown on 16 * 2 LCD. The microcontroller ATMEGA-328P is interfaced with LCD for programming display reason (Fig. 2).

- The amount of mechanical stress of individuals on the floor acquired sensor plate convert in to electrical energy.
- The rural energy to run AC to DC
- It does not contain any pollution.

The relation between electrical voltage and mechanical force is proportional to each other.

As per Hook's Law in $F = -kx$.

x—displacement of springs.

F—resulting force vector.

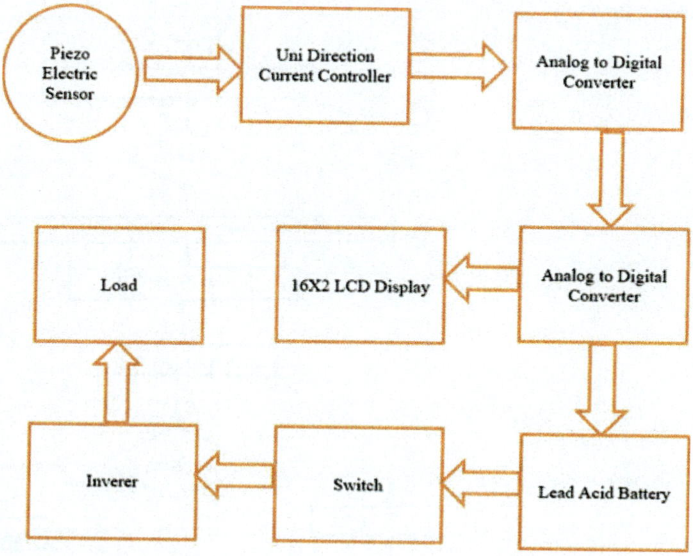

Fig. 2 Block diagram of system

k—spring constant, a constant that depends on the substrate material.

The principle of working of proposed system is primarily based on the piezoelectric sensor plates. For the implementation of this system, we regulate the tile plates above and beneath the sensors and movable spring. The process of non-conventional energy using foot step is changing the acquired mechanical power into equivalent electrical power. Foot step board is consisting of piezoelectric sensors; when strain is applied on the sensors, the sensors will convert mechanical power into electrical energy. The electrical electricity will be storing in the 12v rechargeable battery related to inverter; we are using conventional battery charging unit also for giving supply to the circuitry. The inverter is used to convert the 12 V DC to 230 V AC voltage that is used to run the loads; by the use of this AC voltage, we can function the AC loads.

A flowchart is simply graphical representation of system. Flowchart in Fig. 3 shows the steps in sequential order; generation foot step power through piezoelectric sensor plate process is explained in graphical diagram to reduce the environment pollution using renewable energy.

3.1 Hardware Description

- Piezoelectric sensors
- Led bulb
- Unidirectional current controller
- Lead Acid Battery

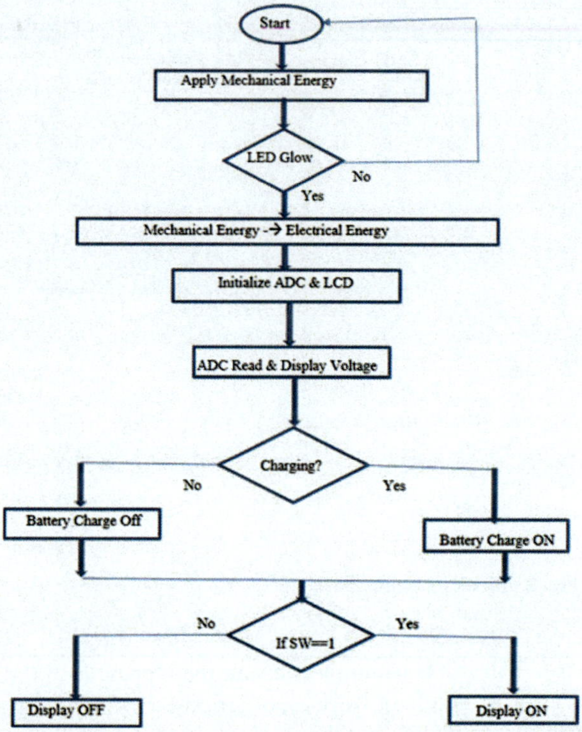

Fig. 3 Working flow of system

- Voltage sampler
- Analog to digital converter
- Arduino Uno
- Inverter
- Switch
- 16 * 2 LCD.

3.2 Software Description

- Arduino IDE
- Embedded C Programming.

4 Results and Discussion

An electric power harvesting system based on piezoelectricity principle is developed and implemented here for electrical energy generation and voltage is obtained from the pedestrian walk pressure applied, the status of voltage is displayed with the help of LCD. A good amount of electrical voltage is produced and depends on the pressure acquired by piezoelectric sensors (Figs. 4 and 5).

The challenge "foot step strength generation" which offers the first-class economical, low-cost electricity answer to frequent people. We have taken one piezoelectric sensor plate and soldered 2-wires on its − ve and + ve sides, also we have connected a single transistor on the + ve end of the module for the purpose of indication and if any rectification required (piezo-sensor plate generates AC voltage and we

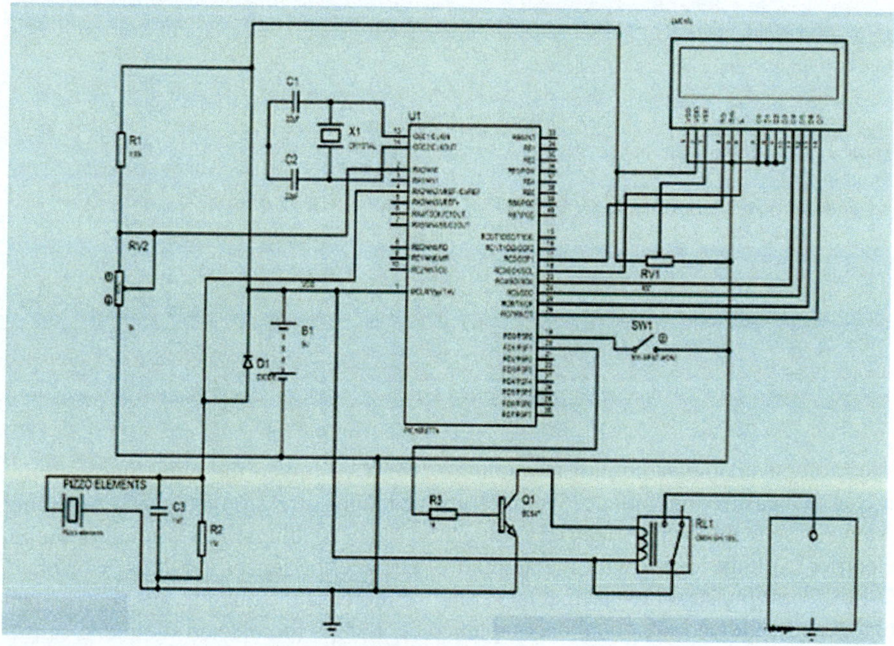

Fig. 4 Schematic diagram of foot step power generation

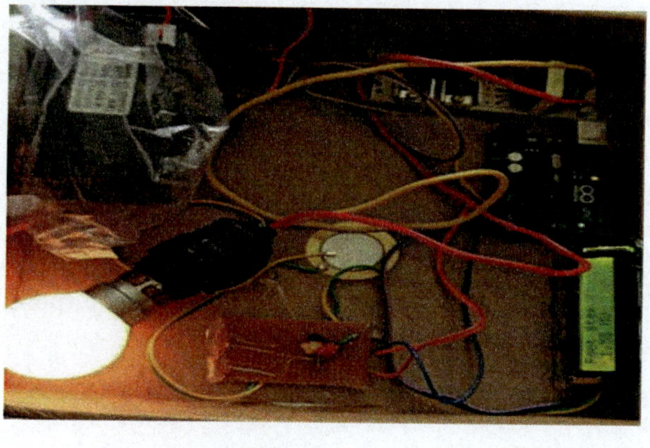

Fig. 5 Output of foot step power generation

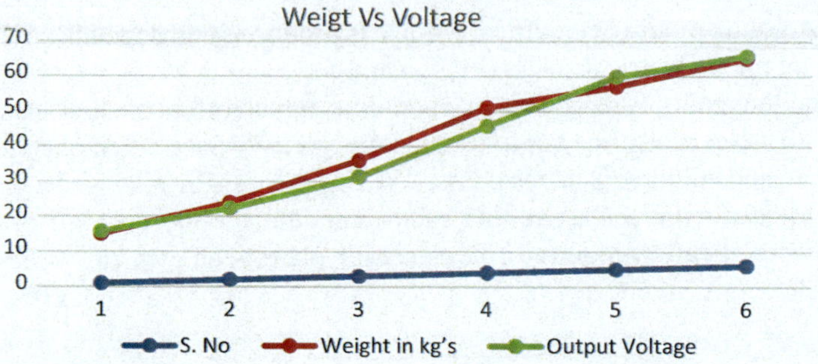

Fig. 6 Graph weight versus voltage

Table 1 Voltage for different weight

S. No.	Weight in kg's	Output voltage
1	15	15.71
2	24	22.34
3	36	31.33
4	51	45.78
5	56	58.84
6	65	66.33

need to convert the same into DC source). We have linked the wires into a multimeter to measure the voltage. We have utilized some stress on the surface of senor plate and seen the output electrical voltage on meter readings. The amount of output electrical voltage varies with the amount of strain experienced on the sensor plate. For different phases of experimental setup, we increase the quantity of piezo-sensor plates and followed the equal accompanied foot-steps (connected the piezo-sensor plates, multimeter, bridge rectifier and wires) to observe how the quantity of produced electrical-voltage would varies and verify the results (Fig. 6; Table 1).

5 Conclusion

A non-conventional, non-polluting energy is achieved when applying a force on piezoelectric sensors. As a result, installation of prosed model in a place where more amount of crowd like malls and entries, where large numbers of people can step on, which in turn would generate more power, thus using this power generated for charging of a connected battery for lightening purpose.

References

1. Prasad PR, Bhanuja A, Bhavani L, Bhoomika N, Srinivas B (2019) Power generation through footsteps using piezoelectric sensors along with GPS tracking. In: 2019 4th International conference on recent trends on electronics, information, communication and technology (RTEICT), pp 1499–1504. https://doi.org/10.1109/RTEICT46194.2019.9016865
2. Quispe JNS, Gordillo AC (2017) Implementation of an energy harvesting system by piezoelectric elements exploiting the human footsteps. In: 2017 IEEE URUCON, pp 1–4. https://doi.org/10.1109/URUCON.2017.8171873
3. Zhu M, Worthington E (2009) Design and testing of piezoelectric energy harvesting devices for generation of higher electric power for wireless sensor networks. In: SENSORS, 2009 IEEE, pp 699–702. https://doi.org/10.1109/ICSENS.2009.5398559
4. Takamura K, Yamada H, Tanaka T, Yada T, Fujiwara H (2018) Equivalent resistance control for maximum power transfer method of piezoelectric element in vibration power generation. In: 2018 International power electronics conference (IPEC-Niigata 2018-ECCE Asia), pp 1381–1385. https://doi.org/10.23919/IPEC.2018.8507750
5. Ghosh J, Sen S, Saha A, Basak S (2013) Electrical power generation using foot step for urban area energy applications. In: 2013 International conference on advances in computing, communications and informatics (ICACCI), pp 1367–1369. https://doi.org/10.1109/ICACCI.2013.6637377
6. Mohammed MA, Mustafa FF, Mustafa FI (2020) Feasibility study for using harvesting kinetic energy footstep in interior space. In: 2020 11th International renewable energy congress (IREC), pp 1–4. https://doi.org/10.1109/IREC48820.2020.9310416
7. Eldho AR, Jimson Varghese TP (2017) Power generation using piezoelectric transduces. Int J Adv Res Electr Electron Instrum Eng 6(3)
8. Mustafa F, Ishak M (2019) Study of footstep power generation using piezoelectric tile. Indonesian J Electr Eng Comput Sci (IJEECS) 15(2)
9. https://www.elprocus.com/footstep-power-generation-system/
10. Meier R, Kelly N, Almog O, Chiang P (2014) A piezoelectric energy-harvesting shoe system for podiatric sensing. In: 2014 36th Annual international conference of the IEEE engineering in medicine and biology society, pp 622–625. https://doi.org/10.1109/EMBC.2014.6943668
11. Datta RK, Rahman S (2014) Power generating slabs: lost energy conversion of human locomotive force into electrical energy. In: 8th International conference on electrical and computer engineering, pp 718–721. https://doi.org/10.1109/ICECE.2014.7026831
12. Triono AD et al. (2018) Utilization of pedestrian movement on the sidewalk as a source of electric power for lighting using piezoelectric censors. In: 2018 3rd IEEE international conference on intelligent transportation engineering (ICITE), pp 241–246. https://doi.org/10.1109/ICITE.2018.8492624
13. Bhaumik A et al. (2017) Non-conventional energy sources using piezoelectric crystal for wearable electronics. In: 2017 1st International conference on electronics, materials engineering and nano-technology (IEMENTech), pp 1–4. https://doi.org/10.1109/IEMENTECH.2017.8077009
14. Sarala T, Shivashankar, Poornima M, Lekhana HD (2020) Generation and utilization of electricity using footsteps as a source of energy. In: 2020 International conference on recent trends on electronics, information, communication and technology (RTEICT), pp 378–382. https://doi.org/10.1109/RTEICT49044.2020.9315633
15. Kamboj A, Haque A, Kumar A, Sharma VK, Kumar A (2017) Design of footstep power generator using piezoelectric sensors. In: 2017 International conference on innovations in information, embedded and communication systems (ICIIECS), pp 1–3. https://doi.org/10.1109/ICIIECS.2017.8275890

A 2.45 GHz Low Noise Figure CMOS-Based LNA Using Constant g_m Biasing Technique

Satyam Saraiyan, Abhinit Saurabh, Shubham Kumar, Shashank Kumar Dubey, and Aminul Islam

Abstract In this paper the first active block of the RF receiver system, i.e., a low noise amplifier (LNA) working at Industrial, Scientific and Medical (ISM) band of frequency 2.45 GHz is demonstrated. Proposed topology of the low noise amplifier is a cascaded type LNA which uses a constant transconductance biasing circuit to reduce the variations in gain and noise figure of LNA while the cascaded stage increases its gain to a higher level. The proposed LNA topology is realized on 180-nm CMOS technology. The simulation result of our proposed LNA topology was able to deliver gain (S_{21}) of 24.64 dB while maintaining a very low value of NF, i.e., 1.723 dB. Good input and output impedance matching have been done with S_{11} of − 32.82 dB and S_{22} of − 6.526 dB and the tradeoff between various parameters has been done in a way to achieve high gain and low noise figure. The proposed circuit exhibits unconditional stability over all ranges of frequency and not only at the operating frequency of LNA. The proposed LNA provides stability factor (K_f) of 1.787 (> 1) and B_{1f} of 727.2 m (> 0). Since the linearity of the RF component is of great importance for the overall performance, so our proposed LNA circuit was able to maintain the linearity in the acceptable range (1 dB compression point of − 21.2763 dB). The demonstrated LNA circuit operates at very low power consumption of 4.68 mW.

Keywords Input matching · Low noise amplifier (LNA) · Noise figure (NF) · Constant g_m biasing · S-parameters

1 Introduction

The first active block of a typical receiver system is the low noise amplifier (LNA). The subsequent stages which follow LNA are mixer and band pass filter. LNA is used to amplify the received signal such that the following signal is able to minimize the

---

S. Saraiyan · A. Saurabh · S. Kumar · S. K. Dubey · A. Islam (✉)
Department of ECE, Birla Institute of Technology, Mesra, Ranchi 835215, India
e-mail: aminulislam@bitmesra.ac.in

J. R. Szymanski et al. (eds.), *Energy Systems, Drives and Automations*,
Lecture Notes in Electrical Engineering 1057,
https://doi.org/10.1007/978-981-99-3691-5_17

noise introduced/added in the latter stages of the receiver system (such as from the mixer, filter, baseband amplifiers) [1–3]. Mixer is used at the receiver side to down convert the high frequency signal to a lower frequency signal which can be further filtered out by using a band pass filter [4].

LNA is one of the most important stages of any wireless communication system. The noise figure and the power dissipation of the LNA is mainly responsible for the sensitivity of the receiver system. Noise figure of the LNA directly adds to the noise figure of the complete wireless receiver system and thus it dominates the overall noise performance of the receiver. For the typical receiver system, noise figure of an LNA must be less than 3 dB to provide an acceptable performance of the receiver system [5, 6]. Power gain of the LNA not only minimizes the noise contribution from the following stages but in modern RF system it also drives the subsequent stage like down conversion mixer without the requirement of impedance matching between the LNA and the mixer [7]. For higher gain of the LNA, minimum reflection loss is required at input and output ports. For that, lower values of input return loss (S_{11}) and output return loss (S_{22}) are required for maximum power transfer to the LNA circuit and hence, for maximum gain. Thus, for typical LNA circuit S_{11} and S_{22} less than −10 dB is enough. Stability of the LNA is defined by the two parameters K_f and $B1_\mathrm{f}$. For the LNA circuit to be unconditionally stable, value of K_f should be larger than 1 and the value of $B1_\mathrm{f}$ (Δ) should be greater than 0 (< 1).

The LNA realized on 180 nm CMOS technology has 24.64 dB gain at 2.45 GHz frequency which lies in the ISM frequency band. Several applications like Cordless phone, Bluetooth devices, Wireless communication networks (Wi-Fi) uses ISM frequencies. Also, this ISM radio band is used for many other industrial, scientific and medical purposes.

In this paper the LNA circuit design and analysis is presented in Sect. 2, simulation results and discussion are discussed in Sect. 3, Sect. 4 presents the conclusion part.

2 LNA Circuit Design and Analysis

2.1 Transmission Gate

Impedance matching is required in LNA to keep the noise figure to a minimum level. To provide an impedance matching to a 50-Ω source without increasing the noise figure beyond the unacceptable region is a challenging task.

The proposed LNA circuit utilizes an inductive degeneration at the source of the common source amplifier which helps in narrowband matching and a common gate stage is connected in cascade on top of it. The cascaded common gate stage helps in improving the gain of the LNA as well as it helps in providing isolation between its input and output. The cascaded stage improves the gain by providing high impedance through which the current flows creating higher voltage at the output thus increasing the gain and it also eliminates the effect of Miller capacitance to a great extent.

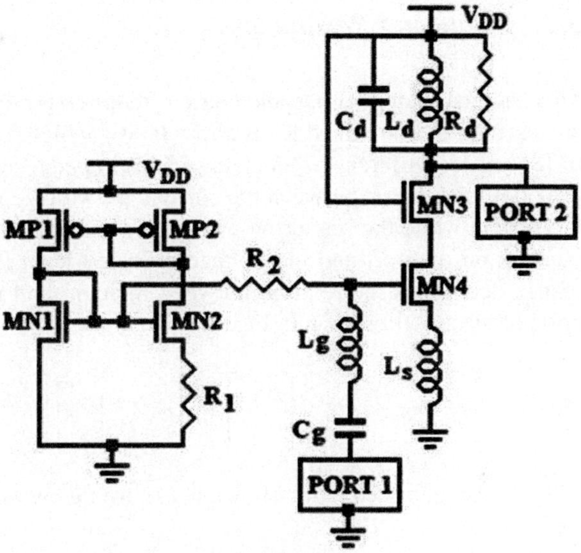

Fig. 1 Proposed LNA circuit

The input impedance matching is done by a gate inductance L_g in series with a capacitance C_g. When looking through the gate of the input transistor, a dominant resistive component is observed which depends on the transconductance, g_m, the source degeneration inductance, L_S, and gate to source parasitic capacitance, C_{gs} of the input transistor and this resistive component is matched using the gate inductance L_g [8]. The transconductance can be controlled externally using the constant g_m biasing circuit's resistance R_1 from Fig. 1.

A high resistance R_g is used to separate the constant g_m biasing circuit from that of the LNA circuit so that the signal cannot enter the biasing circuit. The capacitances C_g and C_o are used as DC blocking capacitances and the resistance R_g is used as AC blocking resistance. The matching was done at the input with a port impedance of 50 Ω and the source degeneration topology provides with a narrowband LNA design at 2.45 GHz frequency with the inductor L_g providing extra variable for the input impedance matching.

3 Results and Discussion

The proposed LNA circuit is shown in Fig. 1. Discussions of the different parameters and their simulated results are presented in the following sections.

3.1 Scattering Parameters

Measurement of magnitude and phase is required for direct measurement of voltages and currents in a circuit which is difficult task at high frequencies such as frequencies of ISM band. Different parameters such as impedances and admittances of a circuit provides the relation between the current and voltage measured at different ports of the system while the scattering parameters provides a relation between the voltage incident on the port and the voltage reflected from the port. Following is the [S] matrix defined as the relation between incident and reflected voltages of the two ports of the system shown in Fig. 2.

$$\begin{bmatrix} V_1^- \\ V_2^- \end{bmatrix} = \begin{bmatrix} S_{11} & S_{12} \\ S_{21} & S_{22} \end{bmatrix} \begin{bmatrix} V_1^+ \\ V_2^+ \end{bmatrix} \quad (1)$$

From the matrix equation shown in (1), for the two-port system as shown in Fig. 2 we can write

$$V_1^- = S_{11} V_1^+ + S_{12} V_2^+ \quad (2)$$

and

$$V_2^- = S_{21} V_1^+ + S_{22} V_2^+ \quad (3)$$

where V_1^+ is the incident voltage at port 1 and V_1^- is the reflected voltage at port 1. Similarly, V_2^+ is the incident voltage at port 2 and V_2^- is the reflected voltage at port 2. S_{11} is defined as the input reflection coefficient, S_{22} is defined as the output reflection coefficient, S_{21} is defined as forward gain coefficient and S_{12} is defined as reverse gain coefficient.

Since for better performance of any system, reflection of power at any port is required to be minimum, thus, very low values of S_{11} and S_{22} is required. For the proposed LNA circuit, value of S_{11} is simulated to be − 32.82 dB and value of S_{22} is simulated to be − 6.526 dB as shown in Figs. 3 and 4, respectively.

For an LNA, the amplification of the received signal should be such that it can withstand the further processing of the signal in the subsequent stages. Thus, S_{21} which is the forward gain coefficient should have a very large value. The S_{21} for the proposed LNA circuit is simulated to be 24.64 dB at 2.45 GHz as shown in Fig. 5. S_{12} which is the reverse gain coefficient should have a large negative value which for our proposed circuit is simulated to be − 37.39 dB at 2.45 GHz as shown in Fig. 6.

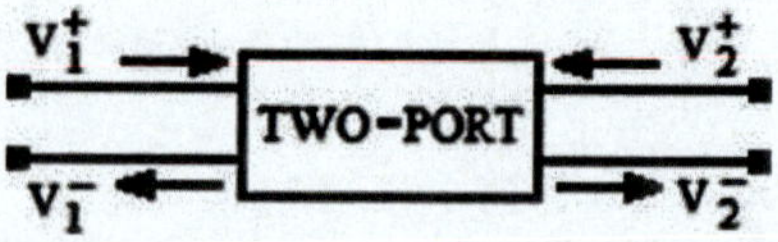

Fig. 2 Two-port network

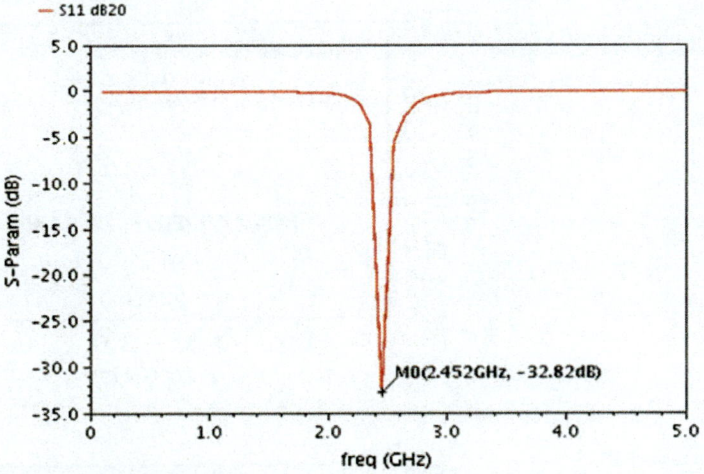

Fig. 3 S_{11} versus RF frequency

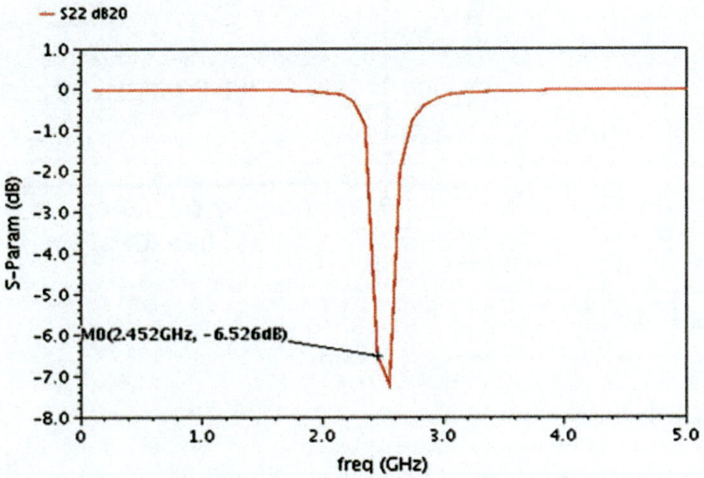

Fig. 4 S_{22} versus RF frequency

3.2 Noise Figure

For a wireless receiver system to have good performance, all the stages in the receiver circuit should add minimum noise to the signal, i.e., each stage should have low noise figure. For this purpose, LNA device is installed as the first stage in the receiver system to minimize the noise additions in the signal during further processing. Noise factor is defined as the ratio of SNR_I (signal to noise ratio at input port) to SNR_O (signal to noise ratio at output port).

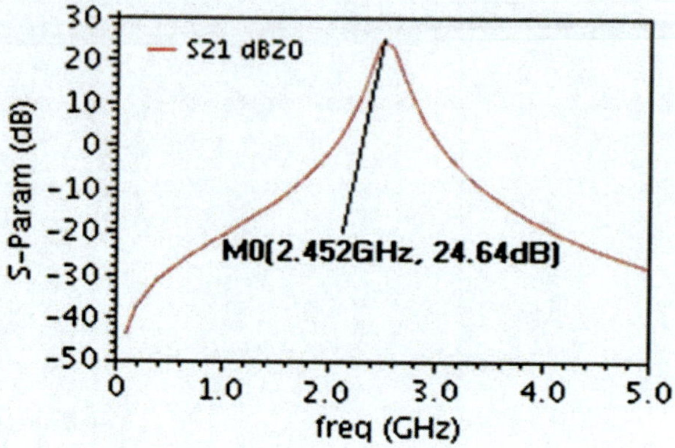

Fig. 5 S_{21} versus RF frequency

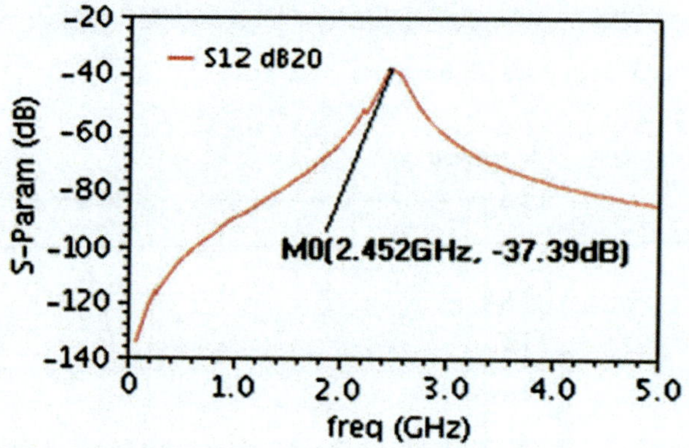

Fig. 6 S_{12} versus RF frequency

$$F = \frac{\mathrm{SNR_I}}{\mathrm{SNR_O}} \tag{4}$$

And noise figure,

$$\mathrm{NF} = 10\log(F) = \mathrm{SNR_{I_{dB}}} - \mathrm{SNR_{O_{dB}}}. \tag{5}$$

For the proposed circuit, simulated result for the NF is 1.723 dB which is less than 2 dB at 2.45 GHz as shown in Fig. 7. The noise figure has a very low value and thus the proposed LNA circuit keeps the noise figure of the receiving system low. NF is the practical noise figure of LNA which includes matching consideration of the circuit.

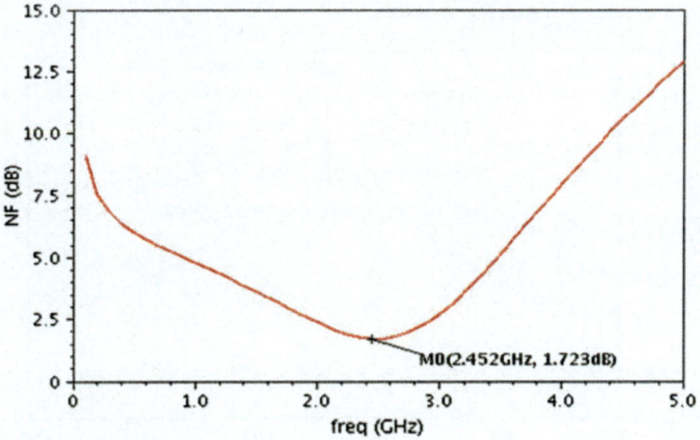

Fig. 7 NF versus RF frequency

3.3 Gain

Three types of gain definitions are used in LNA namely G_T, G_P and G_A. G_T is defined as transducer power gain of the LNA circuit. G_T is the ratio of power delivered to the load and the power available from the source and is given by

$$G_T = \frac{P_L}{P_{avs}} = \frac{|S_{21}|^2(1-|\Gamma_L|^2)(1-|\Gamma_S|^2)}{|1-\Gamma_{in}\Gamma_S|^2|1-S_{22}\Gamma_L|^2} \tag{6}$$

where Γ_L is load reflection coefficient, Γ_S is source reflection coefficient and Γ_{in} is input reflection coefficients. G_P is defined as operating power gain of the LNA circuit. G_P is the ratio of power delivered to the load and the total power inputted to the system and is expressed as

$$G_P = \frac{P_L}{P_{in}} = \frac{|S_{21}|^2(1-|\Gamma_L|^2)}{|1-S_{22}\Gamma_L|^2(1-|\Gamma_{in}|^2)}. \tag{7}$$

G_A is defined as available power gain of the LNA circuit. G_A is the ratio of power available from the network and the power available from the source and is expressed as

$$G_A = \frac{P_{avn}}{P_{avs}} = \frac{|S_{21}|^2(1-|\Gamma_S|^2)}{(1-|\Gamma_{out}|^2)|1-S_{11}\Gamma_S|^2} \tag{8}$$

where Γ_{out} is output reflection coefficient.

Here, we can clearly say that power which is available from the source is always higher than the power which is inputted to the LNA as there is some power loss in

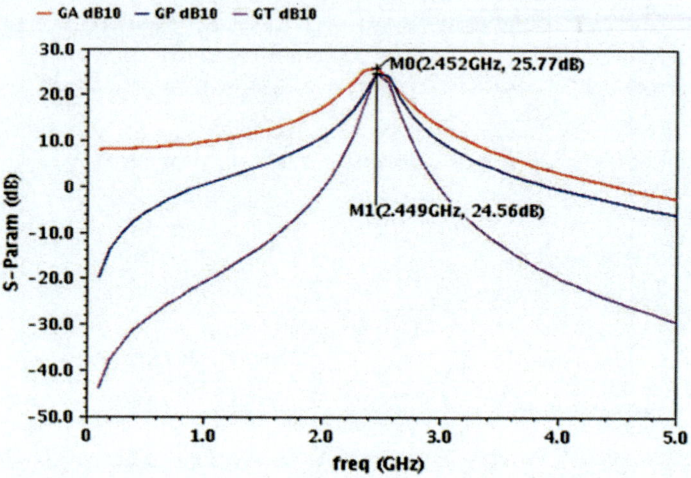

Fig. 8 Gain versus RF frequency

the path. Now from (6) and (7) we can say that $G_P > G_T$ and power available from the LNA is always higher than power available to the load due to power loss in between due to various circuit component. Now from (6) and (8) we can say that $G_A > G_T$. The simulated result of G_T, G_P and G_A are shown in Fig. 8.

3.4 Linearity

Linearity of the LNA is one of the important parameters defining the performance of the LNA. Linearity in simple terms means that output signal is directly proportional to the input signal. But increasing the linearity of the system always comes at the expense of its gain and noise figure, i.e., there is direct tradeoff between the linearity, gain and noise figure of the LNA [9, 10]. There are various parameters defining the non-linearity of the LNA: 1 dB compression point (P1dB), second order intercept point (IIP2) and third order intercept point (IIP3). P1dB lies in the operating point, at which output, and input signal power does not increase by same amount due to non-linearity in the system. System is operated till the region at which actual input–output response is 1 dB lower than the ideal/theoretical response of the system. This point indicates the 1 dB compression point of the LNA and defines the range of input signal's power for the acceptable performance of the LNA. The P1dB obtained for the proposed LNA is − 21.27 dB as shown in Fig. 9.

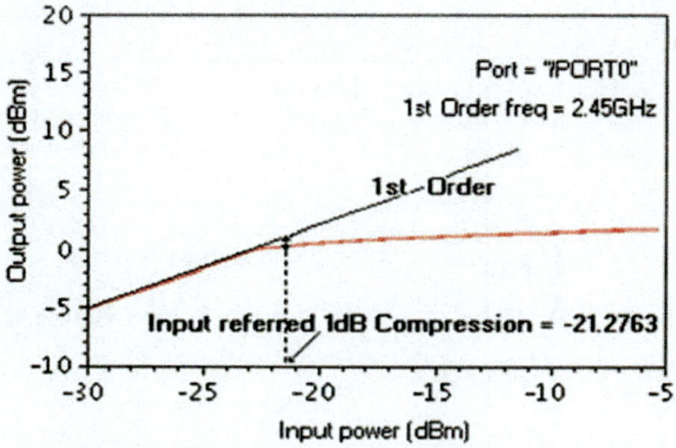

Fig. 9 Plot of output power versus input power to obtain 1 dB compression point

3.5 Stability

Stability of the LNA is an important parameter for defining the proper performance of the LNA circuit. LNA must be stable at all frequencies and not only in the desired frequency range because the input impedances of the receiver circuit can change due to external factor. Stability of the LNA is mainly described by using 'Stern stability factor' (K_f), which is defined as

$$K_\mathrm{f} = \frac{1 - |\Delta|^2 - |S_{11}|^2 - |S_{22}|^2}{2|S_{21}||S_{12}|} \tag{9}$$

where

$$\Delta = S_{11}S_{22} - S_{12}S_{21} \tag{10}$$

Similar to K_f, another factor which defines stability is $B_{1\mathrm{f}}$, which is defined as

$$B_{1\mathrm{f}} = 1 + |S_{11}|^2 - |S_{22}|^2 - |\Delta|^2 \tag{11}$$

For the unconditional stability of the LNA circuit K_f must be greater than 1 and $B_{1\mathrm{f}}$ must be greater than 0. Here unconditional stability means that the circuit will not oscillate for any change in the impedances on either input or output side. Figures 10 and 11 show the plot of stability factor K_f and $B_{1\mathrm{f}}$, respectively.

3.6 Voltage Standing Wave Ratio (VSWR)

VSWR is a measure of how well the impedances are matched at the source side as well as the load side. Impedance mismatch at the source and load side results in

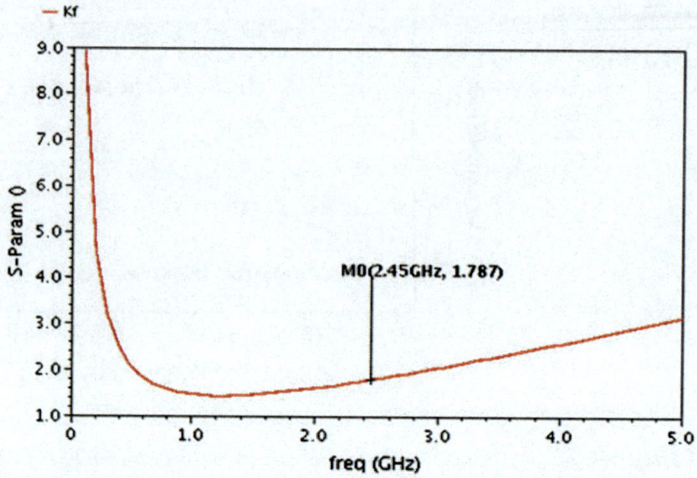

Fig. 10 Stern stability factor (K_f) versus RF frequency

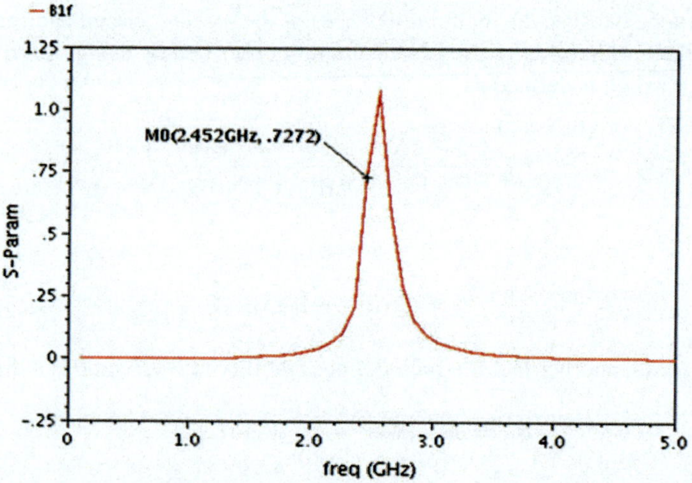

Fig. 11 B_{1f} versus RF frequency

standing waves. The relation of the VSWR with the reflection coefficient Γ is given by

$$\text{VSWR} = \frac{1 + |\Gamma|}{1 - |\Gamma|} \tag{12}$$

and the reflection coefficient is related to matching at the input and output side which is given by S_{11} and S_{22} values.

The value of $VSWR_1$ which is VSWR at the input port as shown in Fig. 12 is close to 1 which shows very good matching at the source/input side. It is also indicated by the S_{11} value which is − 32.82 dB. Similarly, the value of $VSWR_2$ which is VSWR at the output port as shown on Fig. 13 is close to 1. The value of $VSWR_2$ is more than that of $VSWR_1$ since the matching at the input/source side is better as compared to the matching at the load/output side which is also indicated by the value of S_{22} which is − 6.526 dB.

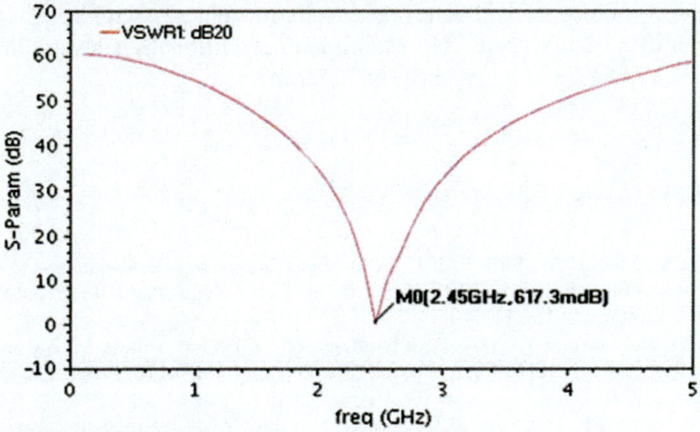

Fig. 12 $VSWR_1$ versus RF frequency

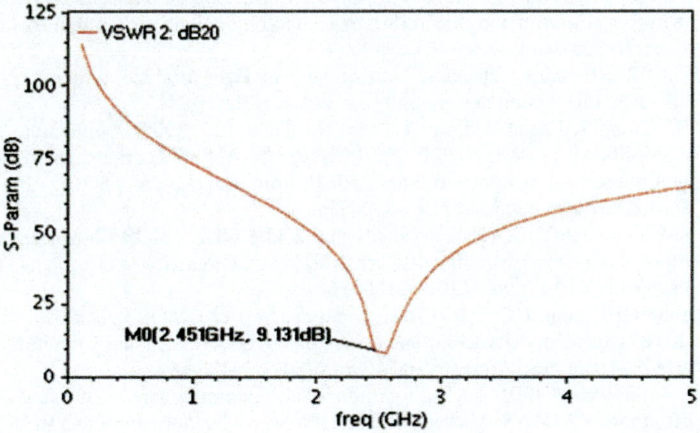

Fig. 13 $VSWR_2$ versus RF frequency

4 Conclusion

This paper presents the design and simulation of a low noise amplifier in 180-nm CMOS technology with very high gain, S_{21} of 24.64 dB and low noise figure of 1.723 dB. Good input and output matching have been achieved (S_{11} of − 32.82 dB and S_{22} of − 6.526 dB) while maintaining the other parameters such as linearity (−21.2763 dBm) in acceptable range. The good input and output impedance matching are also indicated by the VSWR values at the input and output side, i.e., $VSWR_1$ and $VSWR_2$ are close to 1. The proposed circuit operates with very low power consumption of 4.68 mW. The LNA circuit is unconditionally stable with stability factor values K_f of 1.787 (> 1) and B_{1f} of 727.2 m.

References

1. De Souza M, Mariano A, Taris T (2017) Reconfigurable inductorless wideband CMOS LNA for wireless communications. IEEE Trans Circuits Syst I Regul Pap 64(3):675–685. https://doi.org/10.1109/tcsi.2016.2618361
2. Chang C, Chen J, Wang Y (2009) A fully integrated 5 GHz low-voltage LNA using forward body bias technology. IEEE Microw Wirel Compon Lett 19(3):176–178. https://doi.org/10.1109/lmwc.2009.2013745
3. Kumar V, Dubey SK, Islam A (2020) Comparative study of active inductor-based low-noise amplifiers. In: Sikander A, Acharjee D, Chanda C, Mondal P, Verma P (eds) Energy systems, drives and automations. Lecture notes in electrical engineering, vol 664. Springer, Singapore, pp 411–418. https://doi.org/10.1007/978-981-15-5089-8_40
4. Kumar S, Saraiyan S, Dubey SK et al (2020) A 2.4 GHz double balanced down conversion mixer with improved conversion gain in 180-nm technology. Microsyst Technol 26:1721–1731. https://doi.org/10.1007/s00542-019-04718-3
5. Razavi B (2013) RF microelectronics, 2nd edn. Pearson Education India, India
6. Chi-Hsi LR (2012) RF circuit design, 2nd edn. Wiley, Hoboken, NJ
7. Chow YH, Chong T, Hasan Z, Loh CC, Tan TL, Chew EC (2005) A miniature dual-band low-noise amplifier module for IEEE 802.11 b/g/a WLAN applications. In: IEEE MTT-S international microwave symposium digest. IEEE, Long Beach, CA, pp 2071–2074. https://doi.org/10.1109/mwsym.2005.151715
8. Neihart NM, Brown J, Yu X (2012) A dual-band 2.45/6 GHz CMOS LNA utilizing a dual-resonant transformer-based matching network. IEEE Trans Circuits Syst I Reg Pap 59(8):1743–1751. https://doi.org/10.1109/tcsi.2011.2180436
9. Wu L, Leung HF, Luong HC (2017) Design and analysis of CMOS LNAs with transformer feedback for wideband input matching and noise cancellation. IEEE Trans Circuits Syst I Regul Pap 64(6):1626–1635. https://doi.org/10.1109/tcsi.2017.2649844
10. Zokaei A, Amirabadi A (2014) A 0.13 μm dual-band common-gate LNA using active post distortion for mobile WiMAX. Microelectron J 45(7):921–929. https://doi.org/10.1016/j.mejo.2014.03.012

Design of a 2.45 GHz Cascode Low Noise Amplifier with π Matching Technique

Shubham Kumar, Satyam Saraiyan, Abhinit Saurabh, Shashank Kumar Dubey, Santashraya Prasad, and Aminul Islam

Abstract In this paper a low noise amplifier (LNA) working at 2.45 GHz frequency which lies in ISM frequency band is demonstrated. Proposed LNA uses the cascoded topology with constant current source biasing methodology. This biasing technique is used to control the current flowing in the circuit so as to reduce the power consumption and cascoded stage is used to improve the gain of the LNA. Proposed topology of LNA is realized on 180-nm CMOS technology and all simulations are done using cadence virtuoso tool. Since LNA is a first active block of receiver system so we have tried to make the noise figure as low as possible while maintaining other parameters in acceptable range. The simulated results show that the proposed structure is able to provide the very low noise figure of 1.729 dB while maintaining the gain (S_{21}) of 16.34 dB. We are also able to achieve a good input matching (S_{11}) of − 26.82 dB and output matching (S_{22}) of − 21.7 dB at input and output port, respectively. Because of cascoded stage we can achieve good isolation between input and output port (S_{12}) of − 49.04 dB. Our proposed LNA circuit consume very less power of 5.2 mW.

Keywords Low noise amplifier (LNA) · Noise figure (NF) · *S*-parameters · Biasing · Matching

1 Introduction

Advancement in CMOS technology has increased the demand of analog and RF system which plays very important role in today's advanced technology-based system. Because of the demand of nano system technology with low power consumption and fast response the demand of CMOS technology has increased exponentially. All the gadgets and systems working in the range of radio frequency need RFICs

S. Kumar · S. Saraiyan · A. Saurabh · S. K. Dubey · S. Prasad · A. Islam (✉)
Department of ECE, Birla Institute of Technology, Mesra, Ranchi 835215, India
e-mail: aminulislam@bitmesra.ac.in

S. Prasad
e-mail: s.prashad@bitmesra.ac.in

J. R. Szymanski et al. (eds.), *Energy Systems, Drives and Automations*,
Lecture Notes in Electrical Engineering 1057,
https://doi.org/10.1007/978-981-99-3691-5_18

which provide them ease of operation. But the designing of such RFICs is not an easy task since there are lot of trade-offs between its different parameters. Moreover, the power consumption and cost effectiveness make the task more challenging [1, 2].

All the wireless systems have created a new market for receiver system with high performance. This receiver system is composed of different components. In the chain of components LNA is the first component followed by mixer and at the end filters are present. A typical super heterodyne receiver system is shown in Fig. 1. Because of the presence of LNA in first position in a receiver system it plays very important role and all its parameter affect the performance of receiver system more compared to other component's parameter which are present after LNA [3].

In a wireless communication system, the received signal has to overcome a lot of obstacles like trees, buildings, mountains and many other things while traveling from source to destination. Because of such path, generally the power level of received signals are very low around 100 dBm and amplitude level are in the range of μV. To process such a weak signal is really a tough task which is not possible without the help of LNA. Because an LNA does not reduce the noise that is coming with the signal itself but what it does is it amplifies both the signal and the noise without adding further noise to it thereby maintaining the signal to noise ratio of the incoming signal. Because of such operation an LNA is different from a simple amplifier [4]. But an LNA can perform the expected operation only when it satisfies its desirable characteristics which are low noise figure, high gain, low power consumption, large range of linearity, high stability and proper impedance matching at the input port as well as at the output port. Practically an LNA should provide noise figure less than 3 dB and to maximize the gain the reflection losses (S_{11} and S_{22}) should be minimum and less than − 10 dB and its stability which are defined by two parameters K_f and $B1_f$ should also be in acceptable range to provide enough stability.

To provide proper stability the value of K_f should be greater than 1 and the value of $B1_f$ should be less than 0 [4]. The noise figure and the high-power gain are deciding factor for the sensitivity of the LNA because as low as the noise figure will be, more it will contribute toward the low noise amplifying characteristics of LNA and as high as the power gain will be, it will prevent the LNA to degrade the signal to noise ratio due to addition of noise from the subsequent stage of the receiver [5, 6].

Performance of an LNA can be judged based upon different parameter like noise figure, gain, linearity, stability, power consumption and matching of input and output ports. Being a designer one has to focus that on which parameter he has to emphasize

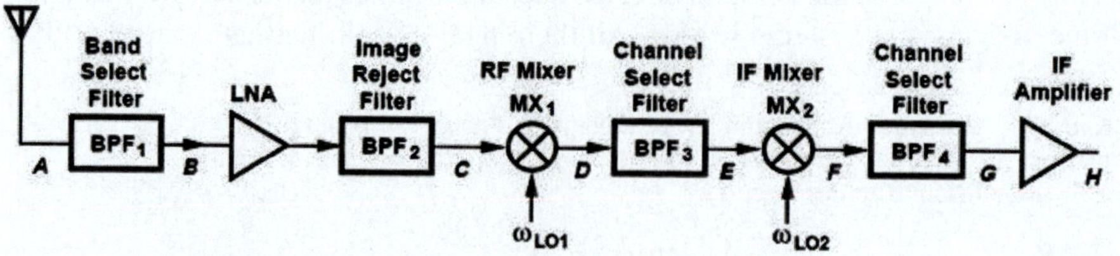

Fig. 1 A typical superheterodyne receiver system

more compared to other parameters since it is not possible to achieve all the parameters in its best form because of the presence of trade-offs between its different parameters. For example, if we increase the gain of LNA then it will also increase its non-linearity and decreases the noise figure. These trade-offs are the most challenging things while designing [7, 8]. In this paper we are mainly focused in increasing the gain and decreasing the noise figure while maintain other parameters in acceptable range. The LNA is designed in 180-nm technology which provides a high gain at a frequency of 2.45 GHz which lies in ISM frequency band so the designed LNA can be used for all application where the operating frequency lies in this range.

This paper is organized into different sections as mentioned below. Section 2 describes the LNA circuit design and analysis. Section 3 describes the results and conclusion is drawn at the end in Sect. 4.

2 LNA Circuit Design and Analysis

A conventional common source amplifier with voltage divider biasing is shown in Fig. 2.

In the shown figure the signal from signal source is coupled into the gate of MOSFET through coupling capacitor C_c which provide DC isolation between amplifier and signal source [9, 10]. To know its amplifying behavior, we need to do AC analysis for which the equivalent small signal model is shown in Fig. 3.

From the above figure we can write the output voltage as

$$V_O = -g_m V_{gs}(r_O \| R_D) \tag{1}$$

Also input gate to source voltage can be written as

$$V_{gs} = \left(\frac{R_i}{R_i + R_S}\right) V_i \tag{2}$$

Now, gain of the circuit can be calculated by simply dividing the output voltage by input voltage.

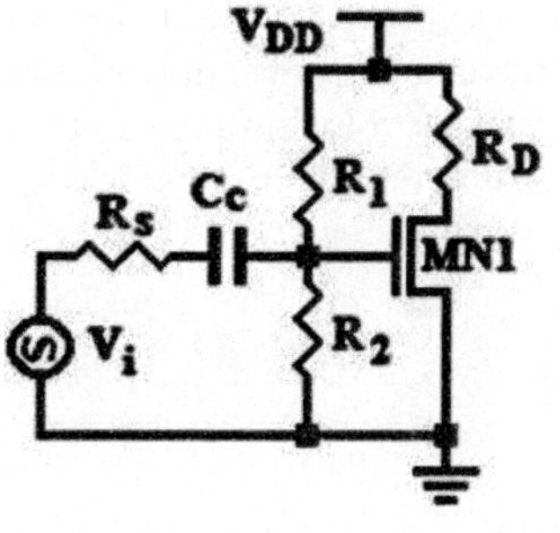

Fig. 2 Conventional common source (CS) amplifier with voltage divider biasing

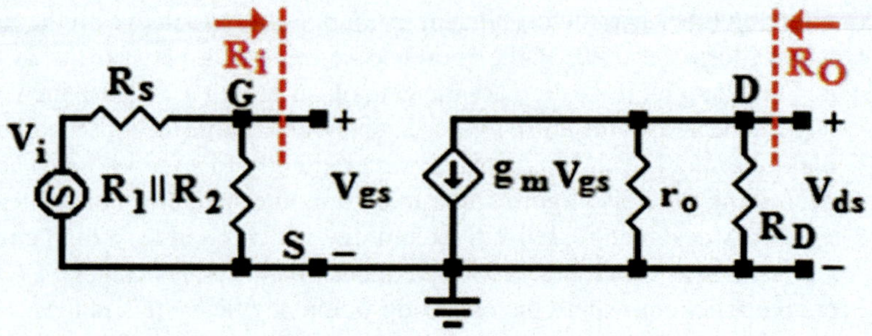

Fig. 3 Small signal model of CS amplifier shown in Fig. 2

$$|A_{\mathrm{V}}| = \left|\frac{V_{\mathrm{O}}}{V_{\mathrm{i}}}\right| = g_{\mathrm{m}}(r_{\mathrm{O}} \| R_{\mathrm{D}})\left(\frac{R_{\mathrm{i}}}{R_{\mathrm{i}} + R_{\mathrm{S}}}\right) \tag{3}$$

In general, the magnitude of R_{i} is far greater than R_{S}. So, gain can be approximated as

$$|A_{\mathrm{V}}| \approx g_{\mathrm{m}}(r_{\mathrm{O}} \| R_{\mathrm{D}}) \tag{4}$$

So, magnitude of output voltage is

$$|V_{\mathrm{O}}| = V_{\mathrm{i}} g_{\mathrm{m}}(r_{\mathrm{O}} \| R_{\mathrm{D}}). \tag{5}$$

This is just a simple approach to show how amplification is done. But to use any circuit as a low noise amplifier, not only gain but many other parameters like noise figure, matching, linearity, stability, etc. are also important which are taken care of by our proposed circuit. All these analyses are described below.

2.1 *Improvement in Matching of Input and Output Port*

In Fig. 4 proposed low noise amplifier is shown. In the circuit MN2 makes common source low noise amplifier with inductive source degeneration by the help of L_{S}. L_{g} and C_{m} are used to make π matching network. MN3 and MN4 are used to create constant current source biasing. R_{g} is used as an AC blocking resistor and C_1 is used as an DC blocking capacitor [11]. MN1 is cascode stage with MN2. C_{d} and L_{d} are used to create load stage [12, 13].

Fig. 4 Proposed design for low noise amplifier

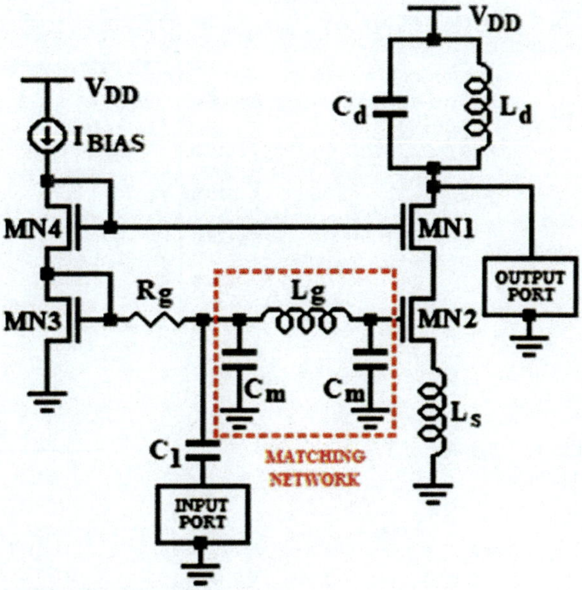

3 Explanation of Simulation Results

The proposed low noise amplifier has been designed and theoretical results have been established with the simulation results. All the simulations have been carried out using industrial standard 180-nm CMOS technology for MOSFET models on virtuoso Analog Design Environment of Cadence. Simulation results of different parameters of low noise amplifier based upon which performance of LNA is judged have been explained below.

3.1 Scattering Parameters

Voltage and current measurements in circuits which are operating at high frequencies is not as simple as compared to the circuits operating at low frequency because of the requirement of magnitude and phase. Parameters like admittance or impedance provides relation between the current and voltage of different port while scattering parameters (*S*-parameters) provides the relation between voltage incident on the port and the voltage reflected from the port. Since the designed LNA is a two-port device so there are four *S*-parameters as given below

$$\begin{bmatrix} V_1^- \\ V_2^- \end{bmatrix} = \begin{bmatrix} S_{11} & S_{12} \\ S_{21} & S_{22} \end{bmatrix} \begin{bmatrix} V_1^+ \\ V_2^+ \end{bmatrix} \tag{6}$$

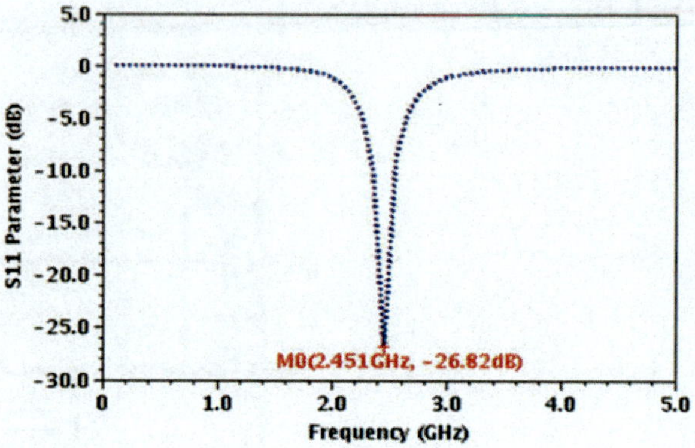

Fig. 5 Response of S_{11} parameter

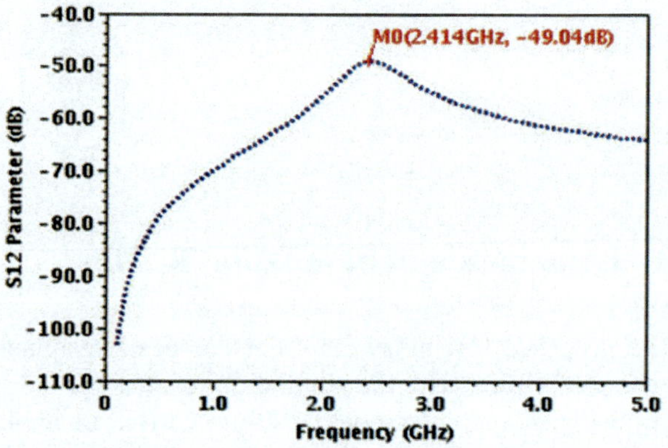

Fig. 6 Response of S_{12} parameter

where V^+ and V^- represent incident and reflected power. The different S-parameters are S_{11}, S_{12}, S_{21} and S_{22}. S_{11} is input reflection coefficient which shows how much power is reflected back at the input. So, its value should be low, S_{12} is reverse gain so its value should be low, S_{21} is forward gain which shows how much power is delivered from input to output so its value should be high, and S_{22} is nothing but output reflection coefficient [14]. After doing the simulation of S-parameters we have got S_{11} as − 26.82 dB, S_{12} as − 49.04 dB, S_{21} as 16.34 dB and S_{22} as − 21.7 dB. Plot of S_{11}, S_{12}, S_{21} and S_{22} are shown in Figs. 5, 6, 7 and 8, respectively.

3.2 Noise Figure and Minimum Noise Figure

The main objective of using a low noise amplifier is to not only amplify the signal but also not to add further noise to the system. So, the analysis of noise figure in case of low noise amplifier is most important parameter for deciding its performance. We can define the noise factor as the ratio of signal to noise ratio at input to signal to

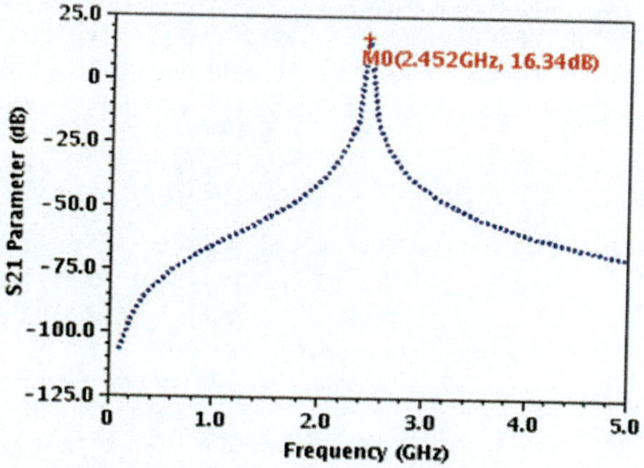

Fig. 7 Response of S_{21} parameter

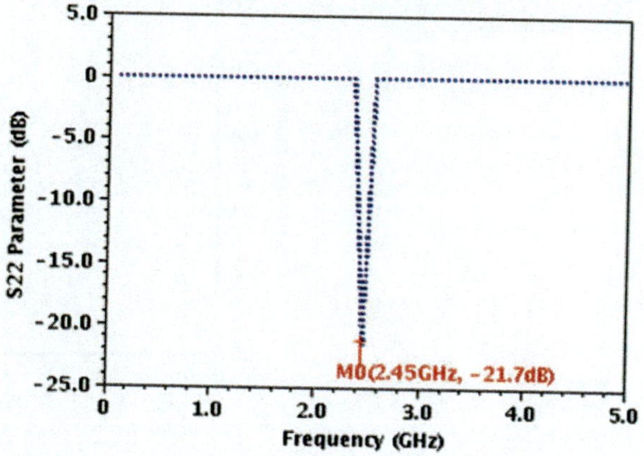

Fig. 8 Response of S_{22} parameter

noise ratio at the output. Mathematically it can be expressed as

$$F = \frac{\mathrm{SNR_I}}{\mathrm{SNR_O}} \tag{7}$$

Hence noise figure of circuit can be expressed as

$$\mathrm{NF} = 10\log(F) = \mathrm{SNR_{I_{dB}}} - \mathrm{SNR_{O_{dB}}}. \tag{8}$$

For the proposed circuit we have done the analysis of noise figure and obtained its value as 1.729 dB which is quite small value and is very suitable for practical use. Simulated plot of noise figure is shown in Fig. 9. To analyze the noise performance of the circuit one more simulation can be done which is known as $\mathrm{NF_{min}}$. $\mathrm{NF_{min}}$ is

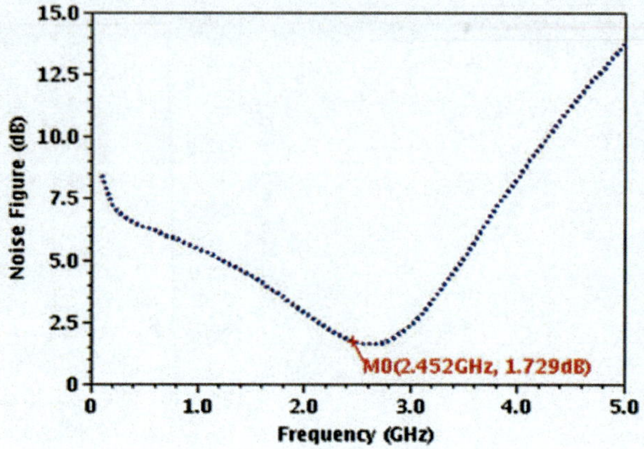

Fig. 9 Plot of noise figure versus frequency

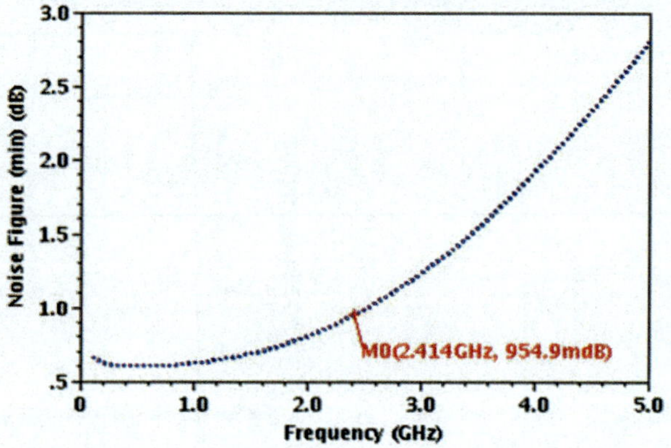

Fig. 10 Plot of NF_{min} versus frequency

the ideal measurement of noise figure without considering matching network. After simulation we have got the NF_{min} as 934.9 dBm. Plot of the same has been shown in Fig. 10.

3.3 Gain

There are different types of gain for which the performance of LNA is judged. These are G_A, G_T and G_P. G_A is the available power gain which is defined as the ratio of power available from network to the power available from the source. Mathematically it is defined as

$$G_A = \frac{1 - |\Gamma_S|^2}{|1 - S_{11}\Gamma_S|^2} |S_{21}|^2 \frac{1}{1 - |\Gamma_{out}|^2}. \tag{9}$$

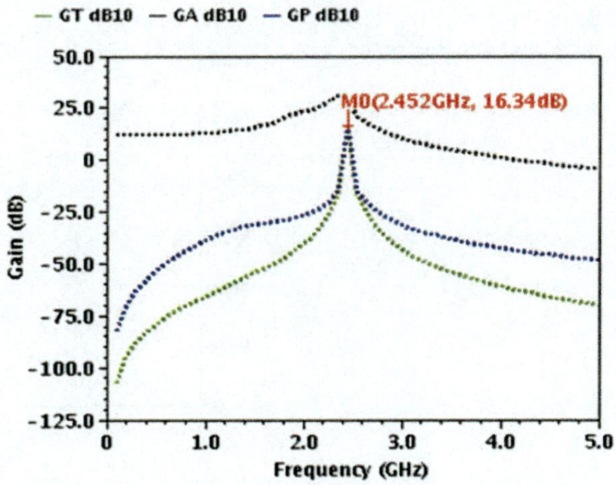

Fig. 11 Plot of G_A, G_T and G_P with respect to frequency

G_T is the transducer power gain which is defined as the ratio of power delivered to the load to power available from the source. Mathematically it can be expressed as

$$G_T = \frac{1 - |\Gamma_S|^2}{|1 - S_{11}\Gamma_S|^2}|S_{21}|^2\frac{1 - |\Gamma_L|^2}{|1 - S_{22}\Gamma_L|^2}. \quad (10)$$

G_P is operating power gain which is defined as ratio of power delivered to the load to power input to the network. We can express it as follows:

$$G_P = \frac{1}{1 - |\Gamma_{in}|^2}|S_{21}|^2\frac{1 - |\Gamma_L|^2}{|1 - S_{22}\Gamma_L|^2} \quad (11)$$

All these gains are simulated and plotted as illustrated in Fig. 11.

3.4 *Voltage Standing Wave Ratio (VSWR)*

VSWR helps to analyze the matching performance of the circuit. As the value of VSWR approaches unity, it indicates better matching. For the input side we have done the simulation of VSWR 1. Similarly, VSWR 2 is done for the output side. These plots are shown in Figs. 12 and 13, respectively.

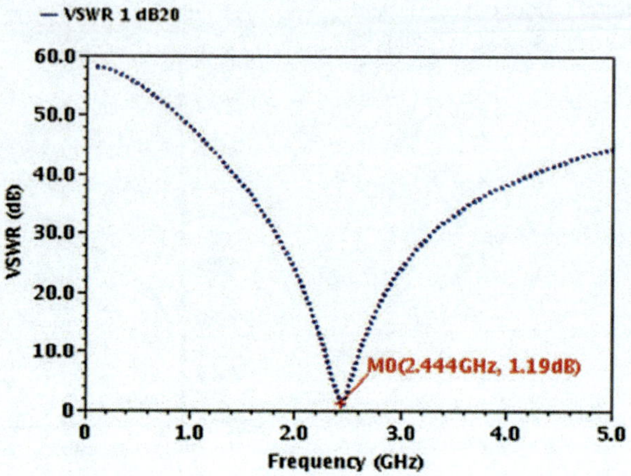

Fig. 12 Plot of VSWR 1 with respect to frequency

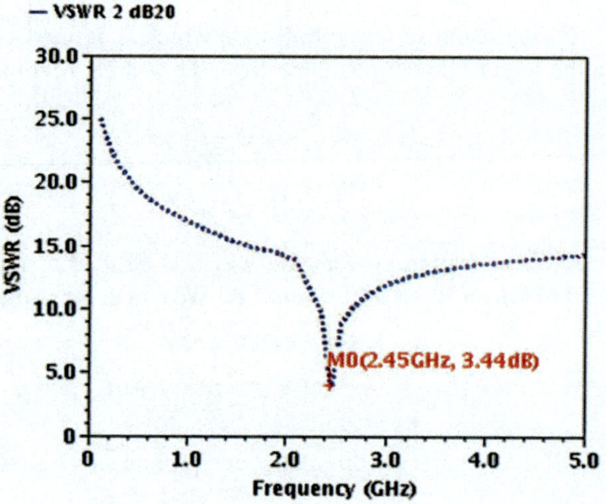

Fig. 13 Plot of VSWR 2 with respect to frequency

4 Conclusion

This paper reported the design and implementation of Common Source cascode amplifier with π matching technique in 180-nm CMOS technology. The proposed design has successfully achieved high gain of the LNA with very low noise figure maintaining all other parameters in acceptable range with low power consumption. The proposed LNA design is suitable for modern RF applications.

References

1. Kumar V, Dubey SK, Islam A (2020) Comparative study of active inductor-based low-noise amplifiers. In: Sikander A, Acharjee D, Chanda C, Mondal P, Verma P (eds) Energy systems, drives and automations. Lecture notes in electrical engineering, vol 664. Springer, Singapore, pp 411–418. https://doi.org/10.1007/978-981-15-5089-8_40
2. Kumar S, Saraiyan S, Dubey SK, Pal S, Islam A (2020) A 2.4 GHz double balanced down-conversion mixer with improved conversion gain in 180-nm technology. Microsyst Technol 26:1721–1731. https://doi.org/10.1007/s00542-019-04718-3
3. Declercq F, Rogier H (2010) Active integrated wearable textile antenna with optimized noise characteristics. IEEE Trans Antennas Propag 58(9):3050–3054. https://doi.org/10.1109/tap.2010.2052555
4. Jou CF, Cheng K-H, Lu E-T, Wang Y (2003) Design of a fully integrated concurrent triple-band CMOS low noise amplifier. In: 46th Midwest symposium on circuits and systems, vol 3. IEEE, Cairo, pp 1239–1242. https://doi.org/10.1109/mwscas.2003.1562519
5. Tsai M-H, Hsu SSH, Hsueh F-L, Jou C-P, Yeh T-J, Song M-H, Tseng J-C (2011) An ultra-low power K-band low-noise amplifier co-designed with ESD protection in 40-nm CMOS. In: 2011 IEEE international conference on IC design & technology. IEEE, Kaohsiung, pp 1–4. https://doi.org/10.1109/icicdt.2011.5783220
6. Parveen F, Rashid AH (2015) Design of a 3.3 V 4.1 GHz narrowband CMOS differential low noise amplifier. In: 2015 IEEE international WIE conference on electrical and computer engineering (WIECON-ECE), Dhaka, pp 218–221. https://doi.org/10.1109/WIECON-ECE.2015.7443901
7. Ding Y, Vehring S, Boeck G (2019) Design of 24 GHz high-linear high-gain low-noise amplifiers using neutralization techniques. In: 2019 IEEE MTT-S international microwave symposium (IMS). IEEE, Boston, MA, pp 944–947. https://doi.org/10.1109/mwsym.2019.8700859
8. Neeraja AR, Yellampalli SS (2017) Design of cascaded narrow band low noise amplifier. In: 2017 international conference on electrical, electronics, communication, computer, and optimization techniques (ICEECCOT). IEEE, Mysuru, pp 1–4. https://doi.org/10.1109/iceeccot.2017.8284642
9. Chang W-M, Cheng K-H, Jou CF (2005) 2.45 GHz/5.2 GHz switched dual-band CMOS LNA with 4 gain control modes. In: 2005 Asia-Pacific microwave conference proceedings. IEEE, Suzhou, p 4. https://doi.org/10.1109/apmc.2005.1606455
10. Yadav N, Pandey A, Nath V (2016) Design of CMOS low noise amplifier for 1.57 GHz. In: 2016 international conference on microelectronics, computing and communications (MicroCom). IEEE, Durgapur, pp 1–5. https://doi.org/10.1109/microcom.2016.7522438
11. Lee L, Jamuar SS, Sidek RM, Khatun S (2006) An improved power consumption circuit of a 5.7 GHz variable-gain low noise amplifier (VGLNA) for RF applications. In: 2006 international RF and microwave conference. IEEE, Putra Jaya, pp 66–70. https://doi.org/10.1109/rfm.2006.331039
12. Wenbo X, Xiaolin Z (2009) A low power low noise amplifier for portable GPS receivers. In: 2009 international conference on communications, circuits and systems. IEEE, Milpitas, CA, pp 793–796. https://doi.org/10.1109/ICCCAS.2009.5250406
13. Tsai J-H, Huang W-L, Lin C-Y, Chang R-A (2014) An X-band low-power CMOS low noise amplifier with transformer inter-stage matching networks. In: 2014 9th European microwave integrated circuit conference. IEEE, Rome, pp 524–527. https://doi.org/10.1109/eumic.2014.6997909
14. Yao T, Gordon M, Yau K, Yang MT, Voinigescu SP (2006) 60-GHz PA and LNA in 90-nm RF-CMOS. In: IEEE radio frequency integrated circuits (RFIC) symposium. IEEE, San Francisco, CA, p 4. https://doi.org/10.1109/rfic.2006.1651107

Characterization of Dynamic C-Gate Using GNRFET @ 16-nm Technology Generation

Rishi Jain, Harsh Malpani, Sanyam Jain, Monalisa Pandey, and Aminul Islam

Abstract This paper proposes a low-power variation—immune GNRFET-based dynamic C-gate. The proposed GNRFET-based dynamic C-gate offers ~ 1.35 × improvement in propagation delay, 2.42 × improvement in average power, ~ 2.37 × improvement in PDP and ~ 4.3 × improvement in EDP. All the simulation results are taken at proposed OEP decided by the optimum results obtained after extensive simulation on SPICE environment.

Keywords Graphene Nano Ribbon Field Effect Transistor (GNRFET) · Power-delay product (PDP) · Energy-delay product (EDP) · Propagation delay (t_P) · C-gate

1 Introduction

Asynchronous circuits are making advancements in the field of digital logic design. In semiconductor industry, these circuits have major contribution in logic design. Hence, the leading design style uses asynchronous circuits [1], because of the silent features of asynchronous circuits which are high performance and low-power applications. Two primary disadvantages of synchronous circuits are: first all signals are

R. Jain · H. Malpani · S. Jain · M. Pandey · A. Islam (✉)
Department of ECE, Birla Institute of Technology, Mesra, Ranchi 835215, India
e-mail: aminulislam@bitmesra.ac.in

R. Jain
e-mail: btech10338.18@bitmesra.ac.in

H. Malpani
e-mail: btech10402.18@bitmesra.ac.in

S. Jain
e-mail: btech10043.18@bitmesra.ac.in

M. Pandey
e-mail: phdec10052.19@bitmesra.ac.in

J. R. Szymanski et al. (eds.), *Energy Systems, Drives and Automations*,
Lecture Notes in Electrical Engineering 1057,
https://doi.org/10.1007/978-981-99-3691-5_19

binary, and another is all signals are discrete [2]. This type of approach greatly simplifies the design process but imposes constraints on the digital circuits which leads to some resolute timing problems when circuit becomes large and complex. Second assumption is rejected by the asynchronous methodology, and therefore can overcome the drawbacks. Present synchronous designs use one or more than one clock signals to control the transfer of data. Such systems may contain large number of clock elements.

The clock signal in synchronous designs has brought consequential challenges like subthreshold leakage, dissipated power, manufacturing variability, large circuit noise sensitivity, or performance improvement and signal reachability, which occurs mostly due to technology scaling. Asynchronous designs eliminate the necessity to have a global synchronization for the circuit and display several advantages when compared with synchronous designs such as electromagnetic compatibility, low-power consumption, robustness and improved noise immunity [3, 4]. Presently, asynchronous designs are used in numerous forms either small or large [4].

C-gate is an important component in the control logic of asynchronous designs [5]. The conventional execution of C-elements is based on precise transistor sizing [6, 7]. As such execution need exact ratios of transistor's driving capacities for correct functionality. Geometric discrepancy (e.g., size of transistor) and other parametric discrepancy, which results to variation in threshold voltage, as a result it affects the driving current of the transistors and also influences the functionality of C-elements.

In the proposed work, we shows analysis of C-gate in asynchronous circuits. We realize a circuit of C-gate using new Field Effect Transistor called Graphene Nano Ribbon Field Effect Transistors (GNRFET). Its various design matrices are studied and compared with that of its CMOS version.

The rest of the paper includes various sections. Section 2 describes the working principle of C-gate in asynchronous circuits. Section 3 discusses about the GNRFET. Section 4 presents the proposed circuit of Muller C-element using GNRFET. Section 5 contains simulated results and comparison. Finally, Sect. 6 concludes the paper.

2 Working Principle of C-Gate

C-gate is also known as Muller C-element or two-hand safety circuit, or coincident flip-flop or hysteresis flip-flop, which is widely used in asynchronous system. In asynchronous designs, handshaking protocol between the sender and the receiver [5] is used in place of global clock synchronization as shown in Fig. 1 and it is well explained from Fig. 2. All four events proceed in cyclic order which are data change, request, data acceptance and acknowledge. Time taken by repeated cycles may be different depending upon the time it takes to produce or consume the data.

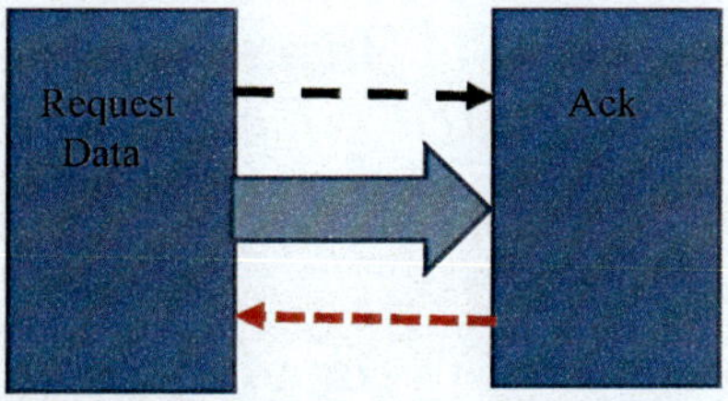

Fig. 1 Handshaking protocol

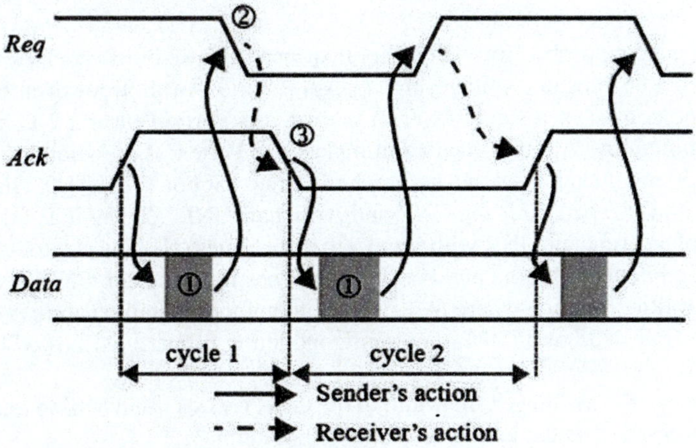

Fig. 2 Timing diagram of handshaking protocol

The basic circuit element C-gate [5] is extensively used in asynchronous circuits. Table 1 depicts the truth table of a 2-input C-gate or Muller C-element and Fig. 3 shows its symbol. It gives output as "0" when it's all inputs are "0" and when it's all inputs are "1" the output is "1". For other combinations of input the output retains its previous value. Consequently, when the output changes from "0" to "1" and "1" to "0", it means that both the inputs are "1" and "0", respectively [6].

Table 1 Truth table of Muller C-element

Input *A*		Input *B*	Output *F*
0		0	0
0		1	F
1		0	F
1		1	1

Fig. 3 Schematic symbol of Muller C-element

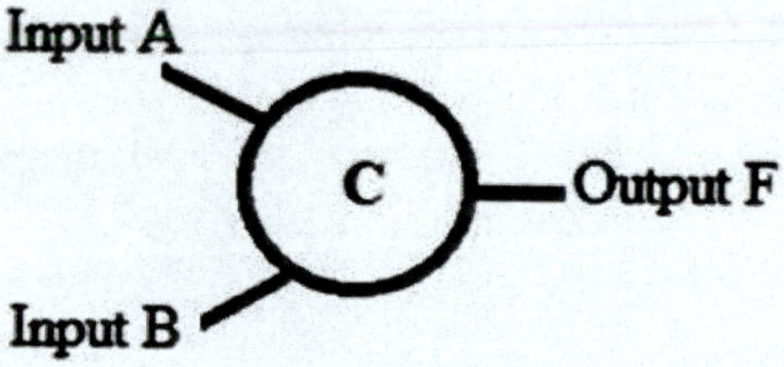

3 GNRFET Structure and Its Working Principle

GNRFET is being probed for various applications because of its eccentric attributes like controllable threshold voltage, high I_{ON}–I_{OFF} ratio. An allotrope of carbon which is graphene consist of a single layer of carbon atom arranged in a 2-D hexagonal lattice in unique way so that each atom makes one vertex. If the width of graphene layer is < 50 nm, then it is known as Graphene Nano Ribbon (GNR) [8]. GNR can be classified into two parts: Zigzag GNR and Armchair GNR. Zigzag Fig. 4 depicts the structure of an armchair GNR with respect to dimer lines (N). The electrical attribute of the ribbon depends on the number of dimer lines for armchair GNR. If the width of ribbon follows relation where $N = 3S + 2$, then more metallic nature is exhibited by ribbon. If $N = 3S$ or $3S + 1$, then semiconducting property is showed by ribbon [9, 10].

In this case, S is an integer, the width of the GNR (W_{GNR}), and N have relationship (1). In this case $a_{\text{c-c}}$ is the lattice constant, is 0.412 nm.

$$W_{\text{GNR}} = (N - 1) \times \left(\sqrt{3/2}\right) \times a_{\text{c-c}}. \quad (1)$$

Fig. 4 Width of armchair GNR with respect to dimer lines [8]

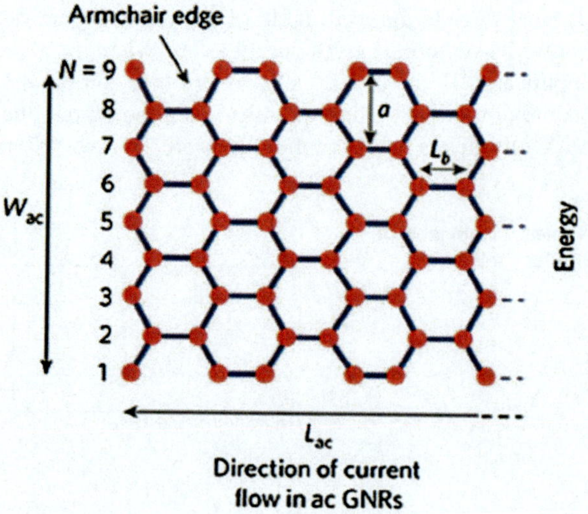

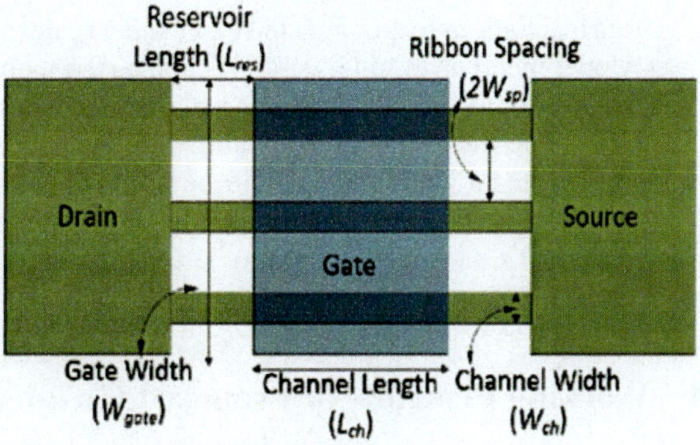

Fig. 5 A multiple ribbon armchair-based MOS type GNRFET [11]

Figure 5 conveys a Metal Oxide Semiconductor (MOS) type armchair GNRFET with multiple ribbons. The GNR channel is intrinsic in nature, laid under the gate. The portion of the ribbon that is outside the gate which connects source and drain part is called the reservoir, and these are having high doping concentration. L_{ch} is channel length, L_{res} is the reservoir length, W_{ch} is the ribbon width, W_{gate} is the gate width and $2W_{sp}$ is the spacing between the ribbons. The transistor can be switched ON and OFF states depending on the voltage given to the gate. The threshold voltage (V_t) of the transistor is given as:

$$V_t = b_g/3e. \tag{2}$$

Here, $b_g = 2|\delta|\Delta E$ is bandgap, and e is charge of an electron. Also, $\delta = 0.27$, 0.4 and 0.66 for $N_a = 3S$, $3S + 1$ and $3S + 2$, respectively, $\Delta E = \pi h v_f / W_{GNR}$, h = Planck's constant, $v_f = 10^6$ m/s. The current through the drain (I_{DS}) in GNRFET can be showed by [11].

$$I_{DS}(\Psi_{CH}, V_D, V_S) = \frac{2qkT}{h}\sum_{\alpha}\left[\ln\left(1+e^{\frac{q(\Psi_{CH}-V_S)-\varepsilon_\alpha}{kT}}\right) - \ln\left(1+e^{\frac{q(\Psi_{CH}-V_D)-\varepsilon_\alpha}{kT}}\right)\right] \tag{3}$$

Here, Ψ_{CH} = Channel Potential, V_D = Drain voltage, V_S = Source voltage, ε_α = Subband edge, α = Subband index ($1 \le \alpha \le N$), k = Boltzmann's constant, h = Planck's constant and T = Temperature.

In this paper, all the GNRFETs have six ribbon GNR channels. The threshold potential is regulated by dimer number which depend on the ribbon width [12]. In GNRFET, multiple graphene nano ribbons are used in parallel which increase the drive strength and make the channel wider and conducting as demonstrated in [11]. Mobility of carrier in GNRFET has been studied by the authors in [11], the mobility with a 2.5 nm wide GNR. They showed 171–189 cm^2/V s, which was estimated based on partial measurements and electrostatic simulations. They estimated mobility using full-band electron and phonon dispersion relations.

Drain of a logic gate is connected to the input (gate) of next logic gate to create data path. Upper metal layer of GNRFET is connected with graphene layer through vias. Due to inadequate knowledge of vertical graphene vias they are considered as metal [13]. Graphene-metal junctions exhibits high contact resistance [13]. To overcome this problem local interconnects will be used between source and drain contact of MOS-GNRFETs. Graphene material will be suitable for these local interconnects to reduce contact resistances and extra vias.

4 Working Principle of Proposed C-Gate Using GNRFET

In integrated circuit design, dynamic logic is a design methodology in combinatory logic circuits. It is distinguished from so called static logic by exploiting temporary storage of information in stray and gate capacitances. The total capacitance comprises of intrinsic and extrinsic capacitance. Dynamic logic circuits require less surface area and are usually faster than their static counter parts.

In Fig. 6 when $A = 0$, $B = 0$, both the P-MOS GNRFET of the first stage are ON and the output F from the second stage comes out to be low. When $A = 1$, $B = 1$, all the N-MOS GNRFET are ON and output F is high. When both the inputs are different, the output of the first stage is in high impedance state as the output of the first stage is not connected to either supply rails. Therefore, the output F will retain its previous value. The output of the first stage is in high impedance state and connected to the drain of transistor XN2 or XP1 at the time when the inputs are different. All the GNRFET used has 12 dimer lines in each ribbon and total 6 ribbons are used. Width of GNRFET is 33.54 nm and that of MOSFET is 32 nm.

5 Simulation Results

This section presents the simulation results, which have been obtained using SPICE simulation environment. Table 2 reports comparison of propagation delay between the MOSFET and the GNRFET realization of C-gate. As can be observed from Table 2, the proposed GNRFET-based C-gate exhibits 1.36 × shorter delay as compared to its MOSFET version at the nominal supply voltage ($V_{DD} = 0.7$ V) as the mobility of electron in GNRFET is higher than that in MOSFET. The comparison of average power (P_{AVG}) dissipation between the MOSFET and the GNRFET realization of C-gate. As expected, the P_{AVG} increases with increase V_{DD}. The proposed GNRFET-based C-gate exhibits its robustness by showing its variability (defined as the ratio of standard deviation (σ) to mean (μ) of a design metric) closer to 0 in the P_{AVG} while varying V_{DD} by 10% around the nominal V_{DD}.

This section presents the simulation results, which have been obtained using SPICE simulation environment. Table 2 reports comparison of propagation delay between the MOSFET and the GNRFET realization of C-gate. The reported t_P data

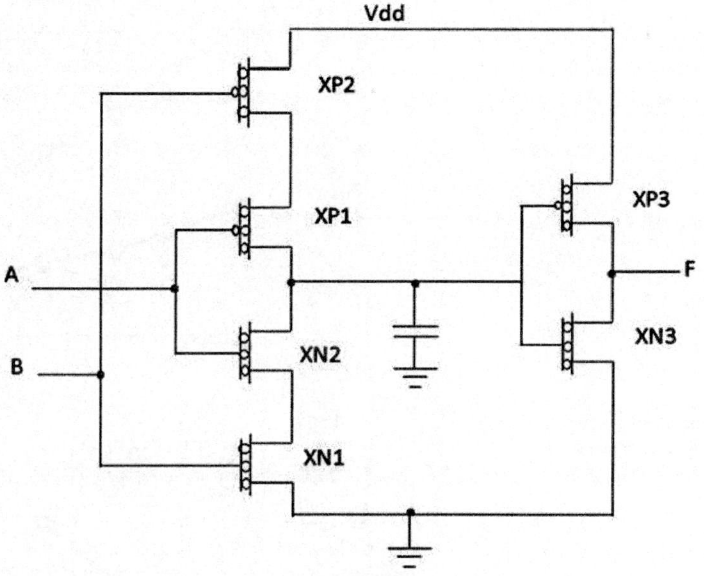

Fig. 6 Proposed circuit diagram of Muller C-element using MOS GNRFET

Table 2 Comparison of propagation delay and its variability with supply voltage (V_{DD})

C-gate design based on	Std. dev. (σ) of t_P (ps)	Mean (μ) of t_P (ps)	Variability (σ/μ)	V_{DD} (mV)
MOSFET	266.6	330.6	0.806	770
	262.7	349.7	0.751	740
	259.3	377.3	0.687	700
	43.11	410.9	0.105	670
	42.41	443.5	0.096	630
GNRFET	6.235×10^{-13}	257.3	2.42×10^{-15}	770
	0	269.4	0	740
	0	278.1	0	700
	0	293.4	0	670
	6.235×10^{-13}	309.2	2.01×10^{-15}	630

of Table 2 have also been plotted in Fig. 7 for making the comparison clearly visible. As can be observed from Table 2, the proposed GNRFET-based C-gate exhibits 1.36 × shorter delay as compared to its MOSFET version at the nominal supply voltage ($V_{DD} = 0.7$ V) as the mobility of electron in GNRFET is higher than that in MOSFET. Figure 7 shows the trend of propagation delay with variation in supply voltage. Figure 8 illustrates the trend of P_{AVG} with the variation in V_{DD}. The proposed GNRFET-based C-gate exhibits 2.42 × lower P_{AVG} dissipation as compared to its

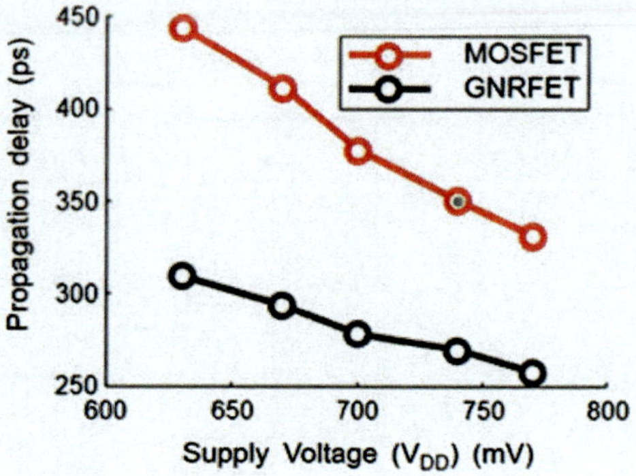

Fig. 7 Variation of propagation delay (t_P) with variation in supply voltage (V_{DD})

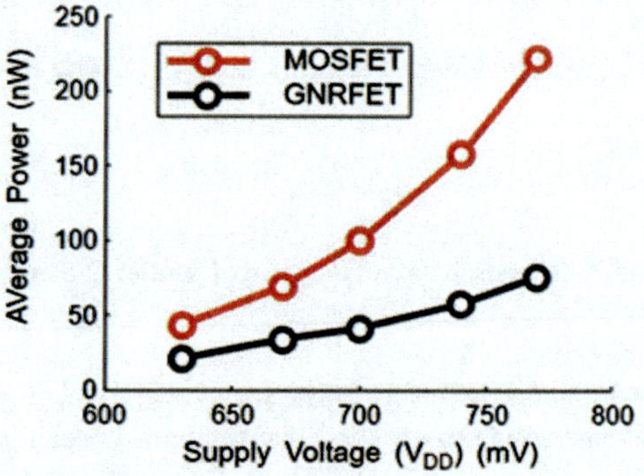

Fig. 8 Variation in average power (P_{AVG}) with variation in supply voltage (V_{DD})

MOSFET counterpart at the $V_{DD} = 0.7$ V as GNRFET shows lesser second order effects compared to that of MOSFET.

The PDP and EDP comparison data of C-gate realized with MOSFET and the GNRFET are plotted in Figs. 9 and 10. As can be observed from Fig. 9, the proposed GNRFET-based C-gate exhibits 2.37 × lesser power-delay product (PDP) as compared to its MOSFET version at the nominal supply voltage ($V_{DD} = 0.7$ V) and Fig. 10 reports the proposed GNRFET-based C-gate exhibits 4.3 × lower energy-delay product (EDP) as compared to its MOSFET version at the nominal supply voltage ($V_{DD} = 0.7$ V).

Figure 11 reports comparison of propagation delay between the MOSFET and the GNRFET realization of C-gate while varying the operating temperature for making the comparison easily comprehensible. As can be observe the proposed GNRFET-based C-gate exhibits 1.36 × shorter delay as compared to its MOSFET version at the room temperature ($T = 25$ °C) as the mobility of electron in GNRFET is higher than that in MOSFET. In MOSFET-based design t_P decreases with increasing temperature as it shows positive temperature coefficient. The proposed GNRFET-based C-gate exhibits its robustness by showing its variability closer to 0 in the propagation while varying temperature from 25 to 100 °C. Figure 12 plot the comparison of average

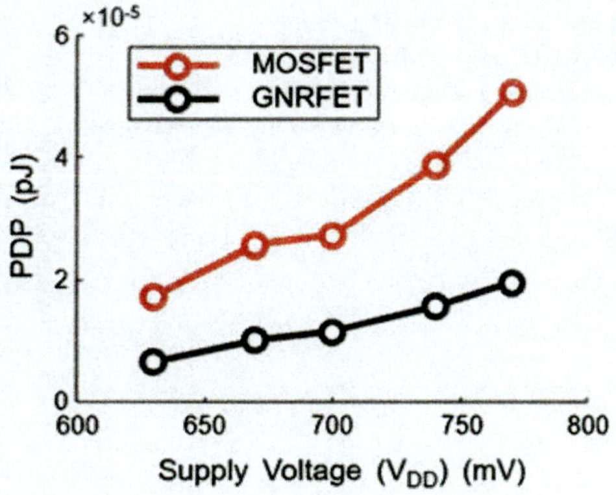

Fig. 9 Variation in power-delay product (PDP) with variation in supply voltage (V_{DD})

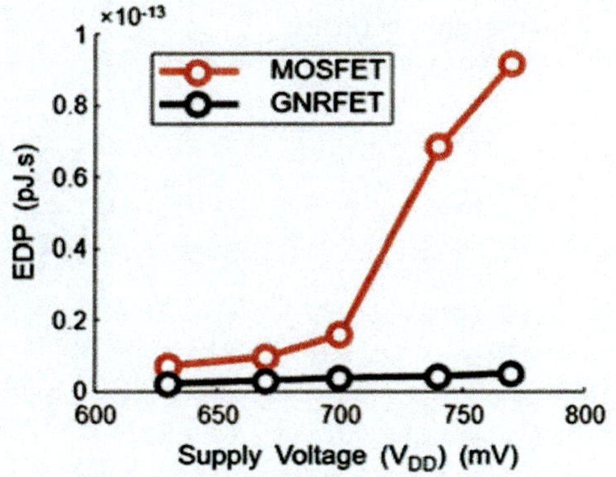

Fig. 10 Variation in energy-delay product (EDP) with variation in supply voltage (V_{DD})

power (P_{AVG}) dissipation between the MOSFET and the GNRFET realization of C-gate for making the comparison clearly perceptible.

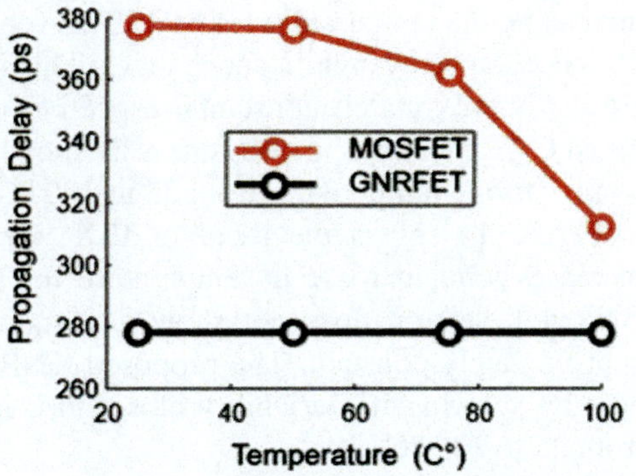

Fig. 11 Variation of propagation delay (t_P) with variation in operating temperature (T)

Fig. 12 Variation of average power (P_{AVG}) with variation in operating temperature (T)

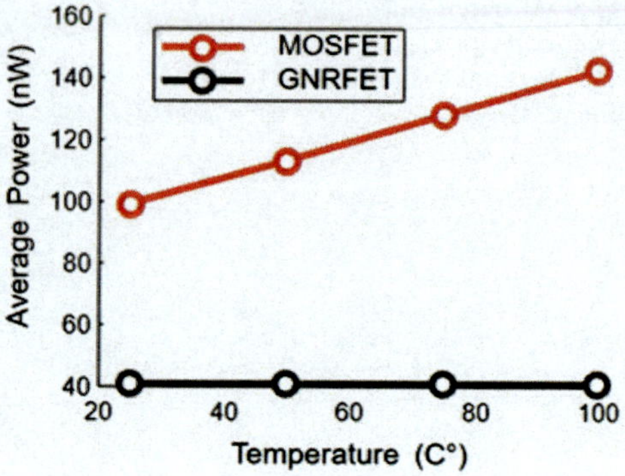

Fig. 13 Variation of power-delay product (PDP) with variation in temperature (T)

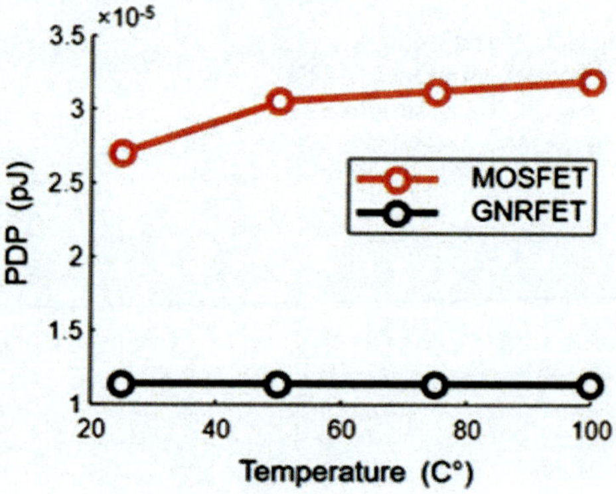

We can also observe from Fig. 13 the proposed GNRFET-based C-gate exhibits 2.37 × lower PDP as compared to its MOSFET version at the room temperature (T = 25 °C).

According to the trend PDP as illustrated in Fig. 13, PDP increases as with increasing the temperature for MOSFET version while GNRFET version of C-gate does not show any significant changes. This is because the delay and average power are also approximately constant irrespective of temperature. The proposed GNRFET-based C-gate exhibits its robustness by showing its variability closer to 0 in the PDP while varying temperature from 25 to 100 °C.

Figure 14 shows the trend of EDP variation with temperature. While EDP increases with increase in temperature for MOSFET version of C-gate however, GNRFET version does not show any significant changes. Rather EDP remains approximately constant. The proposed GNRFET-based C-gate exhibits its robustness by showing its variability closer to 0 in the EDP while varying temperature from 25 to 100 °C.

We have also verify the correctness of voltage transfer characteristics (VTCs) of the proposed C-gate circuit sing SPICE environment. Figures 15 and 16 illustrate the

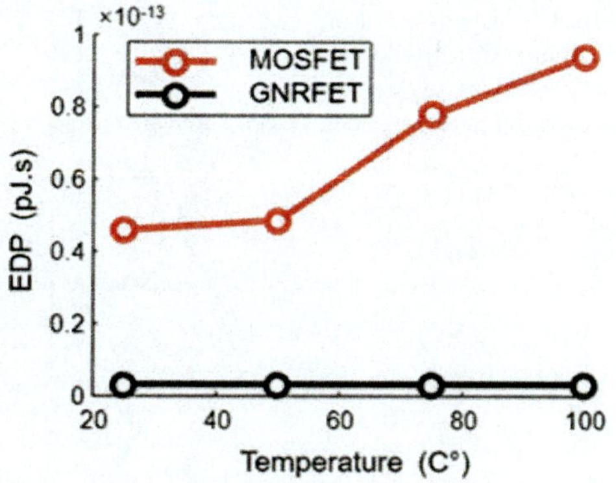

Fig. 14 Variation of energy-delay product (EDP) with variation in temperature (T)

input waveforms and Fig. 17 depicts the output waveform of our proposed C-gate circuit.

Subthreshold leakage current is one of the major concerns in advanced technology generation. Subthreshold leakage current arises when there is weak inversion of

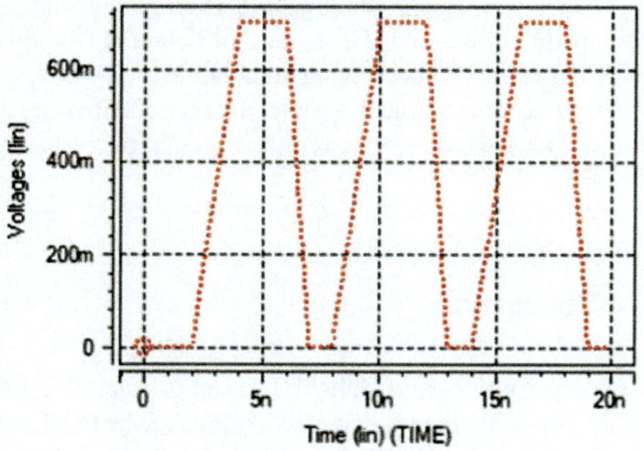

Fig. 15 Waveform of input A, which is applied to the proposed C-gate circuit shown in Fig. 6

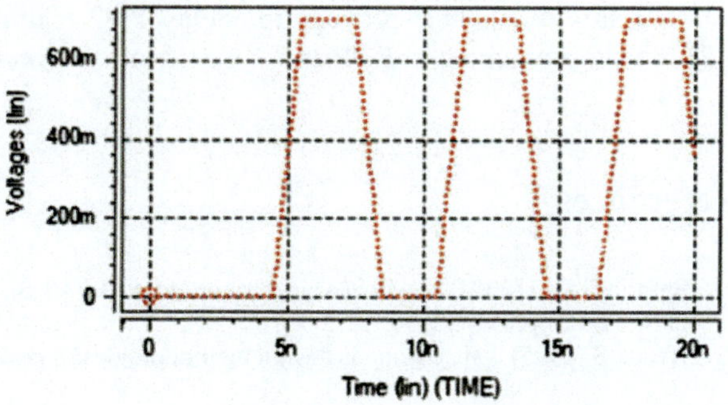

Fig. 16 Waveform of input B, which is applied to the proposed C-gate circuit shown in Fig. 6

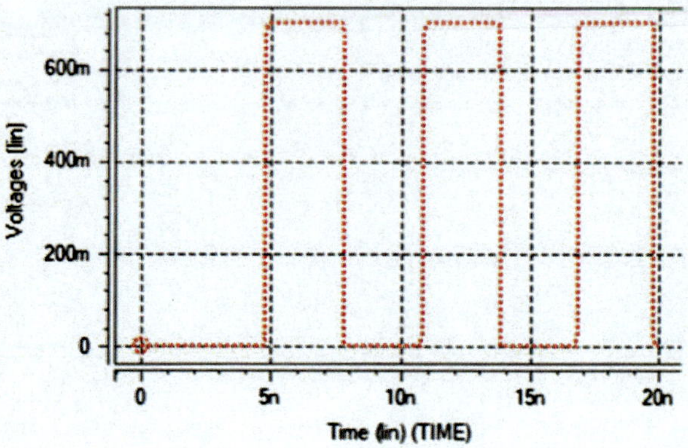

Fig. 17 Waveform of output F, obtained from the proposed C-gate circuit shown in Fig. 6

channel and is given by:

$$I_{\mathrm{SUB}} = I_{\mathrm{s}} \exp\left(\frac{V_{\mathrm{GS}} - V_{\mathrm{t}}}{\eta V_{\mathrm{T}}}\right) \times \left(1 - \exp\left(-\frac{V_{\mathrm{DS}}}{V_{\mathrm{T}}}\right)\right) \tag{4}$$

In (4), I_{SUB} is the subthreshold leakage current, I_{s} is saturation current, V_{t} is subthreshold potential, V_{T} is thermal potential. Subthreshold leakage can be minimized by using devices having higher threshold voltage. Also, MOSFETs do not have the property of multiple threshold voltages and in order to apply multi threshold property, we have to add extra circuitry which in turn consume more area as well as more power.

6 Conclusion

This paper proposes a GNRFET-based C-gate, which exhibits shorter propagation delay, lower average power dissipation, lower PDP and lower EDP. This proposed C-gate circuit is suitable for applications where there is a need to save energy or consume less power because our investigation reveal that the proposed circuit offers ~ 1.35 × improvement in propagation delay, 2.42 × improvement in average power, ~ 2.37 × improvement in PDP and ~ 4.3 × improvement in EDP.

References

1. Smith SC, Di J (2014) Future of asynchronous logic. In: VLSI, circuits for emerging application. CRC Press, pp 311–324
2. Hauck S (1995) Asynchronous design methodologies: an overview. Proc IEEE 83(1):69–93

3. Tang CK, Lin CY, Lu YC (2008) An asynchronous circuit design with fast forwarding technique at advanced technology node. In: Proceedings of international symposium of quality, electronic, design (ISQED'08), pp 769–773
4. Beerel PA (2002) Asynchronous circuits: an increasingly practical design solution. In: Proceedings of international symposium of quality, electronic, design (ISQED'08), pp 367–372
5. Agrawal K, Chowdhury S, Dubey SK, Islam A (2018) Robustness study of Muller C-element. In: Mandal J, Sinha D (eds) Social transformation—digital way. CSI 2018. Communications in computer and information science, vol 836. Springer, Singapore, pp 131–139. https://doi.org/10.1007/978-981-13-1343-1_15
6. Sparso J, Furber S (2001) Principles of asynchronous circuit: a systems perspective. Kluwer Academic Publishers
7. Martin AJ (1989) Formal program transformations for VLSI circuit synthesis. In: Dijkstra EW (ed) Formal development of programs and proofs. UT year of programming series. Addison Wesley
8. Xu W, Lee T-W (2016) Recent progress in fabrication techniques of graphene nanoribbons. Mater Horiz 3:186–207. https://doi.org/10.1039/c5mh00288e
9. Son Y-W, Cohen ML, Louie SG (2006) Energy gaps in graphene nanoribbons. Phys Rev Lett 97(21):216803
10. Mohammed MU, Chowdhury MH (2019) Design of energy efficient SRAM cell based on double gate Schottky-barrier-type GNRFET with minimum dimer lines. In: IEEE international symposium on circuits and systems (ISCAS). IEEE
11. Chen YY et al (2013) A SPICE-compatible model of graphene nano-ribbon field-effect transistors enabling circuit-level delay and power analysis under process variation. In: Design, automation & test in Europe conference & exhibition (DATE). IEEE
12. Sandhie ZT, Ahmed FU, Chowdhury MH (2020) Design of ternary master-slave D-flip flop using MOS-GNRFET
13. Fiori G, Iannaccone G (2007) Simulation of graphene nano-ribbon field-effect transistors as future low-power devices. IEEE Electron Device Lett 28(8)

Study and Analysis of AlGaN/GaN-Based HEMT and MOSHEMT

Shadaf Firdoush, Isha Mishra, Rhea Xalxo, Shashank Kumar Dubey, Santashraya Prasad, and Aminul Islam

Abstract This paper is an extensive study of AlGaN/GaN HEMT and MOSHEMT. Two structures have been simulated, a HEMT and a MOSHEMT structure and their characteristics have been obtained and studied. The gate optimization has been done in this research work. The oxide used beneath the gate contact is made of aluminum oxide with thickness 14 nm. The gate length has been reduced to 250 nm to improve upon the current gain cutoff frequency. The gate leakage current is found to be zero for a greater range of gate voltage in case of MOSHEMT which makes it suitable for radio frequency power applications.

Keywords AlGaN/GaN · Oxide · HEMT · MOSHEMT · Gate leakage current

1 Introduction

Due to advancement in the technology, demand of high-speed devices is increasing drastically. Conventional silicon-based CMOS technology is not suitable for the very high frequency and high-power applications due to various short channel effects (SCEs). III–V group compounds-based devices such as high electron mobility transistors (HEMTs) are most suitable candidate for high frequency and high-power applications [1–5]. HEMT utilizes the concept of heterojunction to improve its performance parameters. Heterojunction is formed at the interface of the wide bandgap and narrow bandgap material. AlGaN/GaN-based HEMT exhibits improved device properties as compared to AlGaAs/GaAs-based HEMT [6–8]. AlGaN/GaN-based HEMTs are most suitable candidate for high frequency wireless communication system [9, 10]. It is also preferable for the low noise applications [11].

S. Firdoush · I. Mishra · R. Xalxo · S. K. Dubey · S. Prasad · A. Islam (✉)
Department of ECE, Birla Institute of Technology, Mesra, Ranchi 835215, India
e-mail: aminulislam@bitmesra.ac.in

S. Prasad
e-mail: s.prashad@bitmesra.ac.in

J. R. Szymanski et al. (eds.), *Energy Systems, Drives and Automations*,
Lecture Notes in Electrical Engineering 1057,
https://doi.org/10.1007/978-981-99-3691-5_20

Gallium Nitride (GaN)-based HEMTs are most preferred for high-power applications and use at high frequency because they possess suitable material properties like high breakdown voltage, low effective mass, high thermal conductivity and high two-dimensional electron gas (2DEG) density of the order of 10^{13} cm^{-2} at the hetero structure face [12–16].

The gate contact in HEMT is a Schottky contact. This Schottky barrier is responsible for controlling and reducing the amount of gate leakage currents [17, 18]. Over time it was found out that by placing an extra layer of GaN on top of AlGaN/GaN structure, this Schottky barrier would increase leading to better control and reduction in gate leakage [19]. However, as the years passed, it was found out that the Schottky barrier is not effective in mitigating gate leakage current to a great extent especially as devices became smaller. Therefore, in recent years, research has gone into placing an additional insulating layer between the metal gate contact and semiconductor layers. In most cases, the insulators used are oxides, creating a metal-oxide semiconductor HEMT (MOSHEMT) structure. Oxide layers on GaN-based HEMTs are created through atomic layer deposition (ALD). ALD makes the fabrication process slightly complicated, but it provides greater freedom into control of the oxide layer that has been chosen. Al_2O_3 has a large band gap (about 8.7 eV), high breakdown field (5 ~ 20 MV/cm), high chemical and thermal stability (amorphous up to at least 1000 °C). Therefore, it is frequently selected as the insulating material beneath gate in MOSHEMT devices. Besides, Al_2O_3 offers ease of deposition [20, 21] although other such materials have been and continue to be explored. MOSHEMTs promise a low gate leakage current (almost zero with optimal oxide thickness) and are therefore strong contenders for low power applications. They have also been employed in bio-sensing applications [22].

There have been several articles discussing the fabrication of MOSHEMT. The addition of an oxide layer surely reduces the gate leakage current, but it also affects the other RF and DC parameters characterizing HEMT and MOSHEMT [23–26]. This paper explores if it is judicious to use MOSHEMT instead of HEMT at extremely high frequencies (EHF). The entire work has been divided into three parts. The first part is concerned with the device description where the simulated structures have been explained in detail. The equations related to important parameters of the device have also been discussed. The second part demonstrates the results that have been obtained because of the simulation performed. The third part draws conclusion on use of MOSHEMT for high frequency and high-power applications.

2 Device Description

2.1 *Structure of HEMT*

HEMT carries current in a two-dimensional electron gas instead of a thick channel. The 2DEG is formed at the heterojunction interface. This heterojunction is formed between a wide and a narrow bandgap material shown in Fig. 1, depicts cross-sectional view of the simulated HEMT device. 6H-SiC is used as substrate material for the proposed device due to its good thermal conductivity. There is a high lattice mismatch between silicon and III-Nitride semiconductors. Hence, on top of substrate is the nucleation layer or wetting layer made of Aluminum Nitride of thickness 40 nm. It accounts for any lattice mismatch between the substrate and the oncoming buffer layer performing what is called wetting of substrate. It also determines the polarity of the device. The buffer layer is made of Gallium Nitride (GaN). It is undoped and has a thickness of 2000 nm. The channel forms at the interface of the buffer and barrier layers. The spacer layer of thickness 2 nm is fabricated on top of buffer layer. It is made of AlGaN. It ensures minimal interaction between the electrons in the channel and the barrier layer. This in turn increases the mobility of the electrons. The barrier layer made of AlGaN is doped with n-type dopants. It has a thickness of 21 nm. Maximum polarization occurs in this layer. Conventional AlGaN/GaN-based HEMTs are generally depletion type because of the polarization charges. By keeping a certain thickness of the barrier layer, the device can be made to operate under a positive threshold voltage.

The next layer is the cap layer. It is made of Gallium Nitride and has a thickness of 3 nm. This layer is used to mitigate the leakage current, and its inclusion depends on the need and specifications of the user. On top of it is the passivation layer made of Silicon Nitride.

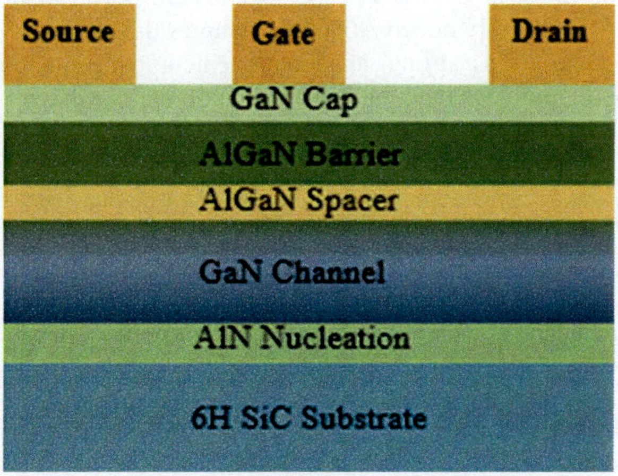

Fig. 1 The cross-section view of the simulated HEMT structure

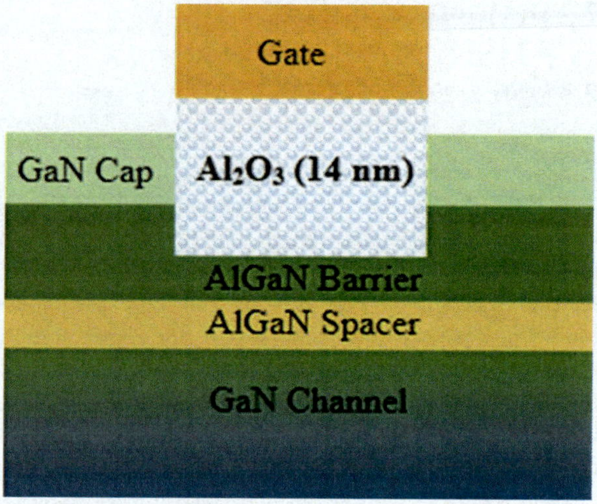

Fig. 2 MOSHEMT with oxide of thickness 14 nm

2.2 *Structure of MOSHEMT*

The layers in the structure of MOSHEMT are same as that of a HEMT structure except for the oxide layer that has been added beneath the gate. Several oxides like Al_2O_3, TiO_2, HfO_2, SiN, SiO_2 and NiO can be used. For the simulated structure Al_2O_3 has been used [25, 26].

Figure 1 shows the cross-section of the AlGaN/GaN-based HEMT structure that has been simulated. It has a gate of length 250 nm and a thickness of 294 nm. The source and drain are of length 800 nm each with a height/thickness of 500 nm each. The source to gate and gate to drain distances are 1400 nm and 4350 nm, respectively. The gate, source and drain are all of gold whose work function is 4.55 eV. Figure 2 shows the AlGaN/GaN-based MOSHEMT. The oxide used is Al_2O_3 of thickness 14 nm. Highly doped AlGaN constitutes the barrier layer which depletes the 2DEG and provides a strong carrier confinement in the quantum well at the hetero interface and minimizes junction leakage and off state leakage.

2.3 *Device Physics*

Unilateral power gain cut off frequency or maximum oscillation frequency is the frequency at which the transistor still provides a power gain. It can be expressed as

$$f_{\mathrm{MAX}} = \frac{f_{\mathrm{T}}}{2\sqrt{(R_{\mathrm{i}} R_{\mathrm{S}} R_{\mathrm{G}})/R_{\mathrm{DS}} + (2\pi f_{\mathrm{T}}) C_{\mathrm{GD}} R_{\mathrm{G}}}}, \tag{1}$$

where R_i is the gate charging resistance, R_S is termed as source resistance, R_G is the gate resistance and R_{DS} is drain to source resistance. f_T is defined as cut off frequency and C_{GD} is the gate drain depletion capacitance.

Cut off frequency or transit frequency is often represented as the inverse of the effective transit time of the carriers through the intrinsic region (region below the gate contact). It can be represented as

$$f_T = \eta v_{sat}/2\pi L_G, \tag{2}$$

where η is defined as modulation efficiency of charge carriers, v_{sat} is defined as the saturation velocity and L_G is defined as device gate length.

Ideally η is 1 for the HEMT but it is always less than 1 for the real HEMTs. We can increase the f_T by reducing the gate length of the HEMT without increasing the gate leakage current.

3 Simulation Results and Discussion

The two structures depicted in Figs. 1 and 2, are simulated using Silvaco and different DC and RF parameters and their trends have been investigated. Figure 3 depicts the I_{DS}–V_{GS} characteristics of both the structures. Figure 4 explores the transconductance extracted from the curve of drain current versus gate voltage by taking the first derivative of drain current and gate voltage. Figure 5 shows the variation of gate leakage current with gate voltage. Figures 6, 7 and 8 are a plot of the output characteristic of the device at different gate biases. Figures 9 and 10 show the trend in RF parameters namely unilateral power gain and Fig. 11 shows the trend in current gain. The threshold voltage (V_T) can be estimated from the I_{DS}–V_{GS} curve (input characteristic) and they are found to be $V_T = -3.9189$ V for HEMT and $V_T = -3.6582$ V for MOSHEMT with oxide thickness 14 nm.

Fig. 3 Drain current versus gate bias for proposed HEMT and MOSHEMT

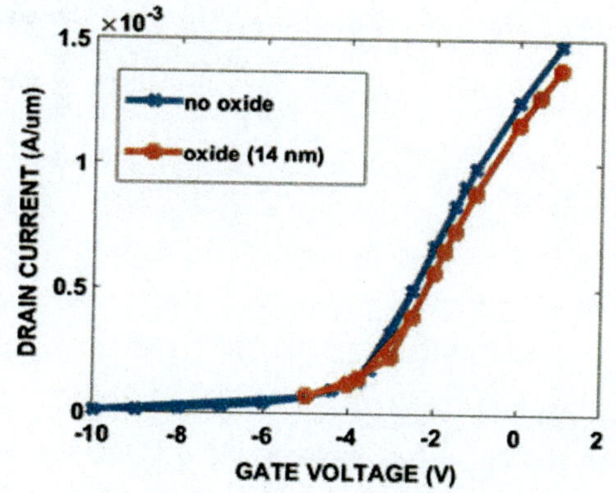

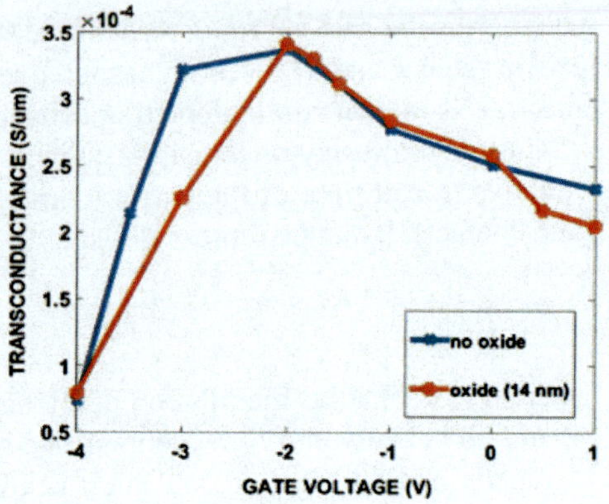

Fig. 4 Transconductance versus gate voltage for proposed HEMT and MOSHEMT

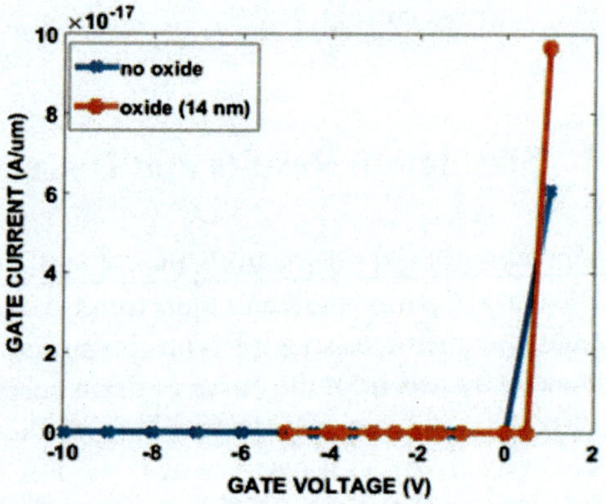

Fig. 5 Gate leakage current versus gate bias for HEMT and MOSHEMT

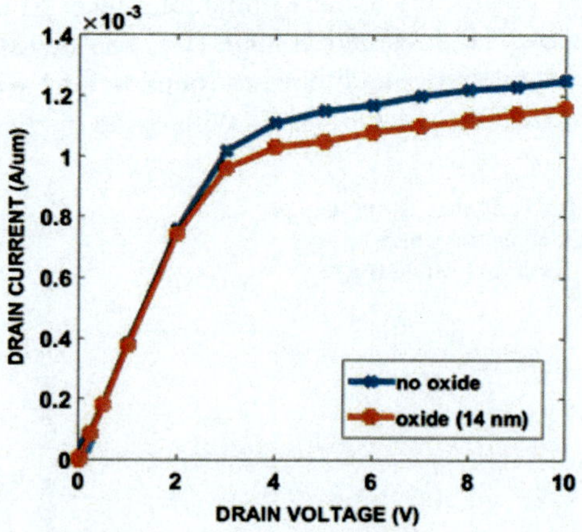

Fig. 6 Drain current versus drain voltage for a gate bias of 0 V

Fig. 7 Drain current versus drain voltage for a gate bias of − 0.2 V

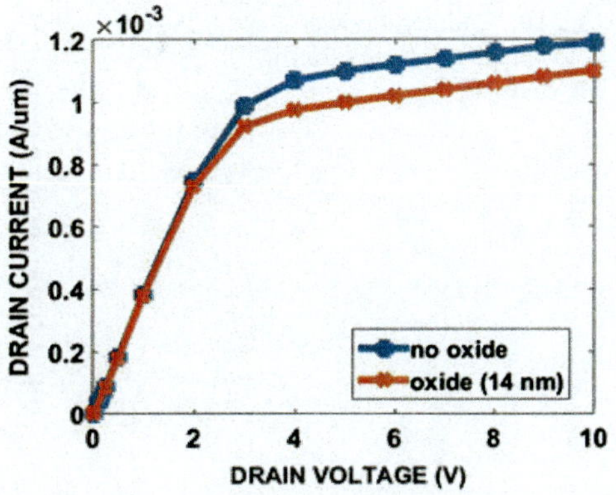

Fig. 8 Drain current versus drain voltage for a gate bias of 1 V

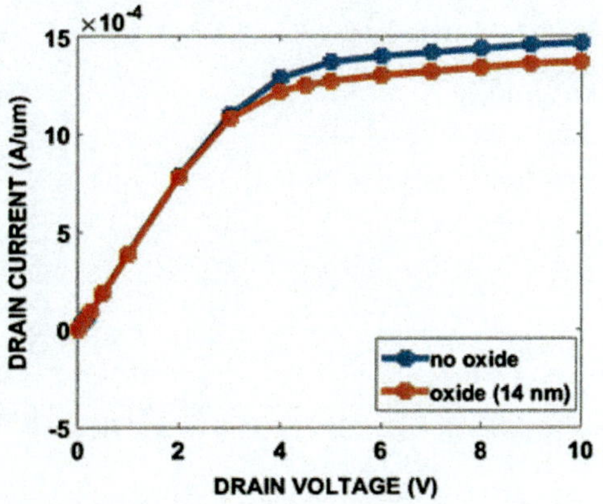

Fig. 9 Unilateral power gain versus frequency for the HEMT structure

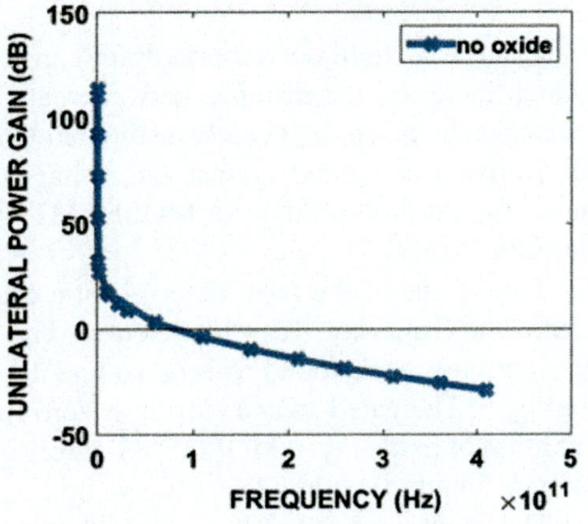

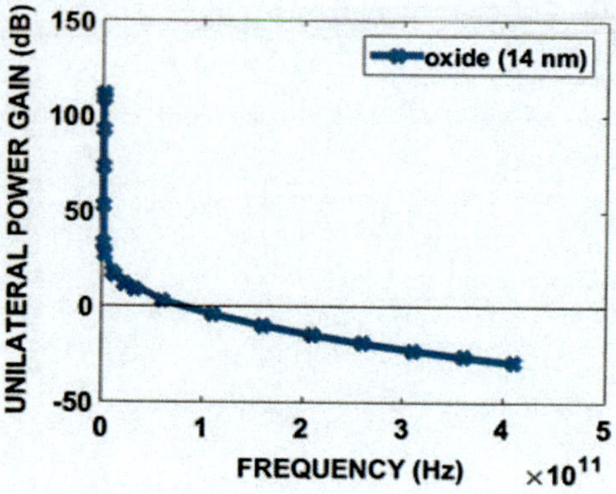

Fig. 10 Unilateral power gain versus frequency for the MOSHEMT

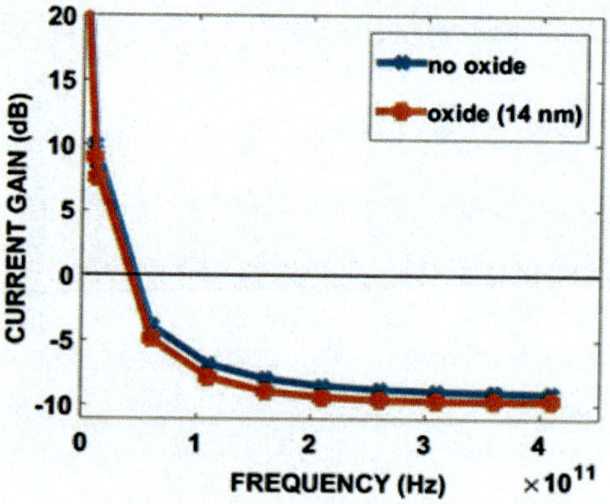

Fig. 11 Current gain versus frequency for the HEMT and MOSHEMT

There is a slight decrease in drain current due to addition of oxide beneath gate which increases the distance between gate and channel which is shown in Fig. 3. Transconductance (g_m) can be estimated from the I_{DS}–V_{GS} curve by taking its first derivative. It is plotted against gate voltage, and it is shown in Fig. 4. Transconductance (g_m) is 349 mS/mm for MOSHEMT with oxide thickness 14 nm and 345 mS/mm for HEMT.

This is one of the most deterministic characteristics which explains up to what depth the oxide layer can be inserted. The minimum and maximum gate leakage current have been found to be zero and 10×10^{-17} A/um, respectively, as shown in Fig. 5. The gate leakage current is zero for a greater range of gate voltage due to addition of oxide layer. MOSHEMT surely gains over HEMT at this miniaturization of gate length providing less leakage up to a greater positive gate voltage.

The drain current has been plotted against drain voltage for a gate bias of 0 V which is shown in Fig. 6. The minimum and maximum drain current has been found to be zero and 1.25×10^{-3} A/um, respectively. The drain current decreases by the usage of oxide layer. This happens because AlGaN above the channel is substituted by the oxide due to very thin spacer. This reduces the amount of AlGaN above the channel thereby decreasing the concentration of 2DEG which causes a decrease in the drain to source current.

To substantiate the point, I_{DS}–V_{DS} is plotted for negative and positive gate biases as well and shown in Fig. 7. For reasons mentioned above the trend is same as obtained for zero gate bias. The minimum and maximum drain current has been found to be zero and 1.2×10^{-3} A/um, respectively, for a gate bias of – 0.2 V. The graph shown in Fig. 8 depicts the output characteristics of the two structures for a positive gate bias. The minimum and maximum current for a positive gate bias of 1 V have been found to be zero and ~ 15×10^{-4} A/um, respectively.

The graph shown in Fig. 9 the variation of unilateral power gain with frequency for the HEMT structure. Unilateral power gain is an important RF parameter. As the frequency increases unilateral power gain decreases. The maximum oscillation frequency is found to be about 82 GHz which lies in the W band. The graph shown in Fig. 10 depicts the variation of unilateral power gain with frequency for the MOSHEMT structure.

The maximum oscillation frequency for the case of MOSHEMT is also found to be around 80 GHz. The current gain drops in case of MOSHEMT because of the addition of oxide layer which increases the distance between gate and channel shown in Fig. 11. The cut off frequency for HEMT and MOSHEMT have been found to be around 46 and 42 GHz. For further reduction in gate length the cut off frequency is expected to rise but leakage current is also expected to soar. Comparative results for HEMT and MOSHEMT are tabulated in Table 1.

The HEMT and MOSHEMT structures are compared against the following parameters, namely the cut off frequency, the maximum oscillation frequency, the oxide thickness and gate voltage up to which no current leakage has been found. As expected, it is in the case of MOSHEMT that the leakage current is zero up to 0.5 V on the gate contact while for HEMT leakage occurs right after 0 V. There is decrease in the cut off frequency because the distance of the gate contact from the channel has increased due to the addition of oxide layer. The oscillation frequency has decreased due to decrease in the cut off frequency.

Table 1 Comparison of HEMT and MOSHEMT

Parameters	HEMT	MOSHEMT
f_T (GHz)	46	42
f_{MAX} (GHz)	82	80
Oxide thickness (nm)	0	14
Gate voltage (up to which zero leakage occur) (V)	0	0.5

4 Conclusion

It can be concluded that the addition of oxide certainly improves device reliability by reducing the gate leakage current but at the same time the thickness must be kept such that it does decrease the drain current and affect other essential device parameters. Hence, the oxide layer should be placed well above the barrier layer. At high frequencies, MOSHEMT also provides a decent current gain and power gain. Hence, it can be used in place of HEMT for high frequency applications.

References

1. Mizutani T, Ohno Y, Akita M, Kishimoto S, Maezawa K (2003) A study on current collapse in AlGaN/GaN HEMTs induced by bias stress. IEEE Trans Electron Devices 50(10):2015–2020. https://doi.org/10.1109/TED.2003.816549
2. Mimura T (2005) Development of high electron mobility transistor. Jpn J Appl Phys 44(12):8263–8268. https://doi.org/10.1143/jjap.44.8263
3. Dubey SK, Sinha K, Sahu PK et al (2020) Characterization of InP-based pseudomorphic HEMT with T-gate. Microsyst Technol 26:2183–2191. https://doi.org/10.1007/s00542-019-04491-3
4. Sinha K, Dubey SK, Islam A (2020) Study of high Al fraction in AlGaN barrier HEMT and GaN and InGaN channel HEMT with $In_{0.17}Al_{0.83}N$ barrier. Microsyst Technol 26:2145–2158. https://doi.org/10.1007/s00542-019-04466-4
5. Endoh A, Yamashita Y, Hikosaka K, Matsui T, Hiyamizu S, Mimura T (2006) Threshold voltage shifts in decananometre-gate AlGaN/GaN HEMTs. Electron Lett 42(8):490–492. https://doi.org/10.1049/el:20060513
6. Dubey SK, Islam A (2020) Study and analysis of AlInN/GaN based high electron mobility transistor. In: Lecture notes in electrical engineering (LNEE), vol 664. Springer Nature, Singapore, pp 449–459. https://doi.org/10.1007/978-981-15-5089-8_44
7. Dubey SK, Islam A (2020) Indium phosphide based dual gate high electron mobility transistor. In: Lecture notes in electrical engineering (LNEE), vol 664. Springer Nature, Singapore, pp 255–264. https://doi.org/10.1007/978-981-15-5089-8_24
8. Dubey SK, Islam A (2020) Effect of source, drain and channel spacing from gate of HEMT. In: Lecture notes in electrical engineering (LNEE), vol 664. Springer Nature, Singapore, pp 81–90. https://doi.org/10.1007/978-981-15-5089-8_8
9. Gupta A, Chatterjee N, Tripathy MR, Pandey S (2016) Design and simulation of GaN HEMT and its application to RF amplifiers. In: 2016 progress in electromagnetic research symposium (PIERS). IEEE, pp 3815–3819. https://doi.org/10.1109/piers.2016.7735432
10. Li H, Yao C, Fu L, Zhang X, Wang J (2016) Evaluations and applications of GaN HEMTs for power electronics. In: 2016 IEEE 8th international power electronics and motion control conference (IPEMC-ECCE Asia). IEEE, Heifei, pp 563–569. https://doi.org/10.1109/IPEMC.2016.7512348
11. Joshin K, Kikkawa T (2006) Recent progress of high power GaN-HEMT for wireless application. In: 2006 Asia-Pacific microwave conference. IEEE, Yokohama, pp 1027–1032. https://doi.org/10.1109/APMC.2006.4429585
12. Kikkawa T (2004) Recent progress and future prospects of GaN HEMTs for base-station applications. In: IEEE compound semiconductor integrated circuit symposium. IEEE, Monterey, CA, pp 17–20. https://doi.org/10.1109/CSICS.2004.1392472
13. Dubey SK, Mishra M, Islam A (2022) Characterization of AlGaN/GaN based HEMT for low noise and high frequency application. Int J Numer Model Electron Networks Devices Fields 35(1):e2932. https://doi.org/10.1002/jnm.2932

14. Dubey SK, Islam A (2021) AlGaN/GaN HEMT with recessed T-gate and floating metal for high power applications. In: 2021 devices for integrated circuit (DevIC). IEEE, Kalyani, West Bengal, pp 216–220. https://doi.org/10.1109/DevIC50843.2021.9455871
15. Dubey SK, Islam A (2022) $Al_{0.30}Ga_{0.70}N$/GaN MODFET with triple-teeth metal for RF and high-power applications. Phys Scr 97(3):034003. https://doi.org/10.1088/1402-4896/ac50c3
16. Khan MA, Simin G, Pytel SG, Monti A, Santi E, Hudgins JL (2005) New developments in gallium nitride and the impact on power electronics. In: IEEE 36th conference on power electronics specialists. IEEE, Dresden, pp 15–26. https://doi.org/10.1109/PESC.2005.1581596
17. Longobardi G, Udrea F, Sque S, Croon J, Hurkx F, Napoli E, Sonsky J (2012) Modelling 2DEG charges in AlGaN/GaN heterostructures. In: CAS 2012 (international semiconductor conference). IEEE, Sinaia, pp 363–366. https://doi.org/10.1109/smicnd.2012.6400760
18. Lenka TR, Panda AK (2011) Characteristics study of 2DEG transport properties of AlGaN/GaN and AlGaAs/GaAs-based HEMT. Semiconductors 45:650–656. https://doi.org/10.1134/S1063782611050198
19. Yu ET, Dang XZ, Yu LS, Qiao D, Asbeck PM, Lau SS, Sullivan GJ, Boutros KS, Redwing JM (1998) Schottky barrier engineering in III–V nitrides via the piezoelectric effect. Appl Phys Lett 73(13):1880–1882. https://doi.org/10.1063/1.122312
20. Eller BS, Yang J, Nemanich RJ (2013) Electronic surface and dielectric interface states on GaN and AlGaN. J Vac Sci Technol A Vac Surf Films 31(5):050807. https://doi.org/10.1116/1.4807904
21. Esposto M, Krishnamoorthy S, Nath DN, Bajaj S, Hung T-H, Rajan S (2011) Electrical properties of atomic layer deposited aluminium oxide on gallium nitride. Appl Phys Lett 99:133503. https://doi.org/10.1063/1.3645616
22. Pal P, Pratap Y, Gupta M, Kabra S (2018) Modeling and simulation of AlGaN/GaN MOS-HEMT for biosensor applications. IEEE Sens J 19(2):587–593. https://doi.org/10.1109/jsen.2018.2878243
23. Alomari M, Medjdoub F, Carlin J-F, Feltin E, Grandjean N, Chuvilin A, Kaiser U, Gaquiere C, Kohn E (2009) InAlN/GaN MOSHEMT with self-aligned thermally generated oxide recess. IEEE Electron Device Lett 30(11):1131–1133. https://doi.org/10.1109/led.2009.2031659
24. Swain R, Jena K, Lenka TR (2016) Modeling of forward gate leakage current in MOSHEMT using trap-assisted tunneling and Poole–Frenkel emission. IEEE Trans Electron Devices 63(6):2346–2352. https://doi.org/10.1109/ted.2016.2555851
25. Basu S, Singh PK, Lin S-K, Sze P-W, Wang Y-H (2010) Effects of short-term DC-bias-induced stress on n-GaN/AlGaN/GaN MOSHEMTs with liquid-phase-deposited Al_2O_3 as a gate dielectric. IEEE Trans Electron Devices 57(11):2978–2987. https://doi.org/10.1109/TED.2010.2071130
26. Liu H-Y, Lin C-W, Hsu W-C, Lee C-S, Chiang M-H, Sun W-C, Wei S-Y, Yu S-M (2017) Integration of gate recessing and in situ Cl−doped Al_2O_3 for enhancement-mode AlGaN/GaN MOSHEMTs fabrication. IEEE Electron Device Lett 38(1):91–94. https://doi.org/10.1109/led.2016.2625304

Implementation and Analysis of CNFET-Based PCRAM Cell Using 32 nm Technology

Vaibhav Gupta, Atharv Kapre, Shashank Kumar Dubey, and Aminul Islam

Abstract This paper demonstrates carbon nanotube field effect transistor (CNFET)-based Phase-Change Random Access Memory (PCRAM) cell. In this design, we have used two CNFETs for the fast switching of resistance in PCRAM cell. The use of CNFET offers far superior electrical characteristics than traditional MOS devices. This paves the way for such major improvements in this design metrics of the proposed cell. By giving high short duration current pulse to PCRAM, we achieved improved read/write access times. Furthermore, the proposed cell consumes negligible hold power.

Keywords Nonvolatile memory (NVM) · Germanium antimony terrarium (GeSbTe) · Phase-Change Random Access Memory (PCRAM) · Carbon nanotube field effect transistor (CNFET)

1 Introduction

CMOS is reaching its scaling limit as there are short channel effects which deteriorate the performance. Taking these things into account the industry needs to find new materials and novel approach to have nonvolatile memory (NVM) developed. Quest in this direction has provided new technologies which include Phase-Change Random Access Memory (PCRAM), Ferroelectric Random Access Memory (FRAM), Magnetic Random Access Memory (MRAM) [1, 2], Resistive Random

---

V. Gupta · A. Kapre · S. K. Dubey · A. Islam (✉)
Department of ECE, Birla Institute of Technology, Mesra, Ranchi 835215, India
e-mail: aminulislam@bitmesra.ac.in

V. Gupta
e-mail: btech10490.18@bitmesra.ac.in

A. Kapre
e-mail: mtec10002.20@bitmesra.ac.in

J. R. Szymanski et al. (eds.), *Energy Systems, Drives and Automations*,
Lecture Notes in Electrical Engineering 1057,
https://doi.org/10.1007/978-981-99-3691-5_21

Access Memory (RRAM) [3, 4], and these provide fast switching speed, and they are expected to be the future memory technologies.

PCRAM has high scalability, fast access time, low power good data retention and low cost. The critical characteristics of PCRAM is its switching property as it switches from amorphous to crystalline states by using electrical pulses. Energy required for operating memory decreases with decrease in cell size. This decreases the writing current with scaling down in size. Data transfer rate and performance of write cycle are the most important criteria. In this paper, we focus on write and read operations, Phase-Change materials, and the solutions to achieve the high speed [5, 6].

As a result of their relatively small dimensions and ability to be applied to many existing and emerging technologies, carbon nanotubes (CNTs) have recently gained significant interest. By controlling the CNT diameter, a CNFET can be controlled to have a different threshold voltage [7]. Since carbon nanotube field effect transistors (CNTFET) are emerging as a viable alternative to silicon transistors by combining low power consumption and high performance, CNFETs have been used to implement the proposed memory.

The article is organized as mentioned below. Section 2 includes overview of Phase-Change memory. Section 3 briefly outlines the CNFET. Section 4 discusses the proposed PCRAM cell and discussion of simulation results. Section 5 concludes this paper.

2 Phase-Change Memory (PCRAM)

Phase-Change materials have a rapid reversible Phase-Change effect. A Phase-Change material usually has two states, a crystalline state with lower resistance and an amorphous state with higher resistance. $Ge_2Sb_2Te_5$ (GST) alloy and $Si_{12}Te_{48}As_{30}Ge_{10}$ (STAG) alloy are the most used Phase-Change materials, which are also used in optical rewriteable disks.

High and low resistances are measured and recorded as data "0" and "1." For switching from disordered or amorphous or high-resistance state to ordered or crystalline or low-resistance state, a PCRAM device needs a short-height long-width SET current pulse, which heats up the amorphous GST or STAG alloy material to a high temperature (< melting temperature), which crystallizes GST or STAG alloy. A PCRAM device needs a tall-height short-width RESET current pulse, which heats up the crystalline GST or STAG alloy material to a high temperature beyond its melting point for switching from ordered or crystalline or low-resistance state to disordered or amorphous or high-resistance state. Due to the short current pulse width the GST or STAG alloy cools down rapidly freezing its atomic structure in a disordered state. The transformation of PCRAM device material (GST or STAG alloy) from crystalline to amorphous and back has been illustrated in Fig. 1. As can be seen from Fig. 1, the duration of the set current pulse is longer than the reset current pulse; however, the height of RESET current pulse is taller than that of SET current pulse. Since phase

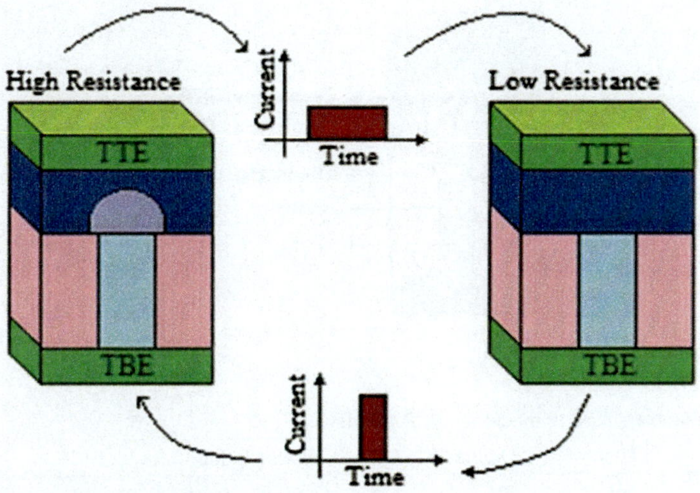

Fig. 1 PCRAM device material (GST or STAG alloy) is sandwiched between to titanium nitride (TiN) or tantalum nitride (TaN) electrodes (TBE and TTE)

transformation is not required for read operation much lower current with less joule heating is needed for reading the cell. A typical PCRAM cell is designed in such a way that the volume of the Phase-Change material is melted and quenched to the amorphous region, so that it completely blocks the current path through the device and allows minimum current to pass [6].

The materials of PCRAM device shown in Fig. 2 are as follows. The material of top and bottom electrodes can be nitride of titanium or *tantalum* [titanium nitride (TiN), *tantalum nitride* (TaN), or alloy of *titanium-tungsten* (TiW)], which can be formed by CVD or sputtering. The material of the isolation layer, which is thicker than bottom and top electrodes, may be borophosphosilicate glass (BPSG), silicon oxide, or silicon nitride, which serves as thermal insulator. The Phase-Change material can be a chalcogenide material composed of In, Ge, Sb, Te, or combinations thereof, such as GeSbTe or InGeSbTe. The thickness of the Phase-Change material can be 20–100 nm. However, 50 nm is preferable [8]. The material for heating element is preferably SiTiNx.

Parameters used for modeling the PCRAM device shown in Fig. 2 are given in Table 1. The SPICE equivalent circuit-level model for is shown in Fig. 3. Basically, it works on four different parts: input energy conversion, temperature transition, Phase-Change, and the geometry/structure/material.

The energy conservation equation, which will further used to calculate the electrical and thermal parameters, where ΔT is the temperature difference between the two electrodes, R_T and C_T are the thermal resistance and thermal capacitance, respectively. A switch is triggered when the temperature reaches melting temperature, and the C_{state} is charged by the voltage source. The equivalent circuit-level model parameters for SPICE simulation are tabulated in Table 2. According to the input energy

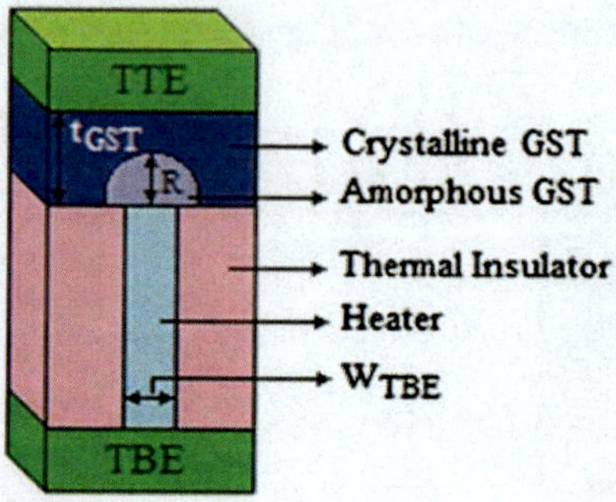

Fig. 2 PCRAM device

Table 1 Parameters used to model the PCRAM device

Symbol	Unit	Values	Description
W_{TBE}	nm	28	Width of bottom electrode contact
t_{GST}	nm	44	Thickness of GST
r_0	Ω	700	Resistance of PCRAM
T_m	°C	570	Melting temperature
T_a	°C	270	Crystalline temperature
E_a	J	8.0×10^{-20}	Crystalline activation region
R_{g0}	Ω	0.02	Resistance of crystalline time
C_{state}	µf	1.0	Capacitance
Heater	Ω	9.30×10^6	Thermal resistance
R	nm	15	Initial amorphous radius

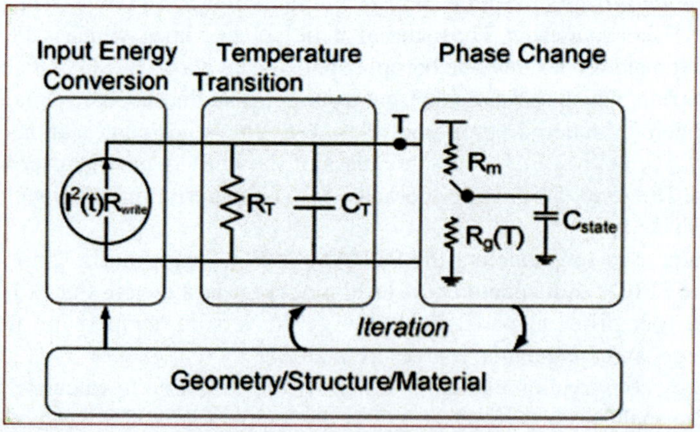

Fig. 3 Equivalent circuit-level model for SPICE simulation [9]

Table 2 Parameters used in circuit-level model for SPICE simulation

Parameters	Description
I	Input current
R_{write}	Resistance of PCRAM during write
R_{T}	Thermal resistance of GST
C_{T}	Thermal capacitance of GST
C_{state}	Storage state of memory cell
$R_{\text{g}}(T)$	Crystallization resistance of GST

block, R_{write} is primarily the resistance of the metallic heater, while I^2R_{write} is the input power. The energy conservation equation for PCRAM is given by

$$\int \left(I^2R_{\text{write}} - \Delta T/R_{\text{T}}\right) = C_{\text{T}}\Delta T. \tag{1}$$

In this equation, R_{write} is the resistance of the metallic heater. I^2R_{write} is the input power in the input energy conversion block. I is the input current passing through the memory device, the ΔT is the temperature difference between the upper electrode and the GST-heater interface, R_{T} is the thermal resistance and C_{T} is the thermal capacitance in the temperature transition block.

The temperature dependent time constant $\tau(T)$ is given by

$$\tau(T) = R_{\text{g}}(T)C_{\text{state}} = T_{\text{m}}C_{\text{state}} \quad \text{for } T > T_{\text{m}} \tag{2}$$

$$\tau(T) = R_{\text{g}}(T)C_{\text{state}} = \text{e}^{E_{\text{A}}/K_{\text{B}}T} \quad \text{for } T_{\text{a}} < T < T_{\text{m}}. \tag{3}$$

In Eq. (2), $R_{\text{g}}(T)$ is the crystallization resistance of GST, C_{state} is the storage state of memory cell, T_{m} is the melting temperature of GST, E_{A} is the active energy during crystallization, and K_{B} is the Boltzmann constant, T is the temperature [9].

3 Carbon Nanotube Field Effect Transistor

Our proposed PCRAM design uses CNFET as shown Fig. 4. We have adopted a semi-empirical model that describes the I–V characteristics in a small carbon nanotube channel material. Since CNFET's behavior is like that of MOSFET, the threshold voltage of the device shown in Fig. 4 is given by

$$V_{\text{t}} = V_{\text{t0}} - \delta V'_{\text{DS}}, \tag{4}$$

where V_{t0} is the threshold voltage at $V_{DS} = 0$ V and in its strong-inversion condition, δ is the is the drain-induced barrier lowering (DIBL) coefficient, and V'_{DS} is the drain-to-source voltage after subtracting the voltage drops in source and drain resistances and is given by $V'_{DS} = V_{DS} - I_D (R_S + R_D)$.

A virtual source concept is used in calculating the drain current. A CNFET has various parameters such as gate length (L_g), contact length (L_c), diameter (d), and gate oxide thickness (t_{ox}), etc., which have been tabulated in Table 3.

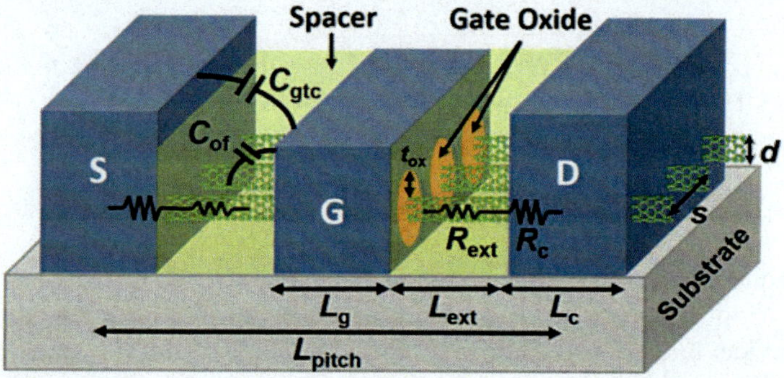

Fig. 4 Structure of CNFET with three CNTs used in a single CNFET for enhancing driving strength of device [10]

Table 3 CNFET with values for simulation

Input	Values	Description
CNFET	1, − 1	1 = NCNFET, − 1 = PCNFET
s	10 nm	Spacing between CNT's
W	1 µm	Width of transistors
L_g	11.7 nm	Gate length
L_c	12.9 nm	Contact length
L_{ext}	3.2 nm	Source and drain length
t_{ox}	3 nm	Gate oxide thickness
k_{ox}	23	Gate oxide dielectric
d	1.2 nm	CNT diameter
Temp	25 °C	Temperature
L_{pitch}	31 nm	Gate pitch

4 Proposed PCRAM Cell

The proposed PCRAM model has two CNFETs and one Phase-Change memory element as shown in Fig. 5. The description of all the symbols used for the proposed PCRAM cell are tabulated in Table 4.

When the radius of mushroom is large the GST material exhibits higher resistance than when the radius of mushroom is less. The crystallization time constant can be given as τ, from Eq. (2).

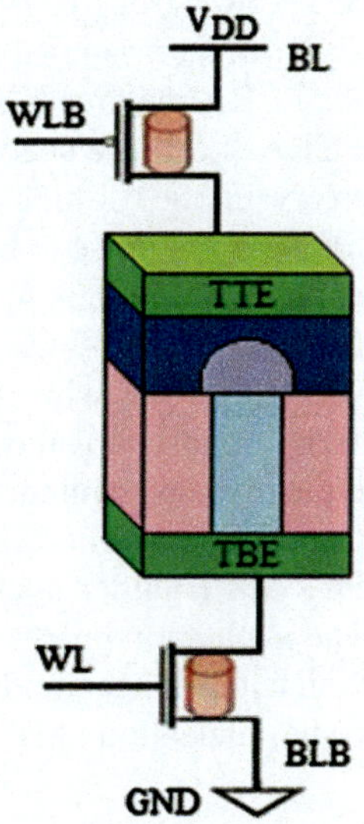

Fig. 5 Proposed PCRAM circuit design

Table 4 Various symbols used for the proposed PCRAM cell

Symbol	Description
W	Transistor's width
L	Transistors gate length
V_{DD}	Supply voltage
V_g	Word line (WL) voltage pulse amplitude
t	Word line (WL) voltage pulse duration
BL	Bit line
BLB	Bit line bar

4.1 Write Operation

It is assumed that the PCRAM cell is initially storing "1" (which means that the PCRAM element is in crystalline or low-resistance state). Now to write a "0" into the PCRAM cell, we need to change the state of PCRAM element from crystalline to amorphous (or low-to-high resistance) state. This needs a tall-height short-width RESET current pulse to be passed through the PCRAM element. For doing so we have applied WL = 0.38 V for a duration of 3 ns only. This could change the GST resistance from 9 kΩ to 3 MΩ as can be seen from Fig. 6. The voltage pulse applied at WL was from 10 ns while performing simulation (not from 0 ns).

Now, it is assumed that the PCRAM cell is initially storing "0" (which means that the PCRAM element is in amorphous or high-resistance state). To write a "1" into the PCRAM cell, we need now to change the state of PCRAM element from amorphous to crystalline (or high-to-low resistance) state. This needs a short-height long-width SET current pulse to be passed through the PCRAM element. For doing so we have applied WL = 0.34 V for a duration of 5 ns. This could change the GST resistance from 1.9 MΩ to 9 kΩ as can be seen from Fig. 7. The voltage pulse applied at WL was from 10 ns while performing simulation (not from 0 ns).

As opposed to amorphization, crystallization generally takes a much longer time, typically tens to hundreds of nanoseconds, and it often occurs at higher temperatures than ~ 700 K but below melting temperature (T_{melt}). The crystalline kinetics of PCRAM is either nucleation or growth oriented at elevated temperatures [11, 12]. The nucleation process occurs when crystalline growth reaches a critical point after which it no longer holds its stability and may become unstable. GST material provides faster nucleation [13].

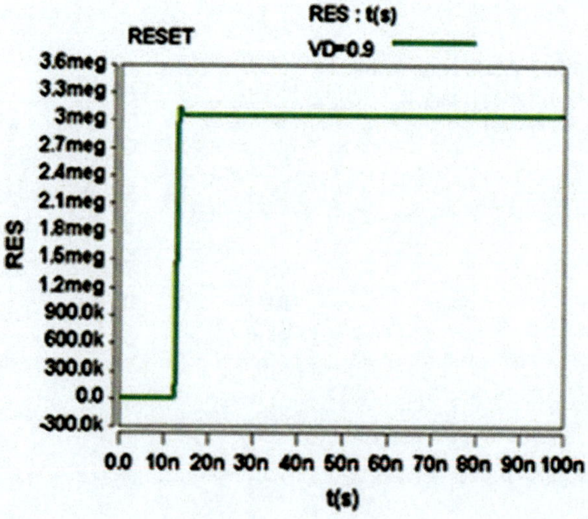

Fig. 6 Change in GST resistance from 9 kΩ to 3 MΩ while writing "0" into the PCRAM cell, which was previously storing "1"

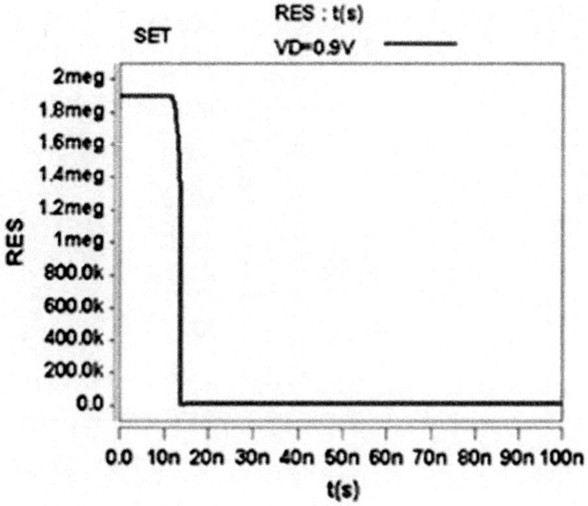

Fig. 7 Change in GST resistance from 1.9 MΩ to 9 kΩ while writing "1" into the PCRAM cell, which was previously storing "0"

The delay for writing "1" into the PCRAM cell is estimated to be 35 ns. This write delay is estimated by observing the time (or x) axis. The time difference is noted when PCRAM temperature falls to 0 K after getting heated (the difference between room temperature and PCRAM temperature becomes 0 at 45 ns of the x-axis) as shown in Fig. 8. Hence, the write delay is 35 ns because the reference time taken for the following simulations is 10 ns. The delay for writing "0" into the PCRAM cell is estimated to be 10.42 ns. This write delay is estimated by observing the time (or x) axis. The time difference is noted when PCRAM temperature falls to 0 K after getting heated (the difference between room temperature and PCRAM temperature becomes 0 at 20.42 ns of the x-axis) as shown in Fig. 9. Hence, getting a write delay of 10.42 ns because the reference time taken for the following simulations is 10 ns.

4.2 Read Operation

For performing read operation, we applied a square pulse of height 0.27 V for 5 ns at WL (WLB is complement of WL) and a ramping down (from 1 to 0 V) V_{DD} at BL. For read operation, one reference PCRAM element along with two CNFETs are used. The reference PCRAM element is initialized with a resistance value of 200 kΩ, and a sense amplifier is connected at the bottom electrode of the PCRAM elements (see Fig. 10). Since WL (WLB is complement of WL) applied to both sides, the FETs are driven with equal drive strength. However, the current through the PCRAM element will be different because of difference is resistance. If the PCRAM cell element is already storing a "1," then its resistance is low (~ 9 kΩ), which will result in higher current through it. This will result in higher voltage developed at bottom electrode (BE) of PCRAM element compared to the bottom electrode (BE)

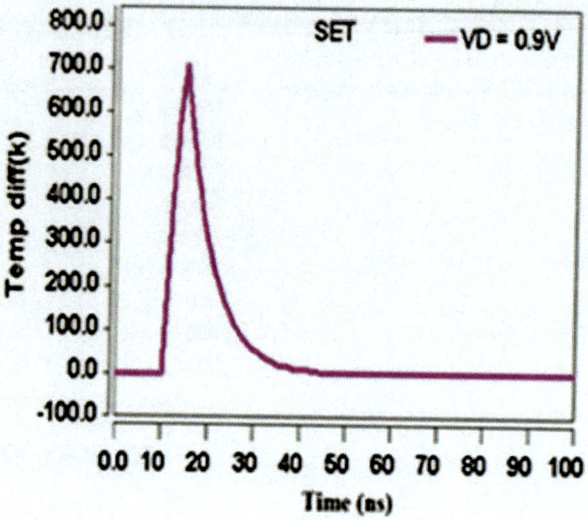

Fig. 8 Delay for writing "1" into the PCRAM cell is estimated to be 35 ns

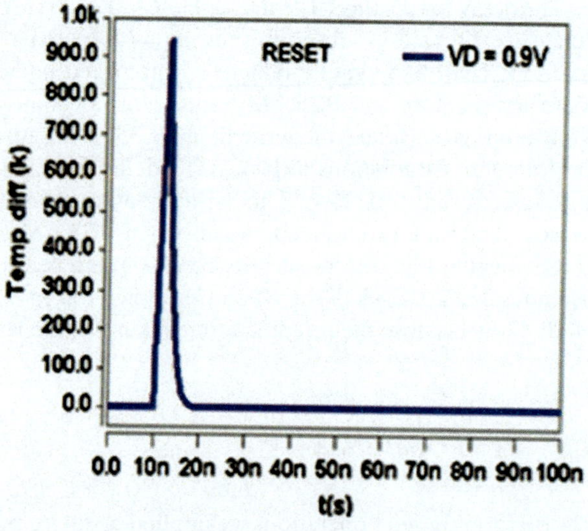

Fig. 9 Delay for writing "0" into the PCRAM cell is estimated to be 10.42 ns

of reference PCRAM element as can be seen from Fig. 11. The potential difference between the BE of the PCRAM element becomes higher than 50 mV, which can be sensed by the sense amplifier only after 11 ns. Therefore, read delay in this case is 1 ns. The reference time taken for the simulations is 10 ns.

If the PCRAM cell element is already storing a "0," then its resistance is high (~ 3 MΩ), which will result in lower current through it. This will result in lower voltage developed at bottom electrode (BE) of PCRAM element compared to the bottom electrode (BE) of reference PCRAM element as can be seen from Fig. 12. The potential difference between the BE of the PCRAM element becomes higher

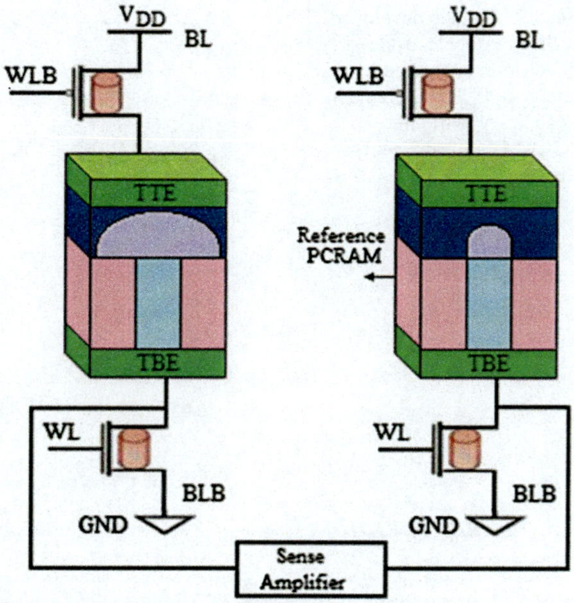

Fig. 10 Read operation for PCRAM cell

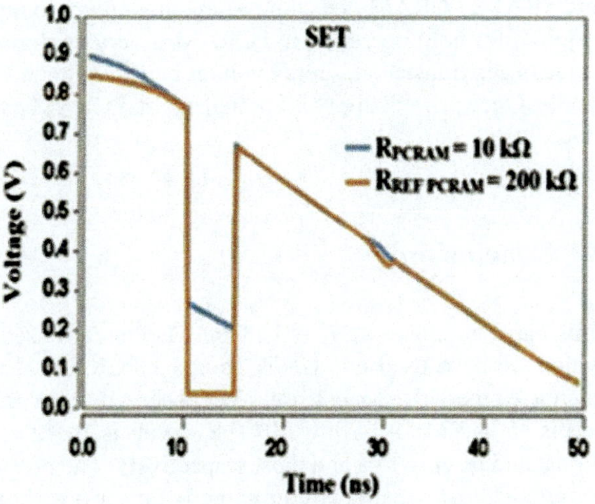

Fig. 11 Voltage developed at the bottom electrode of PCRAM element and reference PCRAM element. Read delay while the PCRAM cell is currently storing (SET) "1" is 1 ns

than 50 mV, which can be sensed by the sense amplifier only after 17 ns. Therefore, read delay in this case is 7 ns. The reference time taken for the simulations is 10 ns.

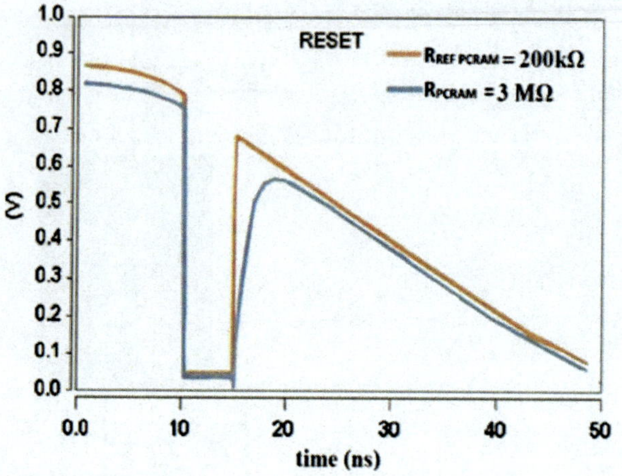

Fig. 12 Voltage developed at the bottom electrode of PCRAM element and reference PCRAM element. Read delay while the PCRAM cell is currently storing (RESET) "0" is 7 ns

4.3 Hold Operation

PCRAM stores data in the form of resistance (Amorphous → high resistance → 0 and crystalline low resistance → 1). In comparison with other nonvolatile memory and SRAM, PCRAM does not require any power to retain its value. So, hold power required for hold operation in PCRAM is very negligible. Using power gating technique (sleep transistor), supply voltage can be switched OFF to save hold power drastically. Once power supply is applied again, PCRAM can wake up with its previously stored data.

5 Conclusion

This paper proposes a $2T - 1$ PCRAM element-based nonvolatile memory cell, in which we have used two CNFETs and 1 PCRAM element. In this work, we have decreased the write access time. We observed write access time for wring "1" (i.e., set) is 35 ns and for writing "0" (i.e., reset) is 10.42 ns. Similarly, read access time for set and reset is 1 ns and 6 ns, respectively. The proposed PCRAM cell is suitable for applications where extreme standby power (i.e., hold power) saving is required.

References

1. Saha S, Dubey SK, Banerjee S, Pal I, Islam A (2018) Nonvolatile write driver for spin transfer torque memory and logic design. In: Mandal J, Sinha D (eds) Social transformation—digital way. CSI 2018. Communications in computer and information science, vol 836. Springer, Singapore, pp 156–166. https://doi.org/10.1007/978-981-13-1343-1_17

2. Gupta N, Thakur P, Dubey SK, Islam A (2017) Design of nonvolatile MRAM bitcell. In: 2017 7th international symposium on embedded computing and system design (ISED). IEEE, pp 1–4. https://doi.org/10.1109/ised.2017.8303917
3. Dubey SK, Reddy A, Patel R, Abz M, Srinivasulu A, Islam A (2020) Architecture of resistive RAM with write driver. Solid State Electron Lett 2:10–22. https://doi.org/10.1016/j.ssel.2020.01.001
4. Dubey SK, Islam A (2020) Design of resistive random access memory cell and its architecture. Microsyst Technol 26:1325–1332. https://doi.org/10.1007/s00542-019-04663-1
5. Shi L, Zhao R, Chong TC (2011) Phase change random access memory. In: Developments in data storage. Wiley-IEEE Press, pp 277–296. https://doi.org/10.1002/9781118096833.ch13
6. Chen X, Hu H, Huang X, Cai W, Liu M, Lam C, Lin X, Zhang L, Chan M (2020) A SPICE model of phase change memory for neuromorphic circuits. IEEE Access 8:95278–95287. https://doi.org/10.1109/ACCESS.2020.2995907
7. Obite F, Ijeomah G, Bassi JS (2018) Carbon nanotube field effect transistors: toward future nanoscale electronics. Int J Comput Appl 41(2):149–164. https://doi.org/10.1080/1206212x.2017.141511
8. Chen Y-C, Wang W-H (2010) Method of fabricating a phase-change memory. Patent No. US 7670871B2, 2 Mar 2010
9. Xu Z, Sutaria KB, Yang C, Chakrabarti C, Cao Y (2012) Hierarchical modeling of phase change memory for reliable design. In: 2012 IEEE 30th international conference on computer design (ICCD). IEEE, Montreal, QC, pp 115–120. https://doi.org/10.1109/iccd.2012.6378626
10. Lee C-S, Pop E, Franklin AD, Haensch W, Wong H-SP (2015) A compact virtual-source model for carbon nanotube FETs in the sub-10-nm regime—part I: intrinsic elements. IEEE Trans Electron Devices 62(9):3061–3069. https://doi.org/10.1109/ted.2015.2457453
11. Kwong KC, He J, Mok PKT, Chan M (2011) Phase-change memory RESET model based on detailed cell cooling profile. IEEE Trans Electron Devices 58(10):3635–3638. https://doi.org/10.1109/ted.2011.2162843
12. Savransky SD (2005) Model of OFF-ON transition and SET process in phase-change memory. In: Symposium non-volatile memory technology. IEEE, Dallas, TX, pp 5–109. https://doi.org/10.1109/NVMT.2005.1541414
13. Kwong KC, Chan M (2009) Circuit implementation to describe the physical behavior of phase change memory. IEEE, Hong Kong, pp 1–4. https://doi.org/10.1109/EDSSC.2008.4760753

FPGA-Based Military Vehicles Detection and Classification from Drone Videos Using YOLOv5

D. Sree Soumya, Ch. Aishwarya, and S. Vasavi

Abstract Drones are prevailing in both military and civil applications. Tracking the rivalry from the surveillance videos recorded by the drone has become a necessity for the purpose of reconnaissance. Object detection is one of the most common tasks performed by drones in order to track and classify the objects in the drone video. The proposed model is built in two modules. In the first phase, the object detection model is developed using YOLOv5 which detects and classifies military vehicles such as Military Tanks and Armoured Personnel Carriers (APCs). In the second phase, the object detection model is deployed into the Field Programmable Gate Array (FPGA). FPGAs are frequently used as implementation platforms for real-time image processing applications since their structure can exploit both spatial and temporal parallelism. The proposed system can be implemented by using a complex detection algorithm in a real-time embedded system using Field Programmable Gate Array (FPGA). Along with providing real-time implementation of object detection, FPGA also improves the performance of the detection model by a high measure.

Keywords Object detection · YOLOv5 · Drones · Military vehicles · FPGA

1 Introduction

Drones are very prominent as they have numerous use cases associated with them. When flying over a lower altitude, drones capture more details on objects, whereas higher altitudes cover a larger area. Many challenges are associated with drone object detection. To discuss one would be the altitude at which images and videos are captured as it greatly affects the size of the required object or the shooting angle. Additionally, weather conditions and lighting changes also have a drastic impact over the visibility of the object of interest. The proposed model is object detection

D. Sree Soumya · Ch. Aishwarya · S. Vasavi (✉)
Department of Computer Science and Engineering, V R Siddhartha Engineering College, Vijayawada, India
e-mail: vasavi.movva@gmail.com

J. R. Szymanski et al. (eds.), *Energy Systems, Drives and Automations*,
Lecture Notes in Electrical Engineering 1057,
https://doi.org/10.1007/978-981-99-3691-5_22

model using YOLOv5 integrated with FPGA. A custom data set is created for this purpose and the data has been collected from various sources for training the model. Due to the present day's global conflicts, detection of enemy movements is considered crucial. Performance requirements should also be noted while designing such systems. FPGAs or Field Programmable Gate Arrays are a solution to this problem. The proposed system gives FPGA implementation of detecting army vehicles from the surveillance videos captured by drone.

1.1 Basic Concepts

(a) **YOLO**. YOLO is a convolutional neural network which is best used for the purpose of real-time object detection. CNNs are classifier-based frameworks that interact with incoming images as organized arrays of data with the goal of recognizing patterns. Compared to other models, it is quick and maintains a good accuracy. YOLOv5 as shown in Fig. 1 is a collection of object detection architectures and models pre-trained on the COCO data set that reflects Ultralytics open-source research towards future vision AI technologies, combining lessons learned and also best practices gleaned from millions of hours of work.

Figure 1 depicts the YOLOv5 architecture. YOLOv5 architecture is comprised of three parts. The backbone is mainly used to extract key characteristics from an image's input. In YOLOv5, cross stage partial network (CSP) is the backbone to

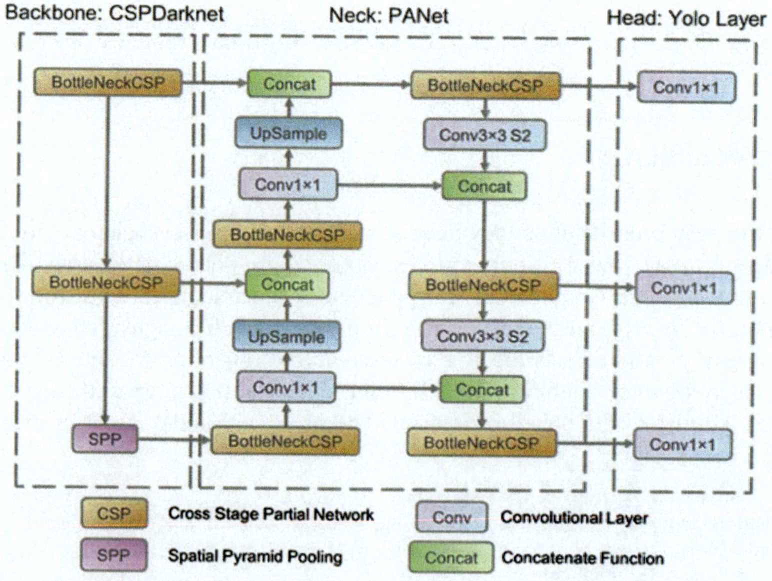

Fig. 1 YOLOv5 network architecture [1]

extract a wide range of important features from an input image. Generally, the neck of the architecture is used to construct feature pyramids. When it comes to object scaling, feature pyramids help models generalize successfully. It makes it easier to recognize the similar object in various dimensions. Feature pyramids are useful for aiding models in performing well on unknown data. On the other hand, PANet is used as a neck to generate feature pyramids. The last detecting process is primarily handled by the model head. It generates the final resultant vectors with class probabilities, objectness scores, and bounding boxes using anchor boxes.

(b) **Intersection over Union (IoU)**. IoU checks if the predicted bounding box will result in an accurate outcome. Intersection over union of the actual bounding box and the predicted bounding box are calculated. Later, IoU will calculate the area of the intersection over union of these two boxes using (1).

$$\text{IoU} = \text{Area of the Intersection}/\text{Area of the Union} \quad (1)$$

Figure 2 depicts the basic concept of intersection over union.

(c) **Non-maximum Suppression**. One of the most problematic challenges with several object detection architectures is that they might detect an object several times instead of simply once. In Non-Max Suppression, we select the predictions with highest confidence and suppress any other predictions with greater than a

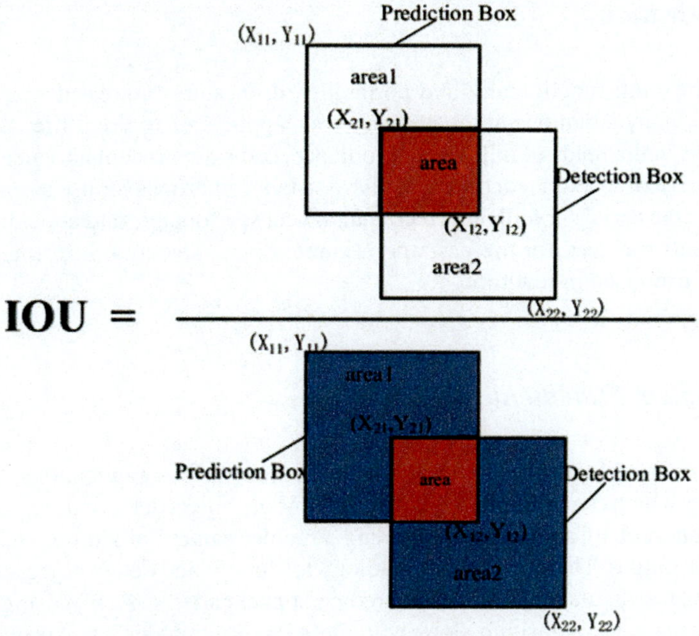

Fig. 2 Intersection over union [2]

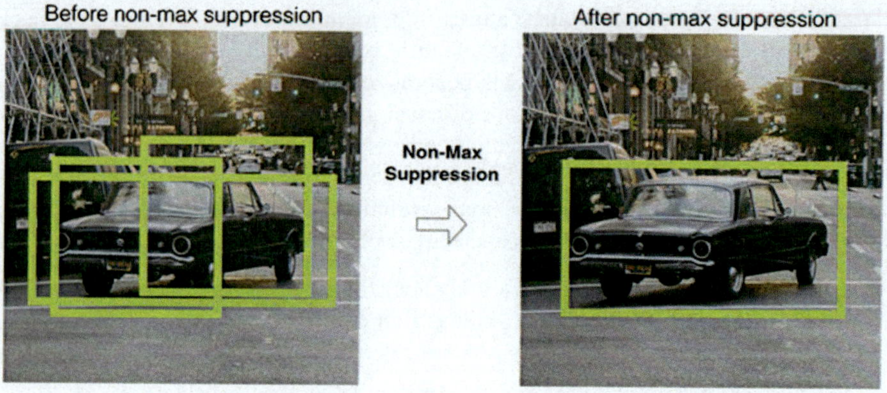

Fig. 3 Non-maximum suppression [3]

threshold of overlap with the selected predictions. To put it another way, we pick the maximum and suppress the ones that are not, hence the name non-maximum suppression.

Figure 3 depicts how non-maximum suppression works.

1.2 *Motivation*

Due to their multi-functionality and portability, drones are in great demand. They are used in many situations and have numerous applications in real life. One such utilization is in the fields of military, surveillance, and safety reconnaissance. People need systems that work in real time with the assistance of drones for use cases such as disaster rescue, aerial surveillance, reconnaissance, spy footage, and attacking roles. This demands the need for the best and accurate object detection implementations that can be exercised in real time.

1.3 *Problem Statement*

The immense need for drones is very high due to their various applications. Some of the use cases which can be implemented in real time are object detection, surveillance, reconnaissance, etc. Capturing videos using a regular camera might not necessarily give the best output. Therefore, a methodology to classify and detect objects military vehicles (such as battle tanks and armoured personnel carriers (APC)) using an efficient object detection algorithm along with Field Programmable Gate Array (FPGA) is required.

1.4 Scope

The scope of the proposed model is limited to drone-based video containing military vehicles such as battle tanks and armoured personnel carriers (APCs). We have included the occlusion as 15%.

1.5 Objectives

The key objective of the proposed model is to detect and classify army vehicles such as battle tanks and APCs from the surveillance video captured by the drone using onboard processing with Field Programmable Gate Array (FPGA).

1.6 Advantages

FPGAs are reprogrammable, and therefore, the proposed model can be extended to the detection of any other type of vehicles. The onboard processing of object detection model with FPGA gives more efficient and accurate results.

1.7 Applications

The proposed model has many applications such as military reconnaissance, aerial surveillance systems, and spy footages, and even more complex tasks such as face recognition are currently a subject of research. Some other object detection applications include detection of pedestrians, people counting, human face detection, text detection, or number plate detection.

2 Literature Survey

Object detection from drone captured images, obtained from VisDrone data set, using improved YOLOv5 has been discussed in [4]. The performance of safety helmet detection model has been analysed based on four different YOLOv5 models in paper [5]. The authors of [6] have discussed about various challenges, issues, and applications involved in vehicle detection from aerial images. The paper provides an overview on conventional algorithms together with machine learning and deep learning concepts to detect objects in aerial images. Embedded system-based object detection system with small drone and deep YOLOv3 has been presented in [7]. In [8], a method for real-time video surveillance in edge computing environments has been produced. The model uses N-YOLO which divides the image into fixed sized images instead of resizing the image in the YOLO algorithm. The most efficient bounding box has been produced by YOLOv3 in comparison with other object detection algorithms. The authors of [9] have presented a unique platform called MPDrone which is based on the MPSoC boards that have FPGA, CPU, and GPU under a single chip. The model has a reconfigurable FPGA that can be utilized to run complicated AI algorithms. Also, the CPU is used to run ROS in order to process communication with the drone controller. In [10], an efficient a real-time object detection framework on limited-resource hardware devices through hardware and software co-design has been highlighted by the authors. The rapid advancement of object detection techniques has gained attention to developing adequate Deep Neural Networks (DNNs). However, the current DNN models do not provide a satisfied solution balancing all attributes model size, speed, and accuracy. In order to compress the YOLOv5 model, a Tensor Train (TT) decomposition has been put forward. An efficient hardware accelerator based on FPGA devices has been developed by unitizing the characteristics of TT decomposition. Experimental results show that the proposed method can significantly reduce the model size and improve the execution time. The classification and detection of objects in satellite using a customized convolutional neural network is discussed in [11]. Three types of objects such as trees, cars, and buildings and cars can be classified in the satellite images through this model. The traditional methods for object classification involve two steps. The regions in which the objects are present are identified in the first step. The objects are classified in the second step. The performance of detection and classification is compared with actual execution of YOLOv3 algorithm for the same data set for other algorithms. The authors of [12] have given the hardware implementation of object detection and recognition using deep learning models. Xilinx PYNQ Z2 and Intel Movidius Neural Compute Stick (NCS) have been used in this paper, and they also increased the accuracy and improved the performance of the system. Single Shot Detector (SSD), Faster Region CNN (FRCNN), and You Only Look Once (YOLO) have been employed for the purpose of object detection. In contrast to the existing models, the results obtained in this paper have shown promising results. Reference [13] gives a clear idea of the existing deep learning techniques for recognition of vehicles from UAV images.

3 Proposed System

3.1 Architecture

The object detection model is trained and developed using YOLOv5 after the data set has been enhanced and preprocessed accordingly. Later, the FPGA is programmed accordingly to deploy the trained YOLOv5 model into it. When the programmed FPGA is integrated with the drone, on-board processing is performed

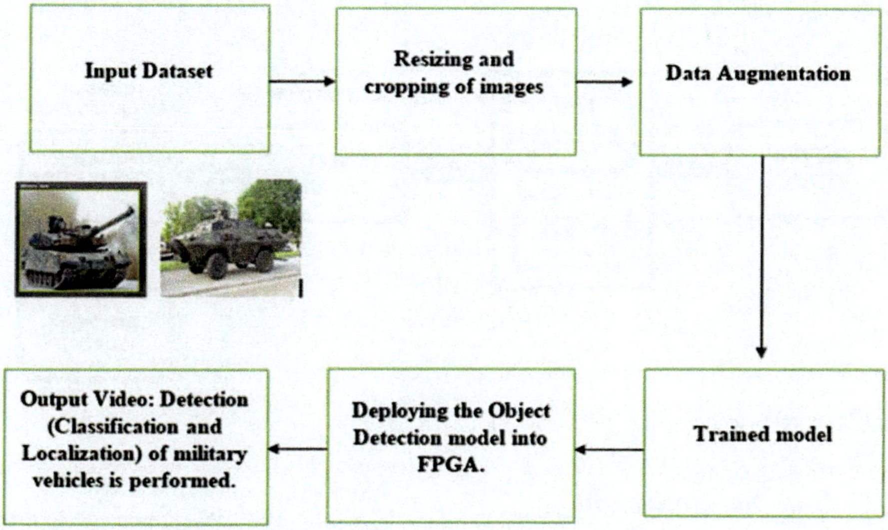

Fig. 4 Architecture of the proposed system

on the captured video and the output video in which the objects are detected is transmitted accordingly.

Figure 4 describes the architecture of FPGA-based object detection for drone video. It describes the different stages involved in the proposed system.

3.2 Methodology

The most powerful object detection method is presently You Only Look Once version 5 (YOLOv5). It is the latest convolutional neural network (CNN) which exactly recognizes and detects objects in real time with a higher accuracy. This method processes the entire image with a single neural network, then divides it into parts and predicts bounding boxes. In this approach, it produces predictions after only one forward propagation through neural network, i.e. it "looks once" at the image. The neck model PANet is used for the fusion of the extracted features. Finally, the YOLO layers produce the output detection results such as score, class, and size. Accurate bounding boxes are predicted through non-maximum suppression technique.

Figure 5 describes the YOLOv5 model which contains CSPDarkNet as its backbone model and is used for feature extraction.

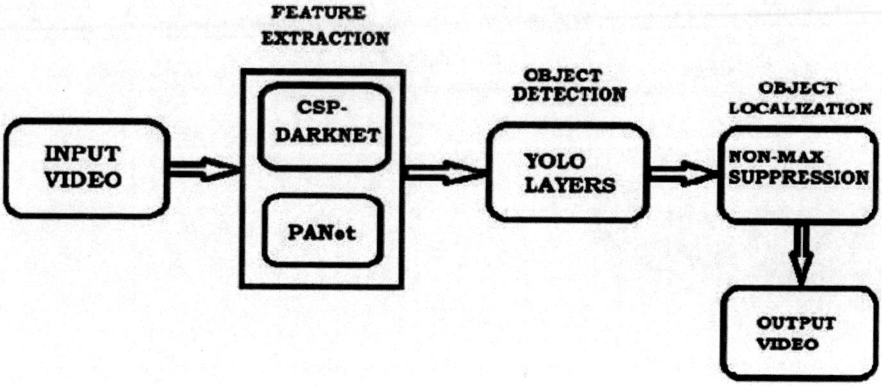

Fig. 5 Object detection using YOLOv5

3.3 Data Set Collection

The object detection model has been trained to detect APCs and battle tanks. However, an open data set containing various classes of military vehicles is not available anywhere. Therefore, a custom data set is created, labelled, and converted according to YOLOv5 format. The images for battle tanks have been taken from Kaggle [14]. Additionally, the images for APCs have been extracted from ImageNet data set. ImageNet [15] is a large image database which consists over 14 million images belonging to different categories. The videos for testing the model have been taking from GitHub.

3.4 Algorithm

In order to train the object detection model using YOLOv5 on a custom data set, first step is to collect the required data and convert it into YOLOv5 format.

The directory format is shown in Fig. 6. Next, select the YOLOv5 model that is appropriate for the proposed system from the available models. YOLOv5 model is used for this purpose of our model. Finally, train the model and run inference on the testing videos and record the observations carefully. The YOLOv5 model can be deployed into the FPGA. For that purpose, to program FPGAs hardware description languages such as VHDL or Verilog, need to be used.

Fig. 6 Directory tree for YOLOv5 model [16]

```
|- workingdir
        |--yolov5
        |--data
            |--images
                |--train
                |--valid
            |--labels
                |--train
                |--valid
```

4 Implementation

This section highlights the output and implementation results of the proposed system. The output results are obtained when the model trained using YOLOv5 object detection algorithm is used to detect tanks and APCs in testing videos. The precision, recall, and mAP of the trained model can be calculated using Eqs. (2), (3), and (4) as described in [17].

$$P = \frac{\mathrm{TP}}{\mathrm{TP} + \mathrm{FP}} \tag{2}$$

$$R = \frac{\mathrm{TP}}{\mathrm{TP} + \mathrm{FN}} \tag{3}$$

$$\mathrm{mAP} = \left(\frac{1}{C}\right) \sum_{k=i}^{N} P(k) \Delta R(k) \tag{4}$$

Here, C denotes the total number of object categories, N denotes the total number of IoU thresholds, k is the IoU threshold, $P(k)$ is the precision, and finally, $R(k)$ is the recall.

The accuracy table can be seen in Table 1.

It can be inferred from the table that the model is giving an accuracy of approximately ~ 90%. The accuracy scores with respect to the classes too can be observed in Table 1.

Table 1 Precision, recall, and mAP of the overall trained model w.r.t. classes

Class	P	R	mAP@0.5
All	0.886	0.829	0.898
APC	0.882	0.746	0.854
Military tank	0.891	0.912	0.941

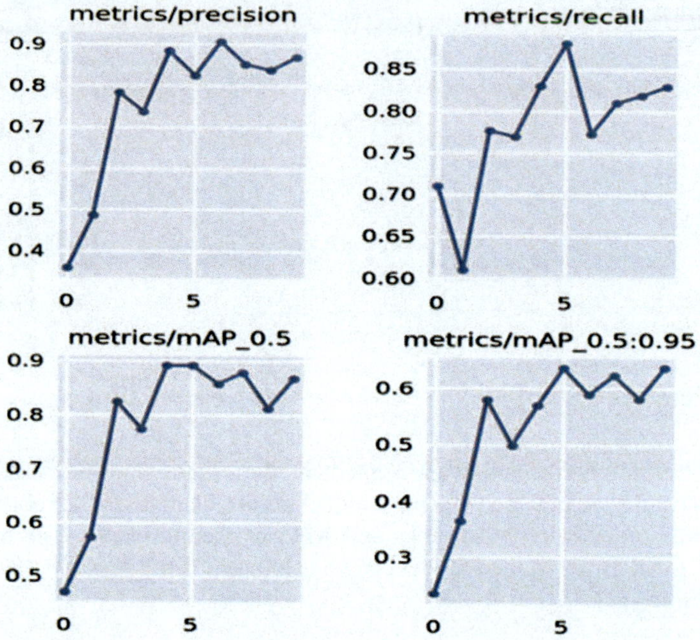

Fig. 7 Graphical representation of precision, recall, and mAP of the trained model w.r.t. number of epochs

Figure 7 gives the graphical representation of precision, recall, and mAP of the trained model with respect to number of epochs. Figure 7 is presenting a clear view of how the metrics are changing with respect to epochs.

The trained model is tested on a video, and in the output the object is classified as a military tank and a bounding box covering the object is drawn.

The output screenshot of the video can be viewed in Fig. 8. From Fig. 8, it can be seen how the model detected the object in the video as a military tank.

5 Conclusion and Future Work

The proposed model FPGA-based object detection for drone video is implemented to classify and detect military vehicles such as battle tanks and armoured personnel carriers (APCs) with an accuracy of ~ 90%. The model is based on FPGA which provides real-time object detection. Results obtained have shown good accuracy. However, the present model can only detect tanks and APCs.

Future work can include expanding and refining the data set in order to obtain better and accurate results in identifying the different classes of military vehicles.

Fig. 8 Results on test video using the trained model

References

1. Xu R, Lin H, Lu K, Cao L, Liu Y (2021) A forest fire detection system based on ensemble learning. Forests
2. Cheng S, Zhao K, Zhang D (2019) Abnormal water quality monitoring based on visual sensing of three-dimensional motion behavior of fish symmetry
3. Jain H, Nandy S (2019) Incremental training for image classification of unseen objects
4. Zhu X, Lyu S, Wang X, Zhao Q (2021) TPH-YOLOv5: improved YOLOv5 based on transformer prediction head for object detection on drone-captured scenarios
5. Zhou F, Zhao H, Nie Z (2021) Safety helmet detection based on YOLOv5. In: 2021 IEEE international conference on power electronics, computer applications (ICPECA), pp 6–11
6. Kumar S, Rajan EG, Rani S (S) A study on "vehicle detection through aerial images: various challenges, issues and applications". In: 2021 international conference on computing, communication, and intelligent systems (ICCCIS), pp 505–509
7. Madasamy K, Shanmuganathan V, Kandasamy V, Lee MY, Thangadurai M (2021) OSDDY: embedded system-based object surveillance detection system with small drone using deep YOLO. EURASIP J Image Video Process
8. Jha S, Seo C, Yang E, Joshi GP (2020) Real time object detection and tracking system for video surveillance system. Multimed Tools Appl
9. Kövari BB, Ebeid E (2021) MPDrone: FPGA-based platform for intelligent real-time autonomous drone operations. In: 2021 IEEE international symposium on safety, security, and rescue robotics (SSRR), pp 71–76
10. Liu M, Luo S, Han K, Yuan B, DeMara RF, Bai Y (2021) An efficient real-time object detection framework on resource-constricted hardware devices via software and hardware co-design. In: 2021 IEEE 32nd international conference on application-specific systems, architectures and processors (ASAP), pp 77–84
11. Deepthi S, Kumar S, Suresh L (2021) Detection and classification of objects in satellite images using custom CNN. Int J Eng Res Technol (IJERT) 10(06)
12. Kaarmukilan SP, Poddar S, Amal Thomas K (2020) FPGA based deep learning models for object detection and recognition comparison of object detection. In: Proceedings of the fourth international conference on computing methodologies and communication (ICCMC 2020). IEEE Xplore part number: CFP20K25-ART. ISBN: 978-1-7281-4889-2

13. Bouguettaya A, Zarzour H, Kechida A, Taberki AM (2021) Vehicle detection from UAV imagery with deep learning: a review. IEEE Trans Neural Netw Learn Syst
14. Military tanks dataset. https://www.kaggle.com/antoreepjana/military-tanks-dataset. Accessed 1 Dec 2021
15. ImageNet dataset. https://image-net.org/index.php. Accessed 1 Dec 2021
16. Train custom data. https://github.com/ultralytics/yolov5/wiki/Train-Custom-Data. Accessed 2 Dec 2021
17. Vasavi S, Kanthi Priyadarshini N, Harsha Vardhan K (2021) Invariant feature based darknet architecture for moving object classification. IEEE Sens Mach Vis Autom Syst 21(10):11417–11426

Design and Implementation of Histogram Estimation Architectures for 8 × 8 Image

Koteswar Rao Bonagiri, Giri Babu Kande, and P. Chandrasekhar Reddy

Abstract Recently, the histogram estimation architectures for analyzing the sampled data have numerous applications in image processing, communications, computer vision, and pattern recognition. In this work, the proposed histogram estimation structure is used to count up the bin values, and also the structure is implemented on Field Programmable Gate Array (FPGA) and also on Application Specific Integrated Circuit (ASIC) platform. The input 8 × 8 image is transformed into binary samples by using MATLAB, and these values are given to the input of Verilog code. The simulation results of the proposed architecture are obtained from the ModelSim software. In this we analyze the performance of both existed and proposed structures for 8 × 8 image and to obtain the performance of ASIC and are implemented in cadence encounter digital IC implementation with 180 nm technology and the percentage decrease of power, delay, and area is 32%, 18%, and 17.6%. And also, the utilization of look up tables, slice registers, and number of flip-flops is reduced for Virtex-6 device.

Keywords Histogram estimation · ASIC · FPGA · Dual port ROM · Optimal bin counter

K. R. Bonagiri (✉) · P. Chandrasekhar Reddy
ECE, JNTU Hyderabad, Telangana, India
e-mail: bonagirikoteswarrao@gmail.com

G. B. Kande
ECE, Vasireddy Venkatadri Institute of Technology, Guntur, AP, India

J. R. Szymanski et al. (eds.), *Energy Systems, Drives and Automations*,
Lecture Notes in Electrical Engineering 1057,
https://doi.org/10.1007/978-981-99-3691-5_23

1 Introduction

Nowadays, the assessment of probability density functions (PDF) based on sampled statistics is an important issue in various applications such as medical imaging, pattern recognition, machine learning, and image processing [1]. One of the fundamental concepts in statistics is PDF which is used for the construction of an approximate of the density function from the observed data samples [2]. Generally, PDF estimation techniques can be classified into two types, namely nonparametric and parametric methods [3]. Nonparametric approaches can be used for all types of density functional forms. When there is an unknown PDF with changing, parametric models can result in unfruitful accuracy [4]. PDFs have an extensive range of uses across an array of applications including image processing, machine learning, and communications. Nonparametric techniques can be fall into two categories: histogram estimation and kernel estimation function [5, 6].

Histogram is a pictorial representation of the sampled data and has many advantages, particularly it can enhance the contrast and choosing the threshold value of an image [7]. In this paper, structure is proposed for nonparametric PDF evaluation utilizing histogram-based method. Generally, histogram estimation methods are plotted using MATALB, where it is not feasible to realize the structure for generation of histograms [8]. The architecture utilizes diversified resources available on current FPGAs. It is possible to execute real-time evaluation on input data samples at higher speeds while extracting a variety of statistical properties [9].

The organization of the work as follows. Related work regarding conventional histogram architectures based on FPGAs is explained in Sect. 2. The proposed work is introduced and explained in Sect. 3. Performance analysis of architectures is demonstrated in Sect. 4. Conclusion and future scope is made in Sect. 5.

2 Related Works

In [10], a structure for histogram estimation was developed to generate the bin structure for any black and white images not only with MATLAB but also the structure was developed in both PsoC and FPGA platforms. The structure consists of two main modules, first module is used for creating the histogram and the second module is for display the histogram bin structure. Due to these two separate modules the FPGA implementation, i.e., the hardware resource utilization of histogram estimation of grayscale images is more.

In this work [11], we analyzed the architectures used to construct the histogram estimations for computer vision application algorithms on FPGAs. In the paper author reviewed various simplification architectures for oriented gradient algorithm for optimizing the design particularly the speed of the architecture. Finally, the author discussed three design guidelines for FPGA-based oriented gradient implementation for various applications.

In this paper [12], they presented an architecture to speed up histogram estimation. The major advantage of this technique is to identify the fast face recognition even though the image has sudden changes due to ambient temperature. The architectures are implemented on Zynq Z-7020 FPGA and the area occupied by the architecture is little bit higher.

3 Proposed DPROM-Based Histogram Structure

The proposed DPROM-based histogram structure is shown in Fig. 1 which contains the memory bank selector, dual port ROM, memory bank selector, and counters.

In this section, 64 samples histogram estimation process is explained clearly. Firstly, an input image of size 8 × 8 which contains 64 samples is read from MATLAB to evaluate the histogram estimation process. The pixel values of the image are transformed into binary data by using "dec2bin" function. The binary sampled data contains 8 bits of sampled data which is stored in a text file and is applied to the hardware to evaluate the histogram estimation process. The 16 pixel sample values are depicted in Fig. 2.

Histogram estimation process for 64 samples is done on FPGA platform. Based on the least significant values, counter passes the samples to dual port ROM which executes the pattern process. The pattern values are given in Table 1.

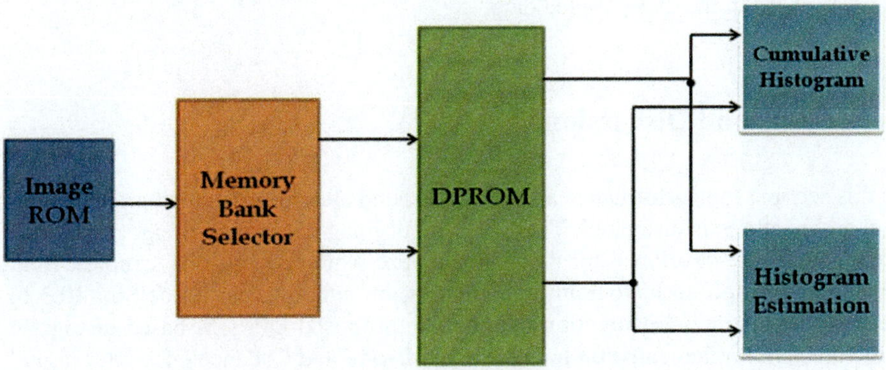

Fig. 1 Block diagram of proposed DPROM based histogram structure

Fig. 2 Input values of image

00000001	00000100	00000000	00000001
00000111	00000110	00000110	00001001
00000101	00000111	00000011	00000101
00000101	00000111	00000011	00001111

Table 1 Address and its pattern values

Address location	Pattern values
0	11111..........111111
1	01111..........111111
2	00111..........111111
3	00011..........111111
4	00001..........111111
⋮	⋮
254	000000.........000011
255	000000.........000001

Firstly, dual port ROM executes the address values of [0] & [1] and its corresponding memory values are given by the following equations.

The output values of dual port ROM are given in Eqs. 1 and 2.

$$16\text{Sample}[0]\_\text{dualportROM}\_\text{out} = 111111 \ldots\ldots\ldots 1111111 \quad (1)$$

$$16\text{Sample}[1]\_\text{dualportROM}\_\text{out} = 011111 \ldots\ldots\ldots 1111111 \quad (2)$$

The pattern process is run for all the input samples based on clock signals, and the dual port ROM output is given to the optimal bin counter, which can count the number of 1's in the 256 sampled data.

4 Results and Discussion

In this section, the performance analysis of existed and proposed histogram estimation architectures is discussed. The 8 × 8 input image is shown in Fig. 3. The two architectures are developed for 8 × 8 image size which contains 64 samples. Both existed and proposed histogram architectures are simulated in ModelSim 10.3 to observe the histogram estimation output. The proposed DPROM-based histogram structures are implemented on both Xilinx software and Cadence encounter digital IC implementation to obtain the performance metrics like area, delay and power of both FPGA and ASIC which are shown in Tables 2, 3 and 4.

The proposed histogram structure is best in terms of speed and delay when compared to the existed architecture and the operating frequency of the proposed structure is 112.335 MHz. The DPROM based histogram structure simulation results are shown in Fig. 4.

Fig. 3 8 × 8 input image

Table 2 ASIC performance of existed and proposed structures using 180 nm technology

Architecture	Area (μm^2)	Power (W)	Delay (ps)	APP ($\mu m^2 \times$ W)	ADP ($\mu m^2 \times$ ps)
Existed Architecture_64 Samples	1,472,301.5	6.24	242	9,187,161.36	356,296,963
Proposed Architecture_64 Samples	1,212,868	4.87	198	5,906,667.16	240,147,864

Table 3 Percentage of reduction of parameters by proposed architecture

Parameters	% of reduction
Power	32
Delay	18
Area	17.6

Table 4 FPGA performance of existed and proposed using Virtex-6 device

Parameters	Available resources	Occupied resources	
		Existed	Proposed
Slice LUTs	46,560	11,974	1672
Slice registers	93,120	1062	1062
Fully used LUT-FF pairs	12,500	536	601
Block RAMs	156	8	8

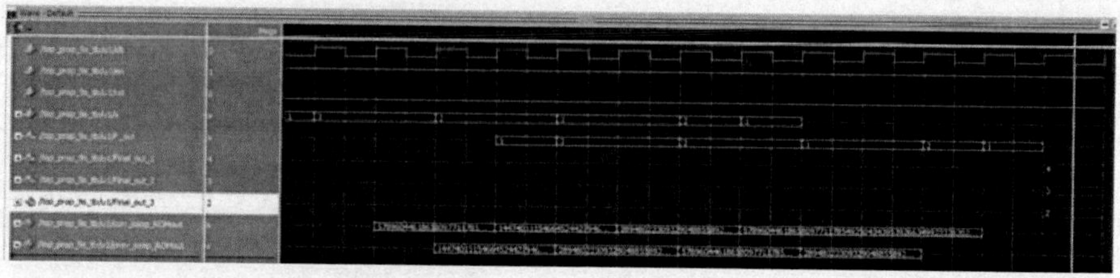

Fig. 4 DPROM-based histogram structure simulation results

5 Conclusion

In this work, the suggested histogram architecture is used to effectively approximate the histogram of the 8 × 8 grayscale image. Due to the usage of dual port ROM and carry select adder, the performance of histogram estimation is improved. The existed histogram architecture uses two ROM blocks, whereas the proposed architecture uses only one dual port ROM to store the old and new samples. Comparing to the existed architecture, the frequency of the proposed architecture is 112.335 MHz means the operating speed of the architecture is improved and delay is reduced. Particularly, the FPGA performance is enhanced in terms its performance metrics like slice LUTs, registers, and fully used LUT flip flop pairs when compared with the existed architecture. The amount of percentage reduction of area, delay, and power by the proposed histogram estimation architecture for 64 samples are 17.6%, 32%, and 18%, respectively.

References

1. Yin S, Ouyang P, Chen T, Liu L, Wei S (2015) A configurable parallel hardware architecture for efficient integral histogram image computing. IEEE Trans Very Large Scale Integr (VLSI) Syst 24(4):1305–1318
2. Pokrajac D et al (2007) Variable width elliptic Gaussian Kernels for probability density estimation. In: 8th ınternational conference on telecommunications in modern satellite, cable and broadcasting services, TELSIKS 2007. IEEE
3. Carbone P, Petri D (2015) Non-parametric estimation of probability density functions via a simple interpolation filter. In: 2015 IEEE international ınstrumentation and measurement technology conference (I2MTC), IEEE
4. Luan H-Y, Hua J (2010) Vicinal risk minimization based probability density function estimation algorithm using SVM. In: 2010 third ınternational conference on ınformation and computing (ICIC), vol 4, IEEE
5. Pană C, Severi S, Abreu GTF (2016) An adaptive approach to non-parametric estimation of dynamic probability density functions. In: 2016 13th workshop on positioning, navigation and communications (WPNC), IEEE
6. Hanebeck UD (2014) Kernel-based deterministic blue-noise sampling of arbitrary probability density functions. İn: 2014 48th annual conference on ınformation sciences and systems (CISS), IEEE

7. Praveen V, Sahitya G, Krishnasree, Naidu CD (2013) FPGA implementation of image enhancement algorithms for biomedical image processing. Int J Adv Res Electrical Electronics Instrumentation Eng 2(11):5749–5753
8. Bittibssi TM, Salma GI, Mehaseb YZ, Henawy AE (2012) Image enhancement algorithm using FPGA. Int J Comput Sci Commun Netw 2(4):536–542
9. Yang Y, Liu YX, Dong QF (2017) Sliced integral histogram: an efficient histogram computing algorithm and its FPGA implementation. Multimedia Tools Appl 76(12):14327–14344
10. Hazra S, Ghosh S, Maity SP, Rahaman H (2016) A new FPGA and programmable soc based VLSI architecture for histogram generation of grayscale images for image processing applications. Proc Comput Sci 93:139–145
11. Ghaffari S, Soleimani P, Li KF, Capson DW (2020) Analysis and comparison of FPGA-based histogram of oriented gradients implementations. IEEE Access 8:79920–79934
12. Bonny T, Rabie T, Hafez AA (2018) Multiple histogram-based face recognition with high speed FPGA implementation. Multimedia Tools Appl 77(18):24269–24288

New Era of Communication: Cognitive Radio a Future Scope

Sanchit Yadav and Ritika Rattan

Abstract Since the advancement being developed in each field, cognitive radio took an important place in communication area. In wireless communication system and mobile technology, cognitive radio plays an important role better in future. It is used in machine learning and with other fields and technologies as well. In this paper, we presented a survey on cognitive radio, associated research issue, its architecture, and various components related to it. Also, we discuss about the efficient utilization of spectrum bands followed by the various effective applications in different fields which has been present.

Keywords Cognitive radio · Wireless communication · Cognitive radio network · Licensed · Unlicensed

1 Introduction

Due to the continuous growth of the communication field, more amount of spectrum bands are required and this finding suitable bands is the main problem in wireless communication [1]. To overcome this problem, cognitive radio is introduced which has the ability to access the spectrum bands and provide them to the various users of various networks and types, by the formation of cognitive radio network [2]. But single spectrum is used by licensed user for specific service, but no unlicensed users are allowed to violate the guidelines, and cognitive radio also overcomes the problem of overloaded spectrums [3]. The cognitive radio has those kinds of capabilities which utilize wireless sensor networks which allocates fixed spectrum [4]. It enhances the quality of information to the users and spectrum can be used effectively by allocating the available band to another user when previous user is not using it, and by this, the utilization of spectrum can be improved.

S. Yadav (✉) · R. Rattan
Department of Electronics & Communication Engineering, Chandigarh University (University Institute of Engineering), Mohali, Punjab, India
e-mail: yadavsanchit97@gmail.com

J. R. Szymanski et al. (eds.), *Energy Systems, Drives and Automations*,
Lecture Notes in Electrical Engineering 1057,
https://doi.org/10.1007/978-981-99-3691-5_24

2 Cognitive Radio Network Architecture

For the development of communication, brief description of the cognitive radio architecture is important and the total architecture is shown in Fig. 2 [5], although the reading of architecture is depended upon the availability of spectrum and sensor nodes, and cognitive radio also may equip the sink. In addition, information also get exchange between sink and sensors including the allocation of spectrum, handoff, and controlled data of used for group formation [6]. The cognitive radio in CR network is able to communicate with the systems by sensing the available networks around it.

The CR networks are type of heterogeneous networks because cognitive radio networks comprise many other kinds of networks and communication systems in it [5]. The architecture design of the cognitive radio network is heading toward the improvement of the whole network for better utilization which means that every user can get whatever they want and fulfill their needs anytime, anywhere while using the network system, and providers are also able to provide better services to their users as proposed in [7] (Fig. 1).

2.1 *Components of Cognitive Radio Architecture*

The primary network and CR network are the two groups in which the components are classified [6]. The networks which already exist are known as primary network or licensed network where its users have the licence to work under certain band of

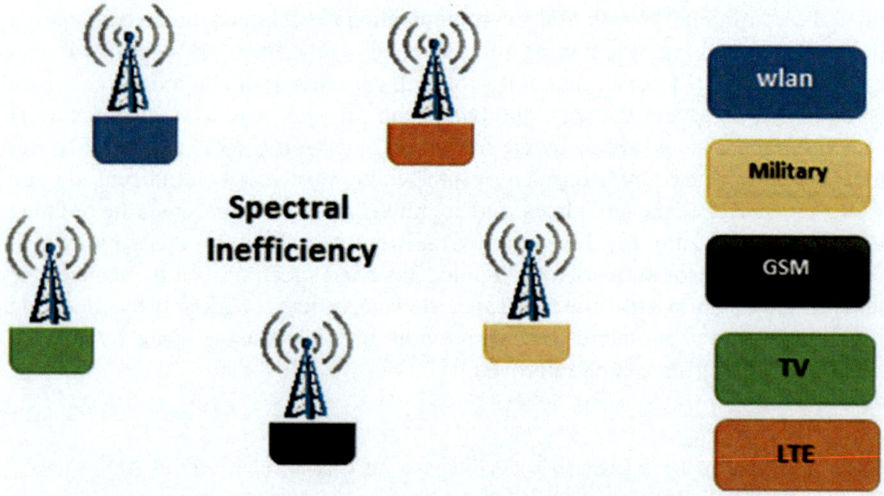

Fig. 1 General structure of cognitive radio

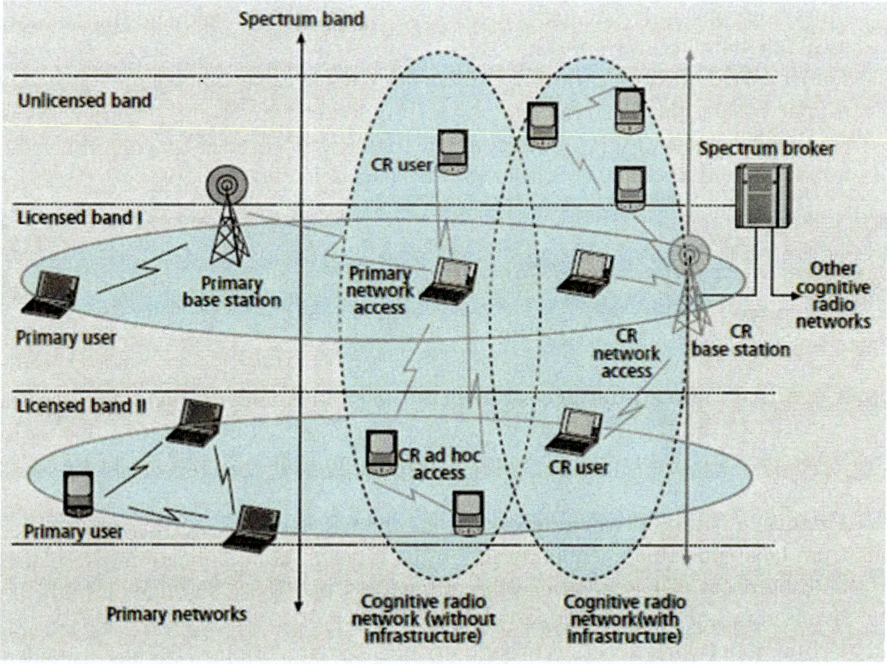

Fig. 2 Architecture of cognitive radio network [8]

spectrum and the users of primary networks are controlled by its primary base station (BS) [5]. As they mainly work in there spectrum, so the unlicensed users would not affect them. On the other hand, the users do not have licence to operate under certain spectrum which are classified as CR network or secondary network or unlicensed network [5]. In the cognitive radio networks, resources of the spectrums on different networks are distributed by the spectrum brokers and the single-hop connections to the user of cognitive radio users are also provided by the base stations of cognitive radio.

2.2 Spectrum Diversity of Cognitive Radio

Both the portions can be accessed by the users of cognitive radio weather it licensed or unlicensed, and also they are divided into two different operation types of bands in cognitive radio.

i. **Licensed band operation**: In this band, cognitive radio mainly focuses on the primary users as this band is basically used by the users of the primary network, and when any primary user appears in the spectrum that is already taken by CR users, then instantly CR users just vacant the spectrum band and find another available band to move in [7].

ii. **Unlicensed band operation**: In this band, CR user just follows the path as of primary users, they also have the same privilege to access the spectrum in the absence of primary users, and to compete for the unlicensed band, cognitive radio users required a refined spectrum for the sharing as shown in [5].

2.3 Network Diversity of Cognitive Radio

The CR users can perform in three different access types as shown in Fig. 2 [8]:

i. **Cognitive radio ad-hoc access**: With the help of cognitive radio ad-hoc access, one user of the cognitive network can communicate with other cognitive radio users directly. Users of CR bands can communicate on both the licensed or unlicensed bands [5].
ii. **Cognitive radio network access**: Inside the CR network, the sharing of spectrum takes place independently, so all the types of communication take place, and the users of CR band can directly access their base stations on both the licensed or unlicensed bands [6].
iii. **Primary network access**: Through the licensed, the primary BS has also been accessed by the cognitive radio (CR) users, and for the access, they require an MAC protocol; also due to this, they can able to access other various primary networks of different technologies [7].

3 Advantages and Disadvantages of Cognitive Radio

From above, it is clear that cognitive radio is miraculously helping technology as it allows constructing networks of radio stations with the help of combinations of cognitive radio networks which are more feasible as compared to the single cognitive radio. Because it improves the capability spectrum sensing and helps in gaining the characteristics of cognitive radio networks [9].

3.1 Advantages

i. The main merit of cognitive radio is to use the spectrum in an efficient way. These types of networks also improve the coverage by relaying technique from a point to another. It helps in reduction of power consumption and does not sacrifice performance of the network.
ii. By sensing channel availability, it avoids intentional radio jamming scenario. They can convert jamming signals into higher quality channels [10].

iii. Cognitive radios can conserve power. These radios can switch to power saving protocol (when the data rate is slow).
iv. Cognitive radio also improves the satellite communication.
v. By sensing environmental man-made interferences, it improves the quality of service (QoS) [11].

3.2 *Disadvantages*

i. In cognitive radios, there is no complete automation as it requires user intervention for any changes to be implemented.
ii. QoS is also affected due to its adverse effects.
iii. Cognitive radios are not secured properly as there are more chances for attackers in comparison with wireless networks [10].
iv. It has a big bandwidth spectrum so it always requires multiband antenna.
v. It requires problems like channel failure, receiver uncertainty for future decision making.
vi. It is used for filtering feature which is required for prior information of primary user [11].

4 Applications of Cognitive Radio

Without interrupting there users and taking care of environment, cognitive radio has some useful application:

i. **Healthcare Application**:

In a medical services framework, for example, telemedicine, wearable body sensors are being utilized progressively. Various remote sensor hubs are put on patients and obtain basic information for far off checking by medical care suppliers [12].

Clinical information is basic, deferral, and mistakenly delicate. Subsequently, the restriction of conventional WSN, as talked about in the past area, limits the possibility of telemedicine [5]. The QoS may not be accomplished at a good level if the working range band is packed in helpful 'telemedicine with BAN'. The utilization of cognitive radio wearable body remote sensors can alleviate these issues because of data transmission, sticking and worldwide operability, hence for the improvement of unwavering quality [13].

Figure 3 presents a model for remote BAN with cognitive radio remote sensors. A lot of examination has been completed in the region of WBASN. BAN stands for body area network. Some underdeveloping countries like India and Nepal are already practising these wireless BAN healthcare systems. It is suitable at the point when the doctors are relatively low.

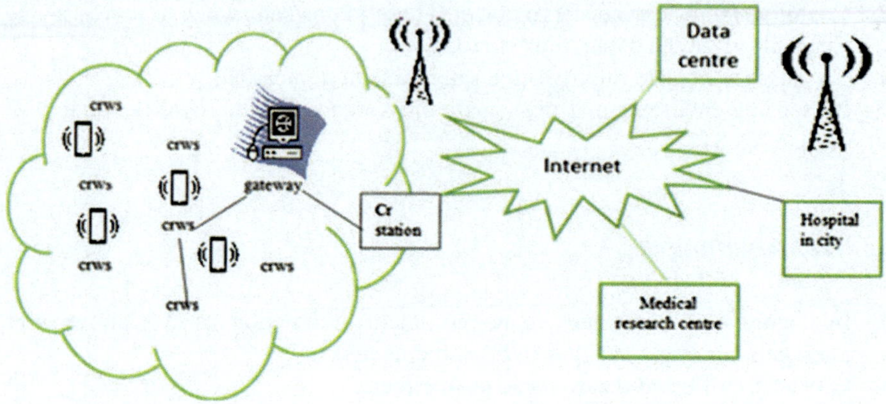

Fig. 3 CRWS in hospital

i. **Indoor Application**:

To achieve adequate QoS, various indoor and potential applications need a WSN climate which is thick can be involved. In achieving reliable correspondence, customary WSN endures possible challenges as the ISM groups are profoundly stuck in indoor regions [13]. Production line mechanization, individual amusement, wise structures, and home checking frameworks are not many WSN indoor application models. The challenges that are looked at through conventional WSN in indoor applications can be decreased through cognitive radio [10].

ii. **Real-Time Surveillance Application**:

An application that is set in observation like ecological checking, biodiversity planning, traffic observing, living space checking, climate observing which influence crops, submerged WSNs, vehicle following, bridge observing, and need decreased communication slow down and channel access. Not much common sense observing applications need upgraded reliability and they are very deferral touchy [11]. At the point when the state of the channel is not of high in multihop WSNs, a postponement to a connection disappointment may occur. At the point when a superior inactive channel is found in CWSN, it bounces to that hub. To improve the channel transmission capacity, the ramifications of various channels at the same time and total of the divert are likely in CWSN [9].

iii. **Diverse Purpose Sensing**:

The ramifications of remote sensors in comparable spaces for different objectives present high. The remote sensors in conventional WSN attempt to get the diversion in no cooperative manner. CWSN may pick different channels for different applications expecting reasonableness and burden offsetting with the help of medium access control (MAC) protocol that is proficient [13].

iv. **Sensing for Diversion**:

The ramifications of remote sensors in comparable spaces for various things present high. The remote sensors in conventional WSN attempt to get the diversion in no cooperative manner. CWSN may pick different channels for different applications expecting reasonableness and burden offsetting with the help of medium access control (MAC) protocol that is proficient [13].

5 Conclusion

The real WSN faces different issues which leads from the compelled transfer speed dispensed to them that are typically unlicensed groups. In the recent year, cognitive radio has proposed various methods for the better and effective utilization of networks' spectrum bands. This paper has investigated different components of the architecture of CR networks then provided the central ideas about its functionality, design, and showed many advantages, applications through which neither users nor environment have to face any problems and harm, and reasonable challenges in CWSN, also proposed the solution to overcome from various problems like the availability of spectrum which arises in cognitive radio due to which this review also promises an effective useful future of cognitive radios technologies in the various fields of wireless communications and mobile networks. This examination clarifies the advantages of CWSN, variety from traditional organizations, and huge application zones.

References

1. Akyildiz IF, Lee WY, Vuran MC, Mohanty S (2008) A survey on spectrum management in cognitive radio networks. IEEE Commun Mag 46(4):40–48
2. Letaief KB, Zhang W (2009) Cooperative communications for cognitive radio networks. Proc IEEE 97(5):878–893
3. Mitola J (2009) Cognitive radio architecture evolution. Proc IEEE 97(4):626–641
4. Akan OB, Karli OB, Ergul O (2009) Cognitive radio sensor networks. IEEE Netw 23(4):34–40
5. Chen KC, Peng YJ, Prasad N, Liang YC, Sun S (2008) Cognitive radio network architecture: part I—general structure. In: Proceedings of the 2nd international conference on ubiquitous information management and communication, pp 114–119
6. Akyildiz IF, Lee WY, Vuran MC, Mohanty S (2006) NeXt generation/dynamic spectrum access/cognitive radio wireless networks: a survey. Comput Netw 50(13):2127–2159
7. Reddy TB, Karthigeyan I, Manoj BS, Murthy CSR (2006) Quality of service provisioning in ad hoc wireless networks: a survey of issues and solutions. Ad Hoc Netw 4(1):83–124
8. Sahai A, Hoven N, Tandra R (2004) Some fundamental limits on cognitive radio. In: Allerton conference on communication, control, and computing, pp 1662–1671

9. Jondral FK (2007) Cognitive radio: a communications engineering view. IEEE Wirel Commun 14(4):28–33
10. Le B, Rondeau TW, Bostian CW (2007) Cognitive radio realities. Wirel Commun Mob Comput 7(9):1037–1048
11. Miller J (2006) An overview of the US and Japanese approaches to cognitive radio and SDR. IEICE Trans Commun 89(12):3168–3173
12. Ileri O, Samardzija D, Sizer T, Mandayam NB (2005) Demand responsive pricing and competitive spectrum allocation via a spectrum server. In: First IEEE international symposium on new frontiers in dynamic spectrum access networks, DySPAN 2005, IEEE, pp 194–202
13. Kaur A, Aryan P, Singh G (2015) Cognitive radio its applications and architecture. Int J Eng Innov Technol (IJEIT) 4(11)
14. Wang B, Liu KR (2010) Advances in cognitive radio networks: a survey. IEEE J Selected Topics Signal Process 5(1):5–23

Fingerprint Door Lock System Using Arduino and Bluetooth Application

Jaspinder Kaur, Harjot Singh Gill, Ajay Kumar, and Jasmeet Singh

Abstract This project is based on door locking mechanism with a fingerprint scanner. The components required are Arduino UNO, fingerprint scanner, Bluetooth module, servo motor, lock and a few different hardware devices. The fingerprint scanner is on the outward panel of the door as a way to get input from person and all of the circuitry can be on the inward panel to keep away from the reach of the consumer. The fingerprint scanner scans the finger and offers output to Arduino. If the print matches with registered print, Arduino sends a call through Bluetooth module to the related tool and asks his permission. If the owner permits the consumer, Arduino will liberate the door in any other case the door could be locked. If the consumer is an unknown individual, the Arduino tells the owner that the individual is unknown and asks if the owner desires to permit the consumer or not. If the owner lets in the consumer to allow in, then Arduino asks the owner to go into the call of the consumer and saves his fingerprints else the door will continue to be locked. For the owner's fingerprint, the Arduino will release the door without asking permission from different devices. In this paper, implementation of door locking mechanism with a fingerprint scanner using Arduino UNO has been discussed in detail. This locking mechanism system ensures precise reliability, fast fingerprint scanning speed and very high protection to the operator that makes this mechanism more reliable.

Keywords Arduino UNO · Bluetooth module · Fingerprint scanner · Servo motor

J. Kaur (✉) · H. S. Gill · A. Kumar · J. Singh
Department of Mechatronics Engineering, Chandigarh University, Mohali, India
e-mail: jaspinder.e10441@cumail.in

H. S. Gill
e-mail: harjot.gill@cumail.in

A. Kumar
e-mail: 19bem1017@cuchd.in

J. Singh
e-mail: 19bem1018@cuchd.in

J. R. Szymanski et al. (eds.), *Energy Systems, Drives and Automations*,
Lecture Notes in Electrical Engineering 1057,
https://doi.org/10.1007/978-981-99-3691-5_25

1 Introduction

Nowadays, people are having a lot of troubles associated with protection whether it is to protect some important things, documentary, house, apartment, locker, etc. [1, 2]. People can't only rely upon the conventional methods of offering safety because in this time they can be easily breakable there are already various methods available in the society to break through old and classic security locks due to this we need something more safer and yet comfortable security system in which we can put our faith that our things and documentaries are totally safe but for that type of security locks they had to be very strong that's why in the market all the good security locks are very bulky and heavier many types of keys and extra systems to protect against any theft activity but the thing is due to all this high security measures they made the lock so complex that in the end it's not comfortable at all as humans. We tend to forget many things including keys, passwords and many other things, so most of the people often complain about getting lost their keys of their master lock and are unable to do anything until the repair man or a company service man comes and helps them, and our concern in this area is to make a lock that is as secure as other heavy and complex lock out there but to keep the complexity of this locking system as simple as possible so one is neither worry about their important things nor there lock keys [3, 4]. They cannot depend upon the keys or cards for locking as they may be misplaced and you need to spend time for arranging new one or they may be stolen and all of your data can visit incorrect hands. Same problems with the password as they may be forgotten or different character can realize approximately the password. Data theft is the biggest thread in today's modern world for the companies and for the individuals data is there life once any of it get stolen they are done for but if we look into fingerprint security lock systems that are invented so far they are pretty good to take care of those thing as they don't need any physical key and so you don't have to care about any key being stolen or misplaced and forgotten as far as the security concerns due to this fingerprint locking system using an software to arrange all the things there is chance of the security system getting hacked but as it's a security lock thus the security measures taken by programmers are very secure and encrypted; thus, there is a small chance of the system getting hacked, and for the solution of this thing, the program of this fingerprint lock is designed in the way that does not allow any type of interaction without matching by fingerprint, so it is pretty secured and very easy to use in this project we made it to go as cheap as possible. Talking about fingerprint security system it is also very useful need to give access to multiple people instead of giving them physical keys we just need to scan the fingerprints it's very convenient and easy compare to old and conventional locking systems. Also inside the company or workplace if the owner of a cabin has the keys, in case his co-employee needs to take something like critical documents, etc., he has to watch for the owner to return back and free up the keys. So, it is very time-eating and irritating trouble and might have an effect on the time table of the workers. But, this is not always the case with fingerprint door lock. By fingerprint, it is so smooth to free up the door and on pinnacle of that it is far a stable manner than traditional methods [5].

In this paper, the door locking mechanism system with a fingerprint scanner using Arduino UNO board has been presented and discussed in detail. This locking mechanism system shows number of advantages such as precise reliability, fast fingerprint scanning speed and very high protection to the operator and reliability.

2 Related Work

Extensive literature reveals that fingerprints are patterns of valleys, ridges and micro-trenches on the finger surface. Similar to the whole lot with inside of human bodies, those trenches shape through aggregate change genetically and environmentally. The inbred code of DNA offers popular arrangement at many manner pores and tissue which have to shape inside growing unborn baby; however, in particular manner, bureaucracy usually ends outcome of disorganized moments. After assist with lining, finger marks are used to create stable and impassable lock systems and numerous locking structures. Lining is technique for setting up verbal exchange among micro-controller and interaction system. Fingerprint interaction systems are well-known and may speak to many microcontrollers. It is an aggregate with gear and program (supply key to speak, additionally referred to as because the Drivers). For easy term, for applying LED into output tool, LED must have been linked with the proper pin in the microcontroller also in that needs to be a software walking in microcontroller to turn it open and shut or flicker or faint. That application may have been evolved the use of many coding language, for example, C, Assembly, Python, etc. [1].

A smart door gets entry to device, the use of fingerprint life science mechanics rates that formerly, for excessive safety regions or into lock territory for banks, conventional locking structures, security codes, have been in work. Nevertheless, those structures had been discovered to be now no longer flawlessly stable [4]. After improvements in generation, RFID playing cards had been used. These playing cards but had been now no longer tons beneficial for the consumer because of possibilities of having absent, theft and buried. Main cause of that observe for offer excessive safety like those excessive quit safety request. Their intention is to take a look at which is to layout a clever door get entry to gadget the usage of fingerprint constituent. Main applications of tool are to offer get entry to most effective legal individuals. Each mechanical and cording key generation will be used for layout them. Small emergency sound beep is supplied toward shield that machine through put up horn when an unauthorized individual violates security of device. Then, a signal suggests some accidental situation. Usually, motors do the locking and unlocking job of the door.

Blueprint and execution of the fingerprint door locking mechanics for split entry for all those thing we says about 'telecellsmartphone' primarily build, totally fingerprint safety machine'. These brought on some smartphone that has modern capabilities that utilize finger marks' identification machine for permit that will get entry for the telecellsmartphone. In accordance with these, especially that device may be useful to establish the connection to the ones telephones and will verify them via their fingerprint sensor at that telecellsmartphone for unlock the door. That machine may seem related with the telecellsmartphone through WiFi, Bluetooth or some softwares may might useful for telecellsmartphone permitting those for interconnection, inshort the result is they joined and finger marks identification system is utilized at maximum latest telephones these days and shortly that finger marks identification primarily works on totally telecellsmartphone will be seen all over, nearly anyone may possess own after which their protection gadget might became so useful.

Publicized in some global bulletin for Modern Analysis of Electrical, Self-regulating and Mechanics Engineering, that the blueprint of finger marks primarily builds on totally igniting structures within cars wherein directory made for legitimate customers is saved with inside that finger marks' constituent. If someone attempts approach function of car at that time Processor fits those finger marks of that individual to saved directory, whenever suit end outcome proves to be a success at that time automobile gets started however in any other case now no longer. Outer gadgets may try to manage via that computer aligned jack. These aligned jacks are easy and cheaper device to constructing pc managed gadgets as well as programs. Any person manner applications are therefore programed for interacting to write down tool programs. Codes that are used for this first model turned into achieved with Visual Basic software. That tell us the technique that is, to begin with consumer is requested to sign up their finger marks. Then enlisting that person's identity is performed. Whenever individual gets permissions, door mechanically unlocks. Later starting their car, that door mechanically locks. If vacation spots are approached, if specific button triggered, that door routinely unlocks. Following some hard and fast clock hold up, that door mechanically locked. In addition, those devices make a specialty of the locking and unlocking the door with use of finger marks. Thus finger marks identification mechanism program permits finger marks only to legitimate customers belongs to car will registered to directory.

A bit of person will able to access car, their fingerprint codes are equivalent resist those fingerprints with inside the directory at the same time as customers without a in shape with inside the database are avoided from the use of the car. that way security of the car increases by a lot and it becomes very easier and convenient to lock and unlock the car and for giving the car access to any other person it only takes a fingerprint scan it can be done via online service of manufacture company that way you have to provide physical keys to anyone and later on we can delete his fingerprint the locking system.

3 Design and Implementation

Figure 1a shows the Arduino UNO board: It is a microcontroller device. It can do various repeated tasks based on the programming [6, 7]. It is the control system of the project. It controls the other devices and makes them work accordingly. Figure 1b depicts the 12 V electronic door lock assembly solenoid (low power consumption) [1]. It works on the principle of electromagnetism. The current in solenoid creates electromagnetic force on latch and push or pull it. Figure 1c displays the finger print sensor. It scans the patterns of valley, pores and ridges on fingers, and based on these patterns, it differentiates among different fingerprints. Figure 1d shows the 12V 2A DC power supply adapter. It gets the AC input and converts it to 12V DC supply. Figure 1e indicates the Bluetooth adapter. It provides the Bluetooth connection with smartphone and Arduino, so that the owner can know who is using the lock [8].

4 Working Flowchart

Figure 2 shows the working of finger print door lock system with the help of flowchart.

When someone will use the fingerprint door lock, the fingerprint scanner will scan the fingerprint. The owner will get the notification from app on phone. When the owner will open the app, it will display the name of person using the door lock if the fingerprint is known. Otherwise, it will display 'Unknown person'. Then, it will ask to let him in. If the fingerprint is known and the owner presses yes button, then lock will open. If owner presses NO, then lock will remain closed. If the fingerprint is unknown and owner presses YES, then lock will open and app will ask owner to enter the name under which the fingerprint will be stored.

5 Performance and Evaluation

Differentiating conventional protected device, this fingerprint door locking device currently at its improves state, green thus extra protected. If any ordinary safety device is produced from locking systems that touch through proper codes, door opens. This machine, a certified thus accurate finger marks are most effective way for opening those protected locked devices.

Locking structures have important roles in each twenty-four hours' life. In order for stable critical and private property in addition to every individual's security, here is not any opportunity for fasten security. However, form fasten gadget carry through, says lots approximately of quantity for what something may become even protected. Fingerprint locking device works on micro-trenches of fingers wherein finger marks' system can be implemented for important thing that opens doors. This system is more secure and extra secured because fingerprints are usually precise and thus

(a) Arduino Uno Microcontroller Board

(b) 12 V Electronic Door Lock Assembly Solenoid

(c) Finger Print Sensor

(d) 12V 2A DC Power Supply Adapter

(e) Bluetooth Adapter

Fig. 1 **a** Arduino UNO microcontroller board, **b** 12 V electronic door lock assembly solenoid, **c** finger print sensor, **d** 12 V 2 A DC power supply adapter, **e** Bluetooth adapter

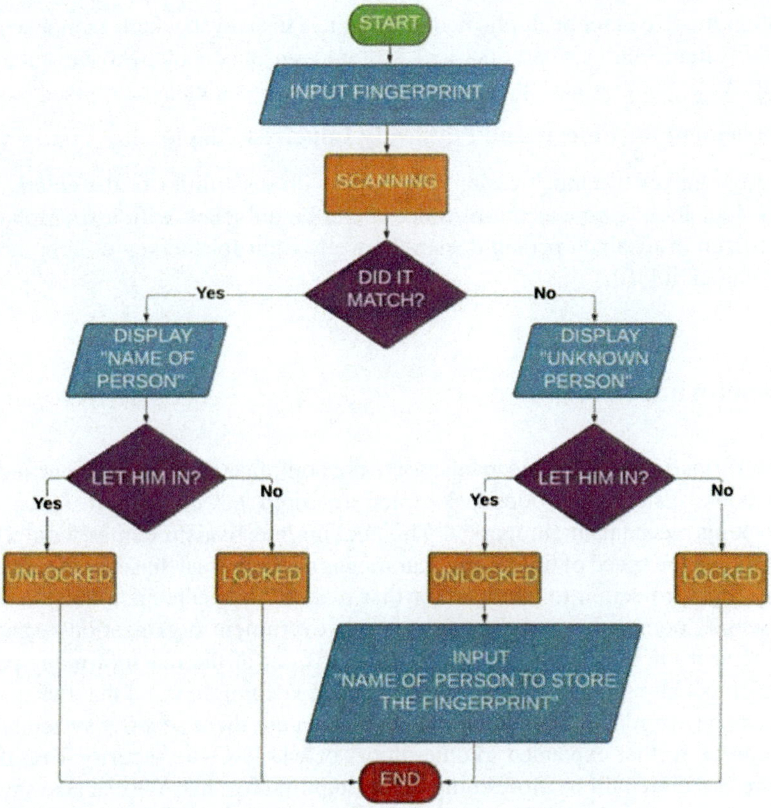

Fig. 2 Working of finger print door lock system with the help of flowchart

cannot duplicated. Many regular similarities and gaps with those security systems of different kinds. Regular key locks, fingerprint code locking mechanism, passcode locking mechanism, life science locking system are simply defense mechanisms that are used for high protection intentions. Every system has their own benefits and drawbacks that make them productive, protected and tough to crack.

A. **Smartphone**

You need to pair the Bluetooth of your smartphone with the application to lock or unlock. Just tap to open your door. It is very good and helpful in case you lose your keys. Also, it gives you the freedom to manage a set of keys.

B. **High Security**

Burglary is a common occurrence for all homeowners. Any unwelcome visitor can easily enter your home, no matter how strong your lock. But a smart lock system is a hard nut to make for it, because it has no buttons. Their approach is based on security agreements included in the banking system. It approaches that their virtual encryption mode cannot be duplicated or broken. This way, hackers cannot select or crash the system. With a smart door lock system, you can decide who can enter your home at any time. It reduces the risk of anything being stolen [9].

C. **Convenient for Elderly and Physically Impaired People**:

Technology makes life much easier for people with disabilities or the elderly. And a clever door lock is no exception! Since it can be unlocked with a smartphone, a senior citizen or disabled person does not have to fight for the key or approach the door to unlock it [10].

6 Conclusion

The fingerprint door locking mechanisms are not complicated for implement and thus it is a flexible system. This door lock system required less financial resources than other locking mechanisms in market. This locking mechanism comes with precise reliability, and the speed of fingerprint scanning is also very fast that enables extra fast and very high protection to the operator that makes this mechanism more reliable. Many people, companies and even private or government organizations want this type of security in their home or offices. But, it is out of the reach of many people because of its extensive cost. This project is made keeping in mind that the security system can be affordable, so much more people can use these security systems. This design can be further expanded to other doors or whole-home security. But, it will cost more and there will be more components required for that. We can also save the print of the users, but it will have a limit. As there is not enough memory to store fingerprints without a computer. For security purpose, the control system is kept inward panel of the door, so that no outsider can access it. We can also provide it energy by solar panels. One of the huge advantages of the fingerprint door lock is that its elasticity. Alternative mechanisms conceivably applied alongside the fingerprint mechanism. Since fingerprints are unique for every individual, so it is a very efficient way of providing security to the important stuff. There are some disadvantages also. We cannot make any changes in hardware easily because now the hardware system is fixed. It needs a continuous power supply so as to keep the equipment on. On power failure, the device stops working. For solving that issue, we can use rechargeable batteries so that we can lower the chances of power failure.

Acknowledgements All the authors would like to thank the Department of Mechatronics Engineering, Chandigarh University, Mohali, Punjab, for providing the opportunity to complete the project.

References

1. Shanthini M, Vidya G, Arun R (2020) IoT enhanced smart door locking system. In: Proceedings of 3rd international conference on smart systems and inventive technology (ICSSIT), no. ICSSIT, pp 92–96
2. Asogwa TC, Fidelis E, Obodoeze C, Obiokafor IN (2007) IJARCCE wireless sensor network (WSN): applications in oil & gas and agriculture industries in Nigeria. Int J Adv Res Comput Commun Eng ISO 3297:3–6
3. Parab S, Mujawar Z, Tandel D, Kuchiwale S (2021) Password-OTP , RFID and android biometric based door lock system using Arduino Mega 2560 microcontroller, pp 2156–2161
4. Prabhakar PAY, Oza PSK, Shrivastava N, Srivastava P, Wadhwa G (2019) Password based door lock system. Int Res J Eng Technol 1154–1157
5. Design and Construction of Smartdoor Security System Using (2020)
6. Louis L (2016) Working principle of arduino and using it as a tool for study and research. Int J Control Autom Commun Syst 1(2):21–29
7. Rathod K, Vatti P, Nandre M, Yenare S (2017) Smart door security using Arduino and Bluetooth application, no 4, pp 2394–0697
8. Morón MJ, Luque R, Casilari E (2014) On the capability of smartphones to perform as communication gateways in medical wireless personal area networks. Sensors (Switzerland) 14(1):575–594
9. Khan R, Khan SU, Zaheer R, Khan S (2012) Future internet: the internet of things architecture, possible applications and key challenges. In: Proceedings of 10th international conference on frontiers of information technology, FIT 2012, pp 257–260
10. Kumar V, Chawda RK (2020) A research paper on smart home. Int J Eng Appl Sci Technol 5(3):530–532

Design and Performance Analysis of DWDM Optical Link for High-Speed Optical Communication

Harpreet Kaur and Ahsanullah Umary

Abstract Developing requests of the net clients is one of the motives which superiority utilizing dense wavelength division multiplexing (DWDM) systems to broadcast optical information. This proposed method has the ability that can transmit numerous wavelengths thru a single optical fiber. In this paper, execution analysis of dense wavelength multiplexing network (DWDM) system has been suggested to evaluate the performance of the suggested network using BER analysis over long distance. DWDM is the advanced type optical network that provide high bandwidth compare to available optical networks that accomplish the need for greater bandwidth. In DWDM, single optical fiber is utilized for transmission of different wavelength. The suggested network model system is implemented by using OptiSystem software, and output of the simulation model is analyzed in term of Q-factor and minimum BER; in general, capacity of the system is improved.

Keywords Q-factor · DWDM · BER analyzer · Optical amplifier · NRZ

1 Introduction

These days are the period of excessive-speed net, small bit fault measure optical fiber systems allowing excessive capability interconnection between transmitters and recipient [1]. To boom the quantity of passage in optical fiber, we are utilizing dense wave length division multiplexing. An optical signal of various wavelengths may be transmitted via the DWDM, and there may be countless range of channel in WDM [2]. Late advancement in transmission limit of DWDM technology alerts in a trunk optical fiber line now in transmission of information beyond 10 Tbits/s over in excess of one hundred km [3, 4]. The populace group has received to depend in media

H. Kaur (✉)
Department of Computer Science Engineering, Chandigarh University, Punjab, India
e-mail: harpreet8307@gmail.com

A. Umary
Department of Electronic and Communication Engineering, Chandigarh University, Punjab, India

J. R. Szymanski et al. (eds.), *Energy Systems, Drives and Automations*,
Lecture Notes in Electrical Engineering 1057,
https://doi.org/10.1007/978-981-99-3691-5_26

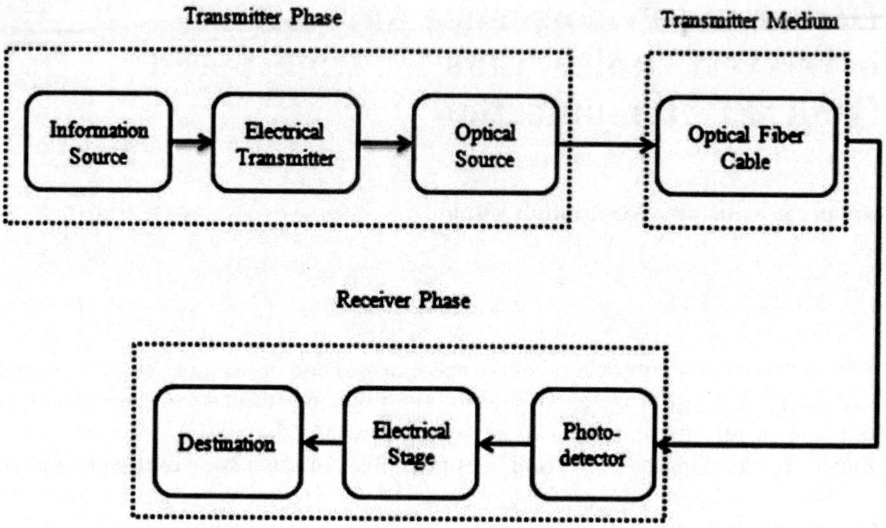

Fig. 1 Schematic illustration of optical communication network

transmission frameworks to fulfill its requirements for data and correspondence [5, 6]. Chromatics scattering of optical fiber and optical intensifier nonlinearities are impacts to limit the transmission limit [7, 8]. On the probability that appropriately controlled a portion of the parameter, at that point improving execution of optical correspondence framework. In optical fiber connection, the info is send from source to target via optical fiber with concept of total internal reflection [9, 10]. Recently, optical communication research is one of the most widely research in telecommunication industry. It lower the total cost of the communication system as well as it gives high-speed data transmission and security [11–13]. Optical communication process includes different phases, input phase transmission phase, and medium as well as receiving phase [14, 15]. The fundamental mechanism of fiber communication scheme is shown in Fig. 1 [16].

Just as it reveal in the above model information such audio, video, speech, or picture is supplied to the electrical transmitter by mean of electrical signal. Electrical phase of transmitter side forces optical source (LED or laser) that generate adjusted light wave carrier [17, 18].

1.1 *Wavelet Division Multiplexing (WDM)*

In WDM, at least two optical signs having various frequencies are joined and at the same time transmitted a similar way over an optical fiber. WDM technology is a free of scale and design, and it can bolster any mix of interface rates. OC-n, OC is representing optical channel and n for what number of channel can be multiplexed

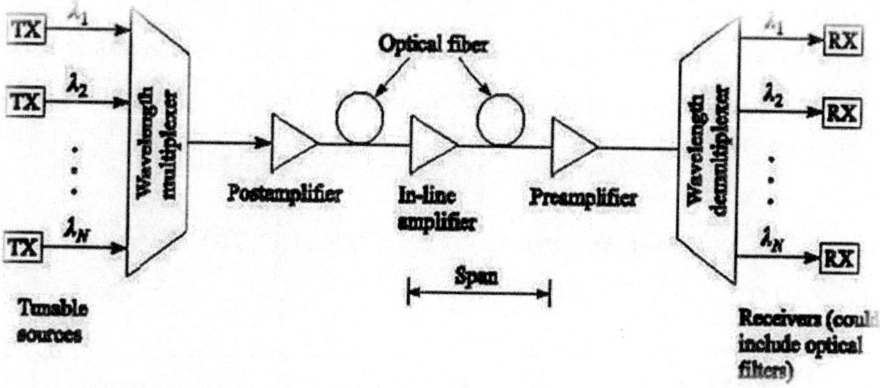

Fig. 2 Typical WDM model

[19]. Figure 7 shows the usage of regular WDM connects [20]. At the transmitter side, a few freely balanced light sources, each discharging signals at various frequency. Here, a multiplexer is expected to join these optical yields into a solitary fiber. At the beneficiary side, a de-multiplexer is required to isolate out these optical signs into channels for signal handling [21, 22] (Fig. 2).

1.2 Dense Wavelet Division Multiplexing (DWDM)

Dense WDM (DWDM) is an innovation with a larger number of frequencies coupled into a fiber than WDM. A prior WDM innovation with few frequencies, superior channel width, and channel separating is named coarse WDM (CWDM) [23]. DWDM frameworks exploit progressed optical innovation, for example, tune-able lasers and narrow-band optical channels to produce numerous frequencies in the range around 1550 nm [24]. Dense wavelength division multiplexing transmits data serially. In today's world of technology, the number of consumers increased that lead to increase bandwidth by using DWDM, this can be achieved [25–28].

Operation of DWDM

Figure 3 shows fundamental schematic diagram of DWDM system model it comprise of following part:

(a) **Transmitter/Receiver**

Transmitters are depicted as DWDM parts due the fact that they offer origin signals which after that multiplexed [29]. The qualities of optical senders utilized in DWDM frameworks are profoundly imperative to framework structure. In DWDM framework, different optical transmitters are utilized as the light sources. Approaching electrical information bits (zero or one) trigger the tweak of a light stream (e.g., a

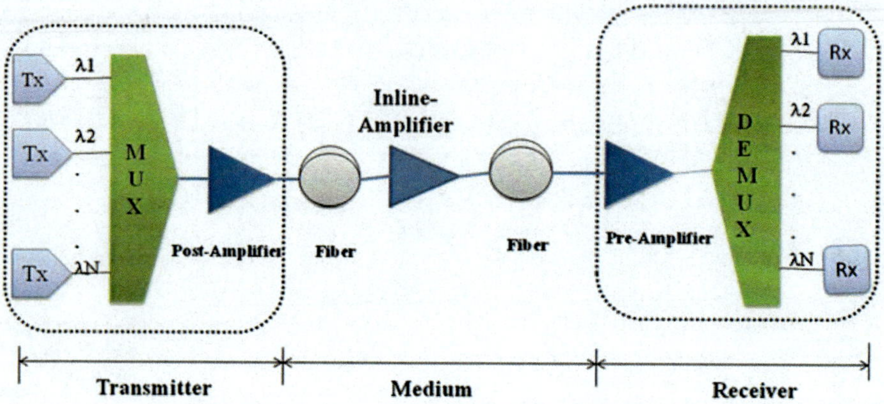

Fig. 3 Schematic illustration of dense wavelength division multiplexing

blaze of light = one, the non-presence of light = zero). Lasers make beats of light. Each light heartbeat has a precise frequency (lambda) communicated in nanometers (nm).

(b) **Multiplexer/De-multiplexer**

The function multiplexer is to join various wavelength into a single fiber or in another words MUX is device that has N inputs and single output (N * 1). On the other hand, de-multiplexer is the opposite of multiplexer; it is 1 into N it means that it accept one input and provide N number of output [31, 32]. Optical multiplexer and de-multiplexer are passive parts. Because the procedure of this multiplexes and de-multiplexing is optical along these lines, there is no need of outer force source [33] (Fig. 4).

(c) **Optical Amplifiers**

Optical amplifiers are utilized here to amplify the light directly without first converting to electrical signal [34]. Optical amplifier can be placed different position within DWDM system to amplify the light employ as booster to boost the light before pass to optical fiber, as in-line amplifiers used to recompense loss in fiber transmission, and as preamplifiers at the destination point to enhance optical signal to required intensity. Optical intensifiers: The optical sign (light) can be enhanced legitimately utilizing optical amplifiers without first changing over it to the electrical sign [35].

(d) **Optical add/Drop multiplexers**

The image describes the functioning of 1 channel optical add-drop multiplexer (OADM). Here, OADM is intended to accept or reject signal alongside unique frequency [36, 37]. Include implies the ability of gadget to add new frequency channel to existing multi-frequency signal. DROP intends to evacuate channel passing sign to another way. The approaching sign is broken into two segments for example drop

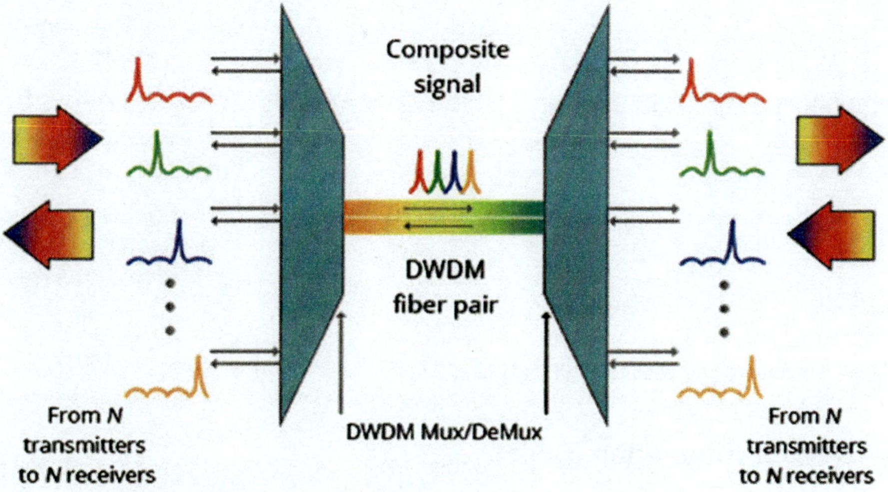

Fig. 4 DWDM multiplexers and de-multiplexers [38]

and go through. The sign that is dropped is passed to the beneficiary of customer gadget, and the go through signal is multiplexed with new signal circulation (Fig. 5).

(e) **Transponder**

Transponder is a tool that is utilized to impart and get signals from fiber, and those are utilized to change over full duplex electrical sign to full duplex optical sign. These proselyte optical signal of one frequency to optical signal of another frequency appropriate for DWDM utilizations. Those are otherwise called optical-electrical-optical (O-E-O) wavelength converters. Subsequently, these are helpful for optical correspondence yet these are massive and devour extra power [38].

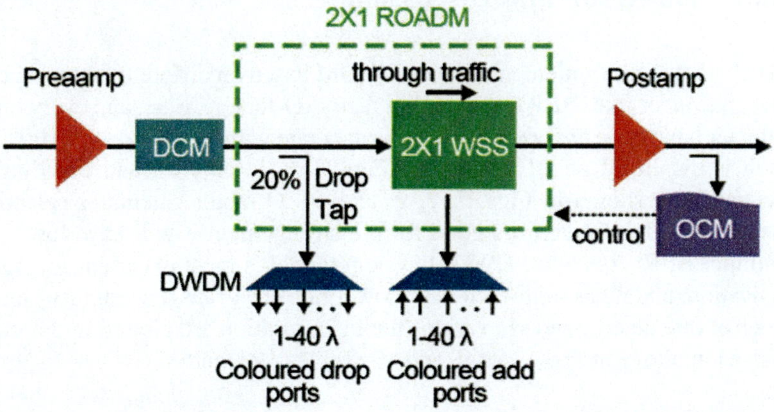

Fig. 5 Optical add-drop multiplexers

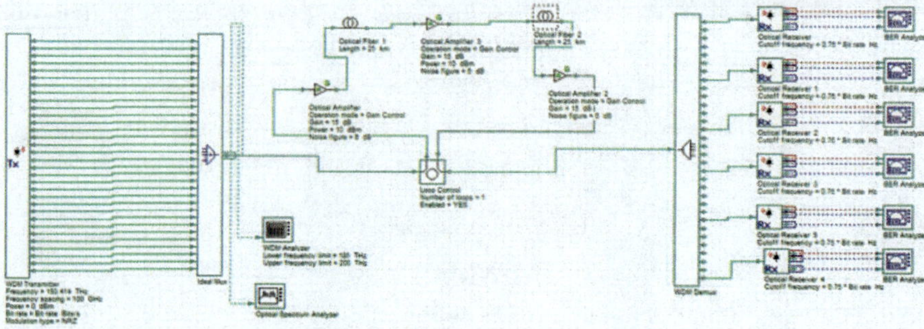

Fig. 6 Structure of a 32 channel DWDM system

2 DWDM Simulation Model

The simulation design of DWDM model is divided into three phase as transmitter (Tx) phase, transmission link, and receiver (Rx) phase as demonstrated in the Fig. 6. Tx part includes of information source, DWM transmitter, optical spectrum analyzer, and WDM analyzer. The transmitter part comprises of a 32-channel wavelength division multiplexing transmitter and multiplexer; the frequency rate is 193.414 THz. To test and confirm that transmitters we use optical spectrum analyzer and WDM analyzer by calculating total power for each channels. At medium, we have utilized optical fiber of total length 50 kms and three amplifiers of gain 15db with power of 10 dbm each that amplify the attenuated signal through the fiber. In that respect different classes of optical amp for instance semiconductor optical booster Raman amplifier and EDFA amp.

3 Simulation Result and Discussion

This part characterizes implementation of DWDM based on different factors such as eye figure, bite error rate (BER), and quality factor (Q-factor). Each and every one of the parameter have been analyzed at 10 Gbps data rate with NRZ and over 100 GHz. The result is displayed on BER analyzer Figs. 7, 8, 9, and 10, and eye diagram for 32-channels is shown in Figs. 7, 8, 9, and 10. Q-factor calculates selectivity of the system. A high Q-factor is good for the arrangement it will have lower loss and minimum BER. As result, DWDM system model is most excellent for optical communication, and it has high-factor and lower bite error rate. The quality element and bite error rate determined via various the fiber duration are plotted in determine as illustrated in above in Figs. 7, 8, 9, and 10 (Tables 1, 2, and 3).

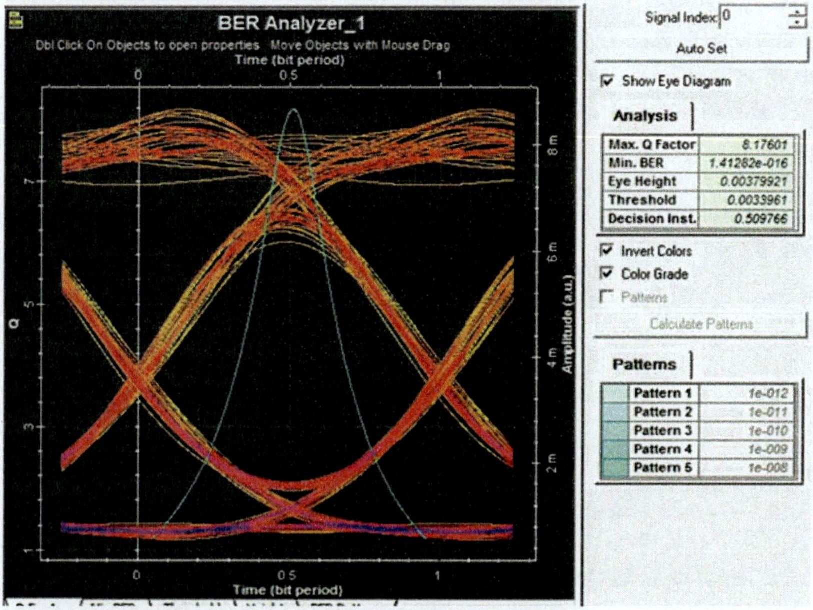

Fig. 7 Eye diagram analyzer for 150 km

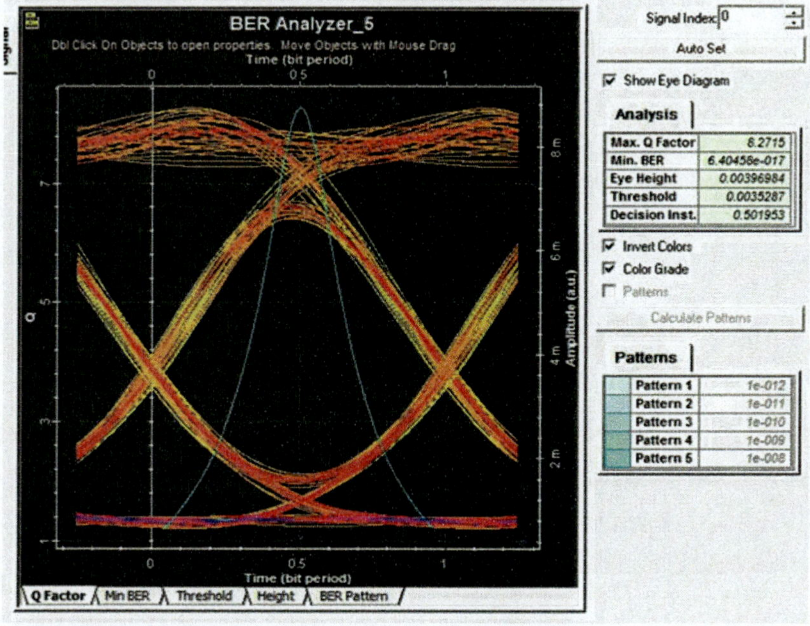

Fig. 8 Eye diagram analyzer for 50 km

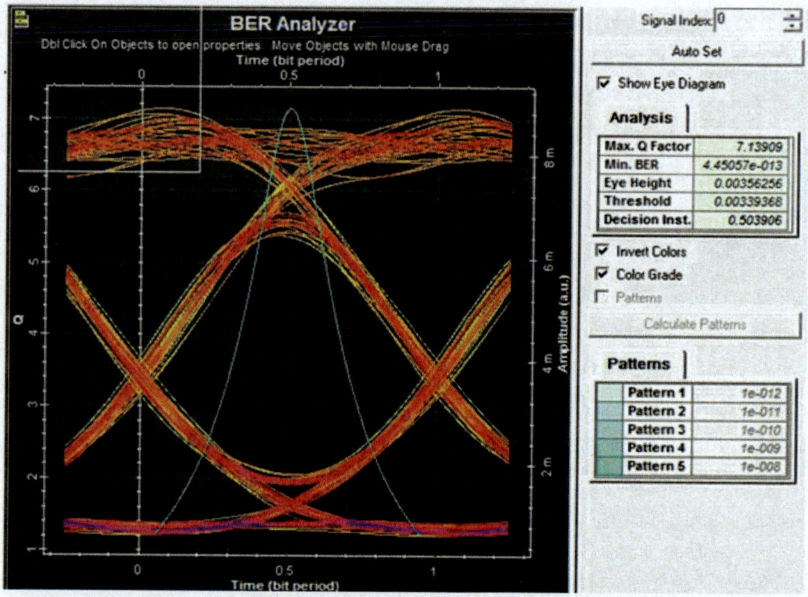

Fig. 9 Eye diagram analyzer for 250 km

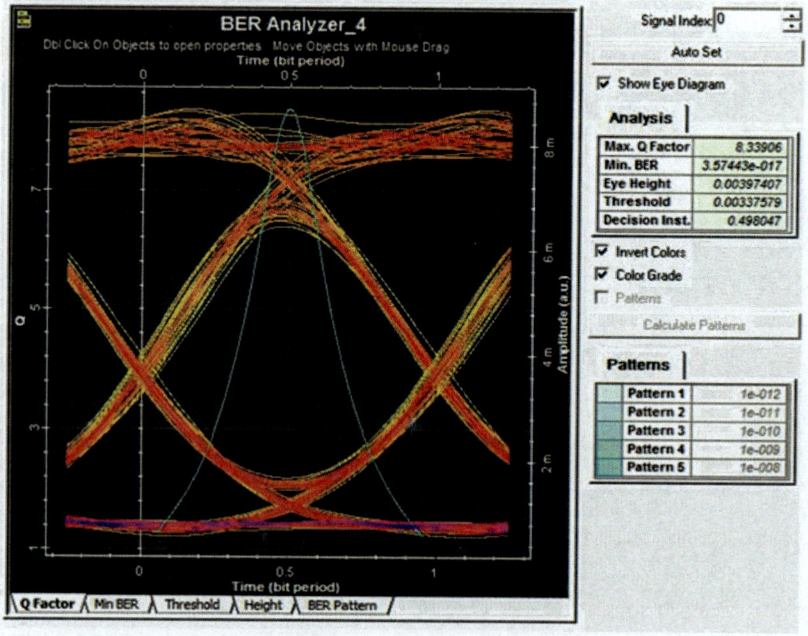

Fig. 10 Eye diagram analyzer for 20 km

Table 1 Simulation parameters of OA

Factors	Value
Span	25 km
Fading	0.2 dB/km
OF extent	25 km
Attenuation	0.2 dB/km

Table 2 Simulation factors of OF

Factors	Value
Process form	Get manage
Achieve	15 dB
Capacity	10 dBm
Noise number	6 dB

Table 3 Simulation variable of DWDM

Factors	Value
Frequency	193.414 THz
Frequency range	100 GHz
Capacity	0 dBm
Extinction ratio	30 dB
Modulation	NRZ

4 Conclusion

In this research article, we have design and implemented DWDM scheme model by means of OptiSystem software. From performance of DWDM, it is obvious that it reform the prospect of optical communication system. A variety of limitation of WDM system could be overcome by utilizing DWDM that provide high bandwidth and transmission losses can be reduce by using amplifier. DWDM network gives high Q-factor with minimum bite error rate (BER) as its mention on the table. This system model has various advantages such as high transmission capacity, flexibility, information transparency, and high reliability, and it is appropriate for long distance transmission.

References

1. Chinenye OD (2017) Enhancing signal production for promulgating information in a fiber optic communication system. Am J Eng Res (AJER) 6(11)

2. Shanker R, Srivastava P, Bhattacharya M (2016) Performance analysis of 16-channel 80-Gbps optical fiber communication system. In: International conference on computational techniques in information and communication technologies (ICCTICT)
3. Tosson M, El-Deeb WS, Abdelnaiem AE (2015) Dispersion compensation techniques for DWDM optical networks. Int J Adv Res Comput Commun Eng (IJARCCE) 4(8)
4. Choi I, Zhou YR, Dynes JF et al (2014) Field trial of a quantum secured 10 Gb/s DWDM transmission system over a single installed fiber, 22 Sept 2014. Optical Express (OSA), vol 22
5. Annapoorani B, Arul Murugan C (2013) DWDM: an innovation in optical communication. Int J Innov Res Comput Commun Eng (IJIRCCE) 1(4)
6. Singh M (2015) Performance analysis of different dispersion compensation schemes in a 2.5 Gbps optical fiber communication link. Int J Technol Enhancements Emerging Eng Res 3(08)
7. Kaur H, Mittal M (2016) Region based image segmentation for brain tumor detection. Int J Eng Manag Res 6:31–34
8. Khair F, Mustika IW, Fahmi, Zulherman D, Hario F (2018) Comparative analysis of dispersion compensating fiber in DWDM system using 10 Gbps and 40 Gbps bit rate. In: 2018 10th international conference on information technology and electrical engineering (ICITEE)
9. Jakborvornphan S (2019) Trend of optical fibre in the next generation. Int J Sci Eng Res 10(3)
10. Sharma P, Arora RK, Pardeshi S, Singh M (2013) Fibre optic communications: an overview. Int J Emerging Technol Adv Eng (IJETAE) 3(5)
11. Ibragimov RZ, Fokin VG (2018) Design of long-haul coherent DWDM optical systems. In: 2018 14th international scientific-technical conference APEIE—44894
12. Kumar R, Srivastva DK, Haider S (2016) Design and analysis of WDM OFDM system using optisystem. Int J Adv Res Comput Commun Eng (IJARCCE) 5(10)
13. Ilic S, Jaksic B, Petrovic M, Markovic A, Elcic V (2013) Analysis of video signal transmission through DWDM network based on a quality check algorithm. In: ETASR—engineering, technology & applied science research, vol 3, no 2
14. Kaur M, Singh S, Kaur G (2019) An efficient method of resource allocation in optical networks using quantum key distribution and OSTBC encoding process. Int J Innov Technol Exploring Eng (IJITEE) 8(12)
15. Idachaba F, Ike DU, Hope O (2014) Future trends in fiber optics communication. In: Proceedings of the world congress on engineering, vol I
16. Biswas D (2017) Modern optical fiber—communication splitter transmitter module. Asian J Appl Sci Technol (AJAST) 1(3):228–231
17. Sharma S, Sharma V, Kaur D (2015) Review paper on DWDM technology. Eur Sci J 2 (Special edition)
18. Gangwar A, Sharma B (2012) Optical fiber: the new era of high speed communication (Technology, advantages and future aspects). Int J Eng Res Dev (IJERD) 4(2):19–23
19. Bhasin A, Kaushik M, Sharma U (2016) Wavelength division multiplexing optical networks. Int J Res Appl Sci Eng Technol (IJRASET) 4(X)
20. Kumar S (2012) Study WDM technology to achieve higher bandwidth requirements in optical fiber based backbone network in communication systems. MERI J Manag IT 5(2)
21. Parul (2014) Brain wavelength division multiplexing: an overview & recent developments. Int J Eng Res Appl (IJERA) 4(5):114–117
22. Rambabu K (2017) Recent development and major implementation channel crosstalk in WDM optical fibre communication link. Int J Recent Trends Eng Res (IJRTER)
23. Sharan L, Shanbhag AG, Chaubey VK (2016) Design and simulation of modified duobinary modulated 40 Gbps 32 channel DWDM optical link for improved non-linear performance. Cogent Eng J 3(1)
24. Kiaee MS, Seraji FE (2016) Design of a 32×5 Gb/s DWDM optical network over a distance of 1000 km. Int J Optics Appl, e-ISSN: 2168-5061
25. Kavitha T, Pavani R (2019) Dense wavelength division multiplexing using dispersion fibre with erbium doped fiber amplifier. Int J Innov Technol Exploring Eng (IJITEE), ISSN: 2278-3075, vol 8, issue 11

26. Paliwal R (2017) Enhancing performance of 10 Gbps DWDM optical link for high speed optical communication. Int J Innov Stud Sci Eng Technol (IJISSET) 3(4)
27. Geerthana S, Preethi Anushiya M, Sasirekha V, Sumitha S, Thasleen Banu A (2017) Performance analysis of 8-channel & 16-channel optical ,fiber using WDM system. Int J Eng Res Technol (IJERT) 5(13)
28. Toda H, Yamashita T, Kitayama K, Kuri-j T (2001) A DWDM MM-wave fiber-radio system by optical frequency interleaving for high spectral efficiency, IEEE
29. Lakshmi DL, Srikanth PC, Sanjaya Kumar C (2017) DWDM in transmission system, a review. Int J Eng Res Electronics Commun Eng (IJERECE) 4(12)
30. Kumar A, Sharma A (2017) A 5x16 Gbps DWDM system for ground-to-satellite using RZ signaling scheme under different turbulences. In: 7th international conference on advances in computing & communications (ICACC)
31. Sarkar C (2013) Interfacing of 8 channel optical MUX DEMUX in DWDM system. Int J Sci Eng Res 4(4)
32. Bakaul M, Lim C, Waterhouse RB (2006) Simultaneous multiplexing and demultiplexing of wavelength-interleaved channels in DWDM millimeter-wave fiber-radio networks. J Lightwave Technol 24(9)
33. Balaj VR, Murugan M, Robinson S (2016) Optimization of DWDM demultiplexer using regression analysis. J Nanomaterials (Hindawi)
34. Wasfi M (2009) Optical fiber amplifiers-review. Int J Commun Netw Inf Security (IJCNIS) 1(1)
35. Jain P, Gupta N (2014) Comparative study of all optical amplifiers. Int J Sci Eng Res 5(11), ISSN 2229-5518
36. Rashed ANZ (2012) Transmission performance evaluation of optical add drop multiplexers in optical telecommunication ring. Am J Eng Technol Res (AJETR) 12(1)
37. Kaur C, Sonia, Singh H (2016) Comparative analysis of DWDM system using OADM (Optical Add/Drop Multiplexer) at different data rate, distance and channel spacing. Int J Eng Dev Res (IJEDR) 4(2)
38. Antil R, Pinki SB (2012) An overview of DWDM technology & network. Int J Sci Technol Res (IJSTR) 1(11), ISSN 2277-8616

A Systematic Review of Automatic Cutting Machine

Sukhmeet Singh and Nitin Sharma

Abstract In a factory or in a kitchen daily we used to cut objects and vegetables. And we have to do this process in an efficient way without damaging the cutting tool. Lots of researchers are doing experiments on cutting of different types of materials like wood, metals, glass, fish, vegetables, etc. They used different types of cutter which can cut the object by using electricity and by using solar light. The main purpose of this paper is to find the gaps from the systematic review on automatic cutting machine. This review was conducted according to PRISMA guidelines. For this we search different libraries, namely Web of science, IEEE and Science Direct. The last search was performed on January 4, 2022, by us. We find 3414 publications from initial search, out of which 17 publications were retrieved. Out of these 17 publications 8 publications were excluded due to duplicate publication. But only 7 publications are of our interest.

Keywords Cutting · Automated

1 Introduction

Nowadays in every field the automation is used. It not only makes our life easier, but we can also do all our work very efficiently. In any factory where we are using automation we have to use the energy. By the help of automation we can automatically turn on and turn off the electricity. For automatic cutting, drilling, painting, etc. we are using automation. So in 2012 C. Postelnicu et al. proposed two models by which we can reduce the consumption of energy and decrease greenhouse gases up to 20%. The Earth has a characteristic nursery impact because of the following measures of

S. Singh · N. Sharma (✉)
Department of Electronics and Communication Engineering, Chandigarh University, Mohali, India
e-mail: nitinsharma.ece@cumail.in

S. Singh
e-mail: Sukhmeet.e9511@cumail.in

J. R. Szymanski et al. (eds.), *Energy Systems, Drives and Automations*,
Lecture Notes in Electrical Engineering 1057,
https://doi.org/10.1007/978-981-99-3691-5_27

nitrous oxide, carbon dioxide, methane and water fume in the climate. These gases ingest infrared radiation transmitted by the Earth which prompts the warming of the outer layer of the planet. The normal nursery impact is because of some parts that are available on the Earth. For example, cutting of plants because of woods fires, nitrogen oxide is available in water and carbon dioxide is available in the seas, Animals and individuals inhale oxygen and delivery carbon dioxide. Aside from that there are a few other man-made reasons of greenhouse gases. Ranchers utilize counterfeit nitrogen in their fields. This delivers nitrogen oxide in the air. Because of deforestation the carbon dioxide level is consistently expanding. This is additionally a significant reason for the expansion in the nursery impact. Consuming of energies, for example, petroleum and coal discharges carbon dioxide in the air which additionally causes nursery impact. By this we can also increase the percentage of renewable energy and efficiency up to 20%. By Cournot and Bertrand models we can find out the energy consumption and total price which we have to pay for it. We can use these models in different categories like in workstation, lightning services, transportation, etc. [1].

Horticulture is a significant wellspring of job in India. In times past the cultivating in the greater part of the towns was crude, when rancher developed grain and cereal harvests. This prompts exceptionally low pay of ranchers. Current horticulture is a way to deal with rural advancements and cultivating rehearses that assist ranchers with expanding productivity. Cultivating is the development of plants, crop, homegrown creatures, fish, birds and so on with the end goal of food, material and different items. It diminishes the utilization of synthetic and other poisonous items, which keeps manor from vermin and infections. With this ranchers can procure better pay by advertising his harvests through co-usable social orders. It is a craft of dealing with the development of plants and creatures for human use. Presently the things have changed because with the utilisation of manufactured manures, cultivating efficiency has expanded a lot. To preserve water numerous techniques are utilized to slow water spill over and trap silt and supplements. In modern farming farmers also use the automation for their good production of the crops, especially in greenhouse where we have to monitor humidity, temperature and pressure. So for this D. Saraswathi et al. from India designed a machine which can monitor temperature, humidity and pressure. By using these designs farmers can monitor humidity level of soil for irrigation purpose. The farmers will get all the values of temperature, humidity and pressure on their mobile app [2].

The safety of all the members in the house is an important factor of our life, especially the elders or old peoples who are living alone in their residence. There are different viewpoints that should be investigated to guarantee that house is a protected spot for every one of the individuals. Security is a wide idea, and its execution is unreasonably difficult, yet it is essential. There is likewise no alternate route with regards to somewhere safe. Mostly kids and elderly folks individuals need neglected care and backing because of the difficulties confronting the formal and casual consideration framework. At the point when we discuss more established individuals close to thirty-three percent of them falls among the sixty-four age bunch bring about shoulder, arm, leg and wounds. One in each five falls among women matured in excess of fifty-four and over brings about a crack. Closed circuit television can monitor what is going on at the premises where they are installed. The presence of a decent security framework including surveillance cameras shows that you care about the wellbeing of your loved ones. Closed circuit television empowers you to observe from a distance, and on the off chance that this is beyond the realm of possibilities, you can watch recorded tapes. Here automation also plays an important role for the safety of the elders. In 2010 T.B.T. Truong et al. from France designed a machine. By the help of this machine one expert or an occupational therapist constantly observes the moment of the old people in the house. In any casualty he will contact any family member and the medical staff [3].

In everybody's life having their own home is one of the fantasies they need to satisfy in their lives. The development business has been surviving a thrilling period of its life. Whenever we are developing a structure then we need to favorite numerous issues. The work deficiency likewise had an impact in blocking efficiency as of late, particularly with laborers not having the great experience. Laborer wellbeing keeps on being an issue tormenting the development business. Correspondence is a significant apparatus in any calling, particularly significant when work is separated in different gatherings. Without clear and viable conveying, significant undertakings can prompt breaks. In the long haul, computerization is probably going to increment efficiency and permit wages to ascend for laborers with cutting edge abilities. The extent of mechanization in development is extremely wide. The new advancements in the fields of software engineering and mechanical technology have assisted with growing new innovations in the field of development industry. We are using the automation not only in the factories, but we are also using it in the construction of the buildings. By using the automation we can increase the quality of the construction as well as the safety of the building. This automation helps labor to do their work efficiency and safety. This automation not only reduces the time but also reduces the fatigue [4].

In this computer era lots of jobs are related to laptop or desktop. Many of these employees have to work with computer and laptop. Computer is a customized gadget with a gathering of guidelines to perform explicit undertakings and create results at an extremely rapid. In spite of the fact that there are many benefits to utilizing a computer, there are likewise risks and inconveniences. The term eye strain is everywhere now and again utilized by individuals to depict a gathering of dubious side effects that are connected with utilization of the eyes. Many individuals invest significant stretches of energy taking a gander at their telephones, tablets, computer screens and TVs and establish a superior workplace to assist with forestalling advanced eye strain. Certain individuals, while focusing on an outwardly extreme errand like perusing fine print, involve the computer for a really long time at a time. Notwithstanding, squinting is a significant natural capacity that keeps the outer layer of the eyes perfect and greased up. Side effects of eye strain are uncommon in offspring of various ages. As youngsters invest more energy than any time in recent memory gazing at computerized screens on computers, tablets, TVs, cell phones and different gadgets, lots of time they forget to take the break when they are working. So E. Beck et al. had designed color bar reminder on your screen. This reminder will give different colors. Up to 1 h continuous working the bar color is white. When the employee works continuously in between 1 and 2 h the bar color is yellow. And continuous work of more than 2 h the color will be red [5].

India is the largest producer of the spices, milk and pulses in the world. As the technology changes, the farming is also converting to the modern farming. Farmers have incredible significance in our general public. They are the ones who give us food to eat. They sow and harvest crops. At the point when the right season shows up, they likewise reap crops. It additionally incorporates developing harvests, vegetables, organic products and blossoms. The economy of any nation just relies upon farming. They are the significant piece of the economy. Natural cultivating is a method, which includes the development of plants and raising of creatures in regular ways. Natural agribusiness joins custom, advancement and science to help the common climate and advance fair connections and a decent personal satisfaction for all individuals. Natural cultivating includes the utilization of natural composts and items acquired that is really great for health. This will help in rationing the proficiency by making its supplement rich, so the jeans and harvests developing that dirt would be sound and supplement rich. Rural water will be water that is utilized to develop new produce and support domesticated animals. Farmers are also using different kinds of apps and weather forecasting information from the internet. So K. Devi et al. had designed artificial intelligence voice bot. The farmer can take help or information about soil, weather, crops as well as use of fertilizers [6].

Wellbeing is a condition of complete physical, mental and social prosperity and not simply the shortfall of infection or sickness. Medical services frameworks are complicated, and there are numerous things you want to be familiar with kinds of emergency clinic frameworks, patient consideration, protection, medical care suppliers and legitimate issues. Wellbeing is a fundamental piece of our life. Being sound and dynamic is generally a benefit for us. Medical services are critical to the general public since individuals get sick, mishaps and crises do emerge. Furthermore, the medical clinics are expected to analyze, treat and oversee various kinds of afflictions and diseases. Health care is portrayed as the recognizable proof, fix, counteraction as well as the executives of a sickness, injury, ailment and the protecting mental and actual prosperity in people. Medical care is characterized as any errand embraced to upgrade the physical, social and mental condition of individuals. Each nation has a huge medical care framework. Automation also changes the world of health care. Nowadays the patient can also take advice from the doctor via the video call. So N. Pereria et al. work on the problems of the healthcare services. The consumer can order the medicines to hospitals or suppliers. In this case he can use the online payment or pay at delivery option. This is also called e-channeling [7].

2 Existing Work

The existing work in the automatic cutting has been searched in different libraries, namely Web of science, IEEE and Science Direct. About 3414 papers were searched and out of which 17 publications were retrieved. Out of these 17 publications 8 publications were excluded due to duplicate publication. But only 7 publications are of our interest. The search process is depicted in Fig. 1.

The existing automatic cutting methods that are searched in the literature can be summed up as given in Table 1.

Nowadays the use of the robot is increasing day by day. Robot is not only used to do our work in home and in offices, but it is also used to attract the customers. We can take help from robot according to our need. Nowadays the robot is also used in medical field. So B. L. Davies had thrown the light to the robot which can do

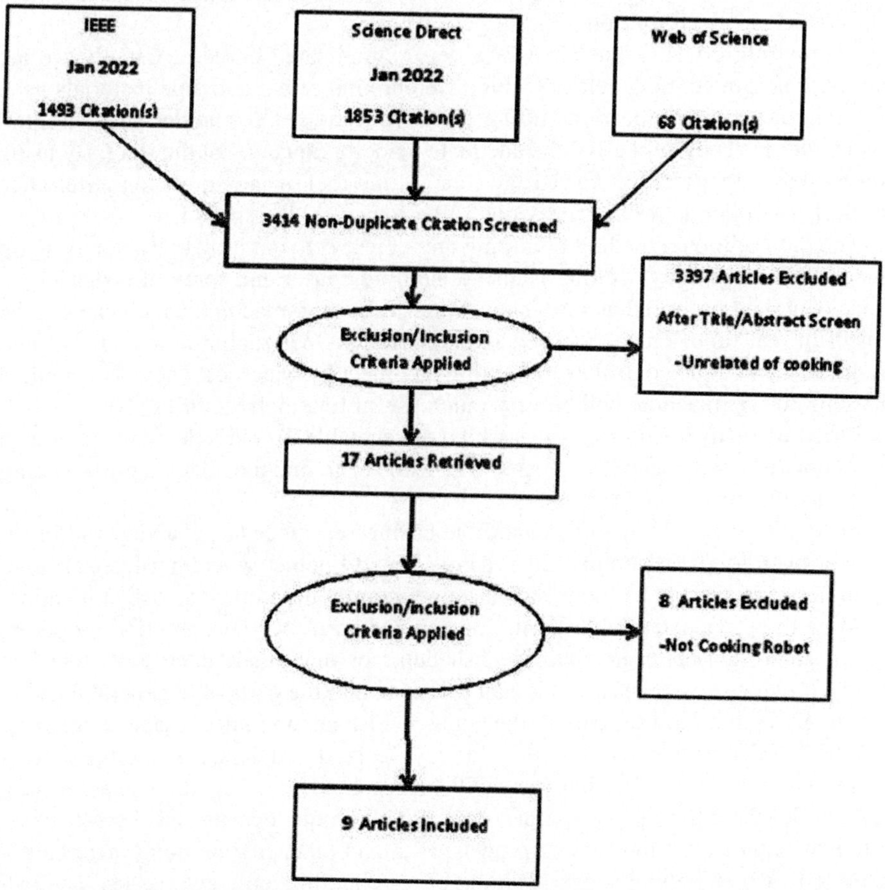

Fig. 1 Search process undertaken to find the suitable publication

Table 1 Various existing object cutting methods

Author	Year	Country	Objective	Object
B. L. Davies	2002	UK	Surgery	Human body
C. Machado et al.	2003	Portugal	Cutting	Pipe
A. Hace et al.	2004	Slovenia	Water jet cutting	Metal sheet
D. Zhou et al.	2006	Atlanta	Cutting vegetable	Potato
H. Lag et al.	2008	Columbia	Automatic cutting	Fish
T. Incirci et al.	2011	Turkey	Cutting using sunlight	Glass
P. Long et al.	2014	France	Cutting at an angle	Soft material
K. M Lee et al.	2018	USA	Find cutting temperature	Sheet
X. Mu et al.	2019	USA	Cutting	Vegetable, fruit

surgery. This robot is having one camera, DC motor, gearbox and different cutting tools which are used for doing the surgery [8].

The cutting-off bars and pipes is a very complicated process. Usually we are cutting it at same cutting velocity. But different kind of bar and pipe materials need different of force and speed of cutting. So C. Machado et al. had designed an intelligent cutting. They suggested the adequate velocity range to cut the steel. By using this cutting velocity range we can save our cutting tool or saw for its long life. They used PI controller as well as fuzzy controller to control the speed [9].

The old traditional method of cutting any object is to cut it by knife or any sharp material. But nowadays cutting is also done by the laser and spray of water jet. A. Hace et al. had proposed an automatic water jet cutting machine which can cut the object in real time. They proposed to use CAD/CAM, graphic user interface and programmable logic controller for cutting an object by water jet. They also claimed that this cutting machine will be very much useful for small factories [10].

There are many methods to cut any kind of vegetable. But in order to make French fry of potatoes we have to slice them. So D. Zhou et al. find the relationship of slicing angle and fracture force [11].

Fishing is one of the world's incredible businesses, since fish is a vital and likely wellspring of food. Fish cultivating is a type of hydroponics wherein fish are brought up in nooks to be sold as food. Fish farming, commonly known as fish cultivation, involves the propagation and rearing of fish, primarily for consumption purposes, within controlled environments like fish tanks or man-made enclosures like fish ponds. It likewise assumes a significant part as one of the parts of in general farming area in giving the food security to the country. Fish are an indispensable wellspring of food, particularly in nations like Norway, Iceland and Japan where the land is depressing or rugged and agribusiness will not be quickly developed. Business fishing segment is where the people or enterprises get all unique oceanic species like stock fish, fish, squid, crab, lobster and prawns. Fish and other marine items are wealthy in proteins which help in weight training or supplant broken down tissues. The fish squanders from canneries are made into fish-suppers, pastes, oils and composts.

In British Columbia millions of dollars was wasted and only because of poor fish processing. So in order to reduce this wastage H. Lag et al. had designed an automatic cutting machine. This machine was made under the collaboration of University of British Columbia with the industrial automation laboratory. This machine is having vertical cutter and horizontal cutter which can cut the fish. There is also one conveyor belt which is used to convey the fish from one side to another. By using this machine not only we can save our time, but all the labors will get rid of the foul smell of cutting fish.

Glass is used in every house. But cutting of glass is a very difficult process. It needs extra care while cutting and handling. So T. Incirci et al. had proposed a cutting machine which can be operated by solar panels. This machine is fully automatic which is controlled by programmable logic controller and motion controller [12].

When we hear the term cutting, then mainly the process which comes in our mind is to fully cut that item. Sometime the requirement is to partially cut an object up to a certain depth. The process of cutting an item up to a certain depth is very difficult process, especially when an item is cut automatically at a certain angle. So P. Long et al. had designed a robotic cutting machine which can cut the soft materials at different angles [13].

When we are cutting any kind of material the temperature of that material goes high. Sometime it also leads to a loss of tool or material which we are cutting. So K. M. Lee et al. had designed a method which takes the image of surface. This images are based on infrared images which can tell about temperature or heat produced at the surface. If the image has red color then the temperature is high at that place. If the images have blue color or light green color then the temperature is low at that area. So by using this technique we can estimate that if the area of the red color increases then the cutting should be stopped. Otherwise the machine can keep on cutting [14].

A human being needs practice of cutting of objects like apple, orange, lemon, fish, etc. If this process can be done by a robotic hand then lots of time will be saved, especially in restaurants where chefs spend lot of time in cutting vegetable and fruit. So X. Mu et al. had proposed knife cutting mechanics. They proposed two degree of freedom robotic arm which is having sensor to sense the force. They proposed slice, press and push sequence to an object [15].

3 Limitation

Many types of research had done for cutting different kinds of objects. Mainly all these cutting are for cutting metals sheets and pipes for both small factories and large factories. But very less amount of work had been done for cutting or slicing different kinds of vegetables like potato, tomato, onion, cauliflower, ginger, etc. We can use all these cut vegetables in automatic cooking machine.

4 Conclusion

The literature review represents the summary of different kinds of cuttings for different kinds of objects. But in all this cutting the temperature of cutting tool is not mentioned. The temperature of the cutting tool plays an important role. Different authors have also used different techniques like water jet spray and knifes. Still we have to do lots of experiment to cut an object by applying minimum power or electricity.

References

1. Postelnicu C, Lastra JLM (2012) Applicability of market-based competition models for energy efficient factory automation systems, pp 1–5
2. Saraswathi D, Manibharathy P, Gokulnath R, Sureshkumar E, Karthikeyan K (2011) Automation of hydroponics green house farming using IOT. In: 2018 IEEE international conference system computation automation networking, pp 1–4
3. Truong TBT, De Lamotte FF, Diguet J, Sa F (2010) Alert management for home healthcare based on home automation analysis, pp 2128–2131
4. Ardiny H,Witwicki S, Mondada F (2015) Construction automation with autonomous mobile robots: a review
5. Beck E, Von Holdt K, Meyer J (2019) Sneaking physical exercise into sedentary work life: design explorations of ambient reminders in opportune moments. In: 2019 IEEE international conference healthcare informatics, pp 1–7
6. Divakar MS, Kumar V, Martina Jaincy DE, Kalpana RA, and Sanjai Kumar RM (2021) Farmer's assistant using AI voice bot, pp 527–531
7. Jayasena S (2006) Automation for service delivery and growth in the Sri Lankan healthcare industry, vol 00, pp 17–20
8. Davies BL (2002) Robotic surgery: at the cutting edge of technology, pp 15–18
9. Machado C, Mendes J, Fonseca J (2003) Intelligent cutting-off of pipes and bars, pp 460–465
10. Hace A, Jezernik K (2004) Control system for the waterjet cutting machine. IEEE/AsME Trans Mechatron 9(4):627–635
11. Zhou D, Claffee MR, Lee K, Mcmurray GV (2006) Cutting, 'by pressing and slicing', applied to the robotic cut of bio-materials , part II: force during slicing and pressing cuts, pp 2256–2261
12. Deri İ, Bölgesi S, Cad O, Istanbul T (2011) This work-in-progress paper was presented as part of the main technical program at IEEE ETFA' 2011 automation of patterned and solar glass cutting and stacking line Taner İncirci, pp 1–4
13. Long P, Khalil W, Martinet P (2014) Robotic cutting of soft materials using force control & image moments. In:2014 13th international conference on control automation robotics and vision (ICARCV), pp 474–479
14. Lee K, Huang Y, Ji J, Lin C (2018) An online tool temperature monitoring method based on physics-guided infrared image features and artificial neural network for dry cutting, pp 1–12
15. Mu X, Xue Y, Jia Y (2019) Robotic cutting: mechanics and control of knife motion, pp 3066–3072

High-Performance Code Compression Using Adaptive Encoding for RISC Processor

D. Lakshminarayana and V. Ganesan

Abstract In this paper, high-performance code compression using adaptive encoding for RISC processor is implemented. The main intent of reduced instruction set computer (RISC) processor is to utilize the small and optimized instructions. Code compression in RISC processor will encode the optimized data to a certain extent. Initially, the data is moved from one place to another using data bus buffer. Adaptive encoding will compress the lossless data and characterize the lossless data. Basically, encoding in RISC processor will improve the usage of look-up tables and slices. It causes the more utilization of area, so to overcome this problem, adaptive encoding in RISC processor is implemented. From results, it is observed that the total delay of conventional encoding in RISC processor is 21.015 ns and adaptive encoding in RISC processor is 4.128 ns. Logic delay of conventional encoding in RISC processor is 12.843 ns, and adaptive encoding in RISC processor is 3.493 ns. Route delay of conventional encoding in RISC processor is 8.172 ns, and adaptive encoding in RISC processor is 0.725 ns. From the simulation results, it is observed that the total delay, logic delay, route delay, memory usage, number of slices, and number of LUTs are reduced effectively compared with the conventional encoding in RISC processor.

Keywords RISC · Code compression · ALU · Instruction adaptive decoder and machine cycle adaptive encoder

---

D. Lakshminarayana (✉)
Department of Electronics and Communication Engineering, Sathyabama Institute of Science and Technology, Chennai, Tamilnadu, India
e-mail: lakshman5504@gmail.com

V. Ganesan
Department of ECE, KG Reddy College of Engineering &Technology, Hyderabad, India
e-mail: ganesan.ece@bharathuniv.ac

D. Lakshminarayana
Bharath Institute of Higher Education and Research, Chennai, India

J. R. Szymanski et al. (eds.), *Energy Systems, Drives and Automations*,
Lecture Notes in Electrical Engineering 1057,
https://doi.org/10.1007/978-981-99-3691-5_28

1 Introduction

RISCs are presently used for all sort of computational errands. In the space of logical registering, RISC workstations are as a rule progressively utilized for process assignment, for example, digital signal processing, digital image processing and so on. RISC ideas help to accomplish given degrees of execution at fundamentally lower cost than different frameworks [1].

Pipelined RISC improves speed and cost viability over the simplicity of equipment depiction language programming and preservation of memory, and RISC-based plans will keep on filling in speed and capacity. The primary highlights of RISC processor are the guidance set that can be designed to speed guidance execution. In the current work, the plan of a 4-bit information width RISC processor is introduced. It has a total guidance set, program and information recollections, broadly useful registers and a straightforward arithmetical logical unit (ALU) for fundamental activities.

In this architecture, the design performance guidelines are of uniform length and comparable design, number juggling activities are limited to CPU registers and just separate burden and store directions access memory. The engineering upholds eighty directions to help arithmetic, logical, shifting and burden–store tasks. Verilog HDL has developed as a standard equipment portrayal language. In this regard, the design approach utilizes Xilinx Vivado to design the architecture.

HDL permits the architecture to be recreated before in the plan cycle to address blunders or examination with various models. Plans depicted in HDL are innovation autonomous, simple to plan and investigate, and are typically more intelligible than schematics, especially for enormous circuits.

Basically, Verilog is utilized as a contribution for combination programs which will create a door-level portrayal for the circuit. The test system which is utilized for the language is Xilinx ISE and Modelsim. Verilog is equipped for portraying basic conduct. Machine cycle directions permit the processor to deal with a few guidelines simultaneously. The processor can work at a high clock recurrence and in this manner yields higher speed. This paper is about plan of a basic RISC processor and blending it. The RISC design adheres to single-cycle guidance execution.

For installed frameworks, there has been an expanded mindfulness as of late on diminishing the memory necessities of uses running in such frameworks. Code pressure lessens the code size of the program to be run in this manner diminishing the general memory prerequisites alongside decreasing the cycles required, framework transport movement and force necessities. Code pressure strategies can be extensively partitioned into two sorts, for example, the compiler-driven improvement strategies and the post-compiler pressure methods.

Overview on the compiler strategies for code pressure, and recorded different techniques utilized by the compiler to diminish code size, e.g., dead code disposal, strength decrease and code-considering. The other class of methods includes object code pressure which is done disconnected and runtime decompression while really executing the code. In this paper, we center around on the latter class of methods.

Pressure of item code varies from ordinary information pressure as it ought to accommodate steady decompression which infers that decompression can begin from any point in the code and not really from the beginning. LZW sort of word reference-based codecs and entropy-based Huffman codes cannot be applied on the grounds that they cannot deal with steady decompression. Besides, such codes are not favored in light of the fact that it will lose the memory arrangement because of the variable length code words prompting debasement in execution. The vast majority of the past work on code pressure has been on fixed length inserted RISC processors.

In this work, we present a code pressure calculation for variable length RISC processors. The feasible pressure for a variable length processor is lesser than that for a fixed length processor as the guidelines are as of now entropy coded somewhat. In this manner even in uncompressed guidelines encoding for variable lengths, the more much of the time happening directions have more limited lengths and less regular ones have longer lengths subsequently lessening the general code size for the variable length processor.

2 Literature Survey

In [1], classical five-stage pipeline structure of RISC architecture is implemented which consists of reconfigurable array. This will interconnect the instructions of data by using the circuits which are related with each other. This RISC architecture consists of an instruction set which is based on the processor performance. Multiplication operation is realized by using ALU in two-level pipelining multiplier. By using the Verilog, HDL process is simulated and prototyped using FPGA development board.

In [2], the RISC processor is designed which executes the very small instructions in same time. In this, complexity of processor will be reduced as well as cost of processor also reduced. In 1970s, 16-bit RISC processor is designed and signified by the number of technical barriers.

In [3], barrel RISC-V processor is designed which controls the deep neural network accelerator. There are eight hardware threads (harts) which is based on the five-stage pipeline data path in this design. By using the strict round-robin scheduler, each thread is executed. As well as this will provide the control of data based on the neural network processing element (PE). General matrix vector (GEMV) operations are performed for the arbitrary precision. This will reduce the area in very effective way.

In [4], arithmetic circuits' usage is increased in redundant number systems (RNS) to increase the speed and performance of systems. Based on the subcircuit level, the arithmetic operations are performed. ALU mainly performs division operation in realization process. Data will be speed up based on the path which is created using CPU. This path will speed up the data to avoid the conversations. Hence in this, RISC-V3 architecture is implemented which uses residue number system. This RISC processor will increase the speed of execution of instructions.

In [5], in this design of RISC-V processor is done based on MCU which consists of Automatic License Plate Recognition (ALPR). Multi-model inference approach is introduced to increase the performance of system. Registration numbers are recognized in this real world. But in this, there will be no restoration process to accelerate the engines for future algorithmic improvements.

In [6], creators executed the 32-cycle RISC MIPS processor on Spartan-6 FPGA. The task includes reenactment and amalgamation of a processor. A RISC is a microchip that had been intended to play out a little arrangement of directions, determined to speed up the processor. The possibility of this undertaking was to make a RISC MIPS processor as a structure block in Verilog HDL.

In [7], a 32-bit MIPS-based RISC core is implemented and executed successfully. The engineering with pipelined control RISC center comprises bring, unravel, execute, pipeline control and memory. The decrease in the force is accomplished utilizing HDL adjustment procedures. This strategy is based on the memory guidance, arithmetic and logic unit and program counter.

In [8], creators introduced the plan of a RISC CPU design dependent on microprocessor interlock pipeline stages (MIPS) utilizing VHDL. It likewise depicts the guidance set, design and timing outline of the processor. Coasting direct number toward fixed number transformation is the principle task, while dealing with these numbers, this change has been accomplished by utilizing float-to-fixed number converter module.

In [9], a five-stage pipelined 32-bit MIPS-based RISC core is implemented and executed successfully. In RISC structure, microprocessor with interlocked pipeline stages is implemented. To speed up the processor, basically a RISC chip is introduced. These all are arranged in parallel format to increase the processor speed. The five modules of MIPS are given as Execution (EX), Instruction Decode (ID), Instruction Fetch (IF), Write Back (WB) and Memory Access (MEM) modules.

In [10], the architecture of five-phase pipeline structure for 32-bit RISC processor is implemented. The Modelsim test system is recreated, and it is finished. A microchip will be there in the RISC processor which will play major role to improve the speed of processor.

3 Adaptive Encoding in RISC Processor Architecture

Basically, the purpose of adaptive encoding technique is to reduce number of LUTs and slices of instructions. Figure 1 shows the adaptive encoding in RISC processor architecture. The main intent of RISC processor is to utilize the small and optimized instructions.

Code compressor in RISC processor will encode the optimized data to a certain extent. Initially, the data is moved from one place to another using data bus buffer. Basically, data bus buffer is a temporary storage which transfers data from main storage. Accumulator will add the data according to the sequence of numbers. ALU will perform the operation at high speed. Instruction register (IR) will control the

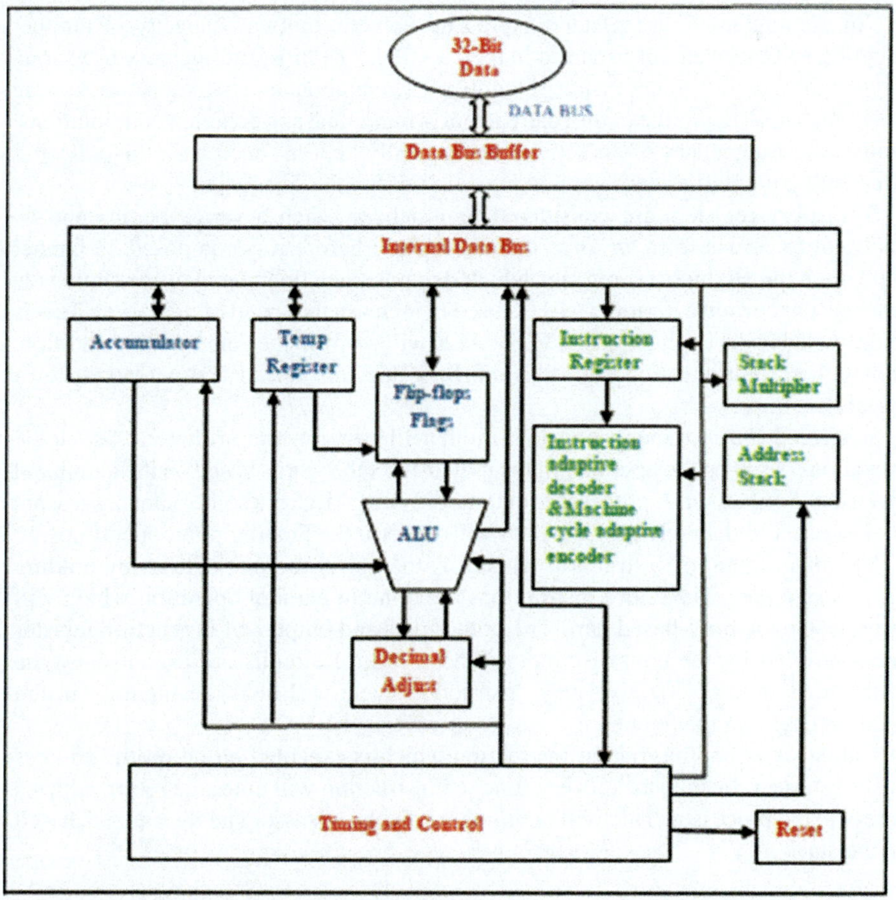

Fig. 1 Adaptive encoding in RISC processor architecture

entire operation by holding the instructions which are decoded. Instruction adaptive decoder and machine cycle adaptive encoder will perform the data according to the request arrival. Stack multiplexer will allow the data in a particular order.

The data bus is a PC subsystem that considers the moving of information starting with one segment then onto the next on a motherboard or framework board, or between two PCs. A commonplace information transport is 32 bits wide. This implies that up to 32 pieces of information can go through an information transport each second.

Data bus buffer (or just buffer) is a district of an actual memory stockpiling used to briefly store information while it is being moved starting with one spot then onto the next. In all cases, the information put away in an information cushion are put away on an actual stockpiling medium.

An accumulator is a register for present moment, halfway capacity of number juggling and rationale information in a PC's CPU. The most rudimentary use for an aggregator is adding a succession of numbers. It goes about as a transitory stockpiling area which holds a middle of the road esteem in numerical and coherent computations. Halfway consequences of an activity are logically kept in touch with the collector, overwriting the past esteem.

Transitory registers are considered uninitialized when a vertex begins and an endeavor to peruse from an impermanent register before it is composed. A banner flip lemon the slightest bit register which demonstrates flood, convey, or sign cycle from past or current activities. ALU is used to increase the speed of operation. This is mainly used in the applications of CPU. ALU will perform the data by using addition, subtraction, multiplication and division. By using this, speed and performance of system are improved.

Basically, the arithmetic and logic unit will perform the arithmetic and logic operations based on the operands. This will instruct the words based on the command given to the processor. A processor contains only one ALU, but some other processors consist of more than one ALUs which will perform the floating point operations.

The main intent of instruction register is to control the instructions by holding the decoded part. The entire instructions are done in parallel operation which will improve the memory-based part. The both inputs and outputs of instruction register are controlled by the control circuits. These control circuits are executed by the instructions in very effective way. Instruction cycle will hold the memory in the location based on the fetches.

Call stack is nothing but an instruction which is executed based on the address this is returned. In call stack, every time the instruction will enter the return address based on the processor. This will continue the fetches to associate the address based on the stack.

4 Results

Table 1 shows the comparison of number of LUTs and number of slices used in conventional and adaptive encoding techniques of RISC processor. From Table 1, it is observed that the adaptive encoding in RISC processor occupies less number of LUTs and slices.

Table 1 Comparison of number of slices and no of LUT's

S. No	Parameters	Conventional encoding in 32-bit RISC processor	Adaptive encoding in 32-bit RISC processor
1	No. of slices	364	223
2	No. of LUTs	660	76

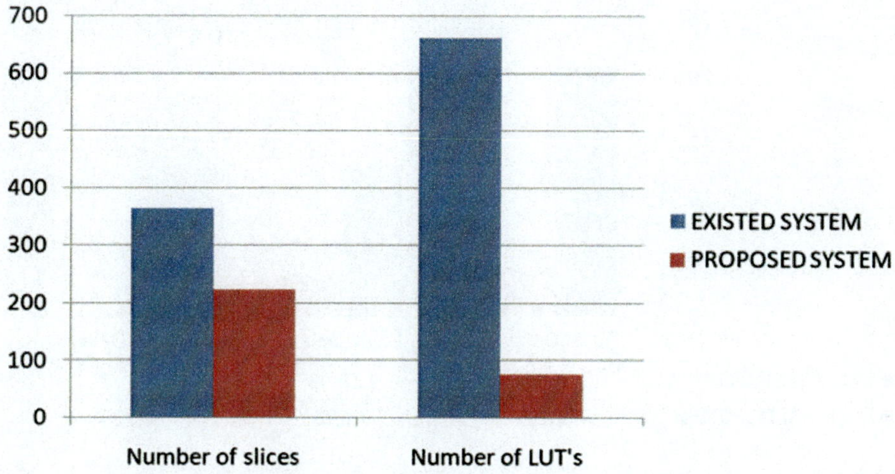

Fig. 2 Comparison of number of slices and no of LUTs

Figure 2 shows the comparison of number of LUTs and number of slices in existed and adaptive encoding in RISC processor. Compared to encoding in RISC processor, adaptive encoding in RISC processor gives effective results.

Table 2 shows the comparison of delay and memory in existed and adaptive encoding in RISC processor. From Table 2, it can observe that adaptive encoding in RISC processor reduces the delay in very effectively compared with encoding in RISC processor. Similarly, usage of memory is also reduced efficiently.

Figure 3 shows the comparison of delay in existed (conventional) and adaptive encoding in RISC processor.

Basically, the total delay is classified into two types; they are logic delay and route delay.

The total delay of conventional encoding in RISC processor is 21.015 ns, and adaptive encoding is 4.128 ns.

Logic delay of conventional encoding in RISC processor is 12.843 ns, and adaptive encoding is 3.493 ns.

Route delay of conventional encoding in RISC processor is 8.172 ns, and adaptive encoding is 0.725 ns (Fig. 4).

Table 2 Comparison of delay and memory

S. No	Parameters	Encoding in 32-bit RISC processor	Adaptive encoding in 32-bit RISC processor
1	Total delay	21.015 ns	4.218 ns
2	Logic delay	12.843 ns	3.493 ns
3	Route delay	8.172 ns	0.725 ns
4	Memory used	4,580,004 Kb	257,644 Kb

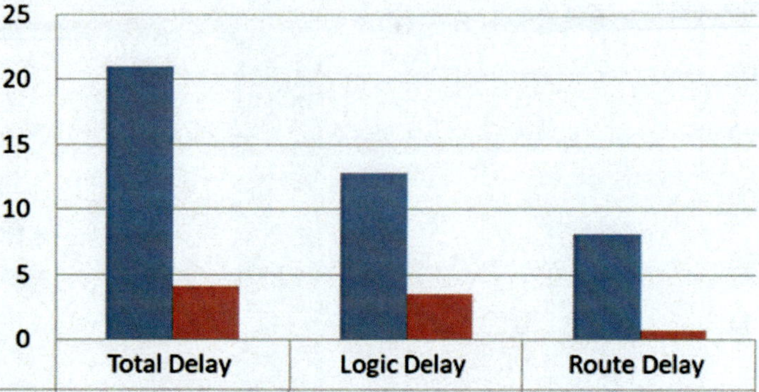

	Total Delay	Logic Delay	Route Delay
EXISTED SYSTEM	21.015	12.843	8.172
PROPOSED SYSTEM	4.128	3.493	0.725

Fig. 3 Comparison of delay

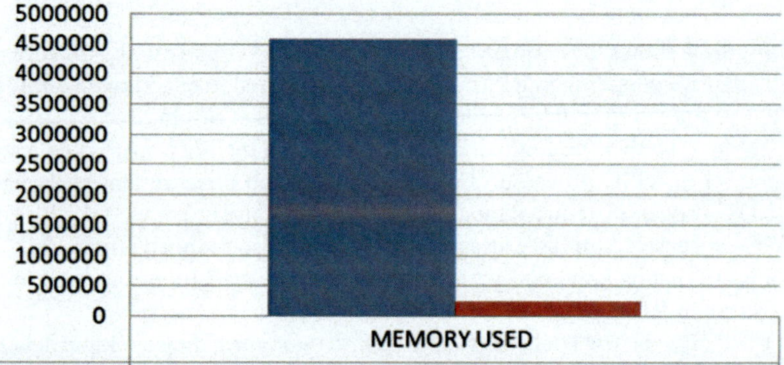

	MEMORY USED
EXISTED SYSTEM	4580004
PROPOSED SYSTEM	257644

Fig. 4 Memory usage

5 Conclusion

Hence in this paper, the design of high-performance code compression using adaptive encoding for RISC processor was implemented. The main intent of RISC processor is to utilize the small and optimized instructions. Instruction adaptive decoder and machine cycle adaptive encoder will perform the data according to the request arrival. Stack multiplexer will allow the data in a particular order. From simulation results, it can observe that the number of slices for encoding in RISC processor is 364 and adaptive encoding in RISC processor is 223. Similarly, the number of LUTs for encoding in RISC processor is 660 and for adaptive encoding in RISC processor is

76. Hence from these simulation results, it can be observed that adaptive encoding in RISC processor gives effective results compared with encoding in RISC processor.

References

1. Lu Y, Liu Y, Liao Y, Liu Y, Xu L (2021) Design of adjacent interconnect processor based on RISC-V. In: 2021 IEEE 4th international conference on electronics technology. https://doi.org/10.1109/ICET51757.2021.9451102
2. Kulshreshtha A, Moudgil A, Chaurasia A, Bhushan B (2021) Analysis of 16-bit and 32-bit RISC processors. In: 2021 7th international conference on advanced computing & communication systems (ICACCS). https://doi.org/10.1109/ICACCS51430.2021.9441873
3. Askari Hemmat MH, Bilaniuk O, Wagner S, Savaria Y, David J-P (2021) RISC-V barrel processor for deep neural network acceleration, 978-1-7281-9201-7/21/$31.00©2021IEEE. https://doi.org/10.1109/ISCAS51556.2021.9401617
4. Reichenbach M, Knödtel J, Rachuj S, Fey D (2021) Risc-V3: a RISC-V compatible CPU with a data path based on redundant number systems, Received 29 Jan 2021, accepted 19 Feb 2021, date of publication March 2, 2021, date of current version 24 Mar 2021. https://doi.org/10.1109/ACCESS.2021.3063238
5. Lamberti L, Rusci M, Fariselli M, Paci F, Benini L (2021) Low-power license plate detection and recognition on a RISC-V multi-core MCU-based vision system, 978-1-7281-9201-7/21/$31.00 ©2021 IEEE
6. Kelgaonkar PS, Kodgire S (2016) Pipelined 32bit RISC MIPS processor on Spartan-6 FPGA. Int J Sci Eng Technol Res (IJSETR) 5(8), ISSN: 2278-7798
7. Ajith Kumar P, Vijaya Lakshmi M (2016) Design of a pipelined 32 bit MIPS processor with floating point unit. Int J Innov Res Sci Eng Technol 5(7), ISSN: 2319-8753
8. Ritpurkar SP, Thakare MN, Korde GD (2015) Design and simulation of 32-bit RISC architecture based on MIPS using VHDL. In: IEEE international conference on advanced computing and communication systems (ICACCS-2015), Coimbatore, India, January 2015. https://doi.org/10.1109/ICACCS.2015.7324067
9. Topiwala MN, Saraswathi N (2014) Implementation of a 32-bit MIPS based RISC processor using cadence. In: IEEE international conference on advanced communication control and computing technologies (ICACCCT), pp 979–983, May 2014. https://doi.org/10.1109/ICACCCT.2014.7019240
10. Katke SP, Jain GP (2012) Design and implementation of 5 stages pipelined architecture in 32 bit RISC processor. Int J Emerging Technol Adv Eng 2(4), ISSN: 2250-2459
11. Sakthikumaran S, Salivahanan S, Kanchana Bhaaskaran VS (2011) 16-bit RISC processor design for convolution application. In: IEEE-international conference on recent trends in information technology, ICRTIT 2011. https://doi.org/10.1109/ICRTIT.2011.5972425
12. Plachno RS, VP of Audio, "A true single cycle RISC processor without pipelining", ESS Design white paper—RISC embedded controller. https://doi.org/10.4218/ETRIJ.04.0804.0003
13. Saambhavi VB, Kanchana Bhaaskaran VS (2011) A 16-bit RISC microprocessor using DCPAL circuits. Int J Adv Eng Technol (IJAET) 2(1):1–9
14. Takahashi Y, Sekine T, Yokoyama M (2009) Design of a 16-bit nonpipelined RISC CPU in a two phase drive adiabatic dynamic CMOS logic. Int J Comput Electrical Eng 1(1), April 2009. https://doi.org/10.1109/APCCAS.2006.342503
15. Samanta S (2009) A contemporary review of adiabatic computing. In: IEEE symposium computers and devices for communication, CODEC 2009

Intrusion Detection and Classification in Wireless LAN Using Data Mining Techniques

K. Rama Krishna Reddy

Abstract The traditional security systems are by no means enough to create a secured IT infrastructure, intrusion detection systems (IDSs), which observe the flow of system works and detect intrusions, which are usually used to complement other defense techniques. However, threats are becoming more powerful and strong, with attackers using new attack methods or modifying existing ones. Furthermore, developing an effective and robust intrusion detection systems is a challenging research platform due to the environment resource restrictions and its constant evolution. This work has a core objective of designing a model for protecting wireless local area network through data mining techniques. Protection from unauthorized access is thus another line of defense for network technologies. The work is to use data mining techniques to assist the IDS construction effort to help mitigate these issues. It has been found that intrusion detection technologies can be implemented jointly with data mining algorithms to spot attacks. An intrusion detection system is used to control network operations, mainly involved to detect and separate unwanted users with classification techniques. As mobile crowd sourcing technologies are applied to smart environments, planners can focus on revolutionizing the world by integrating and coordinating all of the technology resources. An effective IDS must still bring in the technical specifications of the systems being used. Additionally, these algorithms are evaluated for their use in discovering unknown attacks.

Keywords Intrusion detection · Wireless attack · Data mining · Detection algorithms

K. Rama Krishna Reddy (✉)
Department of CSE, KG Reddy College of Engineering and Technology, Moinabad, Hyderabad, Telangana 501504, India
e-mail: krkreddy204@gmail.com

J. R. Szymanski et al. (eds.), *Energy Systems, Drives and Automations*,
Lecture Notes in Electrical Engineering 1057,
https://doi.org/10.1007/978-981-99-3691-5_29

1 Introduction

As the networks have grown, the techniques of communique with inside the wi-fi area have multiplied tremendously. The community connection is shaped among or extra computer systems which can be related to the equal or separate networks using a wi-fi router. Intrusion may be defined because the cap potential to come across and forestall unauthorized access. In the extra traditional enterprise-orientated environment, enterprise networks are going through quite a few barriers with inside the net, wherein intrusion detection is taken into consideration to be a difficult task. Among many famous defenses, intrusion detection has been rounded for pretty a few times. A wide variety of absolutely advanced packages were applied and determined to be functioning. A key difficulty with stopping a higher information of a vulnerability danger is motive that during spite of extra than 25 years of study, this difficulty stays common, attributable to the unexpectedly evolving records generation and the current identity of recent vulnerabilities.

Intrusion mechanisms are referred to for an excessive bit blunders charge and numerous researches have attempted to locate a green discrimination technique, however now no longer but successes. It appears to have an excessive hazard and consequently numerous paintings continue to be going into coming across techniques to pick out intruders. A notably insecure networked gadget necessitates the improvement of higher protection resources. An Intelligent Intrusion Detection Scheme turned into formulated the use of records mining technique. Intrusion detection is the technique of collecting and reading the records from one of a kind reasserts on a tool or community with a view to decide the ability protection weaknesses. Interception is the exercise of tracking sports going on in a computing gadget and seeking to come across any warning signs of protection hazard.

Computational intrusion detection systems (CIDS) has verified to be beneficial on this target. Depending at the records supply taken into consideration, IDS can be both host-primarily based totally or community-primarily based totally. A host-primarily based totally IDS analyzes occasions especially associated with running gadget records consisting of procedure identifiers and gadget calls. Network-primarily based totally IDS analyzes community associated occasions consisting of site visitors volume, IP addresses, provider ports, protocol usage, etc. Depending at the form of evaluation carried out, IDS are categorized as both signature-primarily based totally or anomaly-primarily based totally. Signature-primarily based totally schemes are seeking for signatures or described styles with inside the analyzed records.

This work aims to identify the intruder in WLAN in any organization. Intrusion detection mechanism is employed to screen all incoming users in the network and to watch the exceptional users however aren't extremely trustworthy. The intrusion detection system watches for suspicious anomalies that aims to achieve unauthorized access to the network, like once Associate in Nursing unauthorized users begin to perform functions et al that use their privileges in Associate in Nursing inappropriate manner, the actual fact of the present system works on a firewall so as to secure and also notice the unauthorized person mistreatment Wireless native space Network. A real smart wireless node is characterized by consistency, which may be measured. Non-standard activity in some wireless nodes is also known as security problems. However, so as for a wireless node action to be shapely, it should air Associate in Nursing entity or collective basis specified the model adequately describes it. Data processing ways are often wont to establish and learn wireless network anomalies. Therefore, call tree algorithmic rule is employed to classify the information during this thesis. Parameters like recall, exactness, and accuracy area unit are thought of for analysis of the planned algorithmic rule at the side of existing algorithms like call tree, naive mathematician, and random forest.

2 Related Work

The use of data mining in intrusion discovery is a comparatively ultramodern system. Bracket, association and clustering are all colorful forms of data mining that are extensively used in numerous other sectors. Because algorithms can prize preliminarily retired information from large datasets, it can find preliminarily unknown information that can be used in the intrusion discovery model. This methodology examines the intrusion discovery as a data study, whereas the previous styles were concerned with software engineering. Original input data is converted into ASCII network packet information, which is also converted into connection position information. This access position documents give information about the type of operation, the length of time, and other aspects of connections. Intrusion discovery models are erected from this data using data mining algorithms [1].

Data should be secured during processing, birth, and distribution to insure particular security. As a means of guaranteeing particular protection, all important data should be translated for collection, reclamation, and delivery, and data should be translated for storehouse and transmission. Data encryption offers a good degree of security, but also introduces a number of delicate and fugitive challenges. Big coffers may be booby-trapped with preliminarily undetected trends to support businesses in all feathers of different sectors [2]. These databases also give records on consumers that do not need confidentiality. As mining collects nonpublic data about a reality, it to be misused, performing in a loss to an existent. The particular information isn't at threat of being revealed with sequestration guarding data mining.

The confidentiality of data must be maintained from the moment it is sent to the moment it is retrieved. Not even a small amount of wrongdoing or accidental data failure would be tolerated. Although the danger of mistakes in banking and accounting made them unattractive to big businesses, database data transparency was critical for the small and medium-sized ones. Their high failure rate of error meant that larger corporations shied away from using them, but the small and medium-sized firms absolutely needed data transparency. An analysis of the domain rules resulted in greatly reduced number of mistakes. Created a situation in which "interesting" trends may be inferred from data [3]. Analysts were made aware that these associations were simply conclusions rather than proven facts. Relay bandwidth should be used to enhance professional device functions, and data efficiency is a critical to this. After this, the relay integrity of the data is checked, and there is still some missing data, the conclusion is made. However, if the data continues to be unreadable, the process ends.

In decision tree construction the entropy and information gain are the two statistical methods used to calculate the node information which helps to identify the root node. Entropy is mostly used for iterative steps crafted from three legal guidelines that is lower, i.e., better or same to zero, however additionally from extra than. This is to screen the community and forestall information injection assaults the usage of an intrusion detection system (IDS). One can degree the amount of expertise received through information, the attribute's entropy minus the earlier distribution's entropy. To quantify how a lot of data has been gained, customers must study the attribute's entropy and subtract it from the entropy of earlier expertise [4]. A choice tree is a gadget of nodes (rectangular boxes) and arrows, wherein every node represents a likely desire and every arrow shows a route of action. The desk of capabilities and moves is primarily based totally on the muse of the dataset, as rows and columns. The first node is known as the basis node, the ultimate nodes are leaf nodes which might be taken into consideration because of the inner nodes. An ID3 set of rules includes time and again dichotomizing capabilities at every step, it makes use of a bottom-up approach.

3 Proposed Work

In this proposed work, the authors have used the decision tree algorithm. Generally, the decision tree is used to represent data and elements in terms of diagram. The tree has three elements, namely root, parent, and leaf node. The purpose of this algorithm is to classify the given dataset based on the defined class label. This process is called Iterative Dichotomiser 3 (ID3). Founder of ID3 is Ross Quinlan, who generates a dataset from the decision tree. This concept is used widely in machine learning and natural language processing. ID3 has two components (a) entropy and (b) information gain [5].

i. Entropy and Information Gain

Entropy is the method of representing dataset into decision tree, and information gain is a way to calculate the probability value where every attribute in the dataset having. Let Sr be the set of n data which may be taken from sensor data from any organization or crew, given as Sr1, Sr2, …, Srn. Let S be set of attributes such as s1, s2, …, Sn. Entropy H(S) helps to find an appropriate attribute with minimum values whereas information gain IG(S) which will be applied to optimize the value from the entropy and predict the appropriate result for next step, acts as repetitive iterative process using ID3 algorithm. This iterative process may stratify any one of these cases.

$$\text{Entropy}(S) = -(P_{\oplus} \log_2 P_{\ominus} + P_{\oplus} \log_2 P_{\ominus})$$

where $P_{\oplus}$ is the portion of positive examples and $P_{\ominus}$ is the portion of negative examples in S.

The information gain (Gain(S, A) for any dataset can be defined as (source; Text book)

$$\text{Gain}(S, A) = \text{Entropy}(S) - \sum_{v \in \text{Values}(A)} \frac{|S_v|}{|S|} \text{Entropy}(S_v)$$

where, Values (A) is the all possible values for attribute A, and S_v is the subset of S for which attribute A has value v

ii. Decision Tree

The main objective of this decision tree is to detect and classify the intruder from the dataset. The attack can be classified as: (A) Dos attack, in this category the attackers can hold the information from one node to another (B) Reply attack, in this category the attackers can record the sender's information and also delay the reply information (C) Bias injection, in this type of attack the attackers can modify by adding some unwanted content into our information. The decision tree classifies the data into these three categories.

iii. Steps implemented in this work.

Step 1: Probability of every attribute in the dataset R should be calculated, it has been divided into two; Py (Positive Probability) and Pn (Negative Probability). Entropy (S)= $-[(\text{Py})\text{Log}_2(\text{Py})+(\text{Pn})\text{Log}_2(\text{Pn})]$
Step 2: Calculate the value of entropy for all values in the dataset repeatedly until there is no data in the dataset.
Step 3: Compute the information gain IG(S), that is difference between equation (A)–(B).
Step 4: Create classification matrix with MXN such that rows will be as subset elements and columns have two categories of values. One is entropy value E(S) and second one is information gain IG(S).
Step 5: Decision can be taken with appropriate record based on maximum value of the information gain.
Step 6: Stop the process.
Step 7: The decision tree will be constructed using the above maximum value.

4 Implementation Results

The statistical methodologies like probability has been used to estimate information gain IG (S) to construct the nodes of decision tree for selected sample dataset as given in Table 2. The attributes of sample data consist of IP classes, port number, protocol to be used, to transfer file within an organization.

From the Table 1, it is concluded that the attribute which gains maximum value in information gain should be taken into account for our decision tree. So in the below table, decision should be taken only on port number or protocol rather than IP address.

The result of this work compared with the algorithms like as: decision tree, Naive Bayes algorithm, and random forest algorithm are compared with the proposed IDS, the comparison results are shown in Figs. 1, 2 and 3.

Figure 4 shows the accuracy gain of all three algorithms, in that the proposed algorithm gives 79% of accuracy, 96% of precision, and 65% of recall which was more than the other two existing algorithms.

Table 1 Entropy and information gain

	Entropy	Information gain
IP address	– 0.17816	– 0.35272
Protocol	– 0.15265	**– 0.32825**
Port number	– 0.15265	**– 0.32825**

Table 2 Sample dataset format

Sl. No.	IP address	Protocol	Port number	Intruder
1	O.x.x.x (Class A)	SMTP	25	No
2	128.x.x.x (Class B)	POP3	1886	Yes
3	56.x.x.x (Class A)	UDP	1078	Yes
4	190.x.x.x (Class B)	UDP	110	No
5	45.x.x.x (Class A)	SMTP	25	No
6	146.x.x.x (Class B)	SMTP	2013	Yes
7	75.x.x.x (Class A)	POP3	1886	Yes
8	147.x.x.x (Class B)	UDP	1093	Yes
9	44.x.x.x (Class A)	POP3	68	No
10	76.x.x.x (Class A)	SMTP	25	No
11	172.x.x.x (Class B)	POP3	1886	Yes
12	98.x.x.x (Class A)	UDP	110	No
13	165.x.x.x (Class B)	SMTP	1856	Yes
14	166.x.x.x (Class B)	POP3	68	No

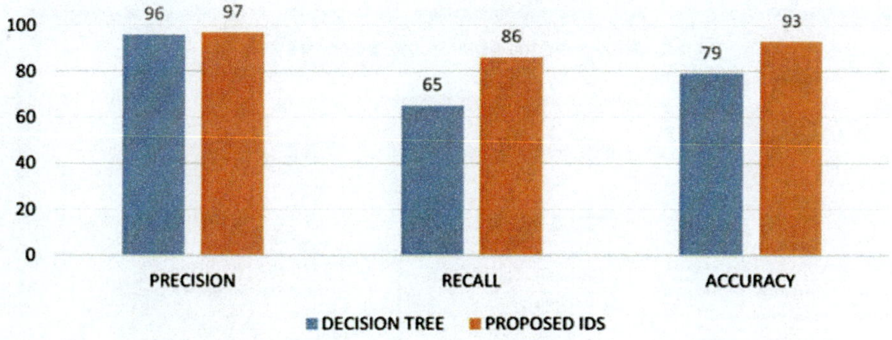

Fig. 1 Comparison result with decision tree and proposed IDS

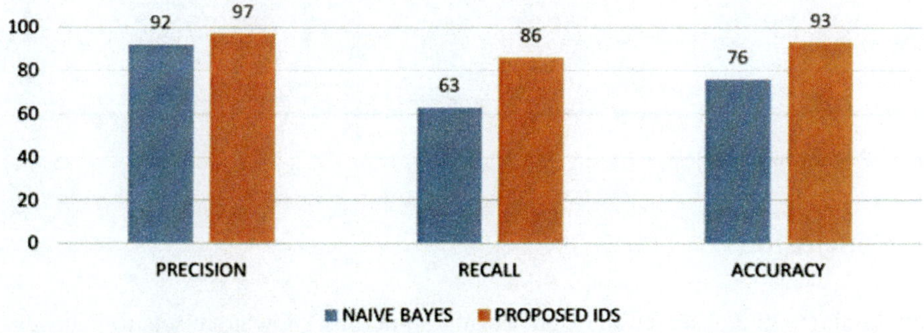

Fig. 2 Comparison result with Naïve Bayes and IDS

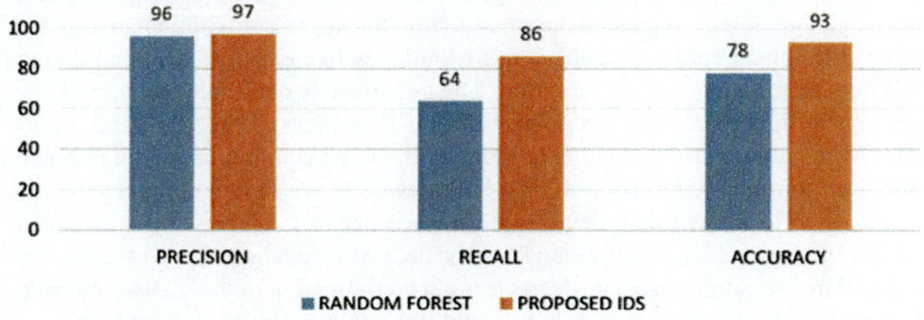

Fig. 3 Results of proposed work with comparison

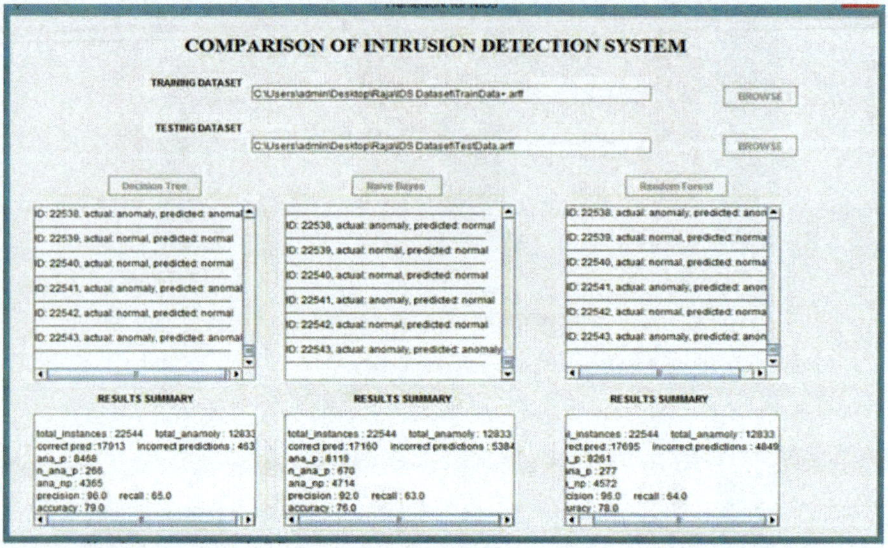

Fig. 4 Performance metrics

5 Conclusion

Data mining is the observation of processing numerous knowledge sets to uncover new facts and summarize the results into sensible information. Distinguishing unexplained attack patterns remains associate degree unresolved issue, though some studies indicate that there is an answer to the current downside. Supporting the knowledge found, it is concluded that so as to find the aggressor with in native space network of a given entity, it ought to use techniques like entropy, ID3, employing a call tree learning algorithmic program. The aggressor is placeable and supports a minimum of three parameters: associate degree IP address, a protocol, and a port range. Its steered that the system is far more correct and economical with reference observance intruders employing a hand-held device like a laptop computer which are connected to the wireless device native space network. This work is meant to estimate entropy and knowledge gain from the decision tree algorithmic program for the aim of investigating associate degree entrant victimization of the ID3 algorithmic program. One step higher than this, it would use one purpose device management and maintenance, creating it terribly user friendly. This result suggests that hand-held devices, personal computing devices connected to the wireless native space network, and laptops that link to the wireless network could also be ready to notice and notice intruders higher. This increased framework has been developed to support the identification of diverse attack vectors through wireless networks. Victimization the confusion matrix aboard this algorithmic program, the planned algorithmic program tried to be associate degree economical performance assessment methodology.

References

1. Jakhar J, Kumar V, Kumar V (2013) Intrusion detection using data mining techniques: a Study through different approach. Int J Appl Sci Technol Res Excell 3(4):1–6
2. Silva J, Cubillos J, Villa JV, Romero L, Solano D, Fernández C (2019) Preservation of confidential information privacy and association rule hiding for data mining: a bibliometric review. Proc Comp Sci 15:1219–1224
3. Mazumdar S (2017) From data integrity to inference integrity. In: 2nd international conference on telecommunication and networks (TEL-NET), pp 1–1
4. Kelley HJ (1969) Entropy of knowledge. Philos Sci 36(2):178–196
5. Goliwale P, Gupta V, Johre A, Bendale S (2018) Intrusion detection system using data mining. Int Res J Eng Technol (IRJET) 05(03):234–238
6. Cahyo AN, Sari AK, Riasetiawan M (2020) Comparison of hybrid intrusion detection system. In: 2020 12th international conference on information technology and electrical engineering (ICITEE), pp 92–97
7. Ajay Kumara MA, Jaidhar CD (2015) Hypervisor and virtual machine dependent intrusion detection and prevention system for virtualized cloud environment. In: 2015 1st international conference on telematics and future generation networks (TAFGEN), pp 28–33
8. Haldorai A, Ramu A, Murugan S (2019) Computing and communication systems in urban development. In: Urban computing. Springer International Publishing, Heidelberg
9. Qu X, Liu Z, Xie X (2009)Research on distributed intrusion detection system based on protocol analysis. In: 2009 3rd international conference on anti-counterfeiting, security, and identification in communication, pp 421–424
10. Sajjad M, Bouk SH, Yousaf M (2015) Neighbor node trust based intrusion detection system for WSN. Proc Comput Sci 63:183–188
11. Raja K, Lilly Florence M (2019) Implementation of IDS within a crew using ID3Algorithm in wireless sensor local area network international conference on inventive computation technologies, Part of the Lecture Notes in Networks and Systems book series (LNNS, vol 98), Springer, ICICIT 2019: Inventive Computation Technologies, pp 467–475.
12. Raja K, Florence ML (2017) Tracking of intruder in local area network using decision tree learning algorithms. Asian J Appl Sci 05(01):46–49

Face Detection and Managing Attendance by Using Raspberry Pi

Bonthala Swathi and Mudigonda RathnaChary

Abstract The management of attendance in our everyday life in schools and colleges is a complicated task for teachers especially for the group of students if it is done by hand. To overcome this problem, the real-time execution of the algorithm with enough correctness is needed, with hardware timing into reflection. There all time execution of the algorithm with enough correctness is needed, with hardware timing into reflection. We proposed face recognition using Raspberry Pi for attendance management system; by using this system, we can reduce the chance of getting proxy attendance. The attendance being marked using live stream video using OpenCV by using the framework comprises four stages database creation, face detection, face recognition, and attendance updating. First, the image of the student is captured through the live stream video then at the point with the images in attendance in the database. Assuming that them at this observed attendance is noticeable as present else absent.

Keywords Raspberry Pi · Face recognition · Face detection · Attendance updates · Database

1 Introduction

Every organization takes attendance either manually, i.e., by using pen and paper by calling their names, or by using an adopted biometrics system such as fingerprint, iris system, RFID card reader, etc. Taking attendance manually by using pen and paper is a time taking process, and there are chances of getting proxy attendance [1]. In this proposed system, we will use an attendance. Attendance will be marked based on facial features. We only use persons' facial features to mark the attendance because facial features differ from person to person. We record the attendance of the students using a live stream video using an OpenCV. Marking of attendance

B. Swathi (✉) · M. RathnaChary
KGReddy College of Engineering and Technology, Hyderabad, Telangana, India
e-mail: swathib@kgr.ac.in

J. R. Szymanski et al. (eds.), *Energy Systems, Drives and Automations*, Lecture Notes in Electrical Engineering 1057,
https://doi.org/10.1007/978-981-99-3691-5_30

using face recognition technology improves two steps. Firstly, detection of faces using live stream video. Secondly, recognition of faces by comparing them among presented images which are stored in the database. The smart way of using managing attendance is by using face recognition. Using this technology, attendance will be recorded accurately. A face recognition system is a time-saving process. We are using facial recognition technology to reduce the flaws of an existing system.

Face recognition is one of the most part use biometrics. It can utilize for security, verification, ID and has lot more benefits. This framework comprises four stages database creation, face detection, face recognition, and attendance updating [2]. Database is created by the images of the students in class. Face detection and identification is performed with the use of Haar Cascade classifier and neighboring binary model histogram algorithm; correspondingly, faces are identified and predictable from real-time video of the classroom [2]. Face Recognition help the management to collect total data and analysis it and show difference if a student is not in class for few min (or) hours. These applications collect the data and make an absent in that class. As teachers cannot take attendance for few minutes so help it and make attendance of student for every few minutes.

As per the past attendance management system, the data which is collected is the biggest issue due to the accuracy of the data found [3]. This is in the view of fact that the attendance not recorded individually by the unusual person. And additionally, the participation of a specific individual be able to in use by mediator not including the recognition of the organization that abuses the exactness of the information [4]. Assuming the institution begin a permission, it might include in the direction of wasted a many individual resource along with time one by one would not be practical in any way. As a result, in the earlier period the recorded of all presence in the system is not dependable for investigation procedure [4].

The third is facing an issue with the ease of access of individual's details with genuine disturbed [5]. For example, the greater part of guardians be exceptionally worried about to follow their kids genuine where about exactly kids go to the classes in school/college [6]. However, earlier there are no methods to find the kids data by guardians. For that reason, advancement should have been done to the past method to get better effectiveness and data correctness [6].

2 Related Works

As indicated by Hanna Mohsin Ahmed, (Hanna Mohsin Ahmed, Rana Talib Rushed, 2020) can have the option to distinguish a Face detection, considerable segment used for the plan of secure as well as monitor, expressly used for disable people via utilize the algorithm of real-time face detection [2]. A Raspberry Pi-base faces identification method with established mechanism of face recognition and detection resolve given as a result the way that image-base biometrics uses a Raspberry Pi is depict [2]. A term paper objective is moving face detection toward an intensity anywhere the organization be capable of alternate by using a key and RFI-Cards designed for getting to security system [3]. Result as to proposed idea know how to functioning method using identification of face using OpenCV, Raspberry Pi, and on an Android app on cell phones [3]. Record establishes a base using the particular 3 fundamentals taking place system know how to apply in a number of more scenario of detection of face into consolidating under modes technology and valuable method on the report of its price.

"Facial Detection and detection System on Raspberry pi with Enhanced security" (D. Sri Sai Mahesh, T. ManeeshReddy, 2020), the previous attendance system's issues is identified based on the identification of face recognition to get solved [5]. Incurrent days our everyday, as we all are depending on the Internet for browsing the web, electronic mail, and sharing the files. The web implies Internetworking of thing show ever a physical entity to facilitate an attribute of Internet protocol deal with the communication between the object and other Internet-enable procedure that makes. The main feature for security such as communicating with the devices is available and provides security without any delay.

The additional security to the face detection as well as identification method used in banks and homes as proposed method will expand the present method [5]. We include a brief description in necessity to create such a system and face detection algorithm used for verification purpose and transfer the information using message [7]. This sort of method just fit in support of marking the attendance of staff as simply relate their essence one time per a day, dissimilar to students which have need to report their attending at each class on a specific day, would be difficult whether the attendance marking system is not convenient [7]. Therefore, to address previously mentioned issue, to complete attendance management system can be advance on fixed plan in such a way the tends to be work basically the same manner with just batteries that makes it manageable [7].

The plan of this article is to capture the video of the students, and it converts into frames, which relate to the database toward the guarantee attendance or nonattendance to mark a specific student to maintain the record. The automatic classroom attendance method helps into expanding the exactness along with speed at last accomplish the high-accuracy real-time attendance in the direction of address the issue designed for programmed classroom valuation.

3 Methodology

For storing each and every student attendance attain the database by capturing the each student face we proposed method to get accurate. The appearance of the student should be captured in such a way that all the characteristic of the students' face should to be detected, even the seating and the posture of the student need to be predictable. There is no requirement for the professor to take the attendance manually in the classroom for the reason that the system records a video and through further handing out steps the face is being predictable and the audience presence folder is updated [7].

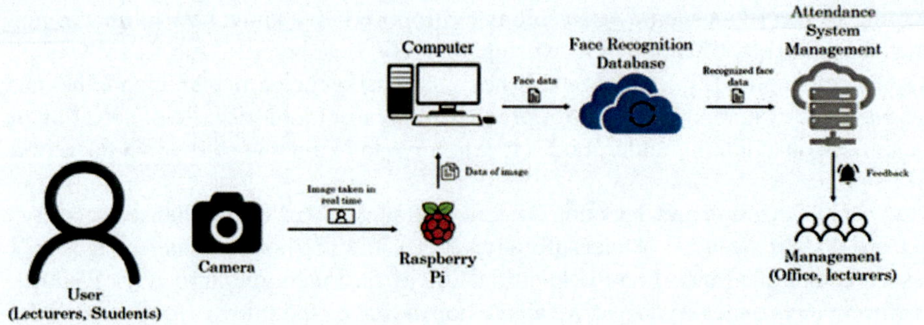

Fig. 1 Face detection method

In this proposed system, attendance will be recorded using live stream video. For face recognition, we will use a person's facial features to mark the attendance [8]. The below Fig. 1 describes the methodology of the face detection method.

In the above figure, we can see that the user is lecturer and students. The lecturer will have a login id and a password to login into the system successfully. The attendance will be marked using live stream video. The front end will have a Raspberry Pi and a camera. Using the camera, image of the students is captured. For detecting the face, the captured images are compared with images of the students who are already present in the database. If the match is found between the captured images and images in the database then attendance is updated. In the end, the attendance will be sent to the lecturers (Fig. 2).

MySQL

It is an open-source relational database management system (RDBMS). Its name is a combination of "My," the name of co-founders Michael Wideness's daughter, and "SQL," the short form for Structured Query Language [8]. A relational database sorts out information into at least one data tables in which data types be related with one another; these relations assist with organizing the data. SQL is language software engineers used to make, alter, and separate data from the relational database, in addition to control user access to the database. Furthermore to relational databases

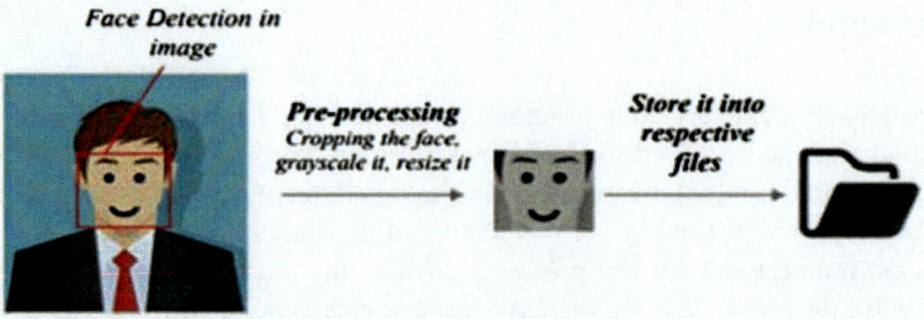

Fig. 2 Image acquisition and preprocessing procedures

and SQL, an RDBMS-like My SQL works with an working arrangement to employ a relational database in a computer's storage system, manages users, allows for network access, and works with testing database integrity and creation of backups [9].

Python

It is an interpreter, object-oriented, high-level programming language with dynamic semantics. Its significant level inherent information structures, combined with dynamic composing and dynamic binding, make it very exceptionally for quick application improvement, just as for use as a pre-arranging or insert language to associate existing works [10]. Python's simple, easy-to-learn syntax emphasizes readability and therefore reduces the cost of program a intendance. The Python translator and the extensive standard library are accessible in source or doubled without charge for every single platform and can be liberally distributed.

Raspberry Pi

There are two most important methods that flow in the software improvement subdivision as follows:

- The design of the faced at a base
- The attendance taking process.

The above-mentioned both the processes are important because they made up the foundation of the attending management method. In this, the development of both flows will be momentarily described. In the meantime, their full functionality, specific requirements.

Installation of Raspberry Pi

The Process of Attendance Taking

Figure 3 illustrates the technique which is used for the creation of face data set, which plans the face identification, the captured image with a knower face in it will then at that point, go through cropping. Training the recognizer, the image in the created list recovered from the csv will followed by pump interested in a identified that is the Eigen force detection to do the training.

Figure 4 illustrates the procedure for taking the attendance during the session such as access the attendance management system website, initiate the python script, and load the training data to acquire portrait. The method will capture student's portrait and handle the undergo the same pre-handling routine and face detection process.

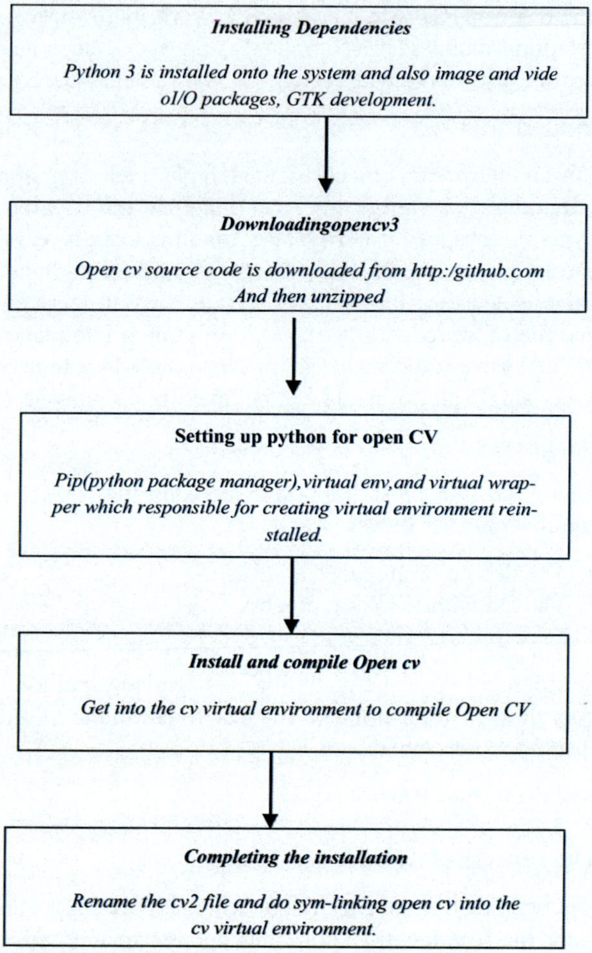

Fig. 3 Flow of the database creation in the framework

4 Implementation

For implementing of face recognition system, there will be the logic behind it. For this, we use python and My SQL. To obtain the Raspberry Pi set, we need to boot as follows:

1. Slot in the Micro-SD memory card into the Raspberry Pi.
2. Associate the Universal Serial Bus device keyboard.
3. Associate the High-Definition Multimedia Interface cables.
4. Associate the Universal Serial Bus Wi-Fi adapter (or Ethernet cable).
5. Skip this step if you are using a Raspberry Pi.
6. Associate the micro-Universal Serial Bus power supply.
7. The Raspberry Pi should now be booting up.

Fig. 4 Procedure for taking the attendance during the session

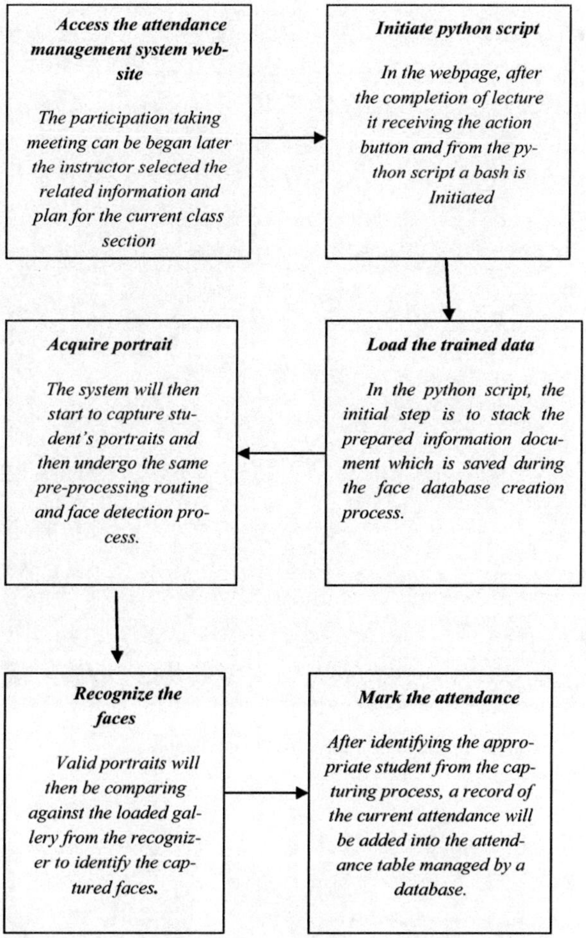

At the point when the Raspberry Pi is done booting up, sign in utilizing username: Pi and secret *t* word: raspberry **Test cases** (Fig. 5).

1. The detailed face database being use is named AT and *T* Face database which know how to be obtain from http://www.cl.cam.ac.uk/research/dtg/attractivefacedatabase.html.

5 Results

This figure representing the captured images of the live stream video in a real time (Fig. 6).

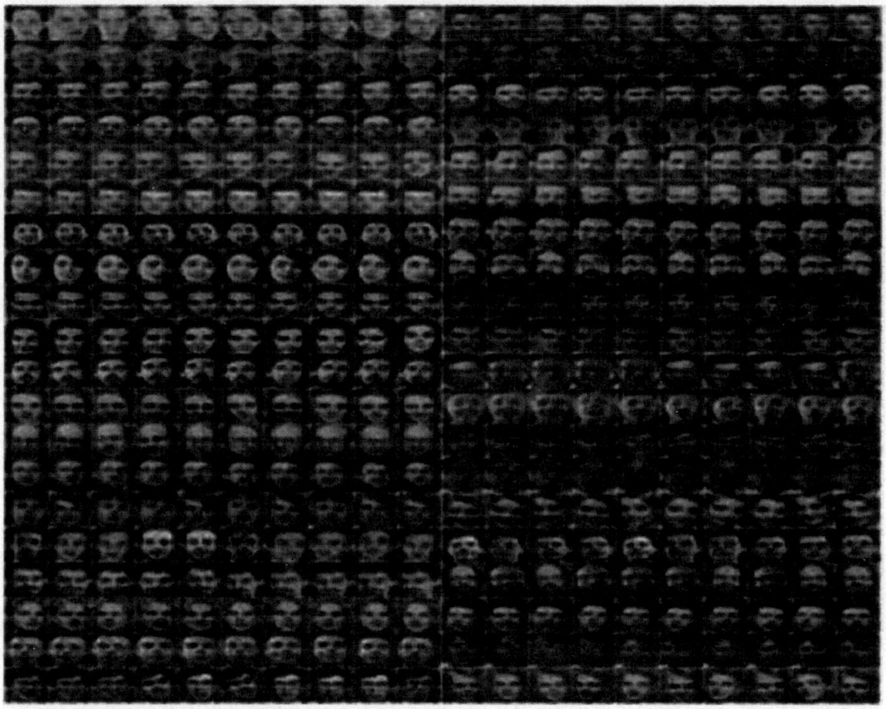

Fig. 5 Sample images captured from the database

Fig. 6 Sample images captured during the live session

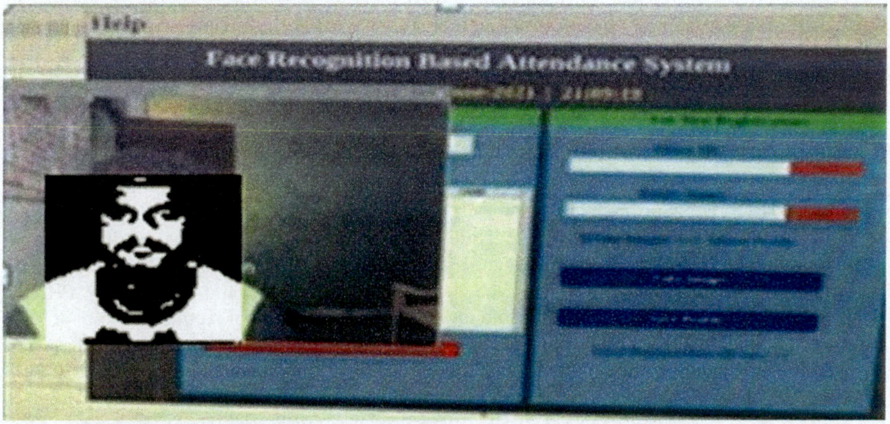

Fig. 7 Attendance recognized system by Raspberry Pi resultant as face identified

Figure 7 representing the output of the program in the form of marking attendance using captured images.

6 Conclusion

This project presents qualified method for number present management method in the classroom location to facilitate knows show to restore the previous manual methods. This process is securing a sufficient amount, dependable, exact, and professional. There is no requirement for specialized equipment for installing the method in the classroom. Attends to be built with a camera and PC. There is a need to use some algorithms that can identify the faces in the cloak to further develop the method performance. Almost all academic institutions have need of the attendance record of students and maintain attendance manually can be a hectic as well last time-consuming task. Consequently, maintaining attendance automatically with the help of face recognition will be very helpful and less lying on your front to mistakes when construed with them annual procedure This will more over control the manipulation of attendance records done by students, and it will save time also. The future scope of the proposed work can be capturing multiple detailed images of the students.

References

1. Danish Gowda HL, Vishal K (2020) Face recognition based attendance system. Int J Eng Res Technol 9(06):70–75
2. Aswin Kumar SV, Chandrasekhar R (2019) Real time face detection using DNN with Raspberrypi. Int J Recent Technol Eng 8(4):177–193

3. International conferences on Information Technology (2019) Information systems and electrical engineering (ICITISEE), pp 304–309
4. Ekenel HK, Stiefelhagen R (2006) Analysis of local appearance-based face recognition
5. Dey S (2015) Effects of feature selection and feature normalization. In: CV PR biometrics work-shop, NewYork
6. Mtshali P, Khubisa F (2019) A smart home appliance control system for physically disabled people. In: Proceedings of the 2019 conference on information communications technology and society (ICTAS), pp 1–5
7. Mohammed MN, Desyansah SF, Al-Zubaidi S, Yusuf E (2020) An internet of things-based smart homes and healthcare monitoring and management system. J Phys Conf Ser 1450:012079
8. Manjula VS, Baboo LDSS (2020) Face detection identification and tracking by PRDIT algorithm using image database for crime investigation. Int J Comput Appl 38(10):40–46
9. Vijay K, Selvakumar K (2021) Brain FMRI clustering using interaction K means algorithm with PCA. In: Proceedings of the international conference on common signal processing (ICCSP), pp 0909–0913
10. Lander K, Bruce V, Bindemann M (2020) Use inspired basic research on individual differences in face identification: implications for criminal investigation and security. Cognit Res Princ Implicat 3(1):1–13
11. Silver D, Schrittwieser J, Simonyan K, Antonoglou I, Huang S, Guez A, Hubert T, Baker L, Lai M, Bolton A, Chen Y, Lillicrap T, Hui F, Sifre L, van den Driessche G, Graepel T, Hassabis D (2019) Mastering the game of go without human knowledge. Nature 550(7676):354
12. Hu Y, An H, Guo Y, Zhang C, Zhang T, Ye L (2020) The development statusand prospects on the face recognition. In: Proceedings of the 4th international conference on bioinformatics and biomedicine engineering, pp 1–4
13. Gottumukkal R, Asari VK (2019) An improved face recognition technique based on modular PC A approach. Pattern Recognit Lett 25(4):429–436
14. Hoyle DC, Rattray M (2019) PCA learning for sparse high-dimensional data. Europhys Lett 62(1):117–123
15. https://wikimili.com/en/Dividend_units. Accessed 23 July 2022

Fully Digitalized Fingerprint and RFID-Based Voting System

Angotu Saida, Yashwanth Rasumalla, and Nikitha Sunkari

Abstract In the current ongoing times, most of the nation's conduct their elections using electronic voting machines (EVM's), where your vote is stored electronically and registered using electronic devices from which little to no wastage of ballot paper takes place. As security is one of the major parameters to be considered during an election that no individual can vote twice or misuse their right to vote, this can be prevented using fingerprint and RFID technology, as the minutiae feature on the finger of each person is dissimilar, the elector can be easily identified and distinguished from others. The machines permit the person to cast the vote with his fingerprint; the fingerprint of the voter is used to uniquely and distinctively identify the person using the fingerprint module; also the system will only allow the user to use the machine once. If the voter tries to use the machine again, it will indicate an error. An admin is assigned to a group of voters standing for the election, the operator only will have the access to delete or add the voters' photo and name. Admin will also authenticate the details of the candidates by proving the voter's identity, and after verification, operator will give the identity card so that the voter can cast his vote at the election booth.

Keywords RFID · EVM · VVPAT · Fingerprint verification

1 Introduction

Elections create opportunities for people to choose their representatives and evaluate them for governance. The central integrity of the elections process invokes the honesty of public democracy. This election process should be robust for selecting

A. Saida (✉) · Y. Rasumalla · N. Sunkari
Department of ECE, KG Reddy College of Engineering and Technology, Hyderabad, Telangana, India
e-mail: saida@kgr.ac.in

N. Sunkari
e-mail: sunkarinikitha@kgr.ac.in

J. R. Szymanski et al. (eds.), *Energy Systems, Drives and Automations*,
Lecture Notes in Electrical Engineering 1057,
https://doi.org/10.1007/978-981-99-3691-5_31

their representatives without any fraudulent behaviors, which should be transparent and comprehensible for voters. The design of a perfect voting machine that is either electronic or based on ballot papers or any other must satisfy various criteria designed by the government. The voter's ballot should be preserved for voter safety purpose. If that evidence exists, then there will be a chance for the candidate to purchase a voter's vote. The voting machine must be tampering resistant.

To deal with such problems, we introduce a fingerprint and RFID-based voting system which sets dual criteria that must be met to utilize the person's right to vote without worrying about the result being manipulated. The system uses a fingerprint scanner to evaluate the identity of the voter. Fingerprints are one of the features in the human body that is unique for every individual around the world; there are no such cases where two persons have the same fingerprints; this include twins too, which makes it a trustworthy feature to consider when it comes to security. The scanner will capture the fingerprints of the person by valleys and ridges on the surface of the finger and digitize them. Each of the fingerprints is stored not by the shape or size of the finger but by the ridge locations (minutiae feature). RFID employs electromagnetic fields to identify and track tags attached to objects. An RFID system consists of a small radio transponder, radio receiver, and transmitter. RFID technology is a system that is bound to give security without worrying about external factors. It consists of two parts, one to be carried by a voter that is an RFID tag and the other at the election booth which is an RFID reader which is a built-in proposed voting system. The tag has an antenna that sends a signal once when it comes in contact with an RFID reader. The RFID reader will evaluate the ID number and details of the tag owner. These systems combined will provide security that is expected for safe and secure voting, and the results will be stored in the cloud server immediately so no tampering with the data is possible. The important factor for excessive-efficient and high-precision HRI is the sensors that help the humanoid robotic to become aware of and apprehend the behaviors of humans [1]. Multi-fingered give up-of-arm-tooling mechanism for dealing with micro-meso scale items. More specifically, the study aims to enhance a compliant prosthetic manipulator designed for rapid 3D manufacturability. The microgripper is comprised of a more than one compliant five beam mechanism arranged symmetrically to provide three-dimensional manipulation [2]. In revealed electronics structures, the roll-to-roll process is a subsequent-generation technique era that may be applied to the manufacturing of diverse electronic merchandise together with bendy digital gadgets, solar cells, and show panels [3]. At gift, it has been advanced in automation technology for decades. Especially, the Internet of Things era is pretty mature. However, inside the Internet of Things technology, there may be a loss of rapid processing of disbursed assignment service architecture [4].

2 Problem Statement and Objective

2.1 Problem Statement

To address the shortcomings of existing ballet machines, EVMs (Electronic Voting Machines), and fingerprint balloting machines, which are susceptible to rigging, there is a need to develop a device that mitigates the risk of hacking, human mistakes and most of all to retailer time.

2.2 Aim and Objective

The major problem with the electronic voting machine is that they are prone to rigging as opposition parties easily influence them. When it comes to data security, it has a minor problem that can cause much trouble, like authenticating the people.

3 Literature Survey

To replace the conventional system used for voting like the ballot box electoral system, an electronic machine was introduced in the year 1998 in India. The ballot box and control unit were two units consisting of robots utilized in India. This makes the voting system hard to manipulate. The system was introduced by Bharat Electronics Limited (BHEL). From 1998 till, now many innovations have taken place in electronic mechanical devices. The system is capable of working for 15 years. As per the suggestion of the committee, the manufacturer has adopted a third-generation design including additional upgrades. Nowadays, electronic machines store votes electronically instead of ballot paper.

Ballot paper voting: During the use of this system, one used to cast his/her vote by writing the name of the nominee they want to elect. VVPAT: The "Voter Verification Paper Audit Trail" the VVPAT system was supposed to independently verify the vote that each person is casting for their candidate correctly. The system has additionally been utilized together with electronic mechanical devices in the 2019 election conducted in India. Electronic voting machines were introduced in the Indian election between the years 1998 and 2001. Before introducing EVM's, the ballot box method was used as the method had many flaws, and it was criticized greatly. Remote Internet Voting: During this technique by the use of the Internet, the user can now access the portal anywhere from his comfort and cast his vote safe and secure.

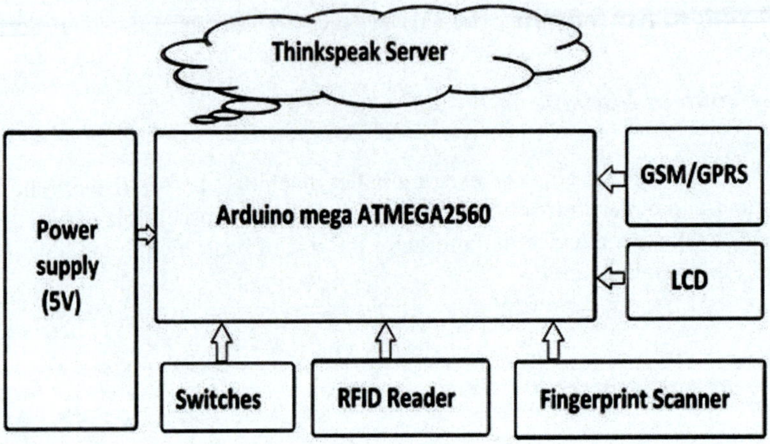

Fig. 1 Block diagram

4 Methodology

The system can be accessed using a login, which requires RFID Card, fingerprint scan, and candidate's name. The system makes use of an online application called Think speak which is supported by all browsers, only legitimate voters whose data has been stored in the database of the election committee will be allowed to vote. Therefore, a fingerprint scanner and RFID system can ensure that only authenticated voters whose data matches with the database can cast their vote (Fig. 1).

5 Existing System

When we come across the election process in our country, there are quite a few systems with fewer complexities, which are secure but practically are not in use, and with the traditional ballot box methods, there is no end to rigging and all other illegal practices that take place during the voting process. With the advancement in technology, many aim to develop a system that would overcome the complexities that are generally faced in the voting process. Few such complexities we have come across are

- Fewer data storage, not being useful for significant elections.
- Complex operation, data not published in the cloud.
- Ballot box is prone to human errors and is time-consuming.

6 Proposed System

The proposed system comprises an Arduino Mega controller (ATmega2560), fingerprint sensor module (R307), RFID reader, I2C LCD, and GSM module. Arduino Mega controller is used because it contains 3 UARTS, I2C, and many GPIO pins available to connect with devices. The proposed system requires three UARTS for GSM/GPRS module, fingerprint sensor, and programming UART. I2C is used for the LCD module to reduce the number of hardware pin connections to connect with the controller. RFID reader reads voters information using RFID Tags. Initially, all voters need to register their details in their RFID tags and scan their fingers to store fingerprints in the database. The current voting system contains voter ID card or Aadhar card for recognition of voters or proof of voter replaced in our paper with RFID. The proposed system initially reads reader data, and if reader data exists in the database, then it allows the person for further process. The voting process is secured using a fingerprint sensor to reduce false votes. The voter scans their finger using a fingerprint sensor; if their fingerprint data exists in the database, then the system sounds their name and allows voters to use their vote pressing button in parallel with the party symbol. Then, automatically voters can check their voting status in LCD. The result button is placed to check results automatically without any manual calculations. Once the polling manager clicks on the result button, results will be displayed on LCD and the same data sent to IoT Things board cloud using GPRS communication technology. The same process continues for all voters.

7 Hardware Requirement and Software Requirement

7.1 Hardware Requirement

- Arduino Mega 2560, GSM module. R307 fingerprint scanner, module, LCD 16 $\times$ 2, RFID reader, RFID card, tactile switches.

7.2 Software Requirement

- Arduino IDE, things speak.

7.2.1 Arduino Mega 2560

Arduino Mega 2560 is a microcontroller that is built around ATmega2560. It packs more memory storage and Input and Output pins compared to other available microcontrollers. It consists of 54 digital Input and Output pins and 16 analog pins allocated onboard. From 54 digital Input and Output pins, 15 pins are used as pulse width modulation (PWM). A 16 MHz of frequency crystal oscillator is included in the board. USB is used to connect and transmit the code from PC onto board a port is given the Arduino to connect the USB. Some versions of the Arduino board lack the feature of DC power jack which is used to power the board like Arduino Pro mini (Fig. 2).

This board packs two voltage regulators; they are 5 and 3.3 V which provide flexibility to work with voltages as per the requirements, unlike Arduino pro mini which has only a single regulator. The major similarity among Arduino Uno and Arduino mega are memory space, bigger size, and more Input and Output pins. Arduino IDE is software used by all versions of the Arduino household. ATmega

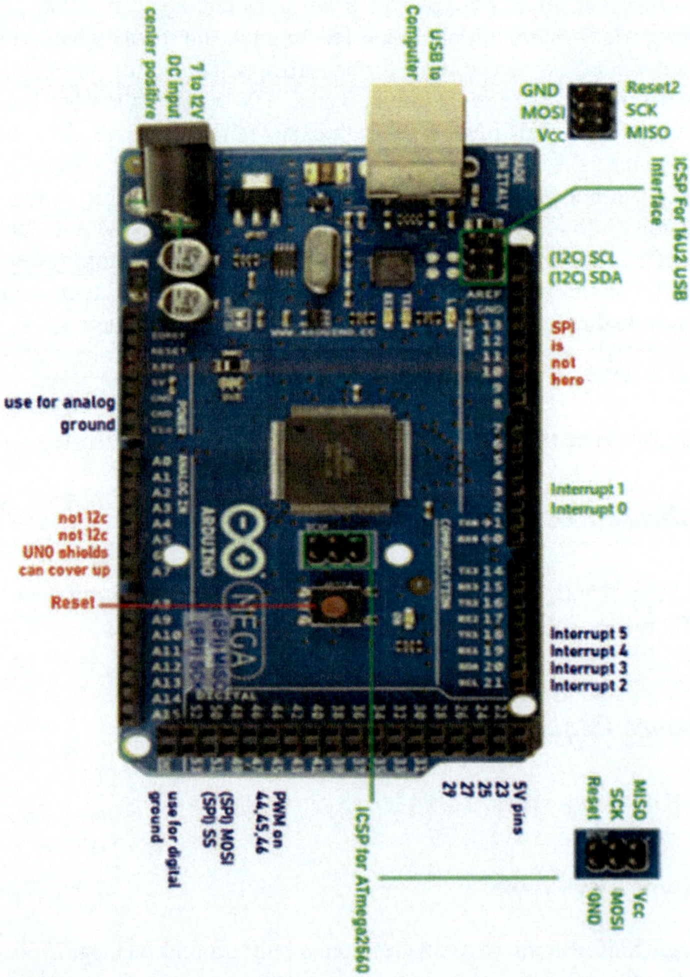

Fig. 2 Pin out of Arduino mega 2560

makes it different from the Arduino pro mini which need a USB cable to serial converter to program the microcontroller. USART consists of a reset button and four hardware serial ports which are used to maintain maximum speed for communication.

Arduino mega is used for projects which require complex circuitry and need large storage space to work; mini projects can be handled using other Arduino boards present in the market but projects like 3D printers or controlling more than one or two motors require additional storage space to store instruction accordingly. There are multiple ports to power up the board such as USB cable, using Vin, or by power jack or by a battery. The Vin and battery are employed to power up the board which is required once you assemble the unit and upload the code to the system via USB.

Arduino Mega 2560 is used for creating standalone projects and combination with other Arduino's can also be designed, but mostly typical projects which require high capacities can be built by this alone.

7.2.2 GSM

The global system for mobile communication is a well-liked standard for communication throughout the countries around. The GSM, the Industry organization of the telephone carrier and producer, has estimated 80 percent of market-making use of the standard. The technology has been used by millions of people throughout the globe, which indicates that subscribers around the globe can communicate through their mobile phones by international standards created among different operators. GSM is a 2nd generation system that is capable of facilitating the widespread implementation of data communications apps into the setup. In addition to the standard AT commands, GSM modems support an additional set of AT commands. These commands are addressed in GSM standards. With the use of these commands, you will be able to perform actions like.

The device is capable of various functions, including viewing, editing, and erasing messages. It also keeps track of signal strength, delivers SMS packages, and monitors charging status (Fig. 3).

The amount of messages that can be processed by given GSM modem per single minute is significantly low, only a few bits between 6 and 10 SMS can be processed per minute.

R307 Fingerprint Scanner Module

Fingerprints are one of the features in the human body that is unique for every individual around the world for us, humans to distinguish between each person is nearly impossible but that is not the case when it comes to computers. They can easily and accurately differentiate between fingerprints of one person from others. To do that we must first capture the fingerprints, it can be done in many ways to capture and digitize them (Fig. 4).

One of the methods to scan a fingerprint is ultrasound scanning. Ultrasound scanning is one of the modern technologies used in many fields.

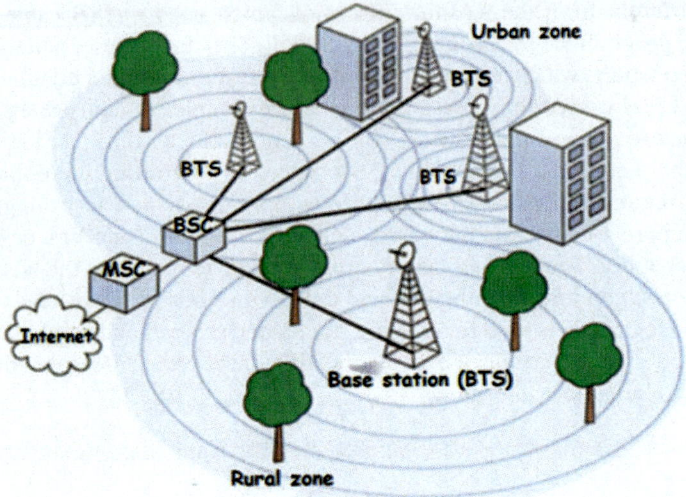

Fig. 3 GSM block diagram

Fig. 4 R307 fingerprint scanner module with cable

7.2.3 Liquid Crystal Display (LCD)

LCD is an even display that makes use of liquid crystal as its main form of operation. LEDs have a huge group of norms for users and organizations, as they are found commonly in television, laptop, smart watches, etc. These took a big leap in order of technological advancements in the field, which include light-emitting diode (LED) and plasma displays; these displays allow the screens to become thinner than the other CRT technology. LCDs require a small amount of power than LED and gas-display. Due to their work on the principle of blocking light instead of discharging it, where LEDs radiate light, crystals in an LCD produce an image with the backlight. As LCDs

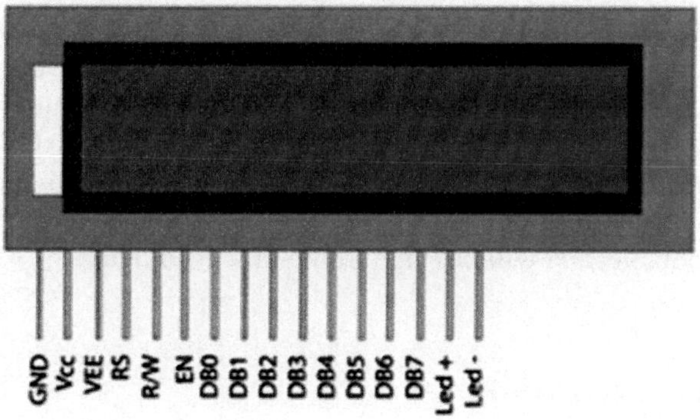

Fig. 5 16 × 2-character LCD

have replaced previous display technology, these have started getting dominated by more advanced versions of the technology in name of organic light-emitting diode (OLEDs).

LCD Pin Out

We come across LCDs every day in the form of bike displays, television set, telephone, personal computers, and smart watches/digital alarm. LCD is an electronic unit that uses liquid crystals to make an image. The 16 × 2 LCD is a basic unit used in basic experimental projects (Fig. 5).

The 16 × 2 indicate 0 displays sixteen characters per line in two lines. In LCD, a character is viewed in a 5 × 7 pixel matrix.

7.2.4 Radio Frequency Identification (RFID)

RFID employs electromagnetic fields to recognize and trace tags linked to things. The system has a radio transponder, receiver, and transmitter. When activated by a pulse nearby reader device, the tag sends the information in the form of digital data, the ID number is sent back to the reader. The number is unique and assigned to different tags, respectively. The tags are powered by readers' radio waves, whereas active tags are powered by a battery and thus can interact at a large distance from the RFID reader. The range can be between on hundred meters. RFID do not need to be in the line of sight like a barcode.

RFID Tag: A RFID comprises two parts—an RFID chip to store the tag's ID and an antenna to transmit and receive signals. These tags are assigned to fixed units to trace them. The tags transmit information via radio frequency waves to be read by the end system. They acquire power from radio waves produced by the reader. These tags receive the transmission from the system. The power goes via a reader built in them to the tags, the power turn on the module, which modulates the energy with the required data and sends the signal toward the reader.

RFID Reader: The reader is considered to be the intellectual part of the whole unit because the reader is responsible to import and export the radio waves to make a connection between RFID tags. The readers vary into two types, fixed RFID is meant to be placed at a constant location such as walls, desks, into the portal, etc. Fixed readers are combined readers, such readers have an antenna that involves one extra antenna port for connections. They are designed to carry around such as fashionable, durable, and without high traffic of tagged items.

8 Algorithm

Step1: Voters' information is stored in their RFID tags and fingerprints in the database. **Step2:** RFID tags are scanned by voters at RFID readers to verify information. If information is available, the voter will be allowed to cast their vote. **Step3:** Voters scan fingerprints and press buttons in parallel to the party symbol. **Step4:** LCD module gives instructions and shows the status to voters. **Step5:** After completion of the voting process the result button is used to check the results (Fig. 6).

9 System Implementation and Result

This project uses RFID technology which requires two separate components: The first one is the RFID reader. It receives and sends radio waves to establish communication among RFID tags. The second one is the RFID tag; this tag is comprised of an RFID chip to store tags ID and an antenna to send and receive signals. Tag gets activated when they acquire energy from radio waves generated from an RFID reader. The voter needs to register his name and thumb impressions in the database. During voting when the voter keeps his thumb in the fingerprint scanner which uses the ultrasound scanning method to capture the fingerprint, the machine checks whether the fingerprint suits the already stored impressions in the database, and if the voter scans his unique RFID TAG given to him while registration matches with the database, then only the person will be allowed to cast his vote, if not he will be stopped from polling. Otherwise the voter can cast his vote by pressing the switches each assigned to a specific candidate. After casting his vote, the voter will see the successful message in the LCD. The admin of the election committee will be able to see the real-time updates in the Think speak web applications by logging into his account. Valid voters will have their name, fingerprint, and different important points in the election committee database. Therefore, a fingerprint scanner will make sure that solely official voters can cast their votes without worrying about the integrity of the system (Figs. 7 and 8).

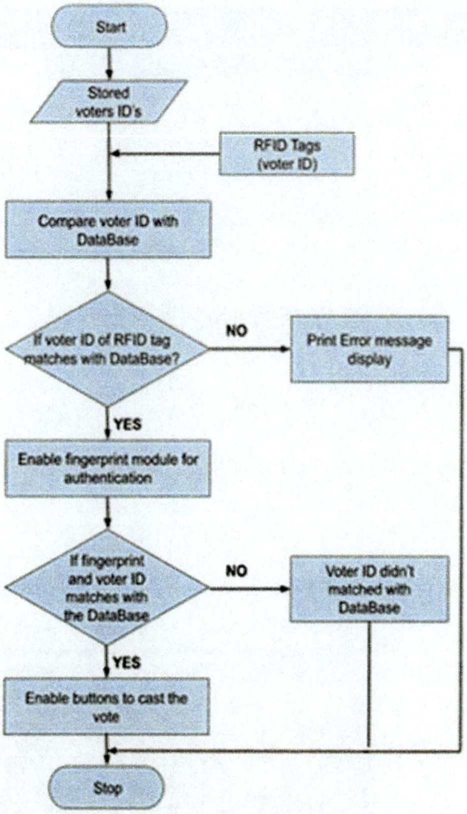

Fig. 6 Flowchart

10 Conclusion

The proposed approach utilizes methods that can prevent illegal votes from taking place. We used technologies that cannot be easily manipulated such as RFID and fingerprint technology. We used RFIDs to check the identity of the voter and compare them with the data stored in the server and employed fingerprint scanners to evaluate the voters if they are legitimate voters or not. If both the criteria are met, then only the voters will be allowed to cast the vote. LCD will display the instructions that are to be followed while voting without the need for another person in the booth. Results will be displayed on the LCD panel. After successfully casting the vote, the data is immediately uploaded onto the server. The online web application that is used to store the data from the voting machine is supported by all browsers so that the admin could log in from anywhere and at any time. The proposed device has been examined several times, and the machine consistently determined that the results were efficient.

Fig. 7 Think speak

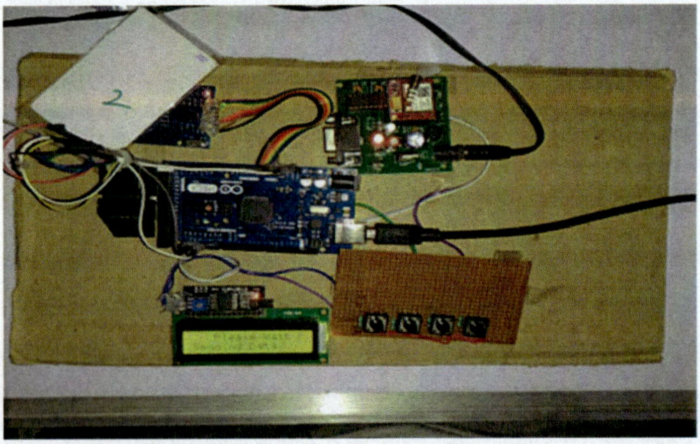

Fig. 8 Fingerprint and RFID-based voting system

11 Future Scope

If the system is improved and funded further in the future, it will become an even better high-quality system for the future world.

The following are some of the strategies that can be used to modify this device:

- The Arduino can be upgraded with a new microcontroller or chip.

- It may be improved even further by designing the PCB unit with VLSI technology. This can help to make the system even smaller.
- Additional sensors can be employed in the future applications.
- Iris recognition, which employs mathematical pattern recognition techniques for increased accuracy and security, can be added.

References

1. Li T, Zhang S (2018) Inkjet-printed proximity sensor for human–robot interaction. Microsyst Technol 24(12):4875–4880
2. Mayyas M, Mamidala I (2021) Prosthetic finger based on fully compliant mechanism for multi-scale grasping. Microsyst Technol 27(5):2131–2145
3. Lee B et al (2020) Preview control of web position in roll-to-roll printing using alignment patterns. Microsyst Technol 26(11):3315–3321
4. Srinivasu G et al (2020) Design analysis of an ultra-wideband antenna for RF energy harvesting in 1.71–12GHz. In: Proceedings of the 2020 11th international conference on computing, communication and networking technologies (ICCCNT). IEEE, New York
5. Khasawneh M, Malkawi M, Al-Jarrah O (2008) A biometric-secure e-voting system for election process. In: Proceeding of the 5th international symposium on mechatronics and its applications (ISMA08), Amman

Paramount Importance of Border Security Smart Robot Using IoT and MATLAB

Angotu Saida and Nehru Jarpula

Abstract In the security border line of control (LoC) most cases are about places where great deal of solidarity like intrusion as violence among group happens. This leads to loss of huge of soldiers, service men as common people. It get difficult to monitor these regions because of harsh weather condition, geographical reason like sand in the deserts, snow clad mountains, forest, water bodies, etc. and fringing LOC. Border security forces are failing to maintain national security and intrusions. To reduce human involvement and their loss we are proposed this system to monitor the LOC using smart robots. The proposed system uses Ultrasonic sensor coupled with a MATLAB processing code for detection of infiltrators and other objects. Ultrasonic sensor coupled with a MATLAB processing code for detection of infiltrators and other objects. Modules are connected with a Blynk app which sends information to the connected device for intimation of the control system. Software program captures intrusion, displays it in the MATLAB visual processor. The controlled-motor laser guns are considered to eliminate different circumstance conditions at the locations. The network integrates the PIR sensor to identify the specific motion and IR sensors to sense doubtful movements. Based on the command decision, a buzzer sensor goes off (Larry Hurwitz, Apparatus and methods for generating alerts, 2010). The sensors have been integrated through IoT for an efficient control of connectivity between sites and large border area.

Keywords LoC · MATLAB · BLYNK APP · IoT · MATLAB visual processor

A. Saida (✉)
KG Reddy College of Engineering and Technology, Hyderabad, Telangana, India
e-mail: angotusaida2@gmail.com

N. Jarpula
Joginpally B R Engineering College, Hyderabad, Telangana, India

J. R. Szymanski et al. (eds.), *Energy Systems, Drives and Automations*,
Lecture Notes in Electrical Engineering 1057,
https://doi.org/10.1007/978-981-99-3691-5_32

1 Introduction

Security of our borders from the illicit development of individuals, medications and weapons while advancing legal travel and exchange, is basic to country security, financial success and power of nation. Border security has the continuously monitoring, evolving policies due to the evolution of terrorism and threats. Robots are being used in every field to increase productivity and efficiency [1]. The Controller, fabrication, design and industrial translation of an Internet of Things (IoT) allow social robot as portrayed through validation of human feelings reaction to its effective interactions. The robot layout facilities on a humanoid hybrid-face that integrates a hard face plate with a digital show to simplify conveyance of complicated facial actions at the same time as supplying the influence of 3-dimensional depth [2]. Wide-location monitoring programs of an Internet of Things (IoT) related wireless sensor network includes sensor nodes (SNs) with restrained hardware and energy sources. The dispensed nature of the sort of network and the problem of far flung get entry to make it greater demands to design an energy-efficient WSN. Moreover, low-power and long-range Wi-Fi connectivity is a mission in IoT applications [3]. Our borders are no different than any field as they employ a large no of troops in the hostile area with a massive risk to human personal. An autonomous robot with be able to do all the things that a normal patrolling protocol demands. With the rapid improvements in technology like IoT, incorporating the technology becomes essential to make it more efficient. It will be able to detect any intrusion and immediately alert the command center over the intrusion detected by the module by processing input from the sensor in a specially designed MATLAB processing code which will not only detect an anomaly but also show its shape to the device connected in the command center. The project which is also coupled with different sensors to detect any changes in the temperature and identify a fire and use water to extinguish it. Usage of autonomous becomes essential in today's border security due to the rising threat to the patrolling officers and country's interest. The MOS technology used in full adder with hybridcomplex Metal Oxide Semiconductor, and the transient response analysis is symmetrical performed, it is occurred in invertor logic gate [4–6]. Also take from design analysis of an Ultra-Wide Band Aerial for radio frequency Energy Harvesting in the frequency range of 1.71–12 GHz and hexagon slotted Ultra-Wide Band monopole for spectrum sensing in the frequency band of 3.1–10.6 GHz [7–10]. The merger of various mixed grained re-configurable solutions permits the energy-efficient implementations of the wireless standards. The work has been completed in the software defined radio (SDR) [11].

2 Literature Review

"Android-based intelligent robot for border security" by Alex et al. [12] Android self-ruling intelligent robot for border security, it recognize the captured image of intruder using camera in android device, alerts security personnel by SMS using GSM and intruder using PIR motion sensor and send this picture to corresponding authorities using android application. This enables the security system to determine an effectively and at low cost to identify an intruder. The development and design of robot car for border security system, the work presents an adaptable smart robot vehicle utilizing remote camera for identifying people, obstacles at main areas, metals, fire and the messages are mail to the concern main systems. This security network makes use of artificial intelligence to produce prompt reaction from the sensors [12]. "SMART SECURITY SURVEILLANCE ROVER" by Suganthi et al. [13] Smart security border surveillance rover is a mobile robot which uses ultrasonic sensor for movement which is connected to wheels by an Arduino. Raspberry pie is also used in the project to work with a webcam that takes detects intrusion and alerts the center. The raspberry is connected to Wi-Fi module. This system has a mobile robot, controlled by Internet, which is consisting of a PIR sensor and camera mounted for detecting the living bodies [13]. "AN ANDROID-BASED INTELLIGENT ROBOT FOR BORDER SECURITY" by Ramesh et al. [14] an android-based intelligent robot security system, it identifies trespassers utilizing the face recognition algorithm. This security network can distinguish a intruder using gatecrasher discovery subsystem which depends on invariant face recognition and it trace the intruder using interloper tracing subsystem depends on streaming technology. Interloper recognition the subsystem captures pictures periodically when it identifies intruder in a secure region and confirm whether the target distinguished is human utilizing invariant face recognition algorithm then robot will be made alert the security system through alert signal using internet. The security system uses the pictures in raspberry pi camera to control robot to identify intruder and motion. Android self-ruling intelligent robot for border security system, which captures image of intruder using Raspberry Pi camera in android device, alerts security personnel by E-mail using GSM, identifies intruder utilizing PIR motion sensor and mail this pictures to corresponding email ID utilizing android application [14].

3 Proposed System

The major components of our project are Node MCU ESP8266, Arduino Nano ATMEGA 328, Ultrasonic sensor, Smoke Senor, Temperature Sensor. The project has two basic modules, one for the detection of the intrusion and the other for take counter measures. The modules are connected to a dedicated device which is used by the officers to check on possible dangers and warnings which will be prompted on the device. This connection is made by the Wi-Fi module present in the dedicated device and in the Node MCU ESP8266. An app called "Blynk" is used to prompt the warning to the security personnel on the device through a graphical interface. The app used due to it being dedicate to emphasize IoT devices which can used in a unison to achieve desired outcome. The most important aspect of the project is to detect intrusion which is carried out by the module which consists of Ultrasonic sensor and Arduino Nano ATMEGA 328.

The Ultrasonic sensor is mounted on a motor which make continuous arcs of rotation like a pendulum to cover the border area. The Ultrasonic sensor coupled with Arduino is used to detect intrusion by using a special dedicated MATLAB processing code which shows visual representation of the intrusion like sophisticated Radar. When the intrusion is detected Ultrasonic sensor picks the signal and starts displaying the intrusion in a red color like radar format along with sending a warning signal to the connected device and the security is intimated about the attempted intrusion, a laser also aimed at the intrusion representing a gun or a countermeasure. The second module which consists of sensors like smoke sensor and temperature sensor, these sensors are used to detect and warn the security of an attack which had resulted in fire and the module is also coupled with a water pump and a tank to act accordingly making autonomous. The Fig. 1 shows the MATLAB module of the security system. An Arduino Nano is used to control the sensors and interface it to a system for the MATLAB Visual Processing code to show the intrusion in the form of a radar like system which can be seen the results section.

The Fig. 2 shows the other module in the security system. A Node MCU is used due to its integrated Wi-Fi module. Other sensors like Temperature sensor, Ultrasonic sensor [15], Smoke sensor, Laser, Buzzer and Relay for the water pump is connected to the Node MCU for the different operations intended in the module. These sensors are used to detect intrusion and automatically take countermeasures to ensure safety.

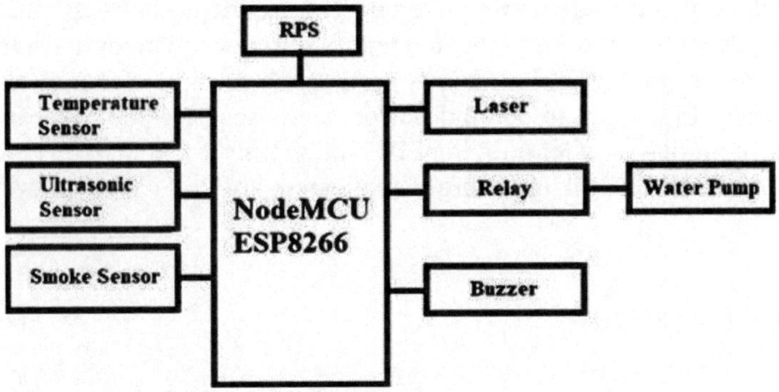

Fig. 1 MATLAB module block diagram

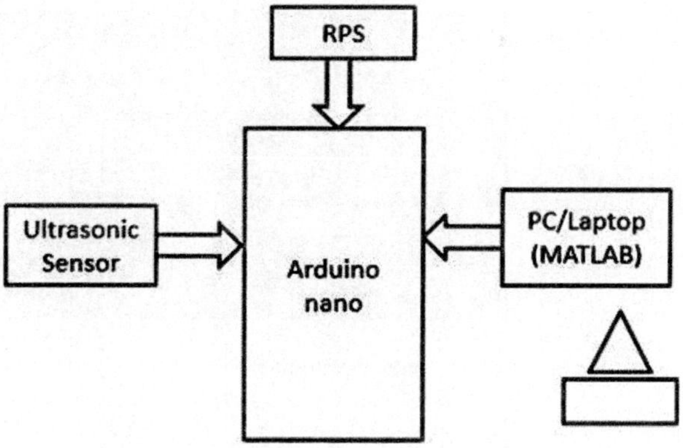

Fig. 2 Counter measure module of the system

4 Result and Discussion

The Ultrasonic sensor used in many applications where measuring distance or sensing objects are required [16]. The module has two eyes like projects in the front which forms the Ultrasonic transmitter and Receiver. The sensor works with the simple formula.

$$\text{Distance} = \text{Speed} \times \text{Time Here the result shown above}$$

The outcome of the project is to detect any intrusion and display the details in the Blynk app in Fig. 3. This helps in knowing intrusion if any by this device without having risk for human loss.

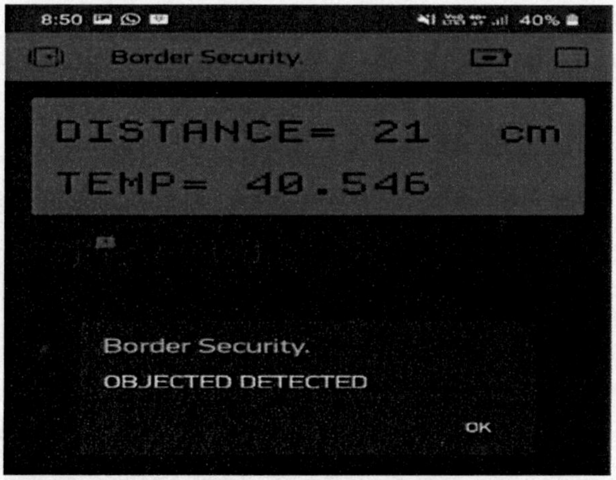

Fig. 3 The object detection output

Fig. 4 The details from sensors output

In Fig. 4 it can be seen that the details from the sensor output is displayed on the screen such as distance and temperature values which intimate the security personnel to act immediately upon it and to overcome the delay.

In Fig. 5 the counter measure can be seen. As a real gun cannot be put into the apparatus by soldiers a laser is installed represent the counter measure actions. This also helps the soldiers to pin point the exact location and react accordingly.

The output from the MATLAB processing code can be seen above. The intrusion and its shape can be seen in the red lines which can be used to determine how big a threat is approaching and what measures can be taken.

Figure 7 it shows front view of the proposed system which will be facing the border for detection of any intrusion. The Ultrasonic sensors can be seen which are connected to a motor to cover large area and monitor effectively.

Figure 8 it shows top view of the proposed system. The Arduino Nano can be seen which connects with a system to provide the MATLAB processing code output. The motor upon which the Ultrasonic sensors are mounted can be seen which rotates, which clear from the top view. Other sensors can also be seen like the smoke sensor and temperature sensor. Water pump along with the tank can also be seen at the top right along with the relay which activates it. Buzzer can also be seen on the left. Clipper switches are used on the either side of the system to maintain the arc like rotation from the motor.

Fig. 5 The counter measure output

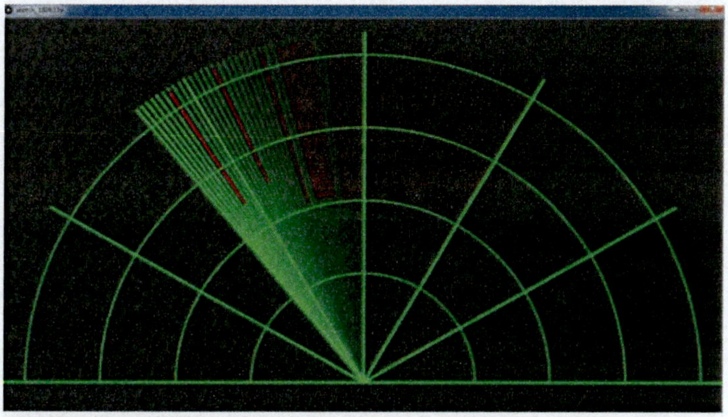

Fig. 6 The MATLAB processing code output

Fig. 7 Proposed system front view

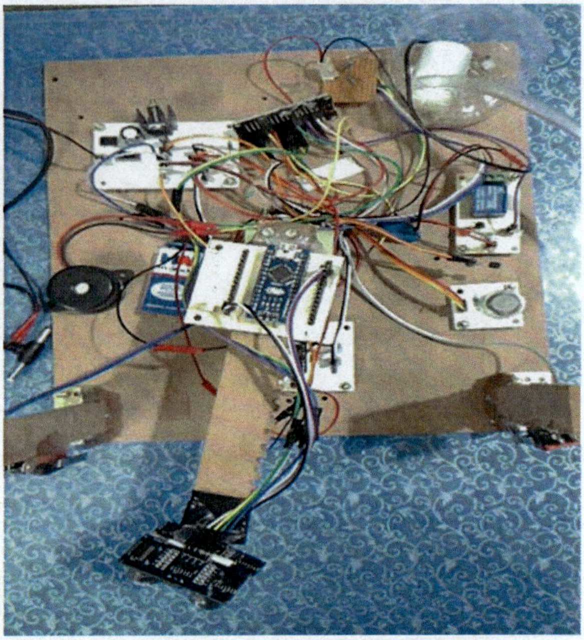

Fig. 8 Proposed system top view

5 Conclusion

Border security is of paramount importance for any country or organization. Current protocols deploy human security personnel on the border for manual patrolling which puts them at high risk for their lives. By this proposed smart robot system we can eliminate the need for human intervention hence reducing the risk of life. It minimizes Collateral Damage by preventing intrusions. The smart robot will be capable of creating a completely interconnected and sophisticated security system for large length borders the extreme environments and conditions. A smart robot that is capable of not only detecting but acting accordingly to the situational conditions. Many bleeding edge technologies can be incorporated in the system like AI and ML coupled with a FLIR camera by simply swapping the existing modules. The system can be incorporated as a chain of robots deployed on the border as the on technology used in IoT. The system very cleverly overcomes the disadvantages of other proposed systems and achieves the main goal of minimizing human casualties.

References

1. Bellavita C (2016) Defence security system: border security. Int J Innov Res Sci Eng Technol
2. Wairagkar M et al (2021) Emotive response to a hybrid-face robot and translation to consumer social robots. IEEE Internet Things J

3. Rahman GME, Khan AW (2021) LDCA: lightweight dynamic clustering algorithm for IoT-connected wide-area WSN and mobile data sink using LoRa. IEEE Internet Things J
4. Guduri M et al (2017) Current-mode circuit-level technique to design variation-aware nanoscale summing circuit for ultra-low power applications. Microsyst Technol 23(9):4045–4056
5. Agarwal VK, Guduri M, Islam A (2014) Power and variability analysis of CMOS logic families@ 22-nm technology node. In: Proceedings of 3rd international conference on reliability, Infocom technologies and optimization. IEEE, New York
6. Guduri M et al (2019) Minimum energy solution for ultra-low power applications. Microsyst Technol 25(5):1823–1831
7. Srinivasu G et al (2020) Design analysis of an ultra-wideband antenna for RF energy harvesting in 1.71–12 GHz. In: Proceedings of the 2020 11th international conference on computing, communication and networking technologies (ICCCNT). IEEE, New York
8. Gayatri T et al (2020) Analysis and design of a planar UWB antenna for spectrum sensing in 3.1–10.6 GHz. In: Proceedings of the 2020 11th international conference on computing, communication and networking technologies (ICCCNT). IEEE, New York
9. Srinivasu G et al (2020) Analysis of a disc-shaped monopole antenna for RF energy harvesting in 1.71–12 GHz. In: Proceedings of the 2020 IEEE India council international subsections conference (INDISCON). IEEE, New York
10. Gayatri T, Anveshkumar N, Sharma VK (2020) A circular planar monopole UWB antenna for cognitive radio applications. In: Proceedings of the 2020 IEEE-HYDCON. IEEE, New York
11. Saida A (2020) An efficient VLSI architecture for software defined radio by using montium processing tile
12. Alex AM, Jose ME, Rinsily KS, Bosco S, Shaji S (2017) Android based intelligent robot for border security. Int Res J Eng Technol 04(04):2041–2043
13. Suganthi N, Arun R, Saranya D, Vignesh N (2017) Smart security surveillance rover 116(12):67–75
14. Ramesh B, Yuvaraj BV, Shankarnag M, Pavan MK, Satwik J (2019) An android based intelligent robot for border security. Int J Comput Sci Mobile Comput 8(2):123–129
15. Alias AT (2016) Department of ECE, MA college of engineering, Kothamangalam ultrasonic vision for border security
16. https://www.migatron.com/ultrasonic-detections-and-control-applications/
17. Larry Hurwitz A (2010) Apparatus and methods for generating alerts

Gap Opening in Graphene by Substrate-Induced Strain Engineering Coupled with Magnetic Spin-Engineering

Sulagna Chatterjee

Abstract Currently, Graphene is emerging as the magic material in the electronics industry. Graphene possesses an excellent range of material properties that include superb mobility, electrical conductivity, high thermal conductivity, mechanical strength as well as flexibility. Besides, Graphene is a transparent material. Further, Graphene is also remarkably suitable for bio-electronic applications due to its antibacterial properties, low Joule heat generation, and high density that allow rapid passage for water evaporation. Also, Graphene is promising for low-power VLSI design as a result of its low drive power consumption. However, its one drawback that prohibits the large-scale, low-cost implementation of Graphene electronics is its non-existent bandgap. Therefore, both research and commercial fraternities are on a quest of suitable method for bandgap opening in Graphene. Several methods including light incidence, impurity doping, etc. have been investigated. But, very little has been availed till date. Moreover, all such methods come with inherent drawbacks that are as significant as decrease in mobility. The current paper proposes a coupled strain–spin engineering technique for inducing bandgap in Graphene. Since Graphene is two-dimensional material, therefore, a supporting substrate is a basic necessity in Graphene electronics. This paper proposes MgO and TiO_2 substrates for mounting Graphene-conducting layer. This, in turn, results in a considerable bandgap in Graphene. Subsequently, magnetic spin effect might be used to tailor such bandgap, according to the requirement of the implementation. Further, by proper application and withdrawal of applied electric field, such strain-induced bandgap might be used for information and memory electronics. ANSYS Multiphysics simulator has been calibrated with experimental observations and has been subsequently used to compare the analytical results.

Keywords Bandgap · Graphene · Mobility · Spin · Strain engineering · Substrate

S. Chatterjee (✉)
Adamas University, Kolkata, India
e-mail: chatterjeesulagna45@gmail.com

J. R. Szymanski et al. (eds.), *Energy Systems, Drives and Automations*,
Lecture Notes in Electrical Engineering 1057,
https://doi.org/10.1007/978-981-99-3691-5_33

1 Introduction

The discovery of Graphene in 2004 [1] in the form of a single layer of Carbon atoms has ushered in a new era in the domain of electronic materials [2]. Graphene has innumerable qualities suitable for application in ultrafast, low-power, flexible electronic devices [3]. Such superior Graphene-based devices find application in every sector of technology, namely biomedical [4], communication, defense, etc. [5–8]. Very high mobility [9], transparency [10, 11], carrier density that may be controlled by applied field [12], excellent mechanical strength, appreciable flexibility, etc. are certain examples of Graphene's remarkable electro-optical and mechanical properties. But in spite of so many unprecedented advantages, the one single fact that Graphene is a gapless material prohibits large-scale, low-cost development and applications of Graphene in next level electronics. Till date, various techniques have been explored with the aim of incorporating a suitable bandgap in Graphene. Doping or chemical substitution is one of such significant and widely used gap-inducing approaches [13]. Nitrogen and Boron are the prevalent dopant types [14–17]. Such substitution-induced gap opening has several drawbacks that includes effectively non-uniform doping, reduction in mobility, and subsequent decrease in conductivity. In order to overcome such drawbacks, various next level structures, namely, Graphene-Nano-Ribbons [14, 16], Graphene hybrids [18–22], patterned and/(or) folded Graphene [23], and Graphene origami [24], have been designed and investigated. But, each of such advanced Graphene-based structures require complicated fabrication process flow and also result in mobility decrement. This research proposes a strain-induced gap opening method that results from the lattice mismatch with the underlying substrate that supports the thin Graphene sheet. Strain engineering is an effective method that tampers with the material properties and if suitably applied results in substantial improvement in device performance [25–33]. Since Graphene is a two-dimensional material therefore, substrate is a basic requirement for Graphene devices. Also, such substrate definitely incurs lattice- and thermo-elastic mismatch with the overlying Graphene film, that in turn results in an inherent process-induced strain in the thin Graphene film. However, such uncontrolled strain rather results in degradation of device performance and gives rise to serious reliability issues [34]. On the other hand, suitably engineered process-induced strain may be used to incorporated energy gap in Graphene [35, 36]. Therefore, this research has proposed a simple yet elegant process-induced strain engineering technique that results from material parameter mismatch with the underlying substrate. Such strain engineering involves minimal process complexities and promises to open desired band gap in Graphene. Further, such strain-induced bandgap might be tuned using spin engineering [37].

2 Theory

The present research has considered the Graphene monolayers to be oriented in such a way so that zigzag and armchair directions lay along *x*- and *y*-axis, respectively. Figure 1 depicts the schematic representation of such Graphene sheet. Study reveals that 11% tensile strain along zigzag and 20% compressive strain along armchair directions result in desired bandgap ~ 1 eV in Graphene [36]. Subsequently, this research has proposed a suitable substrate-induced strain engineering technique that successfully incorporates such desired strain with appropriate magnitude and nature in the Graphene monolayers that lay atop of insulating oxide substrate. Different types of such oxide substrates have been explored. An analytical model has been designed for estimating such substrate-induced strain. Throughout this research article, suffixes f and s have used to demarcate the strained Graphene film and the substrate, respectively. The analytical model that has been used to estimate the stress induced in the Graphene sheet due to lattice mismatch with the insulating substrate is based on the Stoney formula [30, 38].

In this article, a denotes planar hexagonal lattice constant of Graphene. Also, a_f^x and a_f^y denote the lattice constant of the Graphene film along *x*- and *y*-axis, respectively.

$$a_f^y = |2a\cos(120^\circ)| + a = 2a, \tag{1a}$$

$$a_f^x = |2a\cos(30^\circ)| = \sqrt{3}a, \tag{1b}$$

where

$$a = 2.46\text{A}^\circ. \tag{1c}$$

The necessary and sufficient condition under which the mentioned Stoney formulae may be suitably applied is:

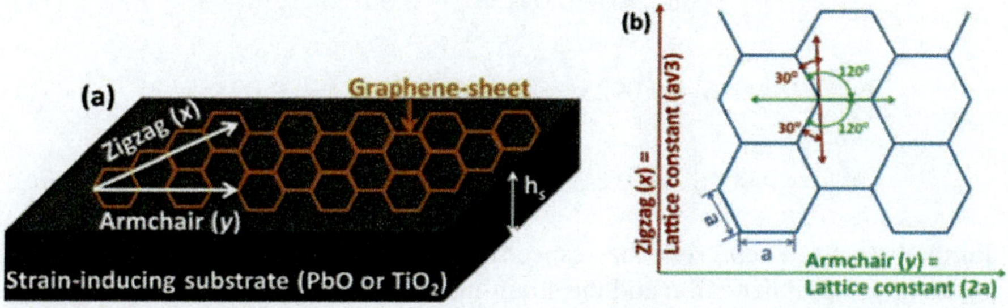

Fig. 1 **a** Schematic diagram of hetero-biaxially strained Graphene sheet lying on suitable substrate that incorporates the desired strain in Graphene for bandgap opening and **b** lattice-orientation for estimation of induced strain

$$r \gg h_s \gg h_f. \tag{2}$$

The Graphene sheet under consideration has the thickness of only single Carbon atoms ~ 3.45A°. Therefore, the above condition is aptly satisfied by the dimensions of the Graphene/substrate composite structure. Subsequently, the resultant stress is denoted by σ_f.

$$\sigma_f = \frac{\varepsilon_m Y_f}{1 + 4hy}. \tag{3}$$

In Eq. (3), ε_m is the inherent strain associated with such stress.

$$\varepsilon_m = \frac{a_s - a_f}{a_f}, \tag{4}$$

where a_f and a_s denote the lattice parameters for the Graphene film and the underlying substrate, respectively. In Eq. (3), Y_f designates the elastic modulus for the Graphene film: Y_f^{SL} for single layer and Y_f^{BL} for bilayer Graphene films.

$$Y_f^{\mathrm{SL}} = (2.4 \pm 0.4) \times 10^{12} = \begin{pmatrix} 2.8 \\ 2.0 \end{pmatrix} \times 10^{12}, \tag{5a}$$

$$Y_f^{\mathrm{BL}} = (2.0 \pm 0.5) \times 10^{12} = \begin{pmatrix} 2.5 \\ 1.5 \end{pmatrix} \times 10^{12}. \tag{5b}$$

For TiO_2 substrate,

$$Y_S = \begin{pmatrix} 0.288 \\ 0.230 \end{pmatrix} \times 10^{12}. \tag{6}$$

From Eqs. (5) to (6), the elastic constants are expressed in units of Pascal.

$$h_f = 0.0046 + 0.0925 n_L - 0.00255 n_L^2, \tag{7a}$$

$$h_f^{\mathrm{SL}} = 0.0046 + 0.0925 \times 1 - 0.00255 \times 1^2 = 0.09455, \tag{7b}$$

$$h_f^{\mathrm{BL}} = 0.0046 + 0.0925 \times 2 - 0.00255 \times 2^2 = 0.1794. \tag{7c}$$

Further, h and m represent the respective ratio(s) of the thickness and elastic modulus of the Graphene film and the strain-inducing underlying substrate.

$$h = \frac{h_f}{h_s}, \tag{8}$$

$$y = \frac{Y_f}{Y_s}. \tag{9}$$

Subsequently, the final complete strain that corresponds to the induced stress, σ_f, is denoted by ε_f and is modeled as:

$$\varepsilon_f = \frac{\sigma_f}{M_f}. \tag{10}$$

3 Results and Discussion

This section presents the induced strain profile in the Graphene sheet due to the application of various oxide insulators as substrate, namely Rutile (TiO_2), PbO, etc., that is estimated as described in the previous section. In this context, it deserves mention that oxide insulators that have tetragonal crystal structure have been chosen for substrate. This is due to the fact that Graphene sheets constitute of hexagonal mesh of Carbon atoms. In order to induce the suitable nature and amount of strain in such horizontal hexagonal atomic mesh, an underlying tetragonal substrate has been found to be the most suitable. The analytically obtained strain values are compared with those obtained from commercial ANSYS Multiphysics software [39] that has been calibrated with experimental results [25, 32]. The simulator works on the basis of Finite Element Mesh (FEM) technique [40]. Therefore, an optimization has been made between mesh fineness and simulation time to obtain superior and exact results. Table 1 presents the material parameters that have been used. Graphene parameters that have used are obtained by aptly considering the projections along *x*- and *y*-axis as well. Each single film of Carbon has been considered to have a single atom thickness of 3.4A°.

The complete hetero-biaxial strain profile for both single and bilayer Graphene has been depicted in Fig. 2. Two types of substrates have been explored, namely PbO and TiO_2. Magnitude of strain along both *x*- and *y*-axis has been plotted. In this context, it deserves mention that tensile strain has been induced along *x*-axis, i.e., zigzag direction, and simultaneously, compressive strain along *y*-axis, i.e., armchair direction, has been incorporated, to suit the requirements of ~ 1 eV gap opening [36]. A range of tensile strain from 4 to 14% for single-layer Graphene on PbO substrate

Table 1 Material parameters for Graphene film and the corresponding underlying substrate

	Lattice constant (a) (A°)	
	X-axis	*Y*-axis
Graphene	4.3	5
PbO	5	4
TiO_2	4.6	3

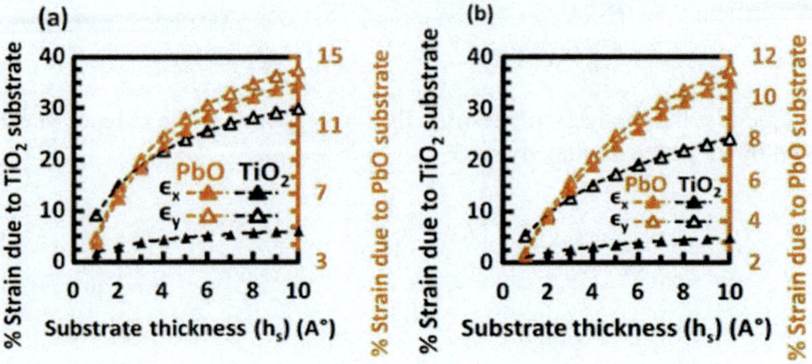

Fig. 2 Variation of the complete hetero-biaxial strain profile that has been induced in: **a** single and; **b** bilayer Graphene film due to underlying PbO and TiO_2 substrates of various thicknesses

and 2–6% for TiO_2 substrate has been obtained. On the other hand, the simultaneous compressive strain counterpart ranges from 4 to 14% and 9 to 30%, under same circumstances, whereas, for bilayer Graphene on the same tetragonal type of substrates, such strain profile sprawls across 2–11% tensile and 2–11.5% compressive for PbO substrate and 1–5% tensile and 5–24% compressive for TiO_2 substrate. This has been observed that as number of layers keep increasing in the Graphene film and substrate thickness remains constant, the induced strain approaches quasi-saturation.

Due to the Dirac-like dispersion feature of Graphene at points K and $K^{/}$ of Brillouin zone, Graphene is quite a suitable material for spin engineering. Experiments have shown that at low temperatures, Graphene has non-trivial topological state. Close to the edges of the Graphene film, at lower temperatures, maximum of hole band at $K^{/}$ and minimum of electron band at K are connected. Bonds with spin polarization might result in such peculiar property. Experiments related to Electron Spin Resonance (ESR) under applied magnetic field have been done with superficially minimized probable sources of externally induced spin and orbit coupling. Such study has revealed that the excitation frequency in the microwave range effectively controls the film resistance of Graphene. This is probably due to coupling effect of carriers with opposite spin. Again, this has subsequently caused due to transitions of the resonant microwave frequencies. Under such circumstances, it deserves mention that Graphene film is capable of blocking transitions even with no external magnetic field. If a magnetic field is applied, then a bandgap ~ 42.2×10^{-6} eV is induced in Graphene film [40]. This, therefore, supports the notion that spin and orbit coupling in Graphene results in desired gap opening. Thus, applied spin engineering in an already strained Graphene film lying on a suitable strain-inducing substrate may result in appropriate bandgap induction in Graphene film.

4 Experimental Feasibility

Graphene is a super thin sheet of Carbon atoms. Therefore, an insulator substrate is a mandatory requirement for the fabrication of any Graphene-based device. The current research has only replaced any possible substrate with PbO or TiO_2 layers of suitable thickness that are capable of incorporating desired amount of strain in the Graphene sheet. This strain in turn opens up the suitable bandgap in Graphene which otherwise constitutes mass-less Fermions. The standard Poly Methyl Methacrylate-assisted (PMMC) wet transfer method technique may be employed for transferring Graphene layer to the desired substrate [34].

5 Conclusion

The current research has for the first time proposed an efficient process-induced strain engineering technique that induces adequate hetero-biaxial planar strain in Graphene film that has been laid on a suitable substrate. Such induced strain, in turn, is capable of incorporating a desired bandgap of around 1 eV in both single and bilayer Graphene films.

References

1. Novoselov KS et al (2004) Electric field effect in atomically thin carbon films. Science 306:666–669
2. Geim AK, Novoselov KS (2007) The rise of Graphene. Nat Mater 6:183–191
3. Novoselov KS et al (2012) Roadmap for Graphene. Nature 490:192–200
4. Hong W, Bai H, Xu Y, Yao Z, Gu Z, Shi G (2010) Preparation of gold nanoparticle/graphene composites with controlled weight contents and their application in biosensors. J Phys Chem C 114(4):1822
5. Biró LP, Nemes-Incze P, Lambin P (2012) Graphene: nanoscale processing and recent applications. Nanoscale 4:1824–1839. https://doi.org/10.1039/C1NR11067E
6. Schwierz F (2010) Graphene transistors. Nat Nanotech 5:487–496
7. Stoller MD, Park SJ, Zhu YW, An JH, Ruoff RS (2008) Graphene-based ultracapacitors. Nano Lett 8(10):3498. https://doi.org/10.1021/nl802558y
8. Wang W, Liu F (2011) Nanopatterned Graphene quantum dots as building blocks for quantum cellular automata. Nanoscale 3:4201–4205. https://doi.org/10.1039/C1NR10489F
9. Peres N (2010) The transport properties of Graphene: an introduction. Rev Mod Phys 82(3):2673. https://doi.org/10.1103/RevModPhys.82.2673
10. Nair RR, Blake P, Grigorenko AN, Novoselov KS, Booth TJ, Stauber T, Peres NM, Geim AK (2008) Fine structure constant defines visual transparency of graphene. Science 6(320):1308. https://doi.org/10.1126/science.1156965
11. Sun R, Zhang Y, Li K, Hui C, He K, Ma X, Liu F (2013) Tunable photoresponse of epitaxial Graphene on SiC. Appl Phys Lett 103:013106. https://doi.org/10.1063/1.4812986
12. Castro Neto AH, Guinea F, Peres NMR, Novoselov KS, Geim AK (2009) The electronic properties of Graphene. Rev Mod Phys 81:109–162

13. Ma X, Wang Q, Chen LQ, Cermignani W, Schobert HH, Pantano CG (1997) Semi-empirical studies on electronic structures of a boron-doped graphene layer: implications on the oxidation mechanism. Carbon 35(10–11):1517. https://doi.org/10.1016/S0008-6223(97)00102-4
14. Martins TB, Miwa RH, da Silva AJR, Fazzio A (2007) Electronic and transport properties of boron-doped Graphene nanoribbons. Phys Rev Lett 98(19):196803. https://doi.org/10.1103/PhysRevLett.98.196803
15. Lherbier A, Blase X, Niquet Y-M, Triozon F, Roche S (2008) Charge transport in chemically doped 2D Graphene. Phys Rev Lett 101:036808. https://doi.org/10.1103/PhysRevLett.101.036808
16. Dutta S, Pati SK (2008) Half-metallicity in undoped and boron doped graphene nanoribbons in the presence of semilocal exchange-correlation interactions. J Phys Chem B 112(5):1333. https://doi.org/10.1021/jp710637c
17. Panchakarla LS, Subrahmanyam KS, Saha SK, Govindaraj A, Krishnamurthy HR, Waghmare UV, Rao CNR (2009) Synthesis, structure and properties of boron and nitrogen doped Graphene. Adv Mater 21(46):4726
18. Liu AY, Wentzcovitch RM, Cohen ML (1989) Atomic arrangement and electronic structure of BC_2N. Phys Rev B 39(3):1760. https://doi.org/10.1103/PhysRevB.39.1760
19. Miyamoto Y, Rubio A, Cohen ML, Louie SG (1994) Chiral tubules of hexagonal BC_2N. Phys Rev B 50:4976. https://doi.org/10.1103/PhysRevB.50.4976
20. Hernandez E, Goze C, Bernier P, Rubio A (1998) Elastic properties of C and $B_xC_yN_z$ composite nanotubes. Phys Rev Lett 80:4502. https://doi.org/10.1103/PhysRevLett.80.4502
21. Golberg D, Dorozhkin P, Bando Y, Dong ZC (2004) Synthesis, analysis and electrical property measurements of compounds nanotubes in the ceramic B–C–N system. MRS Bull 29:38
22. Yuge K (2009) Phase stability of boron carbon nitride in a heterographene structure: a first-principles study. Phys Rev B 79(14):144109. https://doi.org/10.1103/PhysRevB.79.144109
23. Feng J et al (2012) Patterning Graphene. Nanoscale 4:4883
24. Costa AT, Ferreira MS, Hallam T, Duesberg GS, Castro Neto AH (2013) Origami-based spintronics in graphene. EPL 104:47001. https://doi.org/10.1209/0295-5075/104/47001
25. Chatterjee S, Chowdhury BN, Das A, Chattopadhyay S (2013) Estimation of step-by-step induced stress in a sequential process integration of nano-scale SOS MOSFETs with high-k gate dielectrics. Semicond Sci Technol 28:125011. https://doi.org/10.1088/0268-1242/28/12/125011
26. Chatterjee S, Chattopadhyay S (2016) Modeling and estimation of process-induced stress in the nanowire field-effect-transistors (NW-FETs) on insulator-on-silicon substrates with high-*k* gate-dielectrics. Superlatt Microstruct 98:194–202. https://doi.org/10.1016/j.spmi.2016.08.022
27. Chatterjee S, Chattopadhyay S (2017) Analytical modeling of the lattice and thermo-elastic coefficient mismatch-induced stress into silicon nanowires horizontally embedded on insulator-on-silicon substrates. Superlatt Microstruct 101:384–396. https://doi.org/10.1016/j.spmi.2016.12.001
28. Chatterjee S (2018) Chattopadhyay, S: Fraction of insertion of the channel-fin as performance booster in strain-engineered p-FinFET devices with insulator-on-silicon (IOS) substrate. IEEE Trans Elect Dev 65(2):411–418. https://doi.org/10.1109/TED.2017.2781264
29. Chatterjee S, Sikdar S, Chowdhury BN, Chattopadhyay S (2019) Investigation of the performance of strain-engineered silicon nanowire field effect transistors (ε-Si-NWFET) on IOS substrates. J Appl Phys 125:082506. https://doi.org/10.1063/1.5051310
30. Chatterjee S, Mukherjee M (2019) Strain-engineered asymmetrical superlattice $Si/Si_{1-x}Ge_x$ Nano-ATT $\langle p^{++}$-n–n−-$n^{++}\rangle$ oscillator: enhanced photo-sensitivity in terahertz domain. IEEE Trans Elect Dev 66(8):3659–3667. https://doi.org/10.1109/TED.2019.2923108
31. Chatterjee S, Mukherjee M (2020) Direct band gap silicon nanowire avalanche transit time THz opto-electronic sensor with strain-engineering. Opt Quant Elect 52:488. https://doi.org/10.1007/s11082-020-02563-7
32. Mukherjee M, Chatterjee S (2020) Millimeter-wave high-low IMPATT source development: first on-chip experimental verification. Elect Lett 57(7):294–296. https://doi.org/10.1049/ell2.12058

33. Chatterjee S, Mukherjee M (2021) Strained Si/Si$_{1-y}$C$_y$ superlattice based quasi-read avalanche transit-time devices for terahertz ultrafast switches. Appl Phys A 127:155. https://doi.org/10.1007/s00339-020-04187-w
34. Chakraborty D, Chatterjee S, Mukherjee M (2020) Hybrid multi-Graphene/Si avalanche transit time <h-ATT> terahertz power oscillator: theoretical reliability and experimental feasibility studies. IEEE Trans Dev Mater Reliab 20(4):667. https://doi.org/10.1109/TDMR.2020.3018664
35. Gui G, Li J, Zhong J (2008) Band structure engineering of graphene by strain: first-principles calculations. Phys Rev B 78(7):075435. https://doi.org/10.1103/PhysRevB.78.075435
36. Kerszberg N, Suryanarayana P (2015) Ab initio strain engineering of graphene: opening bandgaps up to 1 ev. RSC Adv 5:43810. https://doi.org/10.1039/c5ra03422a
37. Chen X, Liu Y, Gu BL, Duan W, Liu F (2014) Giant room-temperature spin caloritronics in spin-semiconducting Graphene nanoribbons. Phys Rev B Condens Matter 90(12):121403. https://doi.org/10.1103/PhysRevB.90.121403
38. Release 10.0 Documentation for ANSYS ANSYS Inc and ANSYS Europe Ltd 2005 Copyright
39. Freund LB, Suresh S (2004) Thin film materials stress, defect formation and surface evolution. Cambridge University Press, Cambridge, p 94. https://doi.org/10.1017/CBO9780511754715
40. Heinrich B (2019) A spin–orbit gap in Graphene. Phys Rev Lett 122:046403. https://doi.org/10.1038/s41565-019-0409-y

Computer and SoftComputing

A Review of DDoS Evaluation Dataset: CICDDoS2019 Dataset

Mandula China Pentu Saheb, M. Srikanth Yadav, Sallagundla Babu, Jeevana Jyothi Pujari, and Jeevan Babu Maddala

Abstract DDoS assaults are a danger to network security because they overwhelm target networks with malicious traffic, making them unusable as a result. One of the most significant concerns is the lack of a real-time DDoS attack detector that requires low processing power. When it comes to testing novel detection algorithms and approaches, good datasets are vital. After evaluating current datasets, the authors of this paper propose a new DDoS taxonomy. After that, we produce a new dataset, CICDoS2019 that addresses all of our issues. Third, we offer a new family detection and classification technique based on network flow features extracted from the created dataset, which we call FlowNet. Finally, we rank the most relevant feature sets for identifying distributed denial-of-service (DDoS) assaults.

Keywords DDoS · Attacks · Dataset · TCP · UDP

M. C. P. Saheb (✉) · J. J. Pujari · J. B. Maddala
Department of CSE, Vasireddy Venkatadri Institute of Technology, Guntur, Andhra Pradesh, India
e-mail: saheb10thjune@vvit.net

J. J. Pujari
e-mail: jeevanajyothi.p@vvit.net

J. B. Maddala
e-mail: jeevan@vvit.net

M. S. Yadav
Department of IT, Vignans Foundation for Science, Technology and Research (Deemed to be University), Guntur, Andhra Pradesh, India

S. Babu
Department of CSE, V R Siddhartha Engineering College, Vijayawada, Andhra Pradesh, India

J. R. Szymanski et al. (eds.), *Energy Systems, Drives and Automations*,
Lecture Notes in Electrical Engineering 1057,
https://doi.org/10.1007/978-981-99-3691-5_34

1 Introduction

Numerous studies have categorised DDoS attacks in various ways, and each study has its own system of classification. There have been few attacks, despite the fact that everyone has done an excellent job of proposing new taxonomies. There must be more attacks found and new taxonomies developed. A new taxonomy has been designed to identify the distinct attacks that can be carried out utilising the TCP/UDP protocols at the application layer. For the purposes of this chapter, we will use the terms "reflection-based" and "exploitation-based" to describe the taxonomy of distributed denial-of-service (DDoS) attacks, as seen in Fig. 1.

1.1 Reflection-Based DDoS

Reflection-based DDoS assaults hide the perpetrator's identity by using third-party components. Attackers are attempting to flood the victim's system with response packets by sending packets to reflector servers with the source IP address configured to the IP address of the victim. A combination of application layer protocols and transport layer protocols can be used to carry out these attacks, such as TCP or UDP. TFTP, NTP, and CharGen are three examples of UDP-based attacks of this type (as shown in Fig. 1). DNS, LDAP, NETBIOS, and SNMP are just some of the protocols that might be targeted by TCP and UDP assaults.

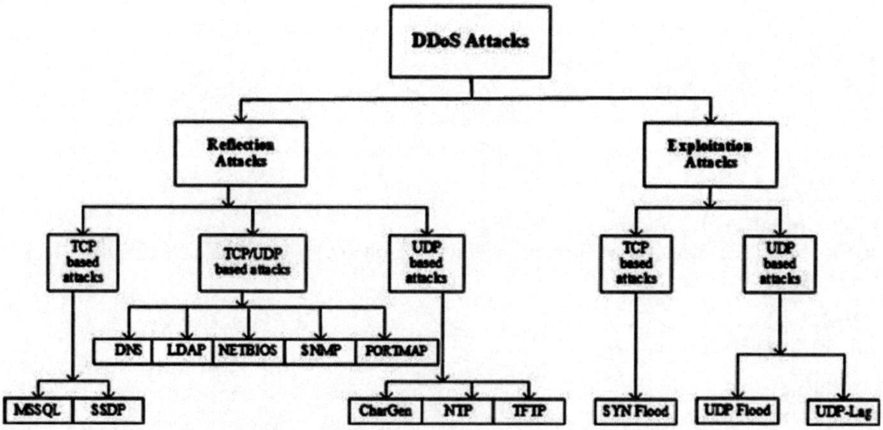

Fig. 1 Taxonomy of DDoS attack

1.2 *Exploitation-Based Attacks*

In exploitation-based assaults, attackers might hide their identities by using lawfully obtained third-party components. Using a reflector server, attackers can flood a target with response packets by sending packets to a reflector server whose source IP address has been set to the IP address of the targeted victim. TCP and UDP are two examples of transport layer protocols that can be exploited in this way. SYN flood, UDP flood, and UDP-Lag are only a few examples of TCP-based exploitation attacks. The host becomes exposed to a UDP flood attack when it receives an excessive volume of UDP packets from a remote host. Hundreds of thousands of UDP packets are being broadcast in rapid succession to multiple ports on the target machine. Consequently, network bandwidth is lowered, the system crashes, and overall performance is adversely affected as a result. It is the SYN flood, on the other hand, that takes advantage of the TCP protocol's three-way handshake and wastes server resources. When the server fails or malfunctions, SYN packets are delivered to the target computer in an attempt to initiate this attack. Attacks like the UDP-Lag attack slow down data transfer between a client and a server. To get an edge over your opponents, it is common practise in online gaming to slow down or halt the movement of other players. Alternatively, network-based software can be used to take advantage of other users' bandwidth by using a lag switch, also known as a hardware switch.

2 Related Work

Businesses and organisations are becoming increasingly concerned about Internet security as the demand for IT services continues to rise. If you want to take down your intranet and the Internet, a DDoS assault is an easy and effective way to do so. Web-based services are unavailable to normal users because of this attack, which has affected a large number of devices. Network, transport and application layers may all be used to carry out DDoS attacks using a wide range of protocols such as TCP/UDP; ICMP; HTTP; and others. Machine learning, knowledge bases, and statistics have been used by many academics to suggest a range of detection and protection strategies. There are a number of drawbacks and issues with each technique, which will be explored in more detail below. Network packet normal distribution cannot be determined with precision using statistical methods, for example. Because they don't have any prior knowledge of the data's distribution, machine learning algorithms have a tough time selecting the ideal feature set.

As a result of these issues, researchers are still unable to find appropriate datasets to test and assess their suggested detection and protection models, such as missing traffic, anonymised data and obsolete attack scenarios. Accurate data collection is a major task in and of itself, as a result [1]. We have two separate additions to offer to this paper in light of these two significant issues. Starting with the present datasets, we assess which have the most significant defects or limitations. A new DDoS dataset, CICDoS2019, was created to solve issues with prior datasets. This dataset will be discussed next. Each benign and denial-of-service flow was analysed using the CICFlowMeter software, which is publicly accessible on the website of the Canadian Institute for Cyber Security [2].

DDoS assaults, such as those that target DNS, LDAP, MSSQL, and the TFTP file system, as well as those that use the UDP protocol, are examined in the second phase of the study (including UDP-Lag and SYN attacks). Aside from these popular methods, we use ID3, random forest, and Naive Bayes and logistic regression to build our models. These are all meant to capture the patterns that we detect. The testing component is used to do this.

3 CICDDoS Dataset

3.1 Description of CICDDoS Dataset

Numerous studies have categorised DDoS assaults in various ways, and each study has its own system of categorization [3]. Identified DDoS assaults and their accompanying mitigation measures, which were then accepted by the community. Automated vulnerability and source address validity were two of the factors utilised to define attacks in this study. Conventional DDoS assaults that are both harmless and up to date will be discussed at the CICDOS2019 conference. Data flow analysis using CICFlowMeter-V3 was used to identify flows based on time stamp, source and destination IP addresses/port numbers/protocols/assault kinds in the network traffic (see Fig. 1). For this dataset, the purpose was to provide an accurate depiction of traffic in the background. The B-Profile system we designed and built was used to profile the abstract behaviour of human interactions and to produce genuine background traffic in the intended testbed. We employed a variety of protocols, including HTTP, HTTPS, FTP, SSH, and email, to produce the abstract behaviour of 25 individuals for this dataset.

There are 50,063,112 entries in the CIC-DDoS2019 dataset, with 50,006,249 rows representing DDoS assaults and 56,863 rows representing benign behaviour. The Center for Internet Security compiled the CIC-DDoS2019 dataset for research purposes only. There are 86 elements in each row. Attacks on this dataset during training and testing are summarised in Table 4. A total of 12 DDoS assaults are included in the datasets for training and testing, including attacks on NTP, DNNS (DNS), MSSQL, NetBIOS, SNMP, SYN, and TFTP (Tables 1, 2).

Table 1 Attack types in CICDDoS2019 dataset

Attack type	Flow count
Benign	56,863
DDoS DNS	5,071,011
DDoS LDAP	2,179,930
DDoS MSSQL	4,522,492
DDoS NetBIOS	4,093,279
DDoS NTP	1,202,642
DDoS SNMP	5,159,870
DDoS SSDP	2,610,611
DDoS SYN	1,582,289
DDoS TFTP	20,082,580
DDoS UDP	3,134,645
DDoS UDP-Lag	366,461

3.2 Experimental Setup

Table 2 includes all of the computers used for training and testing (servers, firewalls, and workstations), along with their operating systems and public and private IP addresses, to illustrate this point. A third party assassinated the relatives of the victims (Attack-Network). A web server, a firewall, two switches, and four personal PCs make up the Victim-Network. All network data, including outgoing and incoming, are collected on one port that serves as a "mirror" on the Victim-main Network. Assaults such as the ones listed below have been reported: (see Table 3).

4 Analysis of CICDDoS Dataset

To begin, we used CICFlowMeter to extract the dataset's 80 traffic attributes. For each DDoS assault, we use the RandomForestRegressor to determine the best detection characteristics. Data from both training and testing are used to evaluate the effectiveness of four common machine learning methods. CICFlowMeter, a flow-based feature extractor that can extract up to 80 features from a packet capture file, was used to extract the network traffic features. The IP address, port number, protocol, and timestamp are all included in the flow label for this application. According on the attack timeline given, we characterise the flows that were established in Sect. 5 (timestamp). The CICFlowMeter website explains and explains in detail all 80 of the retrieved characteristics. A class was discovered via the Scikit-RandomForestRegressor. Prior to this, each feature's relevance is determined and then multiplied by its respective significance value in order to arrive at the final result (Table 4).

Table 2 Attack types and description

Attack type	Description
NTP-based attack	NTP amplification is a type of distributed denial-of-service (DDoS) attack in which the attacker exploits publically accessible Network Time Protocol (NTP) servers to overwhelm the targeted with User Datagram Protocol (UDP) traffic
DNS-based attack	DNS flood is a type of distributed denial-of-service (DDoS) attack in which the attacker targets one or more Domain Name System (DNS) servers belonging to a given zone, attempting to hamper resolution of resource records of that zone and its sub-zones
LDAP-based attack	A DDoS reflection attack is the practice of sending requests using a spoofed source IP address to various servers on the Internet, which in turn will direct their responses to the spoofed address instead of the real sender. The spoofed IP address is that of the intended victim
MSSQL-based attack	The attack occurs when a Microsoft SQL server responds to a client query or request, attempting to exploit the Microsoft SQL Server Resolution Protocol (MC-SQLR), listening on UDP port 1434
NETBIOS-based attack	The primary purpose of NetBIOS is to allow applications on separate computers to communicate and establish sessions to access shared resources and to find each other over a local area network
SNMP-based attack	As the runt iaphs, dis is a vehamesic DDoS attack: sing the Single Novak ManagememProtocol Cane to flood the tries network pipes with handle and of einbin or seconi of attack traffic
SSDP-based attack	A simple service discovery protocol (SSDP) attack is a type of reflection DDoS attacks that exploit the Universal Plug and Play (UPnP) network protocols for sending an amplified traffic stream to the victim's server
UDP-Lag-based attack	A UDP flood is a type of denial-of-service attack in which a large number of User Datagram Protocol (UDP) packets are sent to a targeted server with the aim of overwhelming that device's ability to process and respond. The firewall protecting the targeted server can also become exhausted as a result of UDP flooding, resulting in a denial-of-service to legitimate traffic
WEB DDOS-based attack	This vulnerability compromises a web server or application by exploiting simple HTTP GET or POST request
SYN-based attack	A SYN flood (half-open attack) is a type of denial-of-service (DDoS) attack which aims to make a server unavailable to legitimate traffic by consuming all available server resources. By repeatedly sending initial connection request (SYN) packets, the attacker is able to overwhelm all available ports on a targeted server machine, causing the targeted device to respond to legitimate traffic sluggishly or not at all

Table 3 Hardware used for experimental study

Machine	OS	IPs
Server	Web server: Ubuntu 16.04	192 168 50 1 (day 1) 19216850.4 (day 2)
Firewall	Fortinet	205.174.165.81
PC5 (day 1) 01-11-2019	Windows 7	192.168.50.8
	Windows Vista	192 A68.50.5
	Windows 8.1	192.168.50.6
	Windows 10	192.168.50.7
PC5 (day 2) 03-11-2019	Windows 7	192.168.50.9
	Windows Vista	192.168.50.6
	Windows 8.1	192 A68.50.7
	Windows 10	192 A68.50.8

Table 4 Day-wise attacks were detected with the time of detection

Days	Attacks (detection time)
First day (01-11-2019)	PortMap (9:43–9:51)
	NetBIOS (10:00–10:09)
	LDAP (10:21–1030)
	MSSQL (10:33–10:42)
	UDP (10:53–11:03)
	UDP-Lag (11:14–11:24)
	SYN (11:28–17:35)
	NTP (10:35–10:45)
	DNS (10:52–11:05)
	MAP (11:22–11:32)
	MSSQL (11:36–11:45)
Second day (03-11-2019)	NetBIOS (11:50–12:00)
	SKIMP (12:12–12:23)
	SSDP (12:27–12:37)
	UDP (12:45–13:09)
	UDP-Lag (13.11–13:15)
	WebDDoS (13:18–13:29)
	SYN (13:29–13:34)
	TFTP (13:35–17:15)

5 Conclusion

It is now possible to evaluate intrusion detection systems and methods using the new dataset CICDoS2019 for analysing DDoS assaults. A new DDoS taxonomy for the application layer has been given after evaluating a large number of distinct DDoS attack types and families. Additionally, the most often utilised DDoS datasets have been examined, and the most prevalent vulnerabilities and flaws have been discovered and reported as well. Due to these limitations, we created a new dataset, CICDoS2019, which includes 11 DDoS assaults and was used to evaluate IDS/IPS algorithms and systems. Some of the most crucial factors for identifying distinct DDoS assaults were also included.

References

1. Subbulakshmi T, BalaKrishnan K, Shalinie SM, AnandKumar D, GanapathiSubramanian V, Kannathal K (2011) Detection of DDoS attacksusing enhanced support vector machines with real-time generateddataset. In: Third international conference on advanced computing, pp 17–22
2. Arash Habibi Lashkari MM, Draper Gil G, Ghorbani A (2017) Characterization of tor traffic using time-based features. In: Proceedingsof the 3rd international conference on information systemssecurity and privacy (ICISSP), pp 253–262
3. Mirkovic J, Reiher P (2004) A taxonomy of ddos attack and ddosdefense mechanisms. ACM SIGCOMM Comput Commun Rev 34(2):39–53
4. Jin S, Yeung DS (2004) A covariance analysis model for DDoS attackdetection. In: Proceedings of the 2004 IEEE international conference on communications, vol 4, pp 1882–1886
5. Munivara Prasad AMRK, Rao K (2014) DoS and DDoS attacks: defense, detection and traceback mechanisms: a survey. Global J Comput Sci Technol 14
6. The CAIDA UCSD "DDoS Attack 2007" dataset. http://www.caida.org/data/passive/ddos-20070804 dataset.xml. Accessed Jan 2018
7. DARPA 2000 intrustion detection scenario specific data sets. https://www.ll.mit.edu/r-d/datasets/2000-darpa-intrusion-detection-scenario-specific-data-sets. Accessed Jan 2018
8. Carson Brown AH, Cowperthwaite A, Somayaji A (2009) Analysisof the 1999 DARPA/lincoln laboratory ids evaluation data with netadhict. In: Proceedings of the second IEEE international conference on computational intelligence for security and defense applications, CISDA'09
9. Singh KJ, De T (2015) An approach of DDoS attack detection using classifiers. Emerg Res Comput Inform Commun Appl
10. Yu S, Zhou W, Jia W, Guo S, Xiang Y, Tang F (2012) DiscriminatingDDoS attacks from flash crowds using flow correlation coefficient. IEEE Trans Parallel Distrib Syst 23(6):1073–1080
11. Ali Shiravi MT, Shiravi H, Ghorbani AA (2012) Toward developinga systematic approach to generate benchmark datasets for intrusiondetection. Comput Sec 31:357–374
12. Asosheh A, Ramezani N (2008) A comprehensive taxonomy of DDoSattacks and defense mechanism applying in a smart classification. WSEAS Trans Comput 7(4):281–290
13. Bhardwaj A, Subrahmanyam G, Avasthi V, Sastry H, Goundar S (2016) Ddos attacks, new DDoS taxonomy and mitigation solutions—a survey. In: Proceedings of the 2016 international conference on signal processing, communication, power and embedded system (SCOPES). IEEE, New York, pp 793–798
14. Masdari M, Jalali M (2016) A survey and taxonomy of dos attacks incloud computing. Sec Commun Netw 9(16):3724–3751

15. "Request for comments: 7766 (rfc7766)," in DNS Transport over TCP- Implementation Requirements, 2016
16. Singh K, Singh P, Kumar K (2017) Application layer http-get flood DDoSattacks: research landscape and challenges. Comput Sec 65:344–372
17. Iman Sharafaldin AHL, Gharib A, Ghorbani A (2017) Towards a reliable intrusion detection benchmark dataset. Softw Netw 147:177–200
18. CICFlowMeter 2017. https://github.com/ISCX/CICFlowMeter
19. Pedregosa AGF, Varoquaux G, Duchesnay E (2011) Scikit-learn: machine learning in Python
20. Quinlan JR (1986) Induction of decision trees. Mach Learn 1(1):81–106
21. Breiman L (2001) Random forests. Mach Learn 45(1):5–32

Spotting Irregularities in Infinitesimal Images of Blood via Comparable Fragmentation of Image

Nissankara Lakshmi Prasanna, Koteswararao Velpula, and Narendra Sanam

Abstract Most chronic diseases resulting from the observation of individuals become some specific with their exact clinical picture. These portrayals are of assorted sorts, not out of the ordinary as minuscule pictures, ultrasound, registered tomography, attractive reverberation picture and so on [1, 2]. Barely, a few varieties from the standard can be arranged with horrendous cells in plasma or polyps or breakages or parasitic recognizing, etc. Finding anomalous partial expressions requires doctor's skill; this corresponding distinctive features treated through medication claimed by a doctor or expert. In clinical minuscule portrayals, typical pieces other than sporadic parts are pooled. Random and conventional pieces are not really perceived by sporadic and conventional pieces of portrayals at occurrence; anomalies are debased among ordinary areas of portrayal. Such anomalies stay absentminded in firm parts for a clear area in the portrayals. These anomalies are not really safeguarded nor can cannot escape a thoughtful part genuinely in the portrayal. Irregularities are isolated close by customary pieces of pictures. Recognizing such irregularities is vital. In earlier methods [3–5], these irregularities are recognized and utilized BFS and Shortest Path Algorithm. This paper centers around recognizing irregularities utilizing comparable to discontinuity of clinical minute pictures.

Keywords Irregularities · Image fragmentation · Sickle cell

N. L. Prasanna (✉) · K. Velpula · N. Sanam
Department of Computer Science and Engineering, VasireddyVenkatadri Institute of Technology, Guntur, India
e-mail: nlakshmi@vvit.net

K. Velpula
e-mail: koteswararao@vvit.net

N. Sanam
e-mail: narendracse@vvit.net

J. R. Szymanski et al. (eds.), *Energy Systems, Drives and Automations*, Lecture Notes in Electrical Engineering 1057,
https://doi.org/10.1007/978-981-99-3691-5_35

1 Introduction

Clinical imaging is a technique and practice for creating meaningful representations of the patient's internal parts for clinical review and clinical intervention, comparable to the optical representation of the guaranteed structural capacity of the patient's body is. Various kinds of clinical portrayals are images of changing variety: X-rays, computer tomography, magnetic resonance, positron emission tomography and various microimages.

X-Ray Imaging: This utilizes molecule emanation which is ionized to build portrayals of chronic sickness individual's inside representation by coordinating X-beam shafts into the body. X-beam portrayals are taken advantage of to pass judgment on lungs, breaks in body skeleton and so on.

Ultrasound Imaging: This is likewise called sonography. This utilizes weighty reiteration of sound waves in developing portrayals of inside body organs of an individual. Mainly used for surveying the fetal development, midsection, heart, corridors, veins and so on.

Figured Tomography (CT) Imaging: This imaging unites set of X-beams held from different edges to convey a point-by-point cross-sectional pictures of zones inside persistent's body. Figured tomography pictures grant experts to get outstandingly exact, three-layered viewpoints on explicit areas of inside persistent's body. Registered tomography is utilized to assess closeness, aspect, area of polyps, hairline cracks, tissues and so on.

Attractive Resonance Imaging (MRI): This utilizes an innovation that incorporates radio waves with an attractive field to sort point-by-point portrayals of organs and so on.

Positron Emission Tomography (PET) Imaging: This approach enriches doctors with how inside body organs and the tissues at different areas in body are functioned. This is essentially used to assess neurological defilements likely malignant growth areas, Alzheimer's, Multiple Sclerosis and so on.

Minute Images: Infinitesimal pictures are tiny used in clinical finding [1]. These portrayals are made with an extraordinarily objective of 4096×2160 size (4 K projection), a multiple of the full pixel count as a 1920×1080 pixel high-resolution image. Different sorts of little pictures are:

- Biochemical infinitesimal images.
- Pathological infinitesimal images.
- Micro-natural infinitesimal images.

Observing such anomalous parts of the portrayals requires the expertise of a doctor. This commendable, recognizable evidence promotes and guarantees comprehensive medical care from physicians to patients. This paper includes in observing irregularities probably recognizing sickle cells in minute blood pictures.

2 Related Work

Iron deficiency: Blood contains different unmistakable cells. These are white platelets, platelets and red platelets. Lack of iron is the greatest amount of notable disarray of the plasma. "Lack of iron", the term is subordinate after the outdated Greek term delicacy, which specifies "Lack of Plasma." This is feasible given that the number of red blood cells (RBCs) [4] is reduced or the amount of hemoglobin in the blood is almost below normal. Regardless, it might also consolidate directed oxygen-keeping volume of every hemoglobin iota as deception or need factual development. Lack of iron is definitely an indication of a disease practice as unique to presence an infirmity to self. It is same as outrageous or endless. In debilitating weakness, aftereffects regularly jerk dynamically and evolvement continuously; while in outrageous delicacy, parallel results can be abrupt other than out and out the extra disturbing. Among plentiful components, together dietary (like absence of supplements, minerals and so forth) and non-feeding (defiled) expand to the beginning of shadowiness; inadequacy of iron and celiac contamination embraces a prominent undertaking. Sickliness is ordinarily classified with hemoglobin level under 13.5 g/dl in guys and under 12.0 g/dl in females.

Sickle Cell Anemia: Sickle mobileular ailment is an received blood issue. RBC incorporates hemoglobin which conveys the oxygen in blood [6]. These are curved which enables them to go total minuscule veins to pass oxygen on to out and out bits of the human body. Such cells regularly live around 120 days. Sickle cell disease makes red platelets convert bow formed, which makes them disturbance isolated with practically no issue. This finally causes sickliness considering the way that the phones simply live 10–20 days. Figure 1 shows the state of red platelet and sickle cell. Figure 2 shows blood picture having sickle cells.

Portrayal of Sickle Cell Anemia: Investigation is made when there exists bean formed designs in blood picture.

- Ordinary Cell: Platelets usually have an adaptable round shape and flow efficiently through veins.

Fig. 1 Usual blood cell and sickle cell

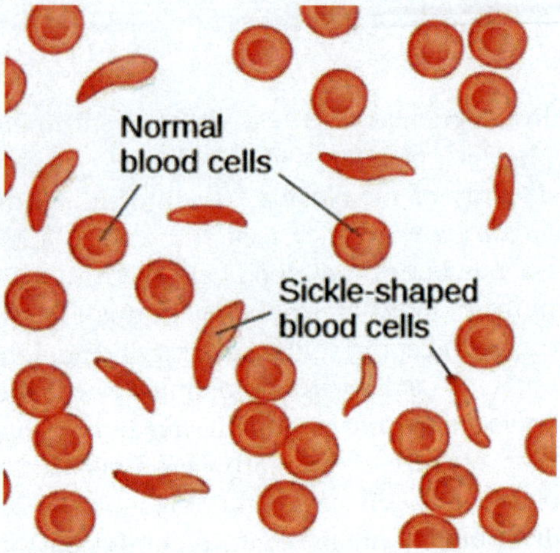

Fig. 2 Blood image with sickle cells

- Sickle Cell: In sickle cell disease, red platelets are shaped as bow formed. Such unusual red platelets, conveying strange hemoglobin known as S shaped hemoglobin, are frangible.

In humanoid wellness, blood being admirably is definitively imperative, and the underlying investigation of disorders is fundamental to keep away from death and the decaying of affliction. With the enhancement of the power of proof, the ability to manipulate images becomes a basic and useful tool in logic testing. It is widely used and can only be used in the field of biomedical exchange. Find the earliest and most effective strategies to classify this disorder using discontinuous imaging, so that sickle cells can be detected in blood imaging, which is fundamental to distinguish this disease [7]. The purpose of this speculation was to decide how expertise in image manipulation might be beneficial in revealing sickle cell disease.

3 Prevailing Methods

3.1 Prevailing Method 1

Figure 3 portrays how sickle id distinguished using structure factor. After taking the image to be examined, it is first changed over to the grayscale image using "rgb2gray." By then, the image is changed over into high differentiation plan for calculating the cutoff using "dark sift." Subsequently, the openings are involved using "imfill". In Fig. 3 the edges are vacant and little articles are emptied using "imclearborder" and "bwareaopen" independently and the objects are checked and the not set in stone to distinguish the sickle cells. Form-factor exploration is done first by recording the cell's large bend length and small center point length [8]. By then, discover the difference between the large and small hatches hatchets. Finally, figure out the construction factor using with help of the condition: The form factor is

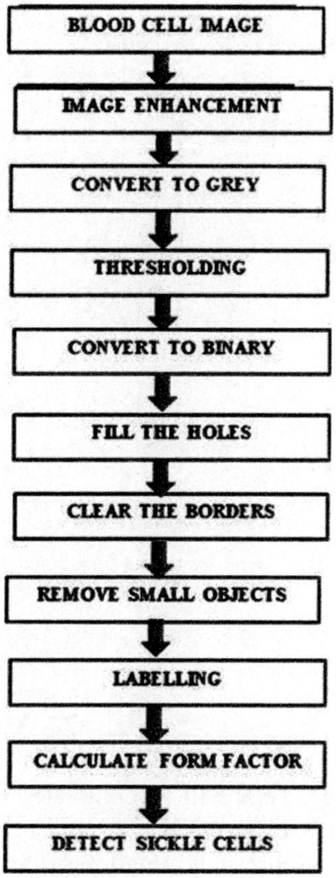

Fig. 3 Sickle cell recognition by form factor

defined by the length of the main midpoint separated by the distinction between significantly small lengths. Perceiving the sickle cells is done by performing testing with environmental elements, sickle cells can be recognized: if the construction factor is not actually the cutoff, by then the cell will be viewed as sickle cell.

3.2 Prevailing Method 2

Figure 4 shows the method for estimation in finding sickle cells using most concise way [9]. Platelet representation is accepted as given and before platelet representation is given to get the basic idea. For better quality, they are replaced by fuzzy representations using facets worked in MATLAB. Ensuing to thresholding, the unclear scale portrayal is changed over to resemble portrayal utilizing the volume im2bw(). The combined portrayal as such procured may fathom restricted introductions. In mandate to procure the specific income, need to fill such holes. This can be done using MATLAB function imfill(). Open elements on edges are removed using Imclearborder() function. Eventually, the borders disappear, and the small items get squeezed

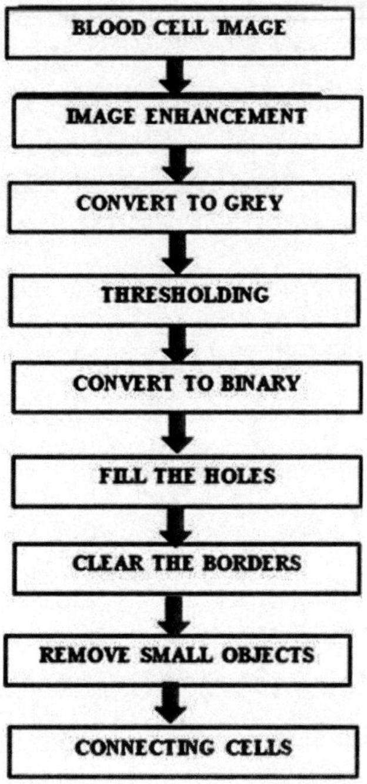

Fig. 4 Sickle cell recognition by clustering

out. They will be completed by separating things by areas that do not necessarily have exact values. This must be achievable making use of the function bwareaopen(). As soon as the small items are destroyed, cells will be associated. This requires all tile parts to be recognized and the volume of pic 2 points is sorted, which is not true for the sorted representation in the outline and extra quests like the counterpart finished now. knnsearch(*X*, *Y*, Name, Value) is utilized to achieve this task. Cells are associated toward the end.

4 Planned Method

The blood representation is expected as data, and then the blood representation is enhanced to join the ideal game plan. Once enhanced, they can be isolated into segments depending on the hidden space. We can then implement representation systems to differentiate between normal cells and foreign cells. After this disturbance, small items are squeezed out. These will be accomplished by recognizing things that have an area of no apparent specific value. Moreover, this should be possible utilizing the volume bwareaopen(). Next, we can actually look at the amount of sickle cells and red platelets and their rates. Figure 5 portrays the proposed strategy.

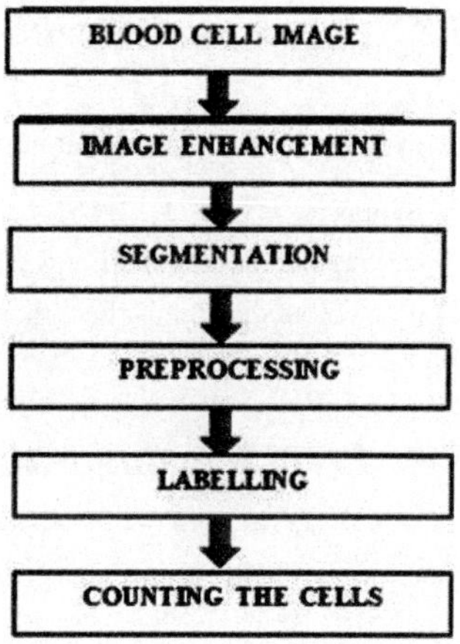

Fig. 5 Planned method for counting cells

5 Implementation

MATLAB is used to complete the new estimation; the "MATLAB" gadget is used. MATLAB has an image planning execute crowd, which involves all limits that are used to explore the image, for instance, scrutinizing, update, fluctuating over from one portrayal type to division, checking and anything which is possible from that point. The limits that used to follow the new estimation are portrayed under [10].

5.1 Reading Image with Imread

MATLAB agrees with the image representation in two-dimensional matrices. The function imread reads the image and returns the matrix [10].

Syntax

Image=imread('file-name');

MATLAB reads the file representing the "filename" in the image with the array. Imread can read a variety of layouts including JPG, PNG, GIF and TIF [10].

5.2 Showing Image with Imshow

Imshow displays the MATLAB on-screen representations.

Syntax

imshow(img)

MATLAB represents img graphically, where img is a grayscale, RGB, or binary depiction. For binary depictions, imshow represents pixels with the 0 for black and 1 for white [11].

5.3 Segmentation with CellSegmentation

cellSegmentation function segments the image into different parts.

Syntax

cellSegmentation(Inputimage,colourspace)

here colourspace must be either RGB, CMYK, YCbCr.

5.4 Preprocessing with Bwareaopen

Noise in an image can be removed using the function bwareaopen.

Syntax

bwareaopen(BW, 100)

it is a filter that can remove noise from a image and performs classification to extract sickle cells [2, 3].

5.5 Labeling with Regionprops

The regionprops is used for adding bounding boxes to cells.

Syntax

regionprops(BW2, 'all')

bounding boxes help in counting the number of sickle and RBC.

5.6 Counting

Having count of sickle cells and red blood cells gives the percentages where the sickle cell percentage and red blood cell percentage are computed as:

Sickle cell percentage:

wbcPercent = (whitecoun/ totalCells).* 100;

RBC percentage:

rbcPercent = (redcount / totalCells).* 100;

6 Observations

Trial input image to planned method is shown in Fig. 6

Sickle cell identification and red blood cell identification in RGB are shown in Figs. 7 and 8.

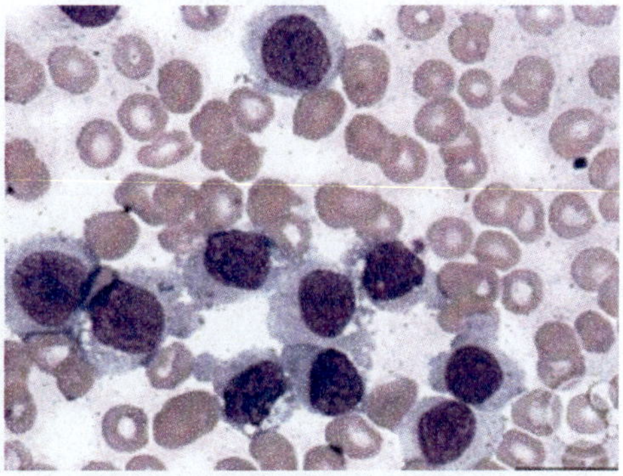

Fig. 6 Results after reading input sample1.jpg file

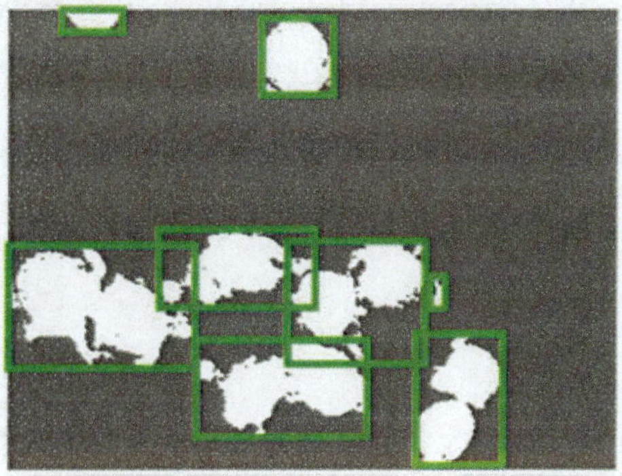

Fig. 7 Results after identifying sickle cells in RGB

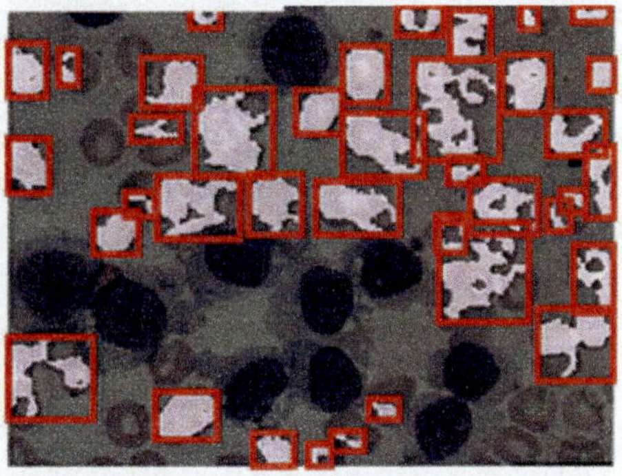

Fig. 8 Results after identifying RBC in RGB

Sickle cell identification and red blood cell identification in CMYK are shown in Figs. 9 and 10.

Red blood cell identification in YcbCr is shown in Fig. 11. There is no identification of sickle cell in YcbCr.

Percentages of identifying sickle cells and red blood cells in RGB, CMYK and YcbCr are shown in Table 1.

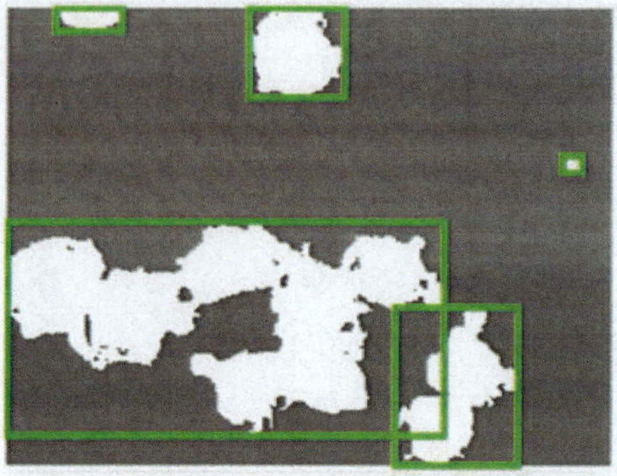

Fig. 9 Results after identifying sickle cells in CMYK

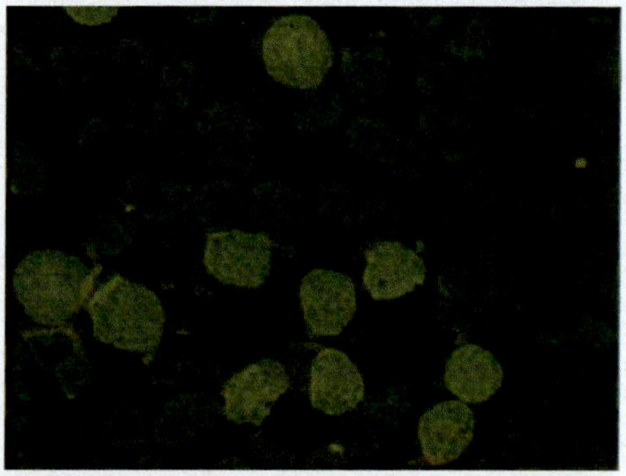

Fig. 10 Results after identifying RBC in CMYK

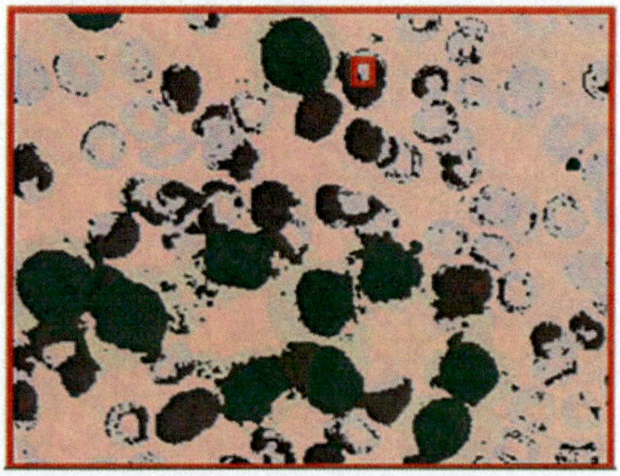

Fig. 11 Results after identifying RBC in YcbCr

Table 1 Percentages of recognizing sickle cells and red blood cells in RGB, CMYK and YcbCr

	RGB	CMYK	YcbCr
Sickle cells	50	52	00
% of sickle cells	29	100	00
Red blood cells	123	00	01
% of red blood cells	71	00	100
Total cells found	173	52	01

7 Conclusion

The proposed computation checks the cells by identifying deviations in microscopic blood images using analogous image fragmentation. Looking at to the past projected arrangements, this planning counts every single under the sun unit without neglecting any unit.

References

1. Medical imaging. https://en.wikipedia.org/wiki/Medical_imaging
2. Mojtaba T, Mona N, Behzad B, Alireza M (2013) New approach to red blood cell classification using morphological image processing. Shiraz EMed J 14(1):23409
3. Chintawar IA, Pravin N, Aishvarya M, Chetan K (2016) Detection of sickle cells using image processing. IJSTE Int J Sci Technol Eng 2(09):2349–2784
4. Sreekumar A, Bhattacharya A (2015) Identification of sickle cells from microscopic blood smear image using image processing. https://Www.Academia.Edu/24101001/Detection_Of_Sickle_Cell_Anemia_In_Red_Blood_Cells_A_Review_Article_In_International_Journal_Of_Advances_In_Engineering_Sciences_And_Applied_Mathematics_March_2015_Reads_166
5. Bacus JW, Weens JH (1977) An automated method of differential red blood cell classification with application to thediagnosis of anemia. J HistochemCytochem 25:614
6. Sickle Cell Disease Symptoms, Causes, Treatments-Web MD. https://www.emedicinehealth.com/sickle_cell_crisis/article_em.htm#what_is_sickle_cell_disease_scd
7. Barpanda S (2013) Use of image processing techniques to automatically diagnose sickle-cell anemia present in red bloodan cells smear. https://ethesis.nitrkl.ac.in/5022/1/109EE0255.pdf
8. Susmitha KA, Tejaswi SG, Kala KL, Pranutha KP, Tirumalasetty S (2018) Classifying deviations in medical microscopic images using evolutionary analysis. Iconic Res Eng J 1(9):233–240
9. Tirumalasetty S, Latha JS, Neeharika J, Pravallika LS, Manasa M (2020) Clustering deviations in medical images using hierarchical clustering and shortest path algorithms. Iconic Res Eng J 9(3):15796
10. Image processing functions in Matlab. https://www.mathworks.com/help/images/ref/regionprops.html
11. Types of Medical Images. https://www.ausmed.com/cpd/articles/medical-imaging-types-and-modalities
12. Poomcokrak J, Neatpisarnvanit C (2008) Red blood cells extraction and counting. In: Proceedings of the 3rd international symposium on biomedical engineering (ISBME2008)

An Approach for Neuronal Cell Segmentation Based on Mask-RCNN

Piyushkumar Hirapara and Afzal Sikander

Abstract Neuronal cell segmentation is an essential subject of research for understanding mental illnesses and developing medications to treat them. It is tough and time-consuming to segment each cell from a microscopic image. Sartorius cell segmentation dataset is newly available and very well-annotated dataset of microscopic images of neuronal cells. For the sartorius cell segmentation dataset, we presented an automatic cell segmentation technique based on the Mask-RCNN model. Mask-RCNN is a state-of-the-art technique for image object detection and instance segmentation. We demonstrate how Mask-RCNN can be utilized to effectively segment neuronal cells. The segmentation results we are getting have a high average precision across a wide range of IoU thresholds. We hope that by employing this technique, we will be able to accelerate the discovery of mental disorder treatments.

Keywords Machine learning algorithms · Image analysis · Medical imaging · Convolution

1 Introduction

Cell instance segmentation is a critical problem in medical image analysis, particularly at the individual neuronal cell level. Alzheimer's disease and tumors are among the most common causes of death and disability worldwide. It is difficult to determine how well this condition responds to treatment and medications. A review of neuronal cells using microscopy could be one way to accomplish this type of endeavor. However, segmenting each cell in a microscopic image can be a difficult and time-consuming task. Deep learning-based algorithms, on the other hand, have demonstrated excellent results for a variety of tasks in the field of computer

P. Hirapara · A. Sikander (✉)
Department of Instrumentation and Control Engineering, Dr. B. R. Ambedkar National Institute of Technology, Jalandhar, Punjab, India
e-mail: afzals@nitj.ac.in

J. R. Szymanski et al. (eds.), *Energy Systems, Drives and Automations*,
Lecture Notes in Electrical Engineering 1057,
https://doi.org/10.1007/978-981-99-3691-5_36

vision, including object identification, object tracking, picture classification, image segmentation, and many more. This task may be difficult because neuronal cells have a distinct, uneven, and concave form, making segmentation difficult. We will look at the Mask-RCNN [1] model for segmentation of sartorius [2] microscopic images of neuronal cells in this paper.

Mask-RCNN was chosen for the challenge of neuronal cell segmentation because the Mask-RCNN model has consistently shown good results on a variety of instance segmentation datasets, and Mask-RCNN is the most intuitive approach to the task.

1.1 Instance Segmentation

The use of deep convolutional neural networks has resulted in significant improvements in image segmentation, starting with a fully convolutional network (FCN) [3], which outperformed all other methods for semantic image segmentation. In semantic segmentation, each object of the same class will be segmented out separately. Over the years, excellent models for semantic segmentation have been proposed, including UNet [4] in 2015, SegNet [5] in 2016, DeepLab family [6] of models in 2017, HRNet [7] in 2019, and many others.

Instance segmentation is a challenging task because it is a combination of two task object detection and segmentation. In instance segmentation, each instance of object will be classified and separately segmented at pixel level. There are a fixed set of classes, but the number of objects from these classes can vary in image.

The approaches provided can be split into two categories. The first is the top-down method, in which we first identify the object, assign a class and bounding box, and then segment the object within the bounding box. We begin with per-pixel labeling or segmentation in the second method and then apply class prediction and bounding box to the output. The Mask-RCNN belongs to the top-down category; it creates bounding boxes for the items in the image first, then performs segmentation on the objects within those boxes.

1.2 The Mask-RCNN

In 2017, the Mask-RCNN model was presented for the job of instance segmentation, and it outperformed any other model at the time. The Mask-RCNN model is based on the Faster-RCNN [8] model, which was created for real-time visual object detection. Faster-RCNN relies on the region proposal network (RPN) [9]. Task of RPN was to propose a region in image where it is more likely to have an object. RPN has improved significantly over time; in the initial RCNN model, RPN was ruled-based and slow; in the Faster-RCNN model, it was replaced with convolutional RPN, which is much faster.

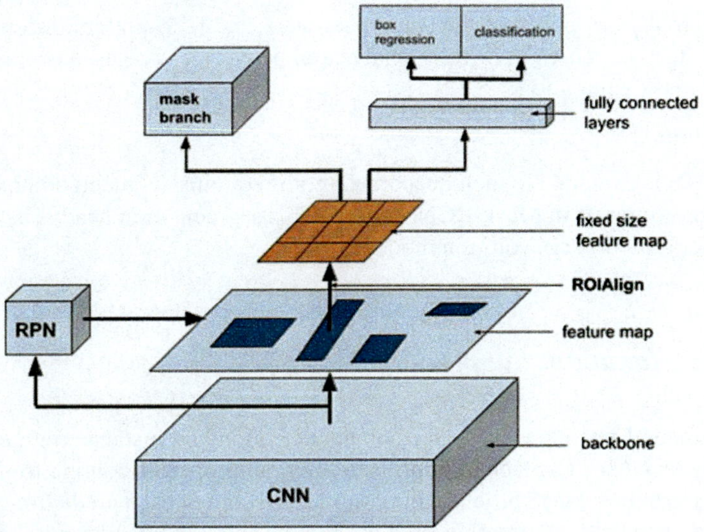

Fig. 1 Mask-RCNN model

The Mask-RCNN model is divided into four stages: feature extractor, region proposal network, region of interest pooling or align, and CNN-based segmentation head (mask branch). Figure 1 depicts all of these stages.

Feature Extractor (Backbone)

The feature extractor's job is to apply a set of convolutional filters or kernels to the input image and extract important features for the model's next stage. Many CNN networks, such as VGG [10], ResNet [11], FPN [9], and others, can be employed for this task.

Region Proposal Network (RPN)

RPN uses a feature map as input from the feature extractor to suggest a region where an object is more likely to be found. The RPN model has two heads: one for classification, which returns true if there is a high probability of an object in a given anchor box and false otherwise; the second head is for bounding box output, and it will provide correction for predetermined anchor boxes, which is then further used to refine the anchor boxes allowing the entire object from input image to be cover in bounding box.

ROI Pool and ROI Align

For class prediction, bounding box prediction, and mask construction, the Mask-RCNN requires a fixed-size vector. The regions proposed by RPN are of different sizes; thus, we require ROI Align.

ROI Align takes input from the feature extractor and RPN and produces a fixed-size feature map on which a convolutional neural network can be used to classify the object and mask generation.

Segmentation Head

Faster-RCNN is extended to include a branch for predicting segmentation masks for each region of interest in Mask-RCNN (ROI). This segmentation head is basically a small mask prediction convolution model.

1.3 Non-Maximum Suppression (NMS)

Many computers' vision tasks, such as object detection and instance segmentation, rely heavily on NMS. The task of NMS is to filter out proposals coming from object detection algorithms. NMS relies on the confidence score of each prediction and IoU of that prediction with every other prediction to do this kind of filtering. There are numerous different types of NMS approaches, which are classified according to their accuracy and speed. There may be several predictions for the same thing during the object detection task, and NMS task is to remove these duplicates based on some threshold. Soft NMS [12], Adaptive NMS [13], Fast NMS [14], and Matrix NMS are just a few of the NMS approaches available.

2 Method and Metrics

2.1 Dataset

The dataset for this task came from the Kaggle competition Sartorius—cell instance segmentation. There are 606 train images in this dataset, each with a resolution of 704 × 520 pixels. Out of 606 training images, 320 images are from Cort cell type, 155 images are from shsy5y cell type, and remaining 131 images are from Astro cell type. A team of professionals has annotated the dataset and verified it on multiple levels. About 80% of the 606 images were used in training of the model and remaining 20% for validation set. It was made sure that the proper portion of images from each cell type ended up in the training and validation sets during the training–validation split.

Data were preprocessed in order to make training easier and more reliable. Dataset had images with broken masks for some cell types which needed to be corrected by some image morphology.

Before training, the dataset was analyzed to determine a measure that could be used during training to create a more robust cell instance segmentation model. There are three types of cells in the dataset: Astro, Cort, and shsy5y.

Table 1 Instance per image per cell type for image data

Cell type	Min	90%	Max	Mean
Astro	5	135	594	80
Cort	4	54	108	33
Shsy5y	49	539	790	337

Table 2 Pixels per mask per cell type for image data

Cell type	Min	90%	Max	Mean
Astro	37	1799	13,327	905
Cort	33	404	2054	240
Shsy5y	30	393	2254	224

The first measure is number of instances per image per cell type and the result of that can be seen in Table 1.

The second important measure is pixels per mask per cell type and the results of that can be seen in Table 2.

2.2 Proposed Approach

The ResNet-50-FPN backbone is utilized with Mask-RCNN. For the purpose of neural cell instance segmentation, an existing version of Mask-RCNN from pytorch was used with significant modifications. The weights of the backbone were pretrained from the MSCOCO dataset [15], whereas the weights of the head were randomly chosen. The model was trained in two stages: first, only the backbone was trained with RPN, and then the model was trained end to end with learning rate scheduler. On a single NVIDIA GPU, the model was trained for more than 30 epochs with a batch size of 2 and a stochastic gradient descent optimizer with a momentum of 0.9. To counteract with overfitting, data augmentation techniques such as random horizontal and vertical flips, as well as some image normalization, were applied.

We have three different forms of cell data, each with a different number of cells per image and varying mask sizes. We will train two different convolutional models to improve the accuracy of the segmentation task. The first model categorizes images by cell type, with three classes (Astro, Cort, and Shsy5y) for each cell type. For classification, this model will have a convolutional feature extractor and a fully connected neural network head. A ResNet-like model was utilized for the feature extraction part. Following classification, we can use the information from the data analysis section to determine the number of instances and mask size for each classed image. Mask-RCNN is the second model, which does instance cell segmentation for each input image. Mask-RCNN output's contains cell class, bounding box, and cell mask. We use the data from the first model, which is image class, to determine how many output predictions to consider during mask formation and the appropriate mask generation threshold from predicted mask. The steps for completing these tasks are outlined below.

The Mask-RCNN model has undergone numerous improvements to improve cell instance segmentation accuracy. We change BOX_DETECTION_PER_IMG to 539, which is 90% of instances per image per cell type for shsy5y cell. By changing BOX_DETECTION_PER_IMAGE, we could set how many region proposals from RPN to carry forward to ROI Align. We could pick how many of this identified object to include in the final mask based on instance per image per cell type data in the postprocessing stage. We will need to make some tweaks to non-maximum suppression for this task (NMS). To filter out predictions, NMS utilizes interaction over union (IOU) and confidence score.

The Mask-RCNN model predicts a probability value for each pixel position, which we use to create a mask by comparing it to a threshold value. The simplest technique for completing this operation is to set MASK_THRESOLD to 0.5. That is, if the likelihood of a pixel in the prediction mask is > 0.5, that pixel is included in the mask. For border scenarios, we will dynamically alter the MAKS_THRESOLD value to ensure that the number of pixels for a particular mask for a specific cell type is between 5 and 95% of the maximum available pixels for that cell type.

3 Experimental Results

We reported the standard COCO metrics including AP (average over IoU thresholds), AP50, and AP75. On the validation dataset, the proposed Mask-RCNN model with ResNet for cell type classification offers an AP of 27.7% which can be seen in Table 3.

According to Table 4, the suggested model produces a high average precision (AP) for Cort cell types and a low AP for Astro and shsy5y cell types. This could be investigated in the future to see what improvements can be made to improve the model's accuracy for all sorts of cells.

The input images, true masks, and predicted masks from the validation set are shown in Figs. 2, 3, and 4.

Table 3 Average precision (AP) for validation dataset

Model	Backbone	AP	AP50	AP75
Mask-RCNN	ResNet-50-FPN	27.7	56.1	23.7

Table 4 Average precision for each cell type

Cell type	Astro	Cort	Shsy5y
AP	14.4	39.8	14.7

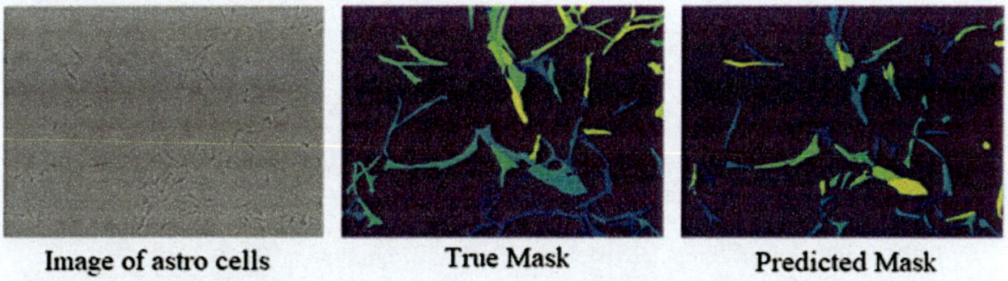

Fig. 2 Segmentation results of astro cell type from validation set

Fig. 3 Segmentation results of cort cell type from validation set

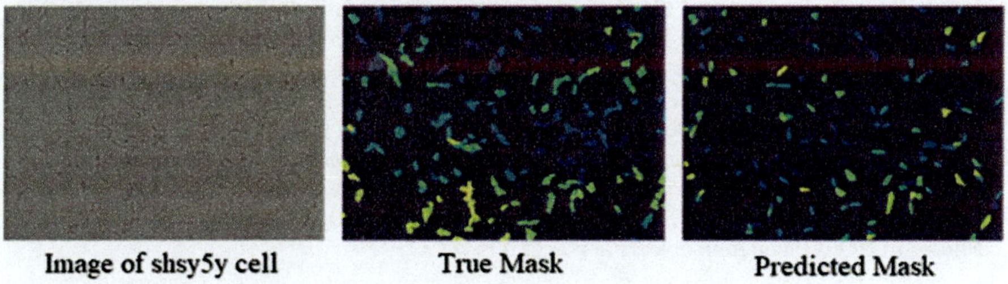

Fig. 4 Segmentation results of shsy5y cell type from validation set

4 Conclusion

Neuronal cell segmentation is an important step in the study of many mental diseases and the development of treatments for them, but it takes a long time and requires expert supervision. Mask-RCNN, a deep learning method, can aid in the cell segmentation process; this model is fast and accurate when compared to other segmentation models. We showed how to use the Mask-RCNN model for cell segmentation in our research. As can be seen in the results' section, the proposed approach produces good segmentation results. We hope that our approach will be useful in future cell segmentation tasks and in the development of mental-health treatments.

References

1. He K, Gkioxari G, Dollar P, Girshick R (2017) Mask R-CNN. In: IEEE international conference on computer vision, pp 2961–2969
2. Sartorius (2021) Sartorius: cell instance segmentation. https://www.kaggle.com/c/sartorius-cell-instance-segmentation/data
3. Long J, Shelhamer E, Darrell T (2015) Fully convolutional networks for semantic segmentation. In: IEEE conference on computer vision and pattern recognition, pp 3431–3440
4. Ronneberger O, Fischer P, Brox T (2015) U-net: convolutional networks for biomedical image segmentation. International conference on medical image computing and computer-assisted intervention. Springer, New York, pp 234–241
5. Badrinarayanan V, Kendall A, Cipolla R (2017) SegNet: a deep convolutional encoder-decoder architecture for image segmentation. IEEE Trans Pattern Anal Mach Intell 39(12):2481–2495
6. Chen LC, Papandreou G, Kokkinos I, Murphy K, Yuille AL (2017) DeepLab: semantic image segmentation with deep convolutional nets, atrous convolution, and fully connected crfs. IEEE Trans Pattern Anal Mach Intell 40(4):834–848
7. Yuan Y, Chen X, Wang J (2019) Object-contextual representations for semantic segmentation. arXiv preprint arXiv:1909.11065
8. Ren S, He K, Girshick R, Sun J (2015) Faster R-CNN: towards real-time object detection with region proposal networks. Adv Neural Inform Process Syst 14:91–99
9. Lin TY, Dollar P, Girshick R, He K, Hariharan B, Belongie S (2017) Feature pyramid networks for object detection. In: IEEE conference on computer vision and pattern recognition, pp 2117–2125
10. Simonyan K, Zisserman A (2014) Very deep convolutional networks for large-scale image recognition. arXiv preprint arXiv:1409.1556
11. He K, Zhang X, Ren S, Sun J (2016) Deep residual learning for image recognition. In: CVPR
12. Bodla N, Singh B, Chellappa R, Davis L (2017) Soft-NMS: improving object detection with one line of code. In: Proceedings of the IEEE international conference on computation vision
13. Liu S, Huang D, Wang Y (2019) Adaptive NMS: refining pedestrian detection in a crowd. In: Proceedings of the IEEE conference on computation on vision pattern recognition
14. Bolya D, Zhou C, Xiao F, Jae Lee Y (2019) YOLACT: real-time instance segmentation. In: Proceedings of the IEEE international conference on computation vision
15. Lin T-Y, Maire M, Belongie S, Hays J, Perona P, Ramanan D, Dollar P, Zitnick CL (2014) Microsoft COCO: common objects in context. In: ECCV

An Overview of MicroRNA Mediated Regulation of TAM and EMT Pathway in Progression of Breast Cancer

Riyanka Shil, Sanmitra Ghosh, Rajib Majumder, and Rudra Prasad Saha

Abstract The most commonly diagnosed female malignancy with a major contribution to global cancer-related mortality is breast cancer. Development and progression of metastatic breast cancer along with its high mortality rate is reported to be attributed to multitude of factors like EMT of cancer cells, resultant enhanced migratory and invasive capacity together with elevated stemness of cells, etc. The breast tumor microenvironment (TME) has elevated inflammation contributed by permeated immune cells, cytokines, and growth factors among which tumor-associated macrophages (TAM) accounts for > 50% of tumor mass. TAM, associated with M2-phenotype of macrophage subpopulation in breast cancer TME, promotes tumor progression and immune suppression by secreting growth factors, pro-angiogenic factors, immunosuppressive factors, and proteases. TAMs communicate with immune cells and tumor cells via cytokines released by it to induce EMT and tumor angiogenesis through JAK2/STAT3/miR-506-3p/FoxQ1 axis or by TGF-β and NFκB. Therefore, TAM targeting has emerged as an attractive strategy for solid tumor therapeutic intervention.

Recently, tumor-derived microRNAs have been implicated as prospective biomarkers for early cancer diagnosis due to their salient contribution to tumor cell-stroma crosstalk through molecular reprogramming of effector functions of stromal components like TAM with pro-tumoral and/or anti-tumoral activities. Various EMT-regulatory factors are targeted by several microRNAs (like miR-200 family) to suppress/promote EMT pathway. Some microRNA (miR-98, miR-27) expressions are reported to be deregulated by TAM-derived cytokines which in turn promote metastasis through EMT pathway. Thus, in breast cancer microRNAs can be a prime mediator of tumor progression and invasion through regulation of TAM activity and EMT pathway.

R. Shil · S. Ghosh (✉)
Department of Biological Sciences, School of Life Science and Biotechnology, Adamas University, Kolkata, India
e-mail: sanmitra.ghosh@adamasuniversity.ac.in

R. Majumder · R. P. Saha
Department of Biotechnology, School of Life Science and Biotechnology, Adamas University, Kolkata, India

J. R. Szymanski et al. (eds.), *Energy Systems, Drives and Automations*,
Lecture Notes in Electrical Engineering 1057,
https://doi.org/10.1007/978-981-99-3691-5_37

Keywords Tumor microenvironment · Tumor-associated macrophage · MicroRNA · EMT pathway · Breast cancer

1 Introduction

Carcinogenesis of breast is one of the most threatening malignancies of women that show a higher mortality rate as compared to all other cancers [1]. Several factors such as environmental, lifestyle, reproductive factors including obesity, high-protein uptake in diet, alcohol consumption, early menarche, age, induced abortion, and others increase the risk to develop breast cancer, among which less than 10% is mediated by the inherited genetic mutations. Therefore, early detection and diagnosis is necessary to decrease the mortality rate [2, 3].

The major cause of its high mortality rate is due to the metastatic capacity acquired by the tumor cells. This metastasis is may be due to the inflammation in the tumor microenvironment, enhanced proliferation, migration of tumor cells, and EMT [4, 5]. Considering the present Indian scenario, the ASR of breast cancer accounts for 22–31%, and the age group may range from 0 to 69 years. In the year 2020, an estimated number of worldwide mortality (A) and new cases (B) of breast cancer in females of all ages have been provided below in Fig. 1, as per World Health Organization.

2 Tumor Microenvironment

A favorable microenvironment is necessary to induce tumor formation and to promote carcinogenesis, apart from the cancer cells and is generally known as the tumor microenvironment (TME) [6]. It constitutes the immediate vicinity of the tumor cells that provide all favorable conditions to promote carcinogenesis. It comprises several components like thriving cancer cells, stroma of the tumor, infiltrating inflammatory cells, blood vasculature, and a number of accompanying tissue cells [7, 8]. The major immune cell component is the TAM, which accounts for more than 50% of the tumor mass [7]. Thus, TME is characterized by inflammation contributed by both cellular and humoral immune responses [6].

3 Tumor-Associated Macrophages (TAM)

The innate immune system harbors the macrophages which are of two types: recruited and tissue specific [9]. Most of these macrophages that migrate to inflammation site arise from the bone marrow and are called recruited macrophages, whereas those macrophages arising from the yolk sac and are associated with that particular tissue, regulating their differentiation and homeostasis are resident [10, 11].

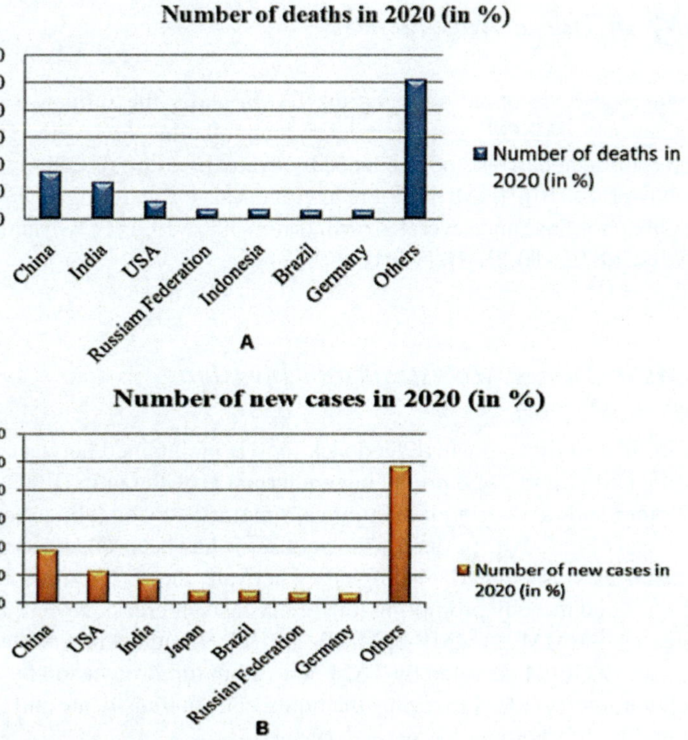

Fig. 1 Bar diagram representing the estimated data for **a** mortality **b** new cases of breast cancer globally

The circulating nascent monocytes are directed toward tumorous zone in the breast TME by various cytokines, chemokines, and growth factors (CCL2, CCL20, CCL18, M-CSF, GM-CSF, VEGF) secreted by stromal and tumor cells which prompts differentiation of monocytes into macrophages [4, 12]. The migrating macrophages together with the resident macrophages participate in polarization process that classifies them as M1 (classically activated macrophages with proinflammatory activities) and M2 (alternatively activated macrophages with anti-inflammatory activities) [10, 12]. Macrophages with M1 phenotype stimulate T_{H1} response and show antitumor activities resulting in reduced tumor aggressiveness [13]. It also shows antimicrobial activities, which produce inflammatory cytokines (IL-1β, IL-6, IL-12, IL-23, TNF-α), chemokines, and RNI/ROI [13, 14]. Macrophages with M2 phenotype can stimulate T_{H2} response showing pro-tumor activities by promoting angiogenesis, immunosuppression, and ability to repair tissues with production of reduced inflammatory cytokines and expresses high levels of IL-10, several chemokines and scavenging receptors [4, 11]. TAM closely resembles the M2 phenotype and plays a crucial role in every stages of tumor progression like angiogenesis, hypoxia, invasion, and metastasis.

3.1 TAMs in Tumor Angiogenesis

During angiogenesis, hypoxic sites recruit TAMs under the influence of various chemoattractants like VEGF-A, angiopoietin-2, endothelins, EMAPII, CXCL12 [4, 15]. TAM recruitment can also be enhanced by overexpression of CSF-1, HIF-1 and HIF-2 and NF-κB [4, 16]. TAMs promote angiogenesis by secreting TNF-α, MMPs, thymidine phosphorylase and several growth factors like VEGF, PDGF and cytokines like TGF-β, CCL8, IL-10, IL-1β [17, 18].

3.2 TAMs in Tumor Metastasis and Invasion

During tumor metastasis, a positive feedback loop is maintained between the tumor cells releasing CSF-1 and TAM producing epidermal growth factor (EGF) by stimulating each other and facilitating the tumor cell migration by enabling them to enter the blood vessels [4, 7, 10]. In one of the studies, it has been observed that CCL2 recruited metastasis associated macrophages activate the CCL2 receptor CCR2, secreting CCL3 and thereby promoting lung metastasis in breast cancer [19]. Secretion of proteases like MMPs (MMP-1, MMP-2, MMP-9, cathepsin), serine proteases (EGF, uPA), and TGF-β1 secreted by TAM carries out tumor invasion by degrading the extracellular matrix (ECM), helping the tumor cells to translocate and begin their deposition and proliferation at the secondary site [4].

4 EMT in Cancer

EMT is the first step in tumor invasion and a major reason for metastasis of breast cancer [21]. During this process, the polarized epithelial cells loose apical-basal polarity, loose expression of epithelial markers (E-cadherin, desmoplakin), shows weakening of cell junctions and transforms into mesenchymal cell phenotype by upregulating the mesenchymal markers (N-cadherin, vimentin, fibronectin, α-SMA) and acquires migratory properties by losing cell–cell adhesion [13]. These cells then invade through the ECM and stromal cell layers followed by penetration into the lumen of the blood vessels and are transported to the site of dissemination as the CTC [10, 22]. As the CTCs reach secondary site, they go through MET and regain the ability to proliferate and differentiate and initiate colonization [23]. For this reversible process, activation of several transcription factors like ZEB-1, ZEB-2, Twist, Snail, Slug, and other signaling pathways including NF-κβ, Wnt, TGF-β, Hedgehog, Notch signaling are required (Fig. 2) [10, 24]. In breast cancer, the CTC level in blood becomes a prognostic factor [25, 26]. EMT often helps CTCs to overcome the problem of "anoikis" or cell apoptosis which is induced due to

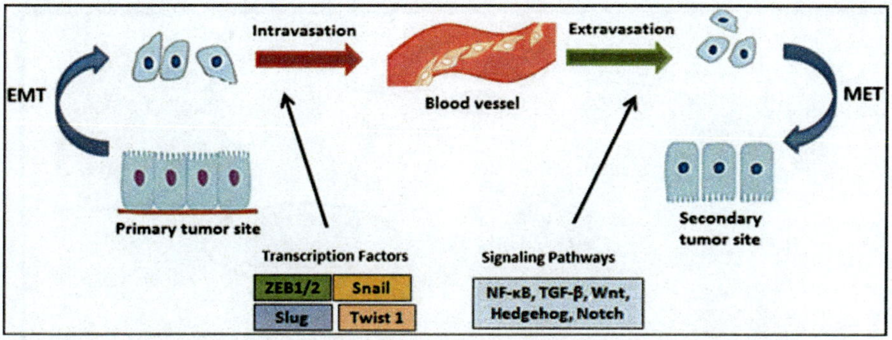

Fig. 2 Reversible process of EMT and MET involving several transcription factors and signaling pathways

the detachment, by facilitating the reattachment of the cancer cells to leukocytes, endothelium, and platelets [27, 28].

5 Crosstalk Between TAM and EMT

The cancer progression and metastasis cannot be carried out solely by the tumor cells, but it also requires an interaction between TAM and EMT to accelerate the process of dissemination and migration of the cancer cells through activation or production of various mediators such as cytokines and growth factors [10]. Some of the interactions between TAM and EMT are mentioned below.

5.1 Positive Feedback Loop Between TAM and EMT

A positive feedback loop as shown in Fig. 3 is maintained between TAM and EMT [29]. As we can see here, TAM secretes cytokines like CCL18 which by down-regulating the expression of miR-98 and miR-27b via the N-Ras/ERK/PI3K/NFκB/Lin28b signaling pathway, enhances the EMT of cancer cells leading to metastasis of lung in breast cancer, whereas in return, the mesenchymal cancer cell activates the macrophages into TAM phenotype by secreting GM-CSF [29, 30].

5.2 Crosstalk Mediated by TGF-B1 and FoxQ1

One of the immunosuppressive cytokine that contributes to differentiation, proliferation and mobility of the cancer cells is TGF-β. TGF-β1, one of its important members,

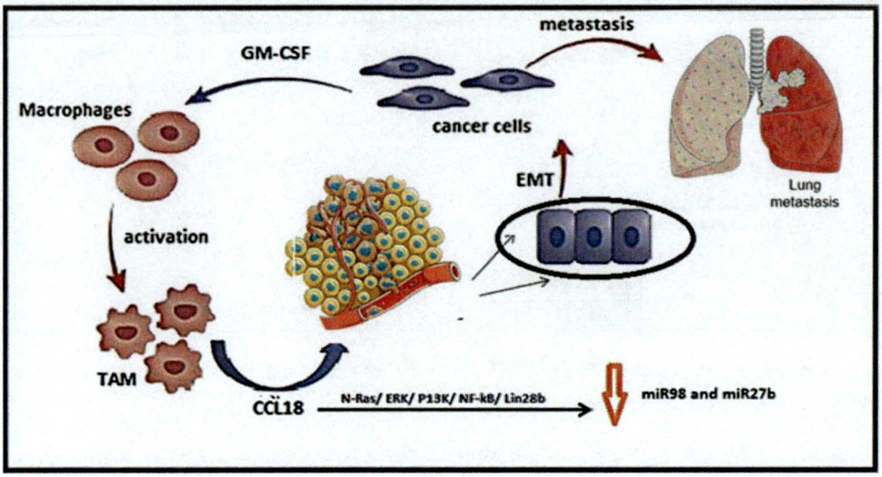

Fig. 3 Positive feedback loop maintained between TAM and EMT

is secreted by TAM to maintain the cancer cells undergoing EMT [10]. TGF-β1 leads to the acquisition of mesenchymal phenotype by TGF-β1/Smad signaling pathway which further activates ZEB-1, ZEB-2, Twist, Slug, Snail that participate in the execution of EMT [13]. Moreover, this crosstalk also involves the NFκB signaling pathway [31]. Recent studies have shown that in the process of TGF-β1 induced EMT, a significant role is played by Ras effector Blimp-1 via downregulation of BMP-5 [32]. FoxQ1, a forkhead TF family member leads to the suppression of E-cadherin and thereby promotes TGF-β induced EMT [33].

5.3 Crosstalk Mediated by IL-6

One of the major causes for poor prognosis and enhanced invasion by the cancer cells is due to IL-6 overexpression in serum [34]. Binding of IL-6 to IL-6 receptor α (IL-6R) and IL-6 receptor β (gp130) activates JAK by phosphorylation which in return phosphorylates the cognate receptor at the cytoplasmic tail and STAT3 [35]. Phosphorylated STAT3 gets translocated into the nucleus as homodimers and by forming a complex with some coactivators such as p68, it promotes the transcription of the target genes [35]. Therefore, by activation of IL-6/JAK/STAT-3 pathway, E-cadherin expression is inhibited, while N-cadherin, Snail, vimentin, and Twist are upregulated inducing EMT [34].

5.4 Crosstalk Mediated by NFκB

NF-κB, a mediator for inflammation, regulates the transcriptional activities in TAM and tumor cells promoting tumor initiation and growth [36]. By the activation of NF-κB, an anti-metastatic microRNA, miR-488 is directly suppressed causing the epithelial cells to acquire mesenchymal phenotype through the activation of two major initiators of EMT, i.e., SATB1 and Twist [36].

The crosstalk between all the components has been summarized in Fig. 4.

6 Role of MicroRNAs in Regulating TAM and EMT

To increase the survival rate for breast cancer in women worldwide, effective management with early diagnosis and proper monitoring of patients is required [37]. Recent studies have shown that circulating microRNA used in the form of a biomarker possess both diagnostic and prognostic value in this context of increasing survival rate [38]. MicroRNAs are small, endogenous, non-coding RNAs that control the gene expression by binding to the 3'UTR complementary sequences of the target mRNA promoting their degradation or translational repression [39]. They can be of two types: oncomiRs and tumor suppressor miRs [40]. OncomiRs promote cell invasion and proliferation by downregulating the tumor suppressor genes, while tumor suppressor miRs negatively regulate the oncogenes inhibiting tumor progression [40].

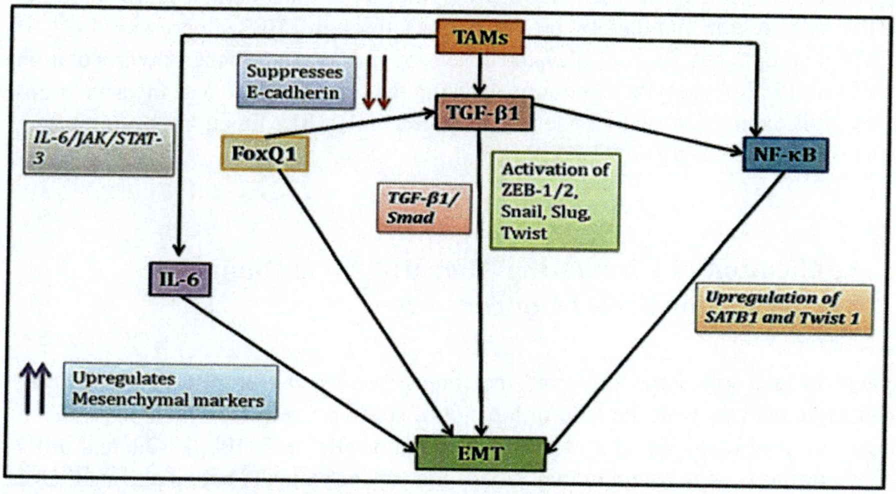

Fig. 4 Crosstalk between TAM and MET

6.1 MicroRNAs as Tumor Suppressors

One of the studies showed that the overexpression of members of miR-200 family suppresses ZEB1/ ZEB2, β-catenin, deactivating TGF-β pathway to promote EMT, as a result the crosslink between EMT and TAM gets interrupted [41–43]. Moreover, in the double negative feedback loop of ZEB1 and miR200s, each of them reciprocally controls their expression [44]. Binding of miR200 family members at 3'UTR of ZEB1 mRNA inhibiting its expression while based on different extracellular signals, ZEB1 can also suppresses miR-200 expression by binding to the promoter directly [44]. Similarly, miR-720, when overexpressed, it inhibits M2 macrophage polarization by targeting GATA3, thereby mitigating the phagocytic ability of M2 macrophage [45]. But, it was also evaluated that even when miR-720 was overexpressed in macrophages, due to an ectopic expression of GATA-3, the M2 phenotype was restored [45]. The expression of miR-98 and miR-27b inhibits EMT-induced cell metastasis by downregulating the CCL18 signaling pathway, otherwise the expression of these miRs gets reduced by TAM secreted CCL18 via the N-Ras/ ERK/PI3K/ NFκB/Lin28b signaling pathway [30].

6.2 OncomiRs as Promoters of Cancer Development

In breast cancer, the most highly expressed oncomiR is miR-21 which elevates cell metastasis by targeting various tumor suppressor genes including TPM1 [46, 47]. The overexpression of miR-221/222 downregulates the E-cadherin level by targeting TRPS1 at 3' region and thereby promotes EMT through ZEB2 suppression [48].

M2 macrophages showed elevated levels of miR-181a, whose downregulation may inhibit M2 phenotype expression, causing the tumor cells to lose invasive properties with expression of M1 phenotype. Here, miR-181a functions by inhibiting expression of C/EBPα and KLF6 [49].

7 Application of Circulating MicroRNAs as Diagnostic Biomarker for Breast Cancer

Heneghan and colleagues observed that breast cancer can be differentiated from other malignancies with the help of a panel of seven oncomiRs, which possessed a diagnostic potential [50]. Out of these seven oncomiRs, miR-10b, let-7a, and miR-155 expressions were found in majority of the cancer patients, but an elevated level of circulating miR-195 were specific to breast cancer [50]. It had a specificity of 91% and sensitivity of 88%, and the latter was further increased to 94% when there was a blend of let-7a, miR-155, and miR-195 [50]. The diagnostic potential of serum miR-214 was investigated by Schwarzenbach and colleagues [51]. In this study, while doing a survey in a cohort of breast cancer patients, four circulating microRNAs, i.e., miR-20a, miR19a, miR-214, and miR-21 have shown to target PTEN [51, 52]. But, only miR-214 had the ability to distinguish between malignant and benign tumors because when a comparative study was conducted between the preoperative and postoperative samples, a significant decline in the level of miR-214 was observed in the latter and a significant rise in the levels of miR-21, miR-20a was observed in the patients having benign tumors [51]. Moreover, a correlation was found between the elevated levels of miR-214 and positive lymph node status, confirming its diagnostic potential for malignancy and metastasis [51]. A study by Cuk et al. showed that a panel of seven oncomiRs (miR-652, miR-148b, miR-409-3p, miR-127-3p, miR-801, and miR-376a/c) possessed a diagnostic potential in a cohort of breast cancer patients of stage I/II [53]. Among them, miR-148b, miR-127-3p, miR-652, miR-801, and miR-409-3p can be used as early diagnostic markers in BRCA because of their elevated levels in sera [53]. In another study by Shen and colleagues with 87 patients of breast cancer and 35 healthy controls, an elevated level of miR-148b, miR-409-3p, and miR-133a was reported in BRCA patients compared to healthy individuals [54]. During the validation phase with these microRNAs, miR-148b and miR-133a were detected as circulating miRNAs, and their potential to be used as a biomarker for BRCA was identified in MDA-231 and MCF-7 breast cancer cell lines [54]. When Zearo and colleagues conducted a study between BRCA cases and healthy controls, although increased levels of seventeen microRNAs were detected in serum of breast cancer patients, most significant increase of serum mir-484 levels was found in breast cancer patients in contrast to healthy controls during the study of the validation cohort of those fourteen microRNAs [55]. Sun and colleagues not only demonstrated miR-155 as a potential diagnostic biomarker but they have also showed the use of miR-155 as a treatment response indicator because after surgery and chemotherapy, it displayed a declined trend [56].

8 Application of Circulating MicroRNAs as Prognostic Biomarker for Breast Cancer

A study on breast cancer serum samples revealed that elevated levels of let-7b could distinguish invasive from non-invasive carcinomas or healthy cases. Again, although increased expression levels of miR-202 were observed in both the cohorts of breast cancer patients, but its elevated level was mostly correlated with a decline in overall survival [57]. In patients of triple negative breast cancer, recently studies showed that some microRNAs like miR-18b, miR-107, miR-103, and miR-652 have the ability to predict the disease outcome [58]. In breast cancer tumor stage is reported to be detected by the elevated levels of miR-34a and miR-21, while breast cancer metastasis can be correlated with overexpression of miR-155, miR-10b, and miR-34a [59, 60].

9 Conclusion

Since breast cancer has become one of the most aggravated diseases disseminating among the females globally, its early detection and treatment is necessary to reduce the mortality rate. It has been found that TAM and EMT in addition with various mediators are often the major players in breast cancer progression, invasion, and metastasis, so targeting these two components can be used as a strategy for therapeutic intervention. Recently, studies have showed that the microRNAs are involved in regulating several transcriptional activities and signaling pathways which are required for upregulating the EMT process along with TAM polarization promoting cancer development and this property therefore paves the way for the microRNAs to be used successfully as a promising diagnostic and prognostic biomarker for future treatments.

References

1. Redig AJ, McAllister SS (2013) Breast cancer as a systemic disease: a view of metastasis. J Intern Med 274(2):113–126. https://doi.org/10.1111/joim.12084
2. Rojas K, Stuckey A (2016) Breast cancer epidemiology and risk factors. Clin Obstet Gynecol 59(4):651–672
3. Benson JR, Jatoi I (2012) The global breast cancer burden. Futur. Oncol 8(6):697–702
4. Chanmee T, Ontong P, Konno K, Itano N (2014) Tumor-associated macrophages as major players in the tumor microenvironment. Cancers (Basel) 6(3):1670–1690
5. Weigelt B, Peterse JL, Van't Veer LJ (2005) Breast cancer metastasis: markers and models. Nat Rev Cancer 5(8):591–602
6. Mantovani A, Ponzetta A, Inforzato A, Jaillon S (2020) Inflammation and tumor progression : double edged swords. Europe PMC Funders Group Innate Immun 285(5):524–532
7. Zhou N et al (2015) Exposure of tumor-associated macrophages to apoptotic MCF-7 cells promotes breast cancer growth and metastasis. Int J Mol Sci 16(6):11966–11982
8. Carter EP, Roozitalab R, Gibson SV, Grose RP (2021) Tumour microenvironment 3D-modelling: simplicity to complexity and back again. Trends Cancer 7(11):1033–1046
9. Boutilier AJ, Elsawa SF (2021) Macrophage polarization states in the tumor microenvironment. Int J Mol Sci 22(13)
10. Song W, Mazzieri R, Yang T, Gobe GC (2017) Translational significance for tumor metastasis of tumor-associated macrophages and epithelial-mesenchymal transition. Front Immunol 8:1–13
11. Cha YJ, Koo JS (2020) Role of tumor-associated myeloid cells contribute. Cells 9
12. Munir MT et al (2021) Tumor-associated macrophages as multifaceted regulators of breast tumor growth. Int J Mol Sci 22(12)
13. Zhang J, Yao H, Song G, Liao X, Xian Y, Li W (2015) Regulation of epithelial-mesenchymal transition by tumor-associated macrophages in cancer. Am J Trans Res 7(10):1699–1711
14. Atri C, Guerfali FZ, Laouini D (2018) Role of human macrophage polarization in inflammation during infectious diseases. Int J Mol Sci 19(6)
15. Leek RD et al (2002) Relation of hypoxia-inducible factor-2α (HIF-2α) expression in tumor-infiltrative macrophages to tumor angiogenesis and the oxidative thymidine phosphorylase pathway in human breast cancer. Cancer Res 62(5):1326–1329
16. Chitu V, Stanley ER (2006) Colony-stimulating factor-1 in immunity and inflammation. Curr Opin Immunol 18(1):39–48

17. Lin EY et al (2006) Macrophages regulate the angiogenic switch in a mouse model of breast cancer. Cancer Res 66(23):11238–11246
18. Lewis CE, Pollard JW (2006) Distinct role of macrophages in different tumor microenvironments. Cancer Res 66(2):605–612
19. Kitamura T et al (2015) CCL2-induced chemokine cascade promotes breast cancer metastasis by enhancing retention of metastasis-associated macrophages. J Exp Med 212(7):1043–1059
20. Micalizzi DS, Farabaugh SM, Ford HL (2010) Epithelial-mesenchymal transition in cancer: parallels between normal development and tumor progression. J Mammary Gland Biol Neoplasia 15(2):117–134
21. Thiery JP, Acloque H, Huang RYJ, Nieto MA (2009) Epithelial-Mesenchymal transitions in development and disease. Cell 139(5):871–890
22. Zhong X et al (2020) Circulating tumor cells in cancer patients: developments and clinical applications for immunotherapy. Mol Cancer 19(1):1–12
23. Chaffer CL, Thompson EW, Williams ED (2007) Mesenchymal to epithelial transition in development and disease. Cells Tissues Organs 185(1–3):7–19
24. Zhang J, Tian XJ, Xing J (2016) Signal transduction pathways of EMT induced by TGF-β, SHH, and WNT and their crosstalks. J Clin Med 5(4):1–18
25. Bidard F, Proudhon C, Pierga J (2016) Circulating tumoe cells in breast cancer. Mol Oncol vol 10
26. Ueo H et al (2015) Circulating tumour cell-derived plastin3 is a novel marker for predicting long-term prognosis in patients with breast cancer. Br J Cancer 112(9):1519–1526
27. Matrone MA et al (2011) Metastatic breast tumors express increased tau, which promotes microtentacle formation and the reattachment of detached breast tumor cells. NIH Public Access 29(22):3217–3227
28. Kumar et al (2011) A pathway for the control of anoikis sensitivity by E-cadherin and epithelial-to-mesenchymal transition. Mol Cell Biol 31(19):4036–4051
29. Su S et al (2014) A positive feedback loop between mesenchymal-like cancer cells and macrophages is essential to breast cancer metastasis. Cancer Cell 25(5):605–620
30. Lin X, Chen L, Yao Y, Zhao R, Cui X (2015) CCL18-mediated down-regulation of miR98 and miR27b promotes breast cancer metastasis. Oncotarget 6(24)
31. Tobar N, Villar V, Santibanez JF (2010) ROS-NFκB mediates TGF-β1-induced expression of urokinase-type plasminogen activator, matrix metalloproteinase-9 and cell invasion. Mol Cell Biochem 340(1–2):195–202
32. Romagnoli M et al (2013) Epithelial-to-Mesenchymal. NIH Public Access 72(23):6268–6278
33. Zhang H et al (2011) Forkhead transcription factor Foxq1 promotes epithelial-mesenchymal transition and breast cancer metastasis. Cancer Res 71(4):1292–1301
34. Sullivan NJ et al (2009) Interleukin-6 induces an epithelial-mesenchymal transition phenotype in human breast cancer cells. Oncogene 28(33):2940–2947
35. Ma JH, Qin L, Li X (2020) Role of STAT3 signaling pathway in breast cancer. Cell Commun Signal 18(1):1–13
36. Li QQ et al (2011) Involvement of NF-κB/miR-448 regulatory feedback loop in chemotherapy-induced epithelial-mesenchymal transition of breast cancer cells. Cell Death Differ 18(1):16–25
37. Youlden DR, Cramb SM, Dunn NAM, Muller JM, Pyke CM, Baade PD (2012) The descriptive epidemiology of female breast cancer: an international comparison of screening, incidence, survival and mortality. Cancer Epidemiol 36(3):237–248
38. Aggarwal V, Priyanka K, Tuli HS (2020) Emergence of circulating MicroRNAs in breast cancer as diagnostic and therapeutic efficacy biomarkers. Mol Diagn Ther 24(2):153–173
39. Ambros V (2004) The functions of animal microRNAs. Nature 431(7006):350–355
40. Zhang B, Pan X, Cobb GP, Anderson TA (2007) MicroRNAs as oncogenes and tumor suppressors. Dev Biol 302(1):1–12
41. Saydam O et al (2009) Downregulated MicroRNA-200a in meningiomas promotes tumor growth by reducing E-Cadherin and activating the Wnt/β-Catenin signaling pathway. Mol Cell Biol 29(21):5923–5940

42. Hamam R et al (2017) Circulating microRNAs in breast cancer: novel diagnostic and prognostic biomarkers. Cell Death Dis 8(9)
43. Zhao M, Ang L, Huang J, Wang J (2017) MicroRNAs regulate the epithelial–mesenchymal transition and influence breast cancer invasion and metastasis. Tumor Biol 39(2):1–8
44. Wu HT et al (2020) Oncogenic functions of the EMT-related transcription factor ZEB1 in breast cancer. J Transl Med 18(1):1–10
45. Zhong Y, Yi C (2016) MicroRNA-720 suppresses M2 macrophage polarization by targeting GATA3. Biosci Rep vol 0, pp 1–9
46. Bautista-Sánchez D et al (2020) The promising role of miR-21 as a cancer biomarker and its importance in RNA-based therapeutics. Mol Ther Nucleic Acids 20(22):409–420
47. Petrović N (2016) MiR-21 might be involved in breast cancer promotion and invasion rather than in initial events of breast cancer development. Mol. Diagn Ther 20(2):97–110
48. Stinson S et al (2011) TRPS1 targeting by miR-221/222 promotes the epithelial-to-mesenchymal transition in breast cancer. Sci Signal 4(177):1–12
49. Bi J et al (2016) miR-181a induces macrophage polarized to M2 phenotype and promotes M2 macrophage-mediated tumor cell metastasis by targeting KLF6 and C/EBPα. Mol Ther Nucleic Acids 5(77)
50. Heneghan HM, Miller N, Kelly R, Newell J, Kerin MJ (2010) Systemic miRNA-195 differentiates breast cancer from other malignancies and is a potential biomarker for detecting noninvasive and early stage disease. Oncologist 15(7):673–682
51. Schwarzenbach H, Milde-Langosch K, Steinbach B, Müller V, Pantel K (2012) Diagnostic potential of PTEN-targeting miR-214 in the blood of breast cancer patients. Breast Cancer Res Treat 134(3):933–941
52. Innocenti GM (2001) The time when the 'Tomte' of evolution was playing with time. Behav Brain Sci 24(2):287
53. Cuk K et al (2013) Plasma MicroRNA panel for minimally invasive detection of breast cancer. PLoS One 8(10)
54. Shen J et al (2014) Circulating miR-148b and miR-133a as biomarkers for breast cancer detection. Oncotarget 5(14):5284–5294
55. Zearo S et al (2014) MicroRNA-484 is more highly expressed in serum of early breast cancer patients compared to healthy volunteers. BMC Cancer 14(1):1–7
56. Sun Y et al (2012) Serum MicroRNA-155 as a potential biomarker to track disease in breast cancer. PLoS ONE 7(10):1–8
57. Joosse SA, Müller V, Steinbach B, Pantel K, Schwarzenbach H (2014) Circulating cell-free cancer-testis MAGE-A RNA, BORIS RNA, let-7b and miR-202 in the blood of patients with breast cancer and benign breast diseases. Br J Cancer 111(5):909–917
58. Sahlberg KK et al (2015) A serum MicroRNA signature predicts tumor relapse and survival in triple-negative breast cancer patients. Clin Cancer Res 21(5):1207–1214
59. Roth C, Rack B, Müller V, Janni W, Pantel K, Schwarzenbach H (2010) Circulating microRNAs as blood-based markers for patients with primary and metastatic breast cancer. Breast Cancer Res 12(6)
60. Si H et al (2013) Circulating microRNA-92a and microRNA-21 as novel minimally invasive biomarkers for primary breast cancer. J Cancer Res Clin Oncol 139(2):223–229

Comparison of Three-Dimensional Conformal Radiotherapy and Intensity Modulated Radiotherapy in Cervix Cancer

Samiul Alim, Md. Masud Parvej, K. A. Khan, Md. Abul Hasnat, Jakir Hosen, Md. Khirul Islam, Sarmin Sultana, Karthick Raj Mani, and M. M. Ahasan

Abstract To investigate the Dose Comparison of Three-Dimensional Conformal Radiotherapy (3DCRT) and Intensity Modulated Radiotherapy (IMRT) underwent Computed Tomography (CT) simulation along with adequate immobilization and positioning devices. Planning Target Volume (PTV) and Organs at Risk (OARs) were delineated slice by slice for all patients. 3DCRT treatment plans were created by photon beam of 15 MV energy using 4-field box technique and Multi-leaf Collimator (MLC) fitted to the Planning Target Volume (PTV). IMRT plans were created by photon beam of 6 MV energy with equally distributed 7 directions gantry angle. We intend to deliver total 50 Gy in 25 fractions for all patients. Dose for the risk organs and planning target were recorded from the Dose Volume Histogram (DVH). *Results*: Target Homogeneity for the 3DCRT and IMRT plan were comparable. Conformity Index shows that over 3DCRT low dose volumes (V_5, V_{10}, and V_{15}) were high in IMRT, but high dose regions were comparatively less. In IMRT plan, the rectal mean doses were reduced by 15% compared to 3DCRT which may result less toxicity in rectum. In IMRT plan bladder mean dose was reduced by 27% compared to 3DCRT plan. *Conclusion*: IMRT shows superior OAR sparing compared to 3DCRT plan. But IMRT treatment plan is very precise so treatment center who does not have IGRT facilities they should not treat with IMRT technique. In stand of IMRT, they can use 4-field box technique using 15MV beams and PTV fitted MLC to reduce the high dose irradiation volume, which may result in low toxicity profile.

S. Alim · Md. M. Parvej · Md. A. Hasnat · J. Hosen · Md. K. Islam · S. Sultana · M. M. Ahasan
Institute of Nuclear Medical Physics, AERE, BAEC, Dhaka, Savar, Bangladesh
e-mail: dr.jakirhosen@baec.gov.bd

K. A. Khan (✉)
Department of Physics, Jagannath University, Dhaka, Jaipur, Bangladesh
e-mail: kakhan01@yahoo.com

K. R. Mani
United Hospital Ltd, Dhaka, Gulshan, Bangladesh

J. R. Szymanski et al. (eds.), *Energy Systems, Drives and Automations*, Lecture Notes in Electrical Engineering 1057,
https://doi.org/10.1007/978-981-99-3691-5_38

Keywords Cervix cancer · Radiotherapy · 3DCRT · IMRT

1 Introduction

Cervical cancer is the fourth common disease for both incidence and mortality. Cervical cancer diagnosed frequently and the leading cause of cancer death. However, substantially vary across countries and within each country depending on the degree of socioeconomic condition including life style factors [1]. The mortality rate in developing countries is high for cervical cancer, with each accounting for 11% of the total female cancer deaths [2]. The rapid evolution of knowledge in cancer biology and the investigation and validation of biologic factors that predict cancer outcome and response to treatment with better accuracy. TNM-based approach in clinical care at an individualized patient level [3]. According to a 2012 systematic review and meta-analysis by Yang et al. based on 13 studies, IMRT significantly reduced the average proportion of irradiated volume of the rectum and small bowel compared with 3DCRT in patients with gynecologic malignancies [4]. IMRT provides a good alternative for treatment of cervix carcinoma with lower acute gastrointestinal and acute genitourinary toxicity with similar target coverage compared to 3D conformal radiotherapy [5]. In IMRT technique, high radiation dose produced significantly less average percent volumes of exposed rectum and small bowel than 3D-CRT, but did not differentially affect the average percent volumes in the bladder and bone marrow [6]. IMRT significantly reduced the rate of toxicity and provided good clinical outcomes compered to 3D-CRT [7]. When we are moving from 3D-CRT to IMRT include more fields, and the dose volume histograms show that, as a consequence, a larger volume of normal tissue is exposed to lower doses. Monitor units (MU) also increased, as a result increasing the total body exposure, due to leakage radiation [8]. It has been pointed out that new technologies, such as IMRT, were responsible for an increase in the doses received out of the field [9]. IMRT technique decreasing some acute complications (GI complication) without worsening outcomes [10]. The IMRT reduces the incidence and severity of acute hematologic toxicity and the volume of bone marrow irradiated at the higher doses for the cervical cancer patient [11]. The modern radiotherapy utilizes extensively the application of medical imaging to determine the extent of diseases, spatial relation between tumor regions and neighboring healthy tissues. Recently advancements in imaging system such as flat panel detector, functional magnetic resonance imaging (MRI), Single Positron Emission Computed Tomography (SPECT), Positron Emission Computed Tomography (PET-CT), Four-Dimensional Computed Tomography, Cone Beam Computed Tomography (CBCT), and ultrasound systems provide complementary information's for precise localization and treatment verification of patient's treatment for cancers sites. Although even with all these resources, implementation toward a real conformal radiotherapy has placed more stringent requirements on the reproducibility of patient positioning, precise target delineation, and accurate beam targeting. Due to the complexity of certain tumor location and shapes, innovative radiotherapy planning techniques have to be applied for the easy and effective sparing of the surrounding normal tissues and organs.

2 Materials and Method

2.1 Materials

At Institute of Nuclear Medical Physics (INMP), we have iX linear accelerator (Varian Medical System, Palo Alto, USA). Which having photon beam of 6MV and 15MV energies, and electron beam of 6, 9, 12, 15, and 18 meV energies.

For treatment plan we used Eclipse Treatment Planning System (TPS) software ver. 13.7 (Varian Medical Systems). For CT simulation we used Phillips Ingenuity TF 128 slice PET/CT scanner (Phillips Healthcare, USA).

2.2 Method

Patient Data: Ten patients with middle age of 45 years (range, 40–55 years) were retrospectively included in this study from our previously treated patient's database. The demographic data which includes age, histopathology, and the staging of patients were listed in Table 1. All the staging of the patients was performed according to the International Federation of Gynecology and Obstetrics (FIGO) classification.

Table 1 Patient data

S. No.	Patient	Age (year)	Histopathological report (HPR)	Stage
1	Patient 1	41	Squamous cell carcinoma	IIB
2	Patient 2	51	Squamous cell carcinoma	IB
3	Patient 3	40	Squamous cell carcinoma	IB2
4	Patient 4	44	Squamous cell carcinoma	IB
5	Patient 5	43	Adenocarcinoma	III (Post Op.)
6	Patient 6	54	Squamous cell carcinoma	IIB
7	Patient 7	47	Adenocarcinoma	IIA
8	Patient 8	49	Squamous cell carcinoma	IIIB
9	Patient 9	55	Adenocarcinoma	IIA
10	Patient 10	47	Squamous cell carcinoma	IIIB

CT Simulation and Delineation

All patients were positioned in supine position with (M/S Civco) the ankle rest. CT simulation was performed using a Phillips Ingenuity TF 128 slice PET/CT scanner (Phillips Healthcare, USA), the institution-based CT protocol with full bladder was used for the data acquisition. The contrast enhanced CT data were acquired in axial mode with a slice thickness of 2.5 mm, field of view of 495 ps timing pixel resolution. Three fiducial markers were placed, one anteriorly above symphysis pubis and two lateral markers in the mid-plane and also a radio-opaque marker was placed at cervical os/vault to identify the caudal border. Once the patient CT data are acquired, the CT images were imported in DICOM format. The structure for body was contoured automatically by the treatment planning system. Several structures and area of interests were delineated. Organs-at-Risk (OARs, i.e., Rectum, Urinary Bladder, Right Femoral Head, Left Femoral Head, Intestine and Cauda Equina) were delineated by the physician slice by slice on the CT images for all patients. CTV and nodal station were contoured according to the consensus guideline of the Radiation Therapy Oncology Group (RTOG). PTV was generated from the CTV (combining the primary and nodal CTVs) by applying an institutional margin recipe of 1 cm in cranio-caudal and 0.7 cm in other direction (Figs. 1 and 2).

Treatment Planning and Optimization

All the treatment plan for Three-Dimensional Conformal Radiotherapy and Intensity Modulated Radiotherapy was planned with iX linear accelerator (Varian Medical System, Palo Alto, USA). The total pelvis was treated including clinically and radiological visible tumor, uterine corpus, upper part of vagina, parametrium, and the lymph nodes. The cranial border was kept at L4–L5 junction, caudal border was kept at below of obturator foramen or inferiorly extended to ensure adequate coverage of vaginal disease (radio-opaque marker placed) extension with proper margins, lateral border was kept at 2 cm from pelvic brim, the anterior border was fixed at anterior cortex of symphysis pubis and posterior border at S2-S3 junction. The conformal four field box technique also planned with 15 MV photons with MLC using a 5 mm fit to PTV. The 3DCRT plans were planned using SAD techniques and normalized at

Fig. 1 3D view of PTV

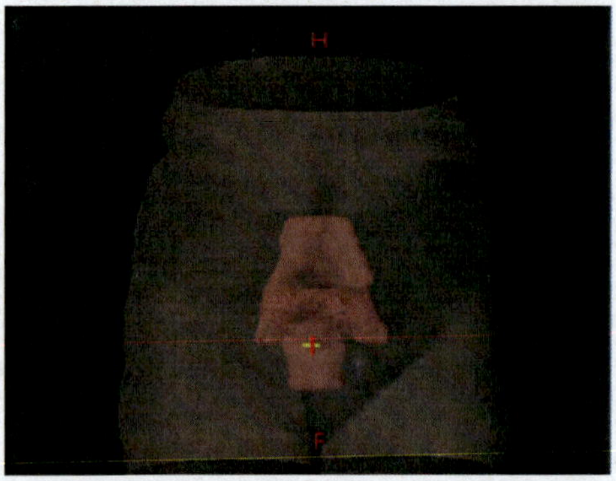

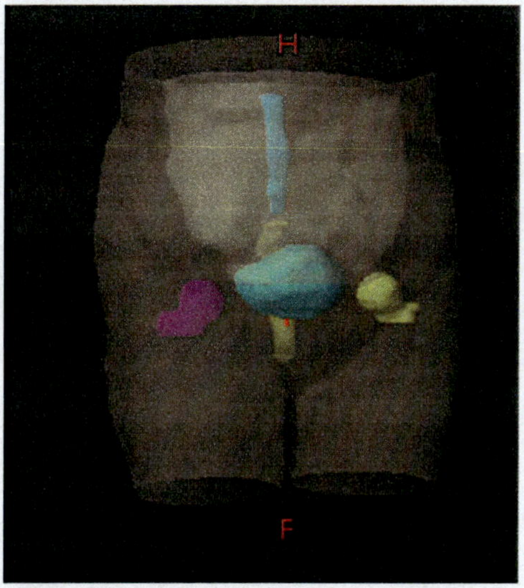

Fig. 2 3D view of organ at risk's

isocenter. Intensity Modulated Radiotherapy plans were generated for all the patients using seven equally spaced coplanar fields. The gantry angles used for the Intensity Modulated Radiotherapy plans were 0°, 51°, 102°, 153°, 204°, 255°, and 306° with collimator rotation if necessary. The 6 MV photon beam along with Millennium 120 leaves was used for the IMRT planning optimization. IMRT plans were optimized by dose volume optimizer algorithm (DVO) ver.13.7 and the smart leaf motion calculator ver. 13.7 were used for calculating the optimal fluence to original fluence. For dose calculation, Analytical Anisotropic Algorithm (AAA) was used, and normalization was done at target mean. All the treatment techniques used in this study were planned with a prescription dose of 50 Gy for PTV with 2 Gy per fraction (Fig. 3).

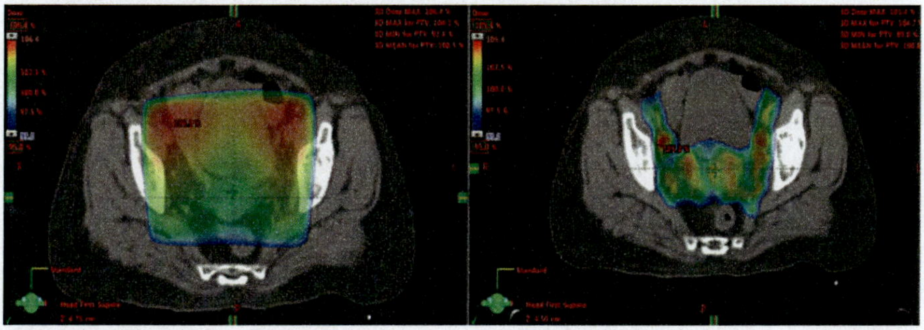

Fig. 3 Typical 95% dose color wash for the 3DCRT and IMRT treatment plan

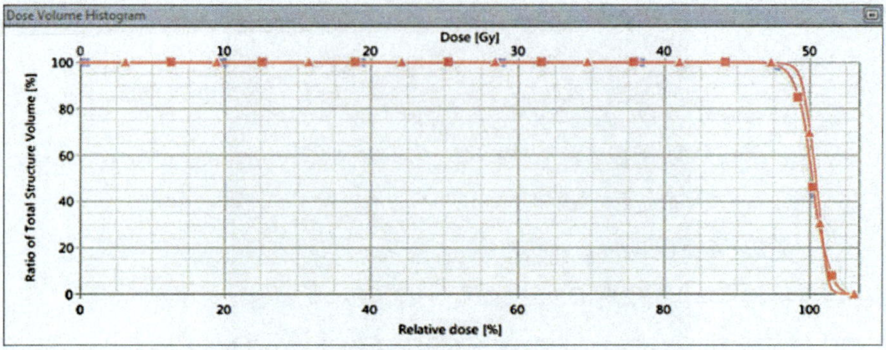

Fig. 4 Dose volume histogram of PTV for 3DCRT and IMRT

Dosimetric Evaluation and Comparison

Quantitative evaluations of plans were performed by taking various dosimetric parameters from dose volume histogram (DVH). For PTV, the values of D_{98}, D_{50}, and D_2 (dose received by the 98%, 50%, and 2% of the PTV volumes, respectively) were defined as metrics for minimum, mean, and maximum doses were documented for bo the treatment techniques. V_{95} (the volume receiving at least 95% of the prescribed dose) and the PTV volumes of all the patients were documented to determine the homogeneity and the conformity indexes.

To compare for these two techniques, ICRU 83 definition was used to determine the dose conformity and dose homogeneity. Dose conformity and homogeneity are independent specifications of the quality of the absorbed dose distribution. Dose conformity characterizes the degree to which the high dose region conforms to the target volume, whereas dose homogeneity characterizes the uniformity of the absorbed dose within the target volume (Fig. 4).

Homogeneity Index (*HI*)

$$HI = \frac{D_2 - D_{98}}{D_{50}} \tag{1}$$

where D_2, D_{98}, and D_{50} are the doses received by 2%, 98%, and 50% of PTV volumes, respectively. HI = 0 (zero) is ideal value.

Conformity Index (*CI*)

In 1993, Radiation Therapy Oncology Group recommended CI as a ratio of the reference isodose (95% isodose volume) volume to the target volume.

$$CI_{RTOG} = \frac{V_{RI}}{TV} \tag{2}$$

where V_{RI} is the reference isodose volume and TV is the volume of the target (PTV).

For OAR minimum, maximum, and mean doses were documented for the parallel and serial structures, respectively, Additional dose parameters such as V_5, V_{10}, V_{15}, V_{20}, V_{30}, V_{40}, and V_{50} (volume receiving at least 5 Gy, 10 Gy, 15 Gy, 20 Gy, 30 Gy, 40 Gy, and 50 Gy, respectively) for Rectum, Bladder, Right Femoral Head, Left Femoral Head, Intestine, and Cauda were recorded. Dose conformity and dose heterogeneity were also calculated using the HI and CI using the Eqs. (1) and (2).

Data Analysis

The statistical data were presented as the average of all the patients followed by the standard deviation ($\bar{x} \pm \sigma\bar{x}$). All statistical analysis was conducted with paired two tailed "T-TEST" with equal variance using Microsoft Word/Excel version 2010 with $p < 0.05$ considered as significant.

Conformity Index and Dose Homogeneity Index

The treatment plan quality has been compared using dose Conformity and dose Homogeneity parameters of two techniques are performed using conformity index (CI) and homogeneity index (HI). The calculated HI and CI of the PTV for both two technique IMRT and 3DCRT are tabulated in Table 2. The average conformity index (CI) ($\bar{x} \pm \sigma\bar{x}$) values were 1.148 ± 0.082423 for IMRT and 2.041 ± 0.400 for 3DCRT. The average homogeneity index (HI) ($\bar{x} \pm \sigma\bar{x}$) values were 0.083 ± 0.024 for IMRT and 0.084 ± 0.012 for 3DCRT.

Low and High Dose Volume

The V_5, V_{10}, V_{15}, V_{20}, V_{30}, V_{40}, and V_{50} low dose and high dose volumes for all patients with 3DCRT and IMRT techniques were tabulated in Table 3. The mean followed by standard deviation ($\bar{x} \pm \sigma\bar{x}$) 3DCRT and IMRT of the V_5, V_{10}, V_{15}, V_{20}, V_{30}, V_{40}, and V_{50} volumes was 45.15 ± 7.026 and 51.870 ± 6.813, 37.716 ± 6.436 and 41.001 ± 5.676, 33.701 ± 6.004 and 35.056 ± 5.098, 31.256 ± 5.626 and 26.344 ± 5.247, 15.605 ± 4.887 and 11.863 ± 4.149, 9.411 ± 3.589 and 6.631 ± 2.820, 4.844 ± 2.173 and 2.41 ± 0.734, respectively. The low dose volumes (V_5 and

Table 2 Conformity index and homogeneity index

Parameter	IMRT	3DCRT
$D_2\%$ (Gy)	51.573 ± 0.316	52.177 ± 0.501
$D_{50}\%$ (Gy)	50.219 ± 0.225	50.426 ± 0.331
$D_{98}\%$ (Gy)	47.361 ± 1.046	47.911 ± 0.945
95% isodose volume (cc)	1039.751 ± 268.364	1800.133 ± 320.004
PTV volume (cc)	903.968 ± 220.553	
CI (RTOG)	1.148 ± 0.082423	2.041 ± 0.400
HI	0.083 ± 0.024	0.084 ± 0.012

Abbreviations: $D_{2\%}$, $D_{98\%}$, and $D_{50\%}$ are the PTV volume received 2%, 98%, and 50% dose, respectively, *PTV* High Risk Planning Target Volume, *CI* Conformity Index, *HI* Homogeneity Index

V_{10}) were higher in IMRT compared to the 3DCRT. A substantial dose reduction in higher dose volumes (V_{50}, V_{40}, and V_{30}) was documented. The mean V_{30}, V_{40}, and V_{50} volumes for 3DCRT for all patients were 31%, 41%, and 100% higher compared to IMRT (Figs. 5 and 6).

Rectum Dose Volume

The mean dose (Gy) of rectum followed by standard deviation ($\bar{x} \pm \sigma\bar{x}$) for 3DCRT and IMRT was 46.485 ± 2.072 and 40.284 ± 2.409. The V_5, V_{10}, V_{15}, V_{20}, V_{30}, V_{40}, and V_{50} volumes for all the patients with 3DCRT and IMRT techniques were tabulated in Table 3. The mean followed by standard deviation 3DCRT and IMRT of V_5, V_{10}, V_{15}, V_{20}, V_{30}, V_{40}, and V_{50} volumes was 100.00 ± 0.000 and 100.00 ± 0.000, 100.00 ± 0.000 and 99.656 ± 0.589, 100.00 ± 0.000 and 98.852 ± 1.779, 99.705 ± 0.784 and 98.071 ± 2.789, 95.007 ± 3.618 and 88.293 ± 7.489, 84.491 ± 8.106 and 54.133 ± 11.406, 43.106 ± 24.524 and 11.038 ± 10.818, respectively. The volumes (V_5, V_{10}, V_{15}, V_{20},) were almost same. A substantial dose reduction in higher dose volumes (V_{50}, V_{40}, and V_{30}) was documented. The mean V_{30}, V_{40}, and V_{50} volumes for 3DCRT for all patients were 7%, 56%, and 290% higher compared to IMRT (Figs. 7 and 8).

Bladder Dose Volume

The mean dose (Gy) of bladder followed by standard deviation ($\bar{x} \pm \sigma\bar{x}$) for 3DCRT and IMRT was 48.372 ± 1.866 and 35.897 ± 3.254. The V_5, V_{10}, V_{15}, V_{20}, V_{30}, V_{40}, and V_{50} volumes for all the patients with 3DCRT and IMRT techniques were tabulated in Table 3. The mean followed by standard deviation 3DCRT and IMRT of V_5, V_{10}, V_{15}, V_{20}, V_{30}, V_{40}, and V_{50} volumes was 100.00 ± 0.000 and 100.00 ± 0.000, 100.00 ± 0.000 and 100.00 ± 0.000, 100.00 ± 0.000 and 99.999 ± 0.003, 100.00 ± 0.000 and 94.600 ± 4.413, 90.272 ± 28.251 and 73.244 ± 10.734, 88.385 ± 8.235 and 46.555 ± 12.935, 69.116 ± 16.562 and 14.916 ± 6.096, respectively. The volumes (V_5, V_{10}, V_{15}, V_{20}) were almost same. A substantial dose reduction in higher dose volumes (V_{30}, V_{40}, and V_{50}) was documented. The mean V_{30}, V_{40}, and V_{50} volumes for 3DCRT for all patients were 23%, 89%, and 363% higher compared to IMRT (Figs. 9 and 10).

Intestine Dose Volume

The mean dose (Gy) of intestine followed by standard deviation ($\bar{x} \pm \sigma\bar{x}$) for 3DCRT and IMRT was 19.224 ± 2.855 and 17.882 ± 1.962. The V_5, V_{10}, V_{15}, V_{20}, V_{30}, V_{40}, and V_{50} volumes for all the patients with 3DCRT and IMRT were tabulated in Table 3. The mean followed by standard deviation 3DCRT and IMRT of V_5, V_{10}, V_{15}, V_{20}, V_{30}, V_{40}, and V_{50} volumes was 62.736 ± 11.012 and 67.390 ± 11.477, 55.056 ± 9.317 and 57.877 ± 51.891, 49.533 ± 8.272 and 51.891 ± 7.4006, 46.847 ± 7.831and 43.483 ± 5.775, 23.503 ± 6.779 and 21.269 ± 6.474, 16.555 ± 6.377 and 11.140 ± 5.391, 10.027 ± 5.114 and 2.948 ± 2.038, respectively. The volumes (V_5, V_{10}, V_{15},) were bit higher in IMRT compere to 3DCRT. A substantial dose reduction in higher dose volumes (V_{30}, V_{40}, and V_{50}) was documented. The mean

Table 3 Static comparison between 3DCRT and IMRT

Organ	Organ volume (cc) ($\bar{x} \pm \sigma\bar{x}$)	Parameter	3DCRT ($\bar{x} \pm \sigma\bar{x}$)	IMRT ($\bar{x} \pm \sigma\bar{x}$)	Difference (%)	p-value
Body	25,080.820 ±6179.332	V_5	45.150 ± 7.026	51.870 ± 6.813	13.713	< 0.001
		V_{10}	37.716 ± 6.436	41.001 ± 5.676	8.7098	< 0.001
		V_{15}	33.701 ± 6.004	35.056 ± 5.098	4.0206	0.020
		V_{20}	31.256 ± 5.626	26.344 ± 5.247	− 15.715	< 0.001
		V_{30}	15.605 ± 4.887	11.863 ± 4.149	− 23.983	< 0.001
		V_{40}	9.411 ± 3.589	6.631 ± 2.820	− 29.539	< 0.001
		V_{50}	4.844 ± 2.173	2.41 ± 0.734	− 50.247	0.001
	Mean (Gy) =		12.997 ± 2.452	12.24 ± 1.942	− 5.0867	0.009
Bladder	196.745 ± 73.077	V_5	100.00 ± 0.000	100.00 ± 0.000	0.000	–
		V_{10}	100.00 ± 0.000	100.00 ± 0.000	0.000	–
		V_{15}	100.00 ± 0.000	99.999 ± 0.003	− 0.001	0.343
		V_{20}	100.00 ± 0.000	94.600 ± 4.413	− 5.400	0.004
		V_{30}	90.272 ± 28.251	73.244 ± 10.734	− 18.862	0.074
		V_{40}	88.385 ± 8.235	46.555 ± 12.935	− 47.327	< 0.001
		V_{50}	69.116 ± 16.562	14.916 ± 6.096	− 78.419	< 0.001
	Mean(Gy) =		48.372 ± 1.866	35.897 ± 3.254	− 21.75	< 0.001
Rectum	85.145 ± 38.14	V_5	100.00 ± 0.000	100.00 ± 0.000	0.000	–
		V_{10}	100.00 ± 0.000	99.656 ± 0.589	− 0.344	0.098
		V_{15}	100.00 ± 0.000	98.852 ± 1.779	− 1.148	0.072
		V_{20}	99.705 ± 0.784	98.071 ± 2.789	− 1.638	0.078
		V_{30}	95.007 ± 3.618	88.293 ± 7.489	− 7.066	0.004
		V_{40}	84.491 ± 8.106	54.133 ± 11.406	− 35.930	< 0.001
		V_{50}	43.106 ± 24.524	11.038 ± 10.818	− 74.393	0.001
	Mean(Gy) =		46.485 ± 2.072	40.284 ± 2.409	− 13.340	< 0.001
Right femur head	58.803 ± 12.857	V_5	100.00 ± 0.000	100.00 ± 0.000	0.000	–
		V_{10}	99.367 ± 1.233	89.371 ± 9.100	− 10.059	0.098
		V_{15}	97.872 ± 3.369	65.540 ± 20.05	− 33.0349	0.072
		V_{20}	96.363 ± 5.296	35.332 ± 24.018	− 63.334	0.078
		V_{30}	43.914 ± 34.890	4.344 ± 7.910	− 90.006	0.004
		V_{40}	10.006 ± 10.081	0.296 ± 0.823	− 97.0417	< 0.001

(continued)

Table 3 (continued)

Organ	Organ volume (cc) ($\bar{x} \pm \sigma\bar{x}$)	Parameter	3DCRT ($\bar{x} \pm \sigma\bar{x}$)	IMRT ($\bar{x} \pm \sigma\bar{x}$)	Difference (%)	p-value
		V_{50}	2.572 ± 4.538	0.000 ± 0.000	– 100.00	0.001
	Mean(Gy) =		30.90 ± 3.219	18.118 ± 3.626	– 41.372	< 0.001
Left femur head	55.334 ± 12.186	V_5	100.00 ± 0.000	100.00 ± 0.000	0.000	–
		V_{10}	99.205 ± 1.433	93.516 ± 6.691	– 5.743	0.010
		V_{15}	97.231 ± 3.888	72.586 ± 18.797	– 25.346	0.001
		V_{20}	95.172 ± 6.085	42.669 ± 23.172	– 55.166	< 0.001
		V_{30}	40.089 ± 10.72	4.810 ± 8.763	– 88.000	0.003
		V_{40}	12.143 ± 13.912	0.057 ± 0.163	– 99.527	0.022
		V_{50}	3.929 ± 7.224	0.000 ± 0.000	– 100.00	0.120
	Mean(Gy) =		31.039 ± 4.049	19.075 ± 3.479	– 38.545	< 0.001
Cauda	25.459 ± 8.479	V_5	61.805 ± 16.134	62.621 ± 12.093	1.320	0.678
		V_{10}	57.079 ± 16.264	52.659 ± 14.004	– 7.743	0.003
		V_{15}	52.763 ± 14.556	47.918 ± 13.402	– 9.182	0.004
		V_{20}	49.998 ± 14.589	44.344 ± 13.446	– 11.308	0.005
		V_{30}	23.038 ± 10.797	21.489 ± 14.118	– 6.723	0.793
		V_{40}	8.058 ± 10.020	2.498 ± 3.156	– 68.997	0.123
		V_{50}	0.000 ± 0.000	0.000 ± 0.000	0.000	–
	Mean(Gy) =		18.103 ± 4.017	16.510 ± 3.498	– 8.799	0.088
Intestine	2668.192 ± 865.802	V_5	62.736 ± 11.012	67.390 ± 11.477	7.418	0.005
		V_{10}	55.056 ± 9.317	57.877 ± 51.891	5.123	0.014
		V_{15}	49.533 ± 8.272	51.891 ± 7.4006	4.760	0.033
		V_{20}	46.847 ± 7.831	43.483 ± 5.775	– 7.180	0.074
		V_{30}	23.503 ± 6.779	21.269 ± 6.474	– 9.505	0.148
		V_{40}	16.555 ± 6.377	11.140 ± 5.391	– 32.709	0.001
		V_{50}	10.027 ± 5.114	2.948 ± 2.038	– 70.599	< 0.001
	Mean(Gy) =		19.224 ± 2.855	17.882 ± 1.962	– 6.980	0.022

V_{30}, V_{40}, and V_{50} volumes for 3DCRT for all patients were 10%, 48%, and 240% higher compared to Intensity Modulated Radiotherapy Plan (Figs. 11 and 12).

Right Femoral Head Dose Volumes

The mean dose (Gy) of right femoral head followed by standard deviation ($\bar{x} \pm \sigma\bar{x}$) for 3DCRT and IMRT was 30.90 ± 3.219 and 18.118 ± 3.626. The V_5, V_{10}, V_{15}, V_{20},

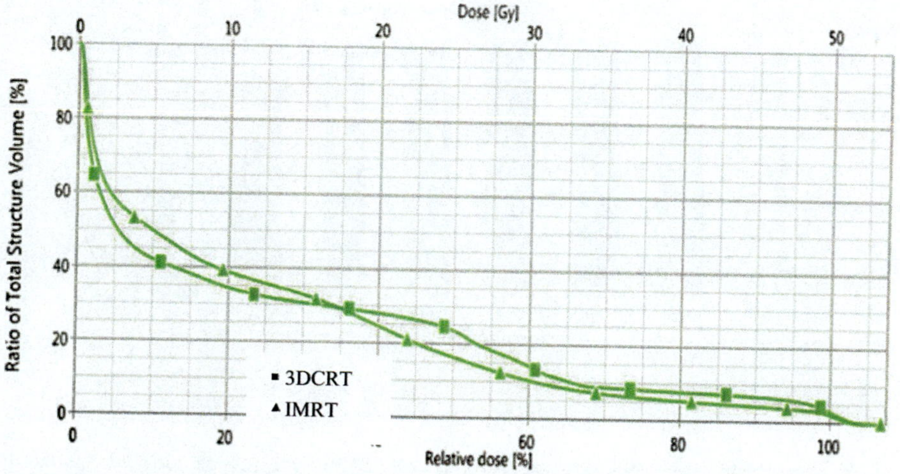

Fig. 5 DVH of low and high dose volume

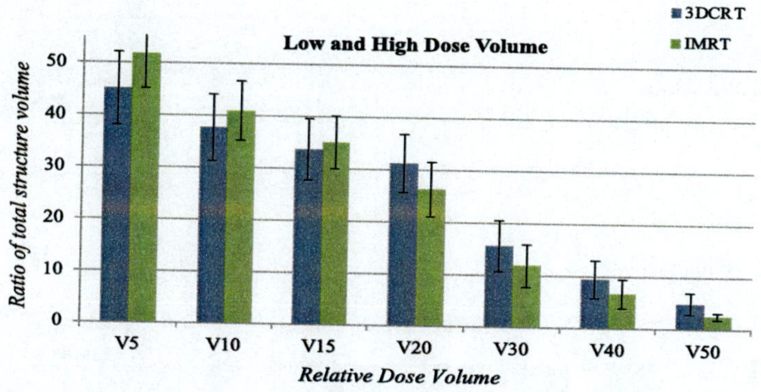

Fig. 6 Graph of low dose and high dose volumes

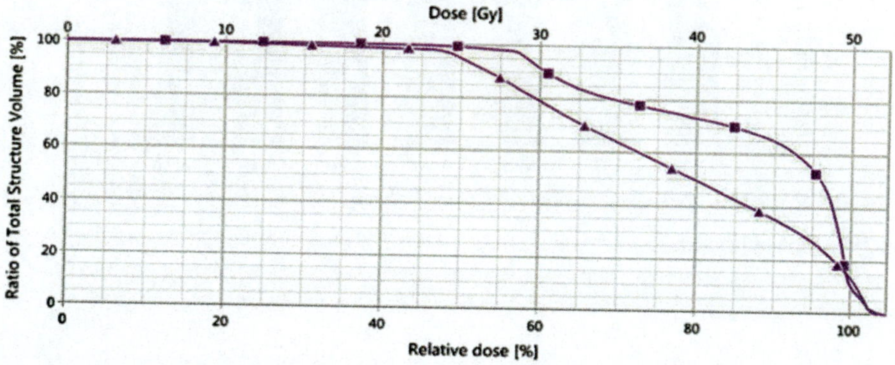

Fig. 7 DVH of rectum

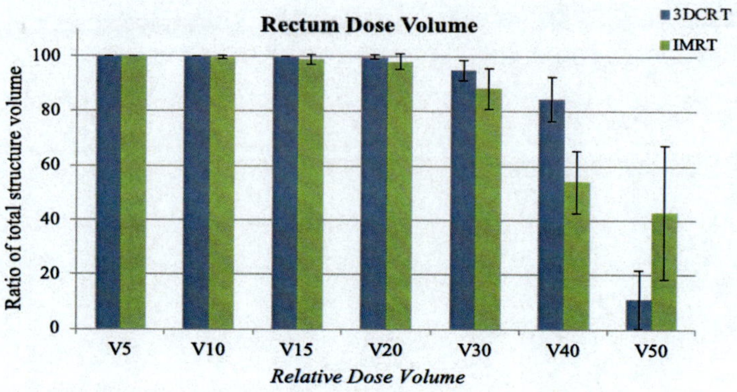

Fig. 8 Graph of rectum

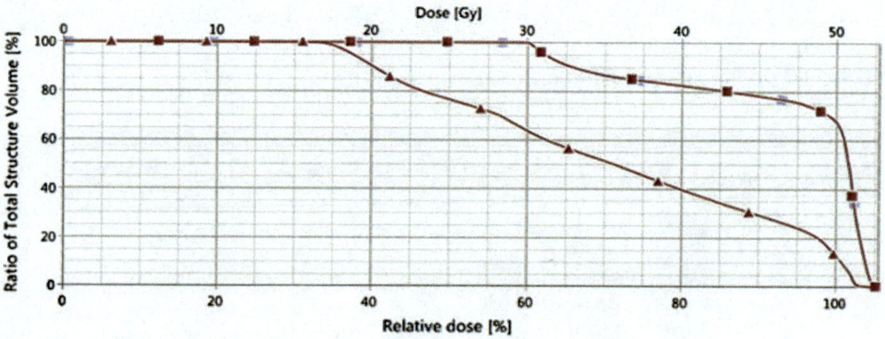

Fig. 9 DVH of bladder

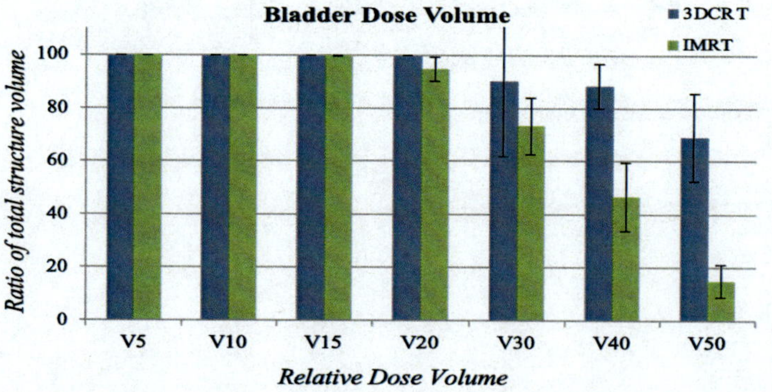

Fig. 10 Graph of bladder

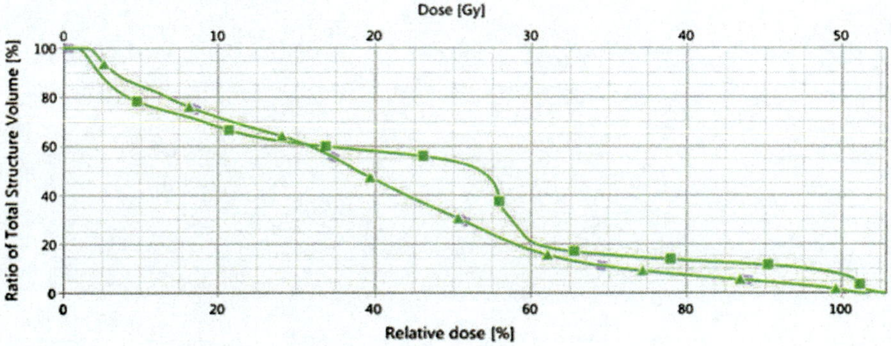

Fig. 11 DVH of intestine

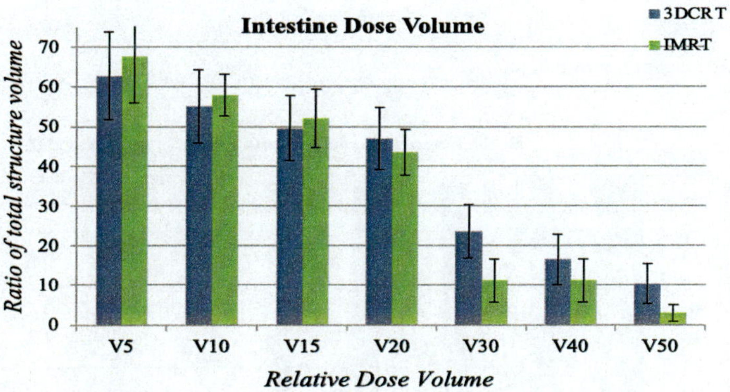

Fig. 12 Graph of intestine

V_{30}, V_{40}, and V_{50} volumes for all the patients with 3DCRT and IMRT techniques were tabulated in Table 3. The mean followed by standard deviation 3DCRT and IMRT of V_5, V_{10}, V_{15}, V_{20}, V_{30}, V_{40}, and V_{50} volumes was 100.00 ± 0.000 and 100.00 ± 0.000, 99.367 ± 1.233 and 89.371 ± 9.100, 197.872 ± 3.369 and 65.540 ± 20.05, 96.363 ± 5.296 and 35.332 ± 24.018, 43.914 ± 34.890 and 4.344 ± 7.910, 10.006 ± 10.081 and 0.296 ± 0.823, 2.572 ± 4.538 and 0.000 ± 0.000, respectively. The volumes (V_5) were almost same, and from V_{10} dose Reduction started. A substantial dose reduction in volumes (V_5, V_{10}, V_{15}, V_{20}, V_{30}, V_{40}, and V_{50}) was documented. The mean V_5, V_{10}, V_{15}, V_{20}, V_{30}, V_{40} volumes for 3DCRT for all patients were 11%, 49%, 172%, 900%, and 3080% higher compared to IMRT. And V_{50} level there is no dose in IMRT plan (Figs. 13 and 14).

Left Femoral Head Dose Volume

The mean dose (Gy) of left femoral head dose (Gy) followed by standard deviation ($\bar{x} \pm \sigma\bar{x}$) for 3DCRT and IMRT was 31.039 ± 4.049 and 19.075 ± 3.479. The V_5, V_{10}, V_{15}, V_{20}, V_{30}, V_{40}, and V_{50} volumes for all the patients with 3DCRT and

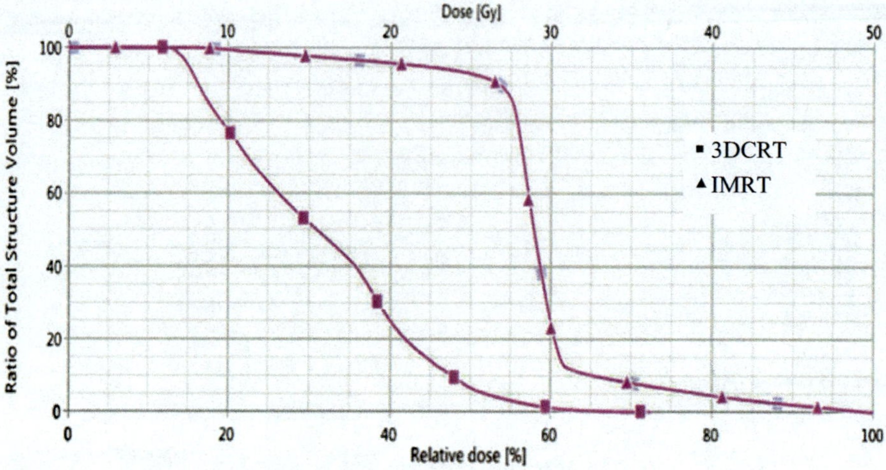

Fig. 13 DVH of right femoral head

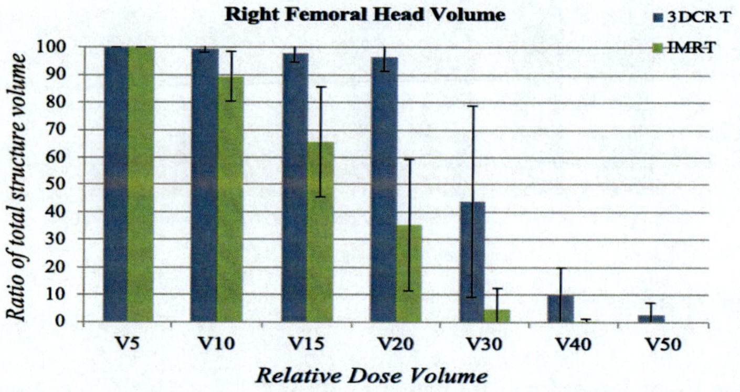

Fig. 14 Graph of right femoral head

IMRT technique were tabulated in Table 3. The mean followed by standard deviation 3DCRT and IMRT of V_5, V_{10}, V_{15}, V_{20}, V_{30}, V_{40}, and V_{50} volumes were100.00 ± 0.000 and 100.00 ± 0.000, 99.205 ± 1.433 and 93.516 ± 6.691, 97.231 ± 3.888 and 72.586 ± 18.797, 95.172 ± 6.085 and 42.669 ± 23.172, 40.089 ± 10.72 and 4.810 ± 8.763, 12.143 ± 13.912 and 0.057 ± 0.163, 3.929 ± 7.224 and 0.000 ± 0.000, respectively. The volumes (V_5) were almost same. And from V_{10} dose Reduction started. Volumes (V_5, V_{10}, V_{15}, V_{20}, V_{30}, V_{40} and V_{50}) were documented. The mean V_{10}, V_{15}, V_{20}, V_{30}, V_{40} volumes for 3DCRT for all the patient were 6%, 33%, 123%, 733%, and 21,055% higher compared to IMRT. And V_{50} level there is no dose in IMRT plan (Figs. 15 and 16).

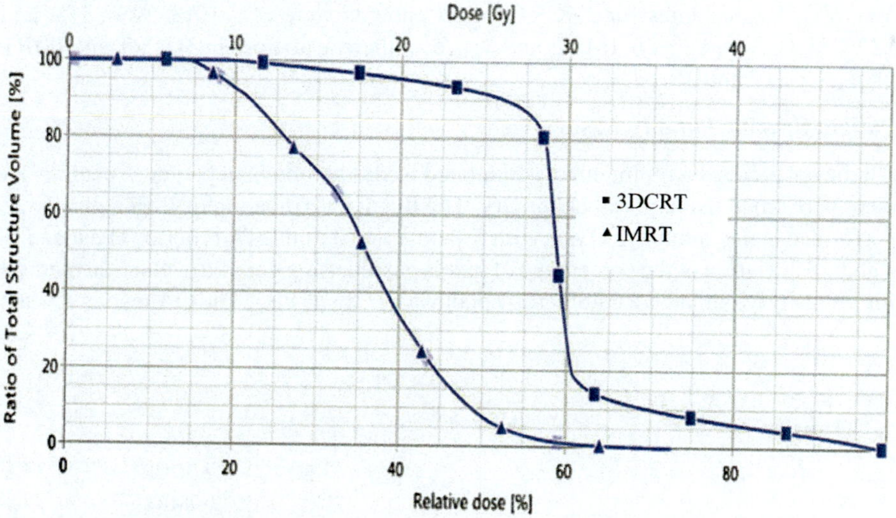

Fig. 15 DVH of left femoral head

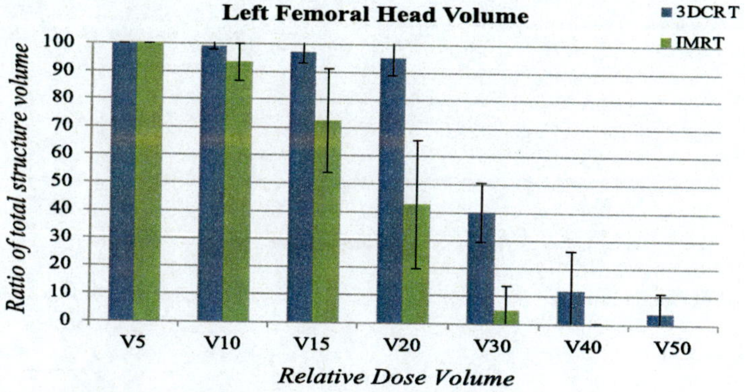

Fig. 16 Graph of left femoral head

Caudal Dose Volumes

The mean dose (Gy) of Cauda Equina followed by standard deviation ($\bar{x} \pm \sigma\bar{x}$) for 3DCRT and IMRT was 18.103 ± 4.017 and 16.510 ± 3.498. The V_5, V_{10}, V_{15}, V_{20}, V_{30}, V_{40}, and V_{50} volumes for all the patients with 3DCRT and IMRT techniques were tabulated in Table 3. The mean followed by standard deviation 3DCRT and IMRT of V_5, V_{10}, V_{15}, V_{20}, V_{30}, V_{40}, and V_{50} and 52.659 ± 14.004, 52.763 ± 14.556 and 47.918 ± 13.402, 49.998 ± 14.589 and 44.344 ± 13.446, 23.038 ± 10.797 and 21.489 ± 14.118, 8.058 ± 10.020 and 2.498 ± 3.156, 0.000 ± 0.000 and 0.000 ± 0.000, respectively. The volumes (V_5) were bit higher in IMRT compered to 3DCRT. Dose reduction started from (V_{10}) in IMRT plan. The mean V_5, V_{10}, V_{15},

V_{20}, V_{30}, V_{40} volumes for 3DCRT for all patients were 8%, 10%, 12%, 7%, and 222% higher compared to IMRT, and V_{50} level there is no dose in 3DCRT and IMRT plan (Figs. 17 and 18).

Patient Specific Quality Assurance

All the ten patients with Intensity Modulated Radiotherapy plan patients specific QA were performed using portal dosimetry. The fluence verification of IMRT plan were performed using gamma analysis with 3% dose and 3 mm DTA criteria. The mean ± standard deviation of the percentage of pixels passed using gamma evaluation method for Intensity Modulated Radiotherapy plan was 97.25 ± 1.35. The QA results reveals

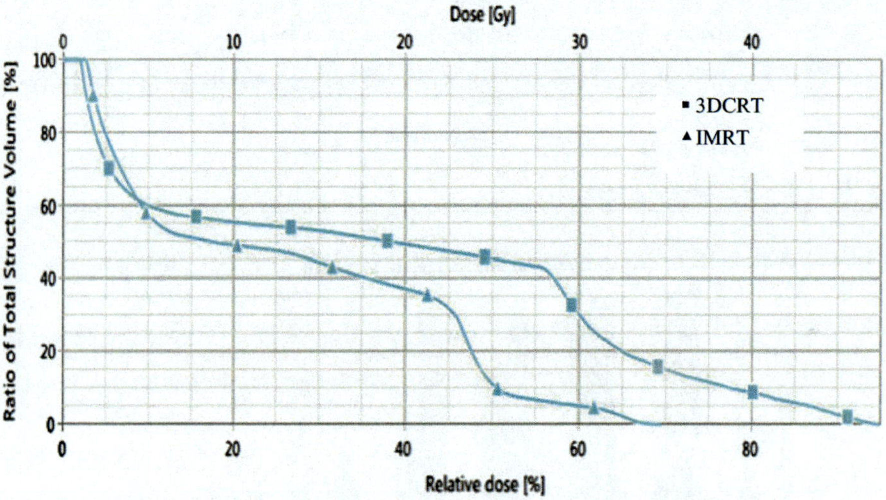

Fig. 17 DVH of Cauda Equina

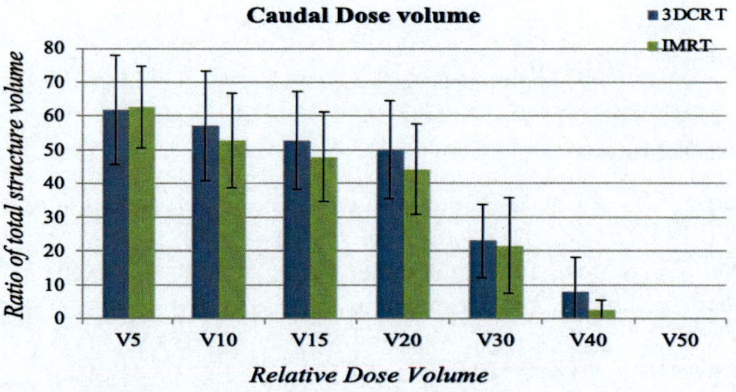

Fig. 18 Graph of Cauda Equina

that the Treatment Planning System predicated fluences and the delivery fluences were well within the clinical acceptance tolerance limits.

3 Results

Target homogeneity for the3DCRT and IMRT were comparable. Conformity Index shows that over 3DCRT low dose volumes (V_5, V_{10}, and V_{15}) were high in IMRT, but high dose regions were comparatively less in IMRT plan; the rectal mean doses were reduced by 15% compared to 3DCRT plan, and bladder mean doses were reduced by 27% compared to 3DCRT plan. Right Femoral Head, Left Femoral Head Cauda, and Body's mean dose reduce by 70%, 62%, 7%, 9% and 6% dose, respectively in IMRT plan. P-value ($p < 0.05$) considered as significant but (bladder V_{10} rectum V_{10}, V_{15}, V_{20} left and right femoral head V_{50} intestine V_{20},V_{30} and cauda V_{30},V_{40}) is higher in 3DCRT plan.

4 Discussion

This study provides a Dosimetric comparison between 3DCRT and IMRT plan in cervical cancer. My data reveals that 3DCRT Mean conformity index clearly illustrates the amount of normal tissue irradiated outside PTV. IMRT resulted in the superior mean CI of 1.148 ± 0.082423, low and high dose volume substantial dose reduction in higher dose volumes (V50, V40, and V30) was seen. But V30, V40, and V_{50} volumes for 3DCRT for all the patient were 31%, 41%, and 100% higher compared to IMRT. The mean dose (Gy) of rectum followed by standard deviation ($\bar{x} \pm \sigma\bar{x}$) for 3DCRT and IMRT was 46.485 ± 2.072 and 40.284 ± 2.409. The volumes (V5, V10, V15, and V20) are almost same, but dose reduction in higher dose volumes (V_{50}, V_{40} and V_{30}) was documented. The mean V30, V40, and V50 volumes for 3DCRT for all patients were 7%, 56%, and 290% higher compared to IMRT. The mean dose (Gy) of bladder followed by standard deviation ($\bar{x} \pm \sigma\bar{x}$) for 3DCRT and IMRT was 48.372 ± 1.866 and 35.897 ± 3.254. The volumes (V_5, V_{10}, V_{15}, and V_{20}) were almost same; dose reduction in higher dose volumes (V_{30}, V_{40}, and V_{50}) was documented. The mean V30, V40, and V50 volumes for 3DCRT for all patients were 23%, 89%, and 363% higher compared to IMRT. The mean dose (Gy) of intestine followed by standard deviation ($\bar{x} \pm \sigma\bar{x}$) for 3DCRT and IMRT was 19.224 ± 2.855 and 17.882 ± 1.962. The volumes (V_5, V_{10}, and V_{15}) were bit higher in IMRT compered to 3DCRT. Dose reduction in higher dose volumes (V_{30}, V_{40}, and V_{50}) was documented. The mean V_{30}, V_{40}, and V_{50} volumes for 3DCRT for all the patient were 10%, 48%, and 240% higher compared to IMRT Plan. The mean dose (Gy) of right femoral head followed by standard deviation ($\bar{x} \pm \sigma\bar{x}$) for 3DCRT and IMRT was 30.90 ± 3.219 and 18.118 ± 3.626. The volumes (V5) were almost same and from V10 dose Reduction started. A substantial dose reduction in volumes

(V_5, V_{10}, V_{15}, V_{20}, V_{30}, V_{40}, and V_{50}) was documented. The mean V_{10}, V_{15}, V_{20}, V_{30}, V_{40} volumes for 3DCRT for all the patients were 11%, 49%, 172%, 900%, and 3080% higher compared to IMRT. And V_{50} level there is no dose in IMRT plan. The mean dose (Gy) of left femoral head dose (Gy) followed by standard deviation ($\bar{x} \pm \sigma\bar{x}$) for 3DCRT and IMRT plan was 31.039 ± 4.049 and 19.075 ± 3.479. From V10 dose reduction started in IMRT plan. The mean V_{10}, V_{15}, V_{20}, V_{30}, V_{40} volumes for 3DCRT for all patients were 6%, 33%, 123%, 733%, and 21,055% higher compared to IMRT, V_{50} level there is no dose in IMRT plan. The mean dose (Gy) of Cauda followed by standard deviation ($\bar{x} \pm \sigma\bar{x}$) for 3DCRT and IMRT was 18.103 ± 4.017 and 16.510 ± 3.498. The volumes (V5) were bit higher in IMRT compared to 3DCRT. Dose reduction started from (V_{10}) in IMRT plan. The mean V_{10}, V_{15}, V_{20}, V_{30}, V_{40} volumes for 3DCRT for all the patients were 8%, 10%, 12%, 7%, and 222% higher compared to IMRT. And V_{50} level there is no dose in 3DCRT and IMRT plan.

5 Conclusion

IMRT shows superior OAR sparing compared to 3DCRT plans. Due to the very tight conformity of IMRT plans, it is recommended to use Image Guidance (CBCT or 2D/2D match) to ensure targets are within the PTV volume. It is not recommended to use IMRT in Carcinoma of Cervix without Image guidance. Centers don't have Image guidance provision can at least use 4-field box technique using 15MV beams to reduce the high dose irradiation volume, which may result in lesser toxicity. Still the concern is the sharp dose fall may result in geographical miss due tumor shrinkage, inter/intra fraction organ motion and volume deformation during radiotherapy.

If we do IMRT plan, then every day we have to take care about Rectum and Bladder filling. Because due to Rectum and Bladder filling, cervix is moving frequently in that condition if we do treat in IMRT technique than we might miss our PTV. Because in IMRT plan, we use very small margin. Though IMRT is having better OAR spearing but if we are unable to ensure bowel preparation and do not have daily imaging system for them better to deliver 3DCRT treatment plan. So during IMRT treatment plan delivery image guidance system can play vital role. Without image guidance, we should not deliver IMRT plan. Otherwise we might miss our target PTV.

In that condition 3DCRT is the best choice. Centers which having large amount of patient for them 3DCRT are best choice because its save huge amount of time.

References

1. Bray F, Ferlay J, Soerjomataram I et al (2018) Global cancer statistics 2018: GLOBOCAN estimates of incidence and mortality worldwide for 36 cancers in 185 countries. CA Cancer J Clin 68:394–424. https://doi.org/10.3322/caac.21492

2. Jemal A, Bray F, Center MM et al (2011) Global cancer statistics. CA Cancer J Clin 61:69–90. https://doi.org/10.3322/caac.20107
3. Amin MB, Greene FL, Edge SB, et al (2017) The eighth edition AJCC cancer staging manual: continuing to build a bridge from a population-based to a more "personalized" approach to cancer staging: the eighth edition AJCC cancer staging manual. CA Cancer J Clin 67:93–99. https://doi.org/10.3322/caac.21388
4. Lin Y, Chen K, Lu Z et al (2018) Intensity-modulated radiation therapy for definitive treatment of cervical cancer: a meta-analysis. Radiat Oncol 13:177. https://doi.org/10.1186/s13014-018-1126-7
5. Naik A, Gurjar OP, Gupta KL et al (2016) Comparison of dosimetric parameters and acute toxicity of intensity-modulated and three-dimensional radiotherapy in patients with cervix carcinoma: a randomized prospective study. Cancer/Radiothérapie 20:370–376. https://doi.org/10.1016/j.canrad.2016.05.011
6. Yang B, Zhu L, Cheng H et al (2012) Dosimetric comparison of intensity modulated radiotherapy and three-dimensional conformal radiotherapy in patients with gynecologic malignancies: a systematic review and meta-analysis. Radiat Oncol 7:197. https://doi.org/10.1186/1748-717X-7-197
7. Lan M-L, Yu X, Xiao H et al (2016) Clinical outcomes and toxicity of postoperative intensity-modulated versus three-dimensional conformal radiation therapy in patients with cervical cancer: pelvic IMRT versus 3D-CRT in cervical cancer. Asia Pac J Clin Oncol 12:430–436. https://doi.org/10.1111/ajco.12476
8. Hall EJ, Wuu C-S (2003) Radiation-induced second cancers: the impact of 3D-CRT and IMRT. Int J Radiat Oncol 56:83–88. https://doi.org/10.1016/S0360-3016(03)00073-7
9. Chargari C, Goodman KA, Diallo I et al (2016) Risk of second cancers in the era of modern radiation therapy: does the risk/benefit analysis overcome theoretical models? Cancer Metastasis Rev 35:277–288. https://doi.org/10.1007/s10555-016-9616-2
10. Isohashi F, Mabuchi S, Yoshioka Y et al (2015) Intensity-modulated radiation therapy versus three-dimensional conformal radiation therapy with concurrent nedaplatin-based chemotherapy after radical hysterectomy for uterine cervical cancer: comparison of outcomes, complications, and dose-volume histogram parameters. Radiat Oncol 10:180. https://doi.org/10.1186/s13014-015-0486-5
11. Hui B, Zhang Y, Shi F et al (2014) Association between bone marrow dosimetric parameters and acute hematologic toxicity in cervical cancer patients undergoing concurrent chemoradiotherapy: comparison of three-dimensional conformal radiotherapy and intensity-modulated radiation therapy. Int J Gynecol Cancer 24:1648–1652. https://doi.org/10.1097/IGC.0000000000000292

An Application of Graph Labeling—Cryptographic Technique with Magic Labeling

Nissankara Lakshmi Prasanna and Nagalla Sudhakar

Abstract Graph theory is an important branch of mathematics which is used to solve variety of problems in computer science by structural models. This structural modeling makes the computer discipline problems to find solution in easier way. Graph labeling is assigning additional information to the graph elements. A lot of research has been surveyed on graph labeling. Magic labeling is a sort of graph labeling. The three types of magic labeling are vertex, edge, and total magic labeling. A lot of research has been surveyed on labeling and its applications in broad. Designing applications over specific labeling algorithm is an interesting and raising domain. Very few authors addressed such applications which can be solved with specific type of labeling. One such application over magic labeling of graphs is addressed here in detail. This article presents a cryptographic technique for data encryption and decryption by using magic labeling of graphs.

Keywords Graph labeling · Magic labeling · Vertex · Edge · Total magic labeling · Cryptographic technique

1 Introduction

The theory of graphs mainly evolved with the rise of computer age. It has several uses in numerous industries, including computer technology, networks of communication, electrical networks, and social media. Graphs have been shown to be an effective mathematical tool for describing the structure of molecules. The use of graph structures can also be used to illustrate control flow.

N. L. Prasanna
Department of Computer Science and Engineering, Vasireddy Venkatadri Institute of Technology, Guntur, India
e-mail: nlakshmi@vvit.net

N. Sudhakar (✉)
Department of Computer Science and Engineering, Bapatla Engineering College, Bapatla, India
e-mail: suds_nagalla@yahoo.com

J. R. Szymanski et al. (eds.), *Energy Systems, Drives and Automations*,
Lecture Notes in Electrical Engineering 1057,
https://doi.org/10.1007/978-981-99-3691-5_39

One of them is graph labeling issues. Numerous variations of labeling have been investigated in the literature. A complete survey on contemporary results, speculation, and open problems in labeling graph is presented by Gallian [1].

The magic squares in number theory include magic labels. In the middle of the 1960s, Stewart [2] and others developed many methods to name the edges of a graph in response to an issue raised by Sedlacek. Stewart outlined the meanings of magic, super magic, and semi magic graphs. All edges of graph are labeled with integers, and each vertex produces a constant and same weight, then it is a semi magic graph. If distinct positive integers are used for labeling, then it can be termed as magic labeling [3].

If vertices considered for calculation of weight function and all vertices are producing same weight, then it can be called as vertex magic total labeling. If edges are considered for calculation of weight function and all edges are producing same weight, then it can be called as edge magic total labeling. If both are included, then it is total magic labeling [4].

Vertex magic labeling [5] is a bijective function that can be written as a mapping of numbers from 1 to n, where n is the number of vertices and edges in the given graph and where the sum of labels for each vertex and each edge incident to it results in a magic constant that is the same for all vertices. A vertex magic complete labeling example on a network of 10 graph elements, or 5 vertices and 5 edges, is shown in Fig. 1a.

Edge magic total labeling may be defined in a similar way. Edge magic total labeling [6] may be written as a bijective function by translating the integers from 1 to n, where n is the number of edges and vertices in the given graph, and the magic constant is the same for all edges. A graph with 9 graph elements, 4 vertices and 5 edges, and an example of edge magic total labeling with magic constant 12 is shown in Fig. 1b. The graph is said to have complete magic labeling [7] if it has both vertex magic total labeling and edge magic total labeling for the magic constants 12 and 9. This is illustrated in Fig. 1c.

Antimagic labeling [8] demands separate weights for each graph piece in contrast to magic labeling. Antimagic labeling of a graph is the process of giving labels to graph components in such a way that the weight of each component varies. There

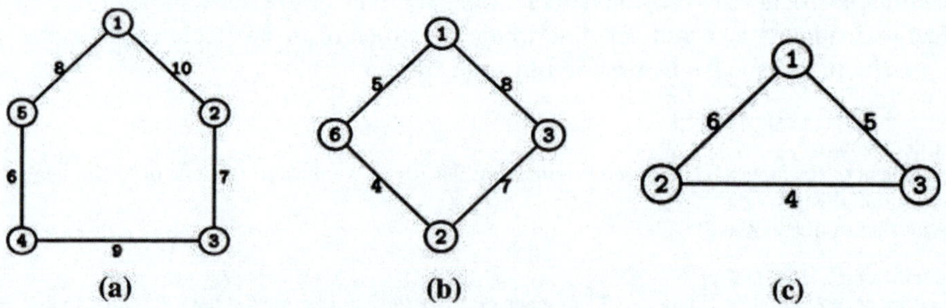

Fig. 1 Examples of VMTL, EMTL, and TML

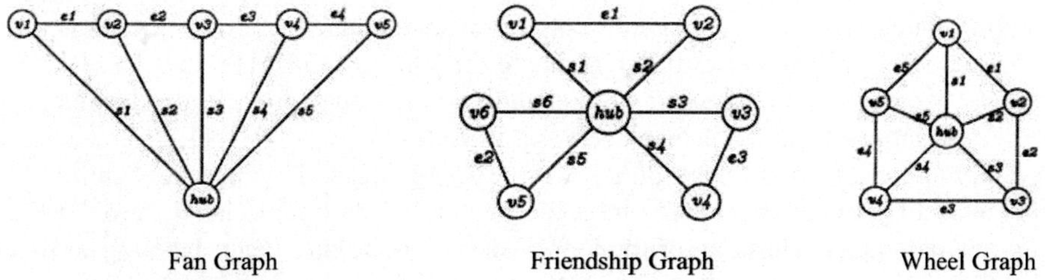

Fig. 2 Families of graph structures

are several different kinds of antimagic labeling, including edge antimagic vertex labeling [10] and vertex antimagic edge labeling [9]. In this region, a lot of work is being done.

Algorithms for graph labeling are created for certain graph architectures. Families of graph structures include those with paths, cycles, wheels, fans, and friendship graphs. Figure 2 depicts the organization of the paths, cycles, wheels, fans, and friendship graphs.

Graph labeling is used in a variety of fields, including coding theory, X-ray crystallography, radar, astronomy, communication network addressing, circuit design, and database administration, to name a few [11]. Magic/antimagic labeling is used in many applications. Unique identification of elements and automatic calculations are main features of magic labeling. For such kind of problems, magic labeling can be applied.

Cryptography is the branch of art and science of keeping messages secret by which unauthorized users cannot access that message. Cryptography [12] is used for security. For data and to restrict unauthorized access, security is needed. Especially for military communications, security is an obvious requirement. Even in many areas like business applications, banking sector and personal information of people need the security among the data. So, security plays a vital role in our day-to-day life. Cryptographers establish a secure system for reliable communication among the parties.

Several cryptographic techniques [13] are proposed, and most of them are based on ASCII values of characters. Magic labeling can be used to improve performance of these techniques which relay on ASCII values. In place of ASCII values, magic labels can be used.

2 Related Work

Jager [14] discussed the basis of VMTL, EMTL, TML, and SAT encodings for 6 combinatorial problems and discuss the labeling clauses and magic clauses. He discussed the basis of VMTL, EMTL, TML, and SAT encodings for 6 combinatorial problems. Jager [15] proposed an encoding for the combinatorial problem magic labeling. They give the SAT cryptographing for magic labeling. And they show the comparison of SAT magic and BACK magic for VMTL and EMTL.

Effective methods are shown by Baker and Sawada [16] to provide all edge magic, all vertex magic total labeling on cycles, and all vertex magic total labeling on wheels. A continuation of this study is done. A more sophisticated labeling algorithm [17] is created to provide any non-isomorphic magic and antimagic label on different families of graphs. These algorithms are used to create the magic labeling in this situation.

For a wide range of applications, including resolving conflicts in social psychology, electrical circuit theory and the energy crisis, coding theory problems, incorporating the design of good radar location codes, synch set codes, missile guidance codes, and helix codes with unmatched autocorrelation properties, and determining ambivalence in X-ray crystallographic dissection, labeled graphs are becoming an increasingly useful family of mathematical models [18]. In the discipline of computer science, designated graphs are significantly resembling the preeminent role. This evaluation demonstrated its impact on networking channels, data mining, cryptography, answering SQL queries, etc. These enormously varied uses lead us to labeled graphs.

Crystallography in particular, and the use of graph labeling in material science [19]. The application of graph labeling in the spectral enactment of material utilizing X-ray crystallography is also made clear in this research. Virtualization, distributed computing, networking, software, and web services are all components of the concept of cloud computing. In this case, the load balancing algorithm is crucial. [20]. The term "load" might relate to delays, network load, memory load, or CPU load. Load balancing is necessary for efficient resource utilization and quick response times. It is a technique for distributing the workload among distributed system nodes so that they all handle the same tasks. This prevents situations when the load is high on certain nodes, low on others, and optimal on others. Load balancing distributes the system's nodes' loads so that they are about equal at all times. In the area of graph labeling, magic labeling is a potent notion that is applied to balance load. With a graph architecture (wheel) that simulates a cloud environment, magic labeling is used, and load sharing between nodes is illustrated using a few instances.

3 Cryptographic Technique

Cryptography is the branch of art and science of keeping messages secret by which unauthorized users cannot access that message. Cryptography is used for security. For data and to restrict unauthorized access, security is needed. Personal information of individuals, corporate applications, banking, and military communications all require data security. Therefore, security is essential to our daily lives. Cryptographers establish a secure system for reliable communication among the parties.

In the present web-connected world, in transactions and identifications numbers are crucial. Some of these details, such as customer social security numbers, tax

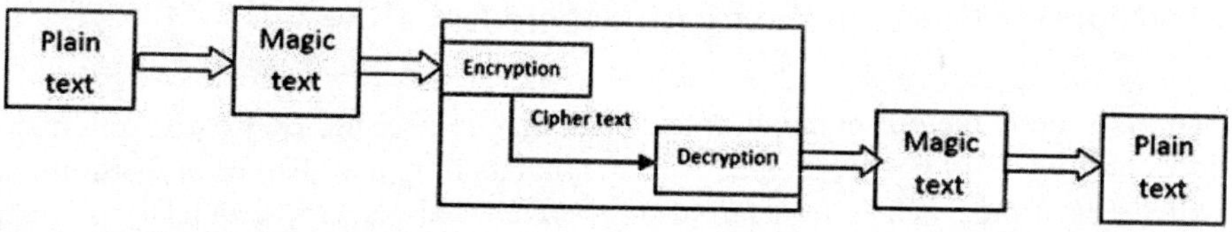

Fig. 3 Proposed encryption algorithm

file numbers, and credit card information, must be kept confidential. This is where cryptography comes into play. Cryptography is a coin with two roles sender and receiver. Sender corresponds to the encryption of the message so that its encrypted disguised form will not reveal the original message to anyone. Receiver will decrypt which inverts the encryption and restore the message. Various methods are available to encrypt numbers.

Several cryptographic techniques are proposed, and most of them are based on ASCII values of characters. Magic labeling can be used to improve performance of these techniques which relay on ASCII values. The primary goal of cryptography is to increase the level of security in message transit over a networked environment.

An innovative encryption method based on magic values is suggested. The idea is shown in Fig. 3. Plain text will be converted into magic text. Now this will be given to any encryption algorithm. Encryption algorithm produces cipher text for magic text. It is observed that the data is secured with double layers: first layer with magic text and second layer formed with encryption technique. So the data can be more secure. The decryption algorithm takes the cipher text and produces magic text from it again by using the same graph plain text can be produced.

4 Proposed System

In this scenario, any paragraph or few sentences or a sequence of words is considered as text. The text will be transferred over the network. The total process is explained in three major steps.

i. Representation of text with magic labeling.
ii. Encryption algorithm for each word by sender.
iii. Decryption algorithm for each word by receiver.

The plain text will be represented as a sequence of magic labeled graphs. Each graph will have its own structure and labeling. A startup message will be prepared based on representation. Sender prepares this and the if receiver decrypts this start up message. It can understand all graphs, their structures, and labeling. Now the second step. By using encryption algorithm, each word of text will be transferred to receiver. Finally in third step, using decryption algorithm for each word produces plain text.

5 Representation of Text with Magic Labeling

The first step is the presentation of text with magic labeling. Entire text or a major portion of text will be represented with magic labeling, and then each word of text will be encrypted and is transferred to receiver. This is very crucial step which improves the overall performance of algorithm. The effective representation leads to few requirements to magic labeling. Representation of a text with magic labeling means identification of two things. The first one is to identify a suitable structure for each word of text. Second is identifying kind of magic labeling. With these two requirements, the given text will be represented with magic labeling. The best design of these two improves the algorithm performance.

Text is composed of a sequence of words. Representation of text with magic labeling is nothing but representation of these words with magic labeling. Each word of text is represented with magic labeling. A suitable graph structures has to be identified for each word. If the same graph structure is used for all words, the word representation may not be in optimal way.

Based on the characters of word, a suitable structure will be identified. For example, all the three different letter words having one character in common will be represented using friendship graph. The word can be represented with minimum number of vertices and edges. If the word has frequent characters, it can be represented with either fan graph or wheel graph. The remaining words are observed with existing (already represented) words. If the characters match is above 50%, then it can be combined with existing graph structure. If the match is 100%, no need of new graph structure. The previous one can be used. All the remaining words can be represented with cycle graph or path graph.

6 Illustration with an Example:

Let the following be the message to be encrypted.

The Sun and the moon are shining in the sky.

The three different letter words in this example are—the, sun, and, are, the, sky, moon.

Edge magic labeling is taken for representation. After representing each word, a suitable edge magic labeling sequence is selected in random and is applied on that graph structure. All these details are used to prepare startup message which will be sent to receiver as first message. This is used to transfer details of graphs, structures, and labeling sequences. Simply the representation in a simple format is sent to the receiver. The agreement between sender and receiver is formed. In the above example, graph structures are identified. Edge magic constant and labeling sequences are randomly chosen and applied on graphs specified in Table 1.

A startup message will be prepared using above information and is sent to the receiver. The receiver decodes startup message and generates all graph structures and

Table 1 Identifying graph structures for words in text

Words	Representation
{the, are}	Structure: Friendship graph size: 2 Magic constant: 15 Vertices: 6(a) 2(r) 1(t) 3(h) 4(e)
{Sun, and}	Structure: Friendship graph size: 2 Magic constant: 15 Vertices: 2(s) 6(u) 3(t) 1(d) 4(n)
{Shining}	Structure: Wheel graph Size: 4 Magic constant: 21 Vertices: 10(s) 5(h) 4(n) 6(g) 1(i)
{in}	Structure: Wheel graph Size: 4 Magic constant: 21 Vertices: 10(s) 5(h) 4(n) 6(g) 1(i)
{moon}	Structure: Cycle graph Size: 3 Magic constant: 10 Vertices Vertices: 5(m) 3(o) 1(n)
{sky}	Structure: Cycle graph Size: 3 Magic constant: 10 Vertices: Vertices: 1(m) 5(o) 3(n)

their labeling. The startup message consists of data, the number of graphs used to represent text, the structure, and labeling sequence of each word of text and the labels which are assigned to characters of word. With this startup message, the receiver immediately prepares required graphs with structure and labeling sequence. So the entire text represented with graph labeling is transferred and decoded by receiver. Now the process of encryption and decryption can be started.

In this way, the given text has to be analyzed and is represented with magic labeling. The kind of analysis and drawing association between the characters in words improves the representation efficiency. This representation is limited to only the graph structure studied in this thesis. But this is left as an exercise for the enthusiastic researchers to give the best results.

7 Encryption Algorithm

After successful representation of the text and startup message is transferred to the receiver, the encryption process can be started. Each word of text will be encrypted using algorithm and is transferred to the receiver.

Table 2 Representing each word of text with graph structures

Word	Encrypted form
The	F1@7:1:1@11:3:1
Sun	F2@7:1:0@11:3:1
And	F2@3:2:0@8:-1:1
The	F1@7:1:1@11:3:1
Moon	C1@4:1:0@7:3:1@2:1:1
Are	F1@11:1:0@8:1:1
Shining	W1@6:1:1@6:4:0@21:3:0@14:3:1@21:3:0@11:1:1@14:1:0
In	W1@6:1:1@6:4:0@21:3:0@14:3:1@21:3:0@11:1:1@14:1:0
The	F1@7:1:1@11:3:1
Sky	F2@3:2:1@11:1:0

Each word is split into a combination of two successive characters because an edge possesses two vertices/characters. To identify an edge, two vertices/ characters are needed. So, every word is split into combination of two successive characters, and the second one will be repeated in order to maintain continuity. For example, the word "the" is split as "th@he". "@" is the delimiter used to separate edges.

The format of encrypted message is "Structureno@edge1@edge2.....". The structure can be cycle, path, wheel, fan or friendship graph. The number indicates the sequence number assigned to the graph. For example, F3 indicates third friendship graph used in startup message.

Structure, number, size, and magic constant can be decided from representation of text. Each edge needs to encrypt now. If there is a direct edge between those two characters, then it decrypted with set of characters labeled with that edge. If no such edge exists, then the path from those two characters is taken. The center label of path will be used as their edge. So each edge requires three things to encrypt. One is label associated with the edge, the path length, and 1 or 0 indicating their alphabetical order. So an edge will be represented as "label: path length: 1 or 0". In case if there is an edge directly between those characters, the path length will be 1 (Table 2).

In this way, each word of text will be encrypted and is transferred to receiver. The following is the encrypted format of words in above example. This encrypted form of message is transferred to the receiver. This kind message cannot be understood by third-party members. So the communication will be secure.

8 Decryption Algorithm

The receiver uses the decryption algorithm. The receiver receives startup message as first message. With this startup message, the receiver has the information regarding number of graphs used to represent text, each graph structure, labeling sequence, and

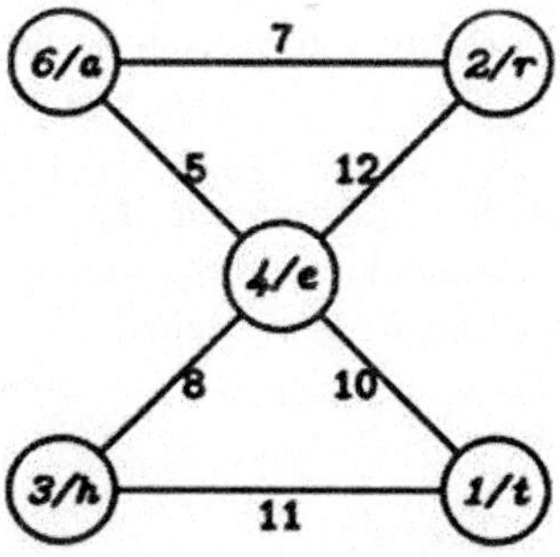

Fig. 4 Example showing encryption of word "the"

the characters on which the labels applied. Whenever the receiver receives startup message, it stores the information and uses it for decrypting each word.

After startup message, the sender sends each word in encrypted format, which has the format "Structureno@edge1@edge2.....". The structure and number used to identify the graph are used to represent the word. Now count the delimiters to find length of the word. The delimiter "@" separates each edge. Each edge label is taken and evaluated to generate text. It consists of three components separated with ":". The first component gives the edge label, second is the path length between those two characters, and finally third is alphabetical order. With this information, the vertices/characters will be identified and produce the original word.

For example, if the message received is "F1@7:1:1@11:3:0", F1 indicates the first friendship graph used for the word. Let the startup message give information of F1, then it is (Fig. 4).

Structure: Friendship graph Size: 2 Magic constant:15 Vertices: 6(a) 2(r) 1(t) 3(h) 4(e).

The first friendship graph size is 2, and magic constant used is 15. The receiver with this information generates the graph.

The first edge is 7:1:1. And 7 is edge label of two vertices (6,2) with path length 1. The characters linked are (a, r). The alphabetical order gives the text as "ar". The second edge is 12:1:0. And 11 is edge label of two vertices (4, 2) with path length 1. The characters linked are (e,r). The alphabetical order is 0, so it gives the text as "re". Finally, all edges will be combined "ar" + "re" = "are". So the word transferred is "are". In this way, the receiver decrypts each word which is sent by the sender.

9 Conclusion

The proposed algorithm produces high security to the text. So far only ASCII values are used to perform encryption/decryption. A new approach is used to represent text. Edge magic labeling sequence is used instead of ASCII values which can be in trivial format. Attacks can state the encrypted message. It will not be in understandable format. This technique can also be combined with any existing cryptographic algorithm to give better performance and high security. The time for encryption and decryption is also very less. Representation of text is the only burden in this process. So this approached method gives high security with dual-layer encryption. New cryptographic technique is designed over magic labeling which provides dual protection for secured transformation of data over networks. Some other applications on magic labeling are also studied.

References

1. Gallian J (2021) A dynamic survey of graph labeling. Electron J Comb DS6
2. Sedlacek J (1963) Problem 27. Theory of graphs and its applications (Smolenice, 1963), 163–164
3. Wallis WD (2001) Magic Graphs. Birkhauser
4. Sudhakar N (2014) Enhanced algorithms for VMTL, EMTL and TML on cycles and wheels. In: Satapathy SC et al (eds.) ICT and critical infrastructure: proceedings of the 48th annual convention of CSI—Volume II, advances in intelligent systems and computing 249, 173. https://doi.org/10.1007/978-3-319-03095-1_19, © Springer International Publishing Switzerland 2014
5. MacDougall JA, Slamin MM, Wallis WD (2002) Vertex magic total labelings of graphs. Utilitas Math 61(3–2)
6. Baskoro ET, Slamin MM, Wallis WD (2000) Edge magic total labelings. Australasian J Comb 22:177–190
7. Wallis WD (2005) Totally magic labellings and graphs. Elsevier Preprint
8. Sugeng KA (2005) Magic and Antimagic Labeling of graphs. AKCE J. Graphs Combin 4(3):261–275
9. Prasanna NL, Sudhakar N, Sravanthi K (2014) Algorithm for vertex anti-magic total labeling on various classes of graphs. Contemp Eng Sci 7(19):915–922. https://doi.org/10.12988/ces.2014.4665
10. Prasanna NL, Sudhakar N (2015) Algorithm for edge antimagic labeling for specific classes of graphs. Int J Exp Algorithms (IJEA) 5(1)
11. Prasanna NL, Sravanthi K, Sudhakar N (2014) Applications of graph labeling in major areas of computer science. Int J Res Comput Commun Technol 3(8), ISSN (Online) 278- 5841 ISSN (Print) 2320-5156.
12. Preneel B (2010) Understanding cryptography: a textbook for students and practitioners. Springer, London
13. Gupta A, Walia NK (2014) Cryptography algorithms: a review. Int J Eng Dev Res 2(2):1667–1672
14. Jager G (2010) A SAT based algorithm for magic labeling. Comb Algorithms 24th Int Workshop IWOCA
15. Jager G (2010) An effective SAT encoding for magic labeling. In: Gerold J, ager, Christian Albrechts University Kiel Computer Science Institute
16. Baker A, Sawada J (2008) Magic labeling on cycles and wheels. In: Combinatorial optimization and applications. Lecture Notes in Computer Science 5165:361–373
17. Prasanna NL, Sudhakar N (2014) Algorithms for magic labeling on graphs. J Theoret Appl Inf Technol 66(1) Issn: 1992-8645 E-Issn: 1817-3195
18. Prasanna NL, Sravanthi K, Sudhakar N (2014) Applications of graph labeling in communication networks. Oriental J Comput Sci Technol Issn: 0974-6471 vol 7(1):139–145
19. Kumar, Vats (2020) Application of graph labeling in crystallography. Mater Today Proc. https://doi.org/10.1016/j.matpr.2020.09.371
20. Prasanna N. Jyothi P (2018) Load balancing strategies in CLOUD an application of magic labeling, 1659–1664. https://doi.org/10.1109/ICCONS.2018.8663105

Architecture, Protocols and Challenges of Internet of Things: An Overview

Srishti Sabharwal and Ashima Shahi

Abstract Internet of Things (IoT) is an emerging technology which transforms the communication over Internet to Machine to Machine basis. It is spanning over leading trends like smart homes, health care, industries, etc. It is result of evolution in communication technologies and aims at enhancing the user experience using various protocols and possible applications. This paper overviews the architecture of the IoT along with the protocols which form the base of communication in IoT. Various application of IoT and challenges of it are discussed to provide insight of its future prospects.

Keywords IOT · Architecture of networks · Layers and protocol · Internet · Data mobility

1 Introduction

Internet of Things connects the physical devices so that those can interact together and deliver the best experience of service to the user. RFID was the first technology to use IoT in 1999 and with the advent of Internet, the number of technologies using IoT increased. IoT basically establishes a network of things which facilitate services in various sectors like transportation, home automation, health care, etc. [1].

IoT makes use of data acquisition techniques and networking technologies to carry out the successful implementation of network of things. Sensors, RFID are some of the data acquisition techniques that allow the smart devices to cumulate data from the surroundings and using Wireless local area networks this data is shared over IoT network and shared accordingly. Smart devices are integral part of IoT

S. Sabharwal (✉)
Electronics and Communication Engineering, Chandigarh University, Chandigarh, Punjab, India
e-mail: Srishtisabharwal17@gmail.com

A. Shahi
Apex Institute of Technology, Chandigarh University, Chandigarh, Punjab, India
e-mail: Ashimashahi.ece@cumail.in

J. R. Szymanski et al. (eds.), *Energy Systems, Drives and Automations*,
Lecture Notes in Electrical Engineering 1057,
https://doi.org/10.1007/978-981-99-3691-5_40

that employ communication protocols like Zigbee, etc. to collect and process the data [2]. IoT paves the way for other technologies and results in innovation that could revolutionize the standard of living. However, the Internet of Things needs to deal with the challenges like security, scalability, mobility which prevent it to efficiently provide the services [3]. In addition to providing Machine to Machine interaction, Internet of Things also provides the facility of interaction between human and machine in more advanced versions.

The 'things' in the IoT refers to array of physical objects including smart phones, tablets, gaming consoles. These equipment's are embedded with sensors and tags like RFID that use gateway to establish connection with Internet. These devices can connect to internet simultaneously and thereby increasing the speed of connection and interaction. The IoT calls for setting the new protocols as the basic protocols of TCP/IP are not apt to deal with the challenges that prevent the efficient use of IoT [4].

2 Architecture

Internet of Things works on layered architecture as it allows establishing connection between large numbers of devices over Internet and it would be more efficient if it will be flexible. It also requires large data capacity to handle the traffic on network due to increasing number of objects on the Internet. Based on the type of architecture, there are various models of the Internet of Things. Challenges like Quality of Service, reliability, privacy, etc. lead to provide more reliable architectures that focus on these factors. Some of the standard architectures are illustrated in Table 1.

Table 1 Different IoT architectures and their layers

Three layer architecture	SOA architecture	Middleware-based architecture		Five layer architecture
Application layer	Application layer	Application layer		Business layer
	Service composition	Middleware layer		Application layer
	Service management	Coordination layer		Service management
Network layer	Object abstraction	Backbone network layer		Object abstraction
Perception layer	Objects	Existing alone application system	Access layer	Objects
			Edge technology	

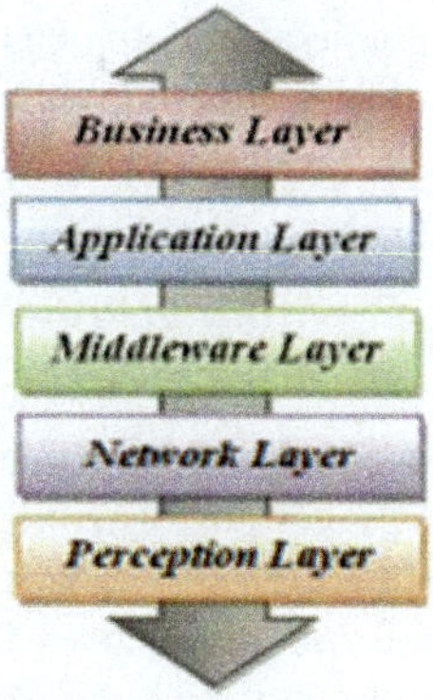

Fig. 1 Five layer architecture [5]

The generic model of IoT consists of 5 layers as shown in Fig. 1:

2.1 *Perception Layer*

Perception layer consists of sensors like RFID, Zigbee, etc. and also some other objects and elements. This layer performs same functions as physical layer of Open Systems Interconnection. The device management of the connected devices is handled by this layer which consists of gathering data from sensors and identification of the data that needs to be collected. The gathered information can be of any kind depending on the application for which the system is set. The collected data is then transferred to the network layer so that it can reach the central system for processing. This is done by perception layer by converting the data to signal or digital form [6].

2.2 *Network Layer*

Secure transfer with confidentiality of information is the responsibility of the network layer. It deploys communication channels like 3G, 4G, Wi-Fi, Infrared, satellite, etc. the choice of which depends on the type of sensors. Therefore, the main function of network layer is to send data to upper layers for further process.

2.3 *Middleware Layer*

This layer is combination of various sub layers acting as interface between different parts of the IoT. The main functions of the Middleware layers are management of service and storing the collected data in the database [7]. It acts as layer that develops a connection between the application layer and the sensors. It can also filter the data and forwards data that is required.

2.4 Application Layer

Based on the information that is filtered and processed by the Middleware layer, this layer performs the application management. The customer can request certain application services which are carried out by application layer. The application layer performs its services for various sectors and technologies like smart homes, smart health, smart transport, etc. [8].

2.5 Business Layer

The bottom four layers are supervised by the business layer. After the careful analysis of data collected from the application layer, business layer makes business models and graphs. Decision making process is supported by business layer and thus aids the managers and executives.

3 Protocols

Whenever data has to be transmitted over network, the communication needs to follow certain rules and regulations that regulate that flow of data, these sets of formats for data communication are called protocols. IoT makes use of protocols for collecting information from the sensors, communication among devices, communication between human and device, etc. [9]. The broad classification of protocol leads to four types of protocols, depicted in Table 2.

Some of the above protocols are discussed below:

3.1 CoAP: Constrained Application Protocol

The main area of application of CoAP is Machine to Machine communication. It is based on Representation State transferred called REST. It is transfer protocol for web that aids in transmitting information among the users and the servers. It is mostly used with constrained nodes/networks. The interaction model of the CoAP is like that of HTTP. There are two sub layers of CoAP: Request Response sub layer and Messaging sub layer. The first layer manages communication based on REST while the messaging layer delivers reliable communication by handling the exchange of

Table 2 Protocols used in IoT

Application protocol	Service discovery	Infrastructure protocols				Influential protocols
DDS	mDNS	Physical device layer	Link layer	Network layer	Routing protocol	IEEE 1883.3. IPSC
CoAP		LTE-A	IEEE 802.15.4	6LoWPAN	RPL	IEEE 1905.1
AMOP		EPC global		IPv4/IPv6		
MOTT	DNS-SD	IEEE 802.15.4				
XMPP		Z-Wave				
HTTP						

messaged between the client and the server. The four types of messaged supported by CoAP are: comfortable, non-comfortable, acknowledge and reset. It also allows multicast communication.

3.2 MQTT: Message Queue Telemetry Transport

It is message transfer protocol with three components: Subscriber, Publisher and Broker. Update in new data is identified by Publisher and the data is broadcasted. Then it is the duty of broker to transmit new information to the subscribers. Quality of Service makes sure that message is delivered by the broker. It executes the process with TCP/IP that orders lossless links. It is lightweight with binary format and fixed header of 2 bytes.

3.3 XMPP: Extensible Messaging and Presence Protocol

This is used for chat, video call and voice-based communication. Instant exchange of messages takes place over the Internet among the users irrespective of the operating system deployed. Authentication, encryption and such factors are taken care by XMPP in instant messaging service. Publish and subscribe communication along with response message system is supported by XMPP. Message stanzas are employed by XMPP to establish client server connection [9]. It is one of the ideal protocols for IoT as it is acknowledged all across the Internet.

3.4 AMQP: Advanced Message Queuing Protocol

This protocol is part of application layer and focuses to create environment which is message oriented. This binary protocol can provide efficient reliability and security. The two sub layers of AMQP are functional layer and transport layer. The former takes care of messaging capabilities while the latter monitors error handling, channel multiplexing. Point to point routing and Publish/Subscribe can also go along with AMQP [10]. Message and exchange queues are integral component of communication using this protocol.

3.5 DDS: Data Distribution Service

This publish/subscribe protocol is developed by Object Management Group (OMG). It is mainly designed for Machine to Machine communication and aims at multicasting with Quality of Service. The sub layers of the DDS are: Data Centric Publish Subscribe (DCPS), Data local reconstruction layer (DLRL). Transmitting the information to the subscriber is the function of DCPS and DLRL acts as interface for functions of the DCPS, it is optional layer. Most of the data transmitted by DDS is done so simultaneously and that also to the various clients/users [11].

4 Application of IoT

Applications of IoT can be distributed in various sectors, like smart home, health care, transportation, lifestyle, etc. Some of which are discussed as follows:

4.1 Smart Health Care

Merging of IoT features with health care objects will upgrade the performances of the assistance of the patients, especially the older ones who need a constant overseeing. In this case it can be required to observe temperature within refrigerator storing medications and to observe patient's health at home and in the hospital. The crisp of IoT nowadays is to invent devices which can be worn to capture the health records of patients and immediately acknowledge right to the doctors and the nurse [12].

Recognition and validation are the two words that express the initial stage of safety in computer systems applicable for health care. Recognition and validation is most often used to control, give permissions and to enhance the confidence of health care workers.

4.2 Smart Homes and Industries

In smart home automation, we have the power to command any appliance of the house through our laptops and mobile phones to communicate with each other and to facilitate us. The energy utilization and temperature can be observed with the help of smart meters.

There are various appliances which makes our home smart using Internet of Things. An appliance may notify the owner about the theft when any valuable article is displaced. Room heater may adapt itself according to the weather; room lighting can adapt to the natural sunlight and could save energy. There are various other aspects like alarming during any unfavorable time or switching of the appliance when not in use. It can also measure quality of air, temperature and ozone present [13].

4.3 Smart Transportation

Motor vehicles are provided with sensors, actuators and processors which gives useful and very important information to the driver and passengers to have knowledge about the navigation and road safety, traffic jams, parking and accidents.

4.4 Smart Agriculture and Environment

IOT could be used in to determine the moisture of the soil and the climate conditions so that the production can be enhanced. It also helps in determining the humidity and temperature required to have control over fungus. IoT can protect us and the environment from natural disasters like earthquake, etc. by measuring the environment conditions [14].

5 Challenges

The key challenges in the IoT need to be addressed to make use of IoT efficiently so that it can lead more innovations in the future. The most prominent challenges of the IoT are as follows:

5.1 Mobility

Mobile users are one of the most possible users of the IoT, which means more mobile devices to be connected on network. This becomes difficult when two mobile users in communication with each other are in different location and have different access methods [15]. However, there are various measures that can be implemented to tackle this obstacle. For, e.g., use of distributed service lifestyle management to monitor the mobility, Web of things, etc.

5.2 Reliability

The networking system is denoted as most reliable if it provides security to its users. It is must in all the layers of the IoT to ensure that the complete information is transmitted without compromising with its loss [16]. There are ways to deal with it, one of which is Adaptive Joint Protocol.

5.3 Availability

In order to get smooth experience of the IoT services, various hardware equipment's and software are required. Unavailability of which renders the whole process null. Power design tools need to be used to ensure that availability does not cause any issue in the IoT process. In addition to this various mathematical model can also be used to monitor it.

5.4 Scalability

The demand of IoT devices is continuously increasing but increasing traffic may affect the quality of service. So, this is another challenge to increase the scale of the use of IoT without compromising with the efficiency of the system [17]. To keep check on this, the existing architectures need to be optimized and may even involve introducing new architecture models with tools and space to manage a greater number of devices.

5.5 *Authentication:*

With increasing devices connected on network, the authentication of devices also needs to be managed. There is need of policies and proxy rules to identify the trusted devices and keep out those which may pose threat to the network.

6 Factors Affecting Performance

In this column, we observed features dominating the performance in IoT. Normally we study the effect of mobile gateways and numerous traffic sources on presenting energy consumption in IoT. IoTs are thought of with many communication patterns. The patterns differ being dependent on the requirements of the applications.

6.1 *Gateway Mobility Effect*

Throughput for a periodic monitoring application running on top of a UDP protocol can be observed by varying the length among sensors and gateway. Then calculate its conforming throughput.

The consequences direct that as length is increased throughput decreases. It is because due to increase in time to receive packet.

6.2 *Power Consumption*

Power consumption can be measured by using a simple battery-based energy model developed using ns-3. The planned energy model works with a battery volume of 9720Watt per second and uses around 0.35 W, 0.025 W and 0.2 W of idle, receive and transmit power, respectively.

6.3 *Impact of Multiple Applications*

Apart from using one gateway, we imagined numerous gateways can be used to question multiple sensors liable on application necessities.

Accordingly, we can have multiple number of traffic sources with diversity in packet size and data rates. Therefore, we can have multiple applications for IoT networks.

7 Conclusion

Internet of Things is revolutionizing technology that widened the horizon of communication by enabling transmission of data among devices. The data can be collected, processed and used to provide the best services to the user. Its use is not only increasing but also inspiring. The challenges arising can be dealt with numerous ways and thus IoT will continue to evolve and introduce new features and technologies. This paper overviews the architecture, protocols and challenges of IoT for those who want to dive deep into this technology and make best of it.

References

1. Atzori L, Iera A, Morabito G (2010) The internet of things: a survey. Comput Netw 54(15):2787–2805
2. Tripathi G, Singh D, Loo KK (2011) EOI: entity of interest based network fusion for future services. In International conference on hybrid information technology, pp 39–45. Springer, Berlin, Heidelberg
3. Choudhary G, Jain AK (2016) Internet of things: a survey on architecture, technologies, protocols and challenges. In: IEEE international conference on recent advances and innovations in engineering (ICRAIE-2016)
4. Florea I, Rughinis R, Ruse L, Dragomir D (2017) Survey of standardized protocols for the Internet of Things. In: 2017 21st international conference on control systems and computer science (CSCS). IEEE, pp 190–196
5. Kraijak S, Tuwanut P (2015) A survey on internet of things architecture, protocols, possible applications, security, privacy, real-world implementation and future trends. In: 2015 IEEE 16th international conference on communication technology (ICCT). IEEE, pp 26–31
6. Yang Z, Yue Y, Yang Y, Peng Y, Wang X, Liu W (2011) Study and application on the architecture and key technologies for IOT. In: 2011 international conference on multimedia technology. IEEE, pp 747–751
7. Wu M, Lu TJ, Ling FY, Sun J, Du HY (2010) Research on the architecture of Internet of Things. In: 2010 3rd international conference on advanced computer theory and engineering (ICACTE). IEEE (vol 5, pp V5–484)
8. Al-Fuqaha A, Guizani M, Mohammadi M, Aledhari M, Ayyash M (2015) Internet of things: a survey on enabling technologies, protocols, and applications. IEEE Commun Surv Tutorials 17(4):2347–2376
9. Giri A, Dutta S, Neogy S, Dahal K, Pervez Z (2017) Internet of Things (IoT) a survey on architecture, enabling technologies, applications and challenges. In: Proceedings of the 1st international conference on internet of things and machine learning, pp 1–12
10. Schneider S (2013) Understanding the protocol behind the internet of things, online electronicdesign.com, embedded, understanding-protocols-behind- internet-things
11. Elsaleh T, Gluhak A, Moessner K (2011) Service continuity for subscribers of the mobile real-world Internet. In: Proc IEEE ICC workshops, pp 1–5
12. Coetzee L, Eksteen J (2011) The internet of things—promise for the future an introduction. 1ST-Africa 2011 conference proceedings (CSIR)
13. Committtee (1987) Global system for mobile communication. A report on GSM, pp 1–2
14. Wu M, Lu TJ, Ling FY, Sun J, Du HY (2010) Research on the architecture of Internet of Things. In: Proc 3rd ICACTE, pp V5–484–V5–487
15. Khan R, Khan SU, Zaheer R, Khan S (2012) Future internet: the internet of things architecture, possible applications and key challenges. In: 10th international conference on frontiers of information technology, pp 257–260

16. Atzori L (2010) Antonio lera, Giacomo Morabito. Internet of things: a survey, ELSEIVER 54(15):2787–2805
17. Macedo D, Guedes LA, Silva I (2014) A dependability evaluation for internet of things incorporating redundancy aspects. In: Proceedings of the 11th IEEE international conference on networking, sensing and control, pp 417–422

Sensor Technology for Fuel Tank Monitoring. A Bibliometric Review

Arshdeep Singh Kalsi and Harjot Singh Gill

Abstract Sensor technology for fuel tank monitoring is still evolving. This technology has diversified applications and different types of sensors are used for the diversified applications. Scopus database was used for this review. This review of 554 documents identifies the dominant authors, funding sponsors, research institutes, future research themes, and regions. Detailed bibliometric analysis was conducted by "R" and results were visualized. The findings of this paper can be beneficial for future researchers to evaluate the themes, future publications, and collaborations. Similarly, future researchers can focus on the mentioned funding sponsors for funded projects in the future.

Keywords Fuel tank · VOS viewer · Bibliometrics · Wireless sensor · Sensor · Review · Fuel efficiency

1 Introduction

Sensor technology has diversified applications in fuel tank monitoring. Fuel tank sensing is applied in industries, automobiles, etc. Sensors have been extensively used in the automobile industry for comfort, safety, and driving efficiency. Monitoring of fuel level is essential in vehicles to perform the timely refilling of fuel and planning the drive. This will become crucial in the case of long drives, especially in the case of trucks. However, all the vehicles are now having fuel tank sensors as there is no other practical way of understanding the fuel level on a real-time basis.

There are several fuel tank monitoring technologies are being applied, distributed fluid level sensors are used in vehicles for monitoring fuel level. These are

A. S. Kalsi (✉)
Department of Mechanical Engineering, Baba Farid College of Engineering and Technology, Bathinda, India
e-mail: arshdeepkalsi@gmail.com

H. S. Gill
Department of Mechatronics Engineering, Chandigarh University, Mohali, Punjab, India

J. R. Szymanski et al. (eds.), *Energy Systems, Drives and Automations*, Lecture Notes in Electrical Engineering 1057,
https://doi.org/10.1007/978-981-99-3691-5_41

temperature-insensitive fiber Bragg grating sensing arrays for monitoring fuel tanks [1]. The ultrasonic sensors are highly effective in monitoring the CNG tanks levels [2]. Non-contact fiber optic fuel level sensors are also used for monitoring automobile fuel tanks [3]. The optic fiber grating sensors are used for monitoring the vibration of aerofil fuel tanks [4].

The H2 fuel tanks can be monitored by conformal skin sensors with safety [5]. Similarly, the anisotropic thermal expansion characteristics of lead metaniobate ceramics are applied for the low-liquid level sensors are used for fuel monitoring in space launch vehicles [6]. The fuel tank levels of aircraft are monitored by multiple microbending optical-fiber sensors [7–13]; the Hydrogen FBG sensor using Pd/Ag film is also used for monitoring fuel tanks of aircraft [14]. The IR sensors and magnetic sensors are also used for the monitoring of petrol fuel tanks in cotton spinning mills [15].

Very few bibliometric studies and systematic reviews were observed in the literature on body weight sensors in the database of Scopus. This review fills this research gap, by a detailed bibliometric analysis of body weight sensors. This paper spots out the active authors, regions, research institutes, document sources, and sponsors on research related to body weight sensors and their applications.

2 Review Methodology

This bibliometric review on fuel tank sensors is based on papers drawn from Scopus and the results are visualized. The resources for this paper were downloaded on 07/09/2021, using the Boolean "Fuel tank sensor". The research model for this paper was inspired by the works [16–25].

3 Discussion

Five hundred and fifty-four documents had been used in this paper. Five hundred and twenty-four documents were in the English language and other documents were in Chinese, German, Japanese, Korean, Russian, Slovenian, and French. The details of the document published are shown in Fig. 1. The document types are shown in Fig. 2 (Figs. 3, 4, 6, 8 and 9).

The most influential countries of this research domain were shown in Fig. 5.

The author details of this research domain were shown in Fig. 7.

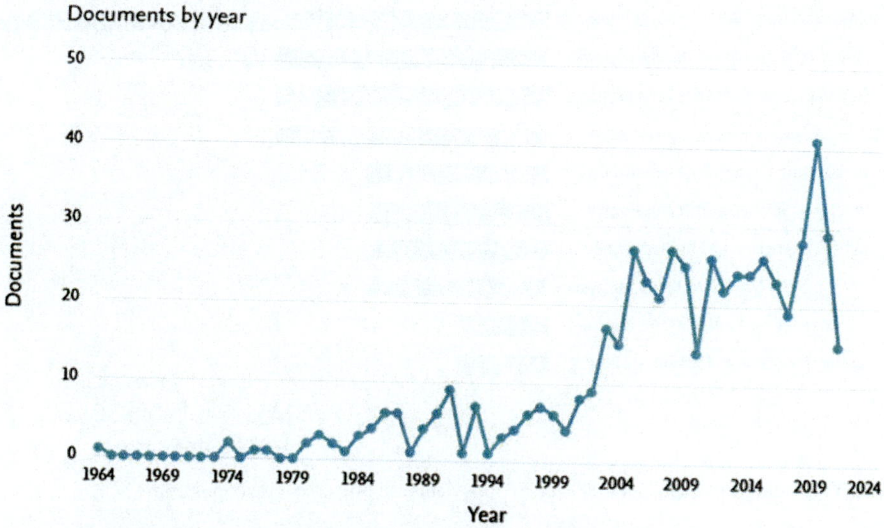

Fig. 1 Documents published on yearly basis. *Source* Scopus database https://www.scopus.com

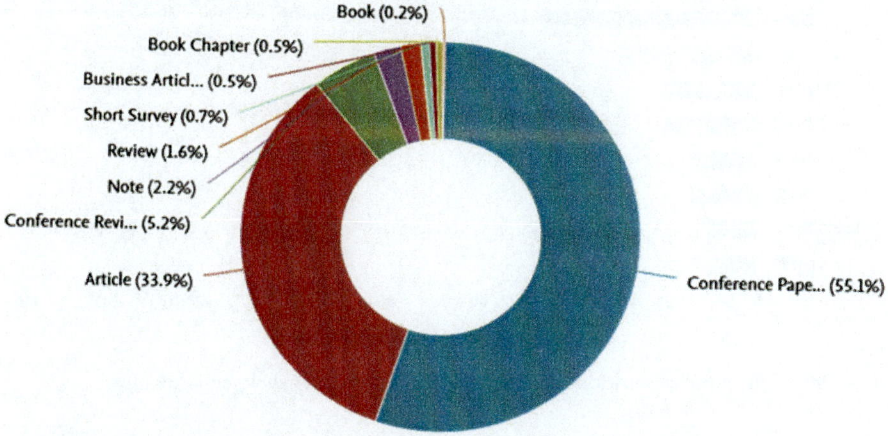

Fig. 2 Types of documents published. *Source* Scopus database https://www.scopus.com

4 Conclusions

Fuel tank sensors are used in industries, automobiles, and aircraft. The various types of sensors used are the distributed fluid level sensors [26]. Ultrasonic sensors [2]. Non-contact fiber optic fuel level sensor [3]. The optic fiber grating sensors are used for monitoring the vibration of aerofil fuel tanks [4]. Fuel tank levels of aircraft are monitored by multiple microbending optical-fiber sensors [7–13]. The Hydrogen FBG sensor [14]. The IR sensors and the magnetic sensors [15].

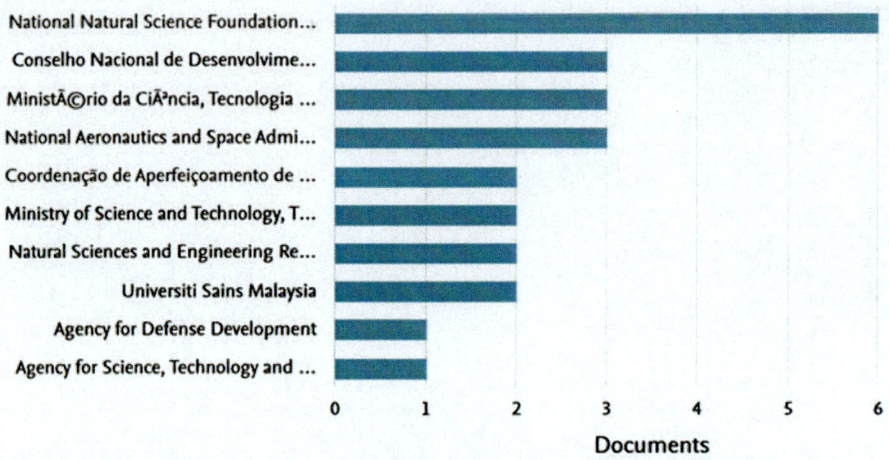

Fig. 3 Leading fund sponsors. *Source* Scopus database https://www.scopus.com

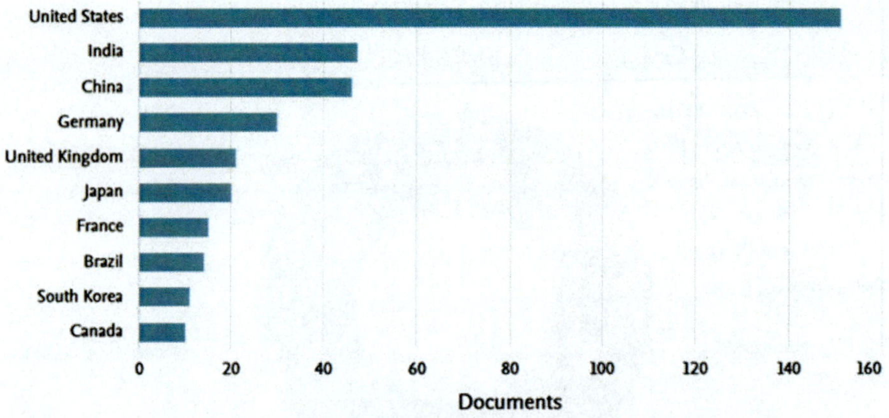

Fig. 4 Most influential countries. *Source* Scopus database https://www.scopus.com

The research domain of sensor technology for fuel tank monitoring, especially for complicated applications are evolving. This fact is evident from the annual publication data in Fig. 1. The majority of the documents are articles and conference proceedings. National Natural Science Foundation of China is the leading fund sponsor of the research domain with 6 funded documents. The leading affiliations connected with the research on the fuel tank sensors are Boeing NASA Marshall Space Flight Center Corporation and Beihang University. The leading countries engaged in the research related to fuel tank sensors are the United States of America, India, and China. The United States of America had strong collaborations with India and Australia. But the lack of collaboration between countries was observed in this research domain.

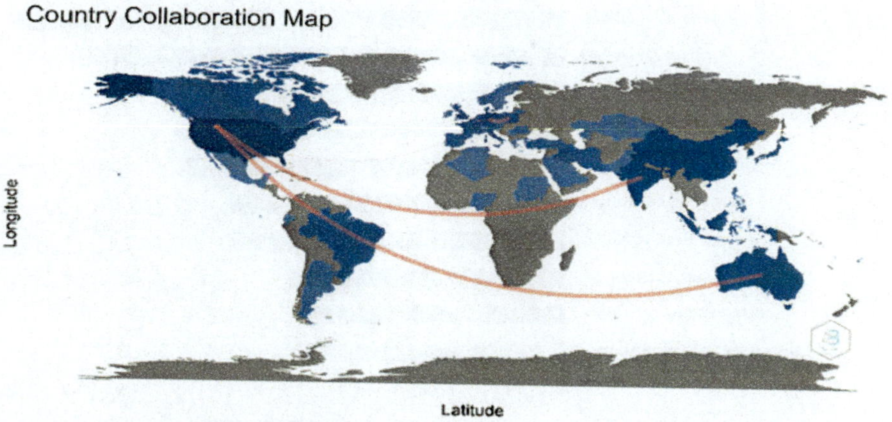

Fig. 5 Research collaboration of top countries. *Source* Scopus database https://www.scopus.com

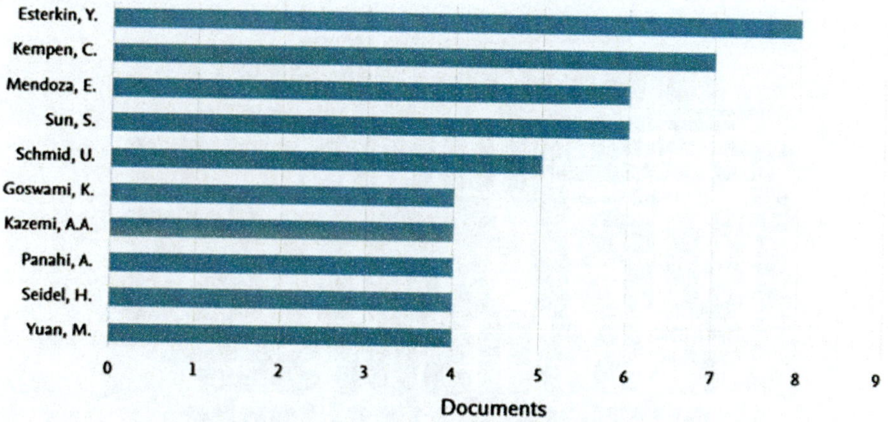

Fig. 6 Most influential authors. *Source* Scopus database https://www.scopus.com

The leading keywords that occurred in the related research on fuel tank sensors are fuels, fuel tanks, sensors, and automobile fuel tanks. The leading authors of this research domain are Esterkin Y, Kempen C, Mendoza E, and Sun S. The leading document sources of this domain are SAE Technical Papers (57 documents), and Proceeding of SPIE (43 documents). The driving themes appearing on the top right corners are sensors, hydrogen fuel cells, fuel storage, digital storage monitoring. The marginal themes are fiber optic sensors, optical fibers, and fiber optics. The emerging or disappearing themes are tanks, aircraft, aircraft fuel tanks, liquids, and level measurement. The generic and transversal themes are Fuel tanks, fuels, and automobile fuel tanks. Future research can be on new materials for fuel tank sensors and new methods of fuel tank sensors.

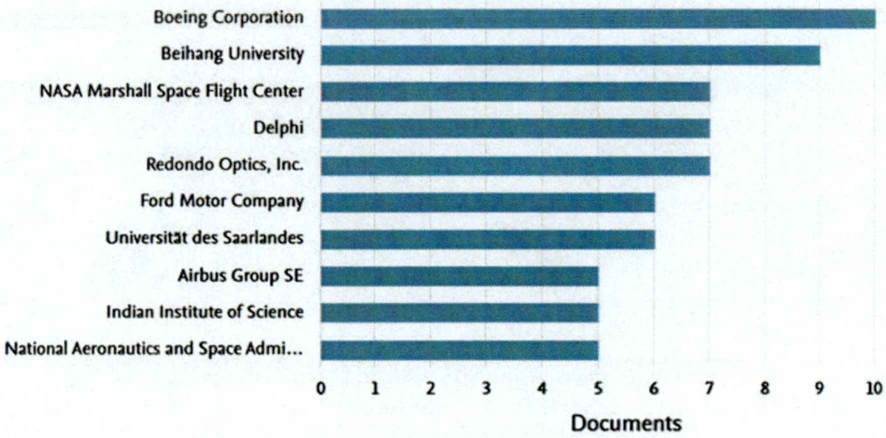

Fig. 7 Most influential research institutions. *Source* Scopus database https://www.scopus.com

Fig. 8 Word cloud

The United States of America and China dominate in funding research and the number of active authors of this research domain is also from these countries. Future researchers and academicians can concentrate on these domains for future publications and collaborations. The leading journals can be contacted for future publications.

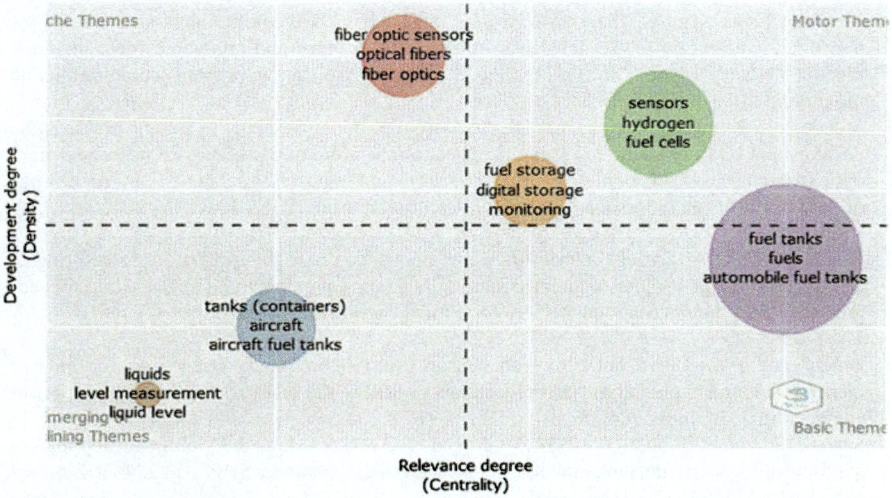

Fig. 9 Keyword analysis

References

1. Arkwright JW, Parkinson LA, Papageorgiou AW (2018) A distributed fluid level sensor suitable for monitoring fuel load on board a moving fuel tank. https://doi.org/10.1117/12.2288050
2. Kim N, Choi D (2020) Analysis according to the change in structure of ultrasonic sensors and the change in pressure inside the tank for measurement of fuel amounts in compressed natural gas (CNG) tank [Analyse der Strukturänderungen von Ultraschallsensoren und Druckänderung i. Materwiss. Werksttech. 51(5):645–653. https://doi.org/10.1002/mawe.201900227
3. Laddha RS, Buchade PB, Shaligram AD (2021) Development of non-contact fiber-optic fuel level sensor for automobile fuel tank. In: 4TH International conference on emerging technologies; MICRO TO NANO (ETMN 2019) 2335. https://doi.org/10.1063/5.0043582
4. Lu G, Liang D-K (2012) Vibration inspection of aerofoil fuel tank with composite material surface based on optic fiber grating sensors. Zhendong yu Chongji/Journal Vib. Shock 31(4):34–37+66
5. Dong S et al (2004) Conformal sensor skin approach to the safety-monitoring of H 2 fuel tanks. Appl Phys Lett 84(21):4153–4154. https://doi.org/10.1063/1.1753651
6. Ibrahim AM, Cozzo TF, Pilgrim SM, Djordjevic BB (1991) Anisotropic thermal expansion characteristics of lead metaniobate ceramics used in the low-liquid level sensors for fuel tanks of space launch vehicles. Int J Thermophys 12(4):751–755. https://doi.org/10.1007/BF00534230
7. Knowles SF, Jones BE, Purdy S, France CM (1998) Multiple microbending optical-fibre sensors for measurement of fuel quantity in aircraft fuel tanks. Sensors Actuators A Phys 68(1–3):320–323. https://doi.org/10.1016/S0924-4247(98)00030-2
8. Mendoza EA, Esterkin Y, Kempen C, Sun, Susko K, Goglia J (2011) Advances towards the qualification of an aircraft fuel tank inert environment fiber optic oxygen sensor system. In: Proceedings of SPIE—the international society for optical engineering 8026. https://doi.org/10.1117/12.887128
9. Mendoza EA, Kempen C, Sun S, Esterkin Y (2014) Highly distributed multi-point, temperature and pressure compensated, fiber optic oxygen sensors (FOxSense^{TM}) for aircraft fuel tank environment and safety monitoring. In: Proceedings of SPIE—the international society for optical engineering 9202. https://doi.org/10.1117/12.2064482

10. Mendoza E, Kempen C, Esterkin Y, Sun S, Susko K (2010) Multipoint fiber optic sensor network for detection of oxygen leaks in aircraft fuel tanks and cryogenic fuel tanks and pipeline delivery systems. In: Proceedings of the 5th European workshop—structural health monitoring 2010, pp 627–632
11. Mendoza E, Kempen C, Esterkin Y, Sun S, Susko K, Goglia J (2012) Progress towards the development and qualification of an all optical temperature and pressure compensated fiber optic oxygen sensor for monitoring ullage environment in aircraft fuel tanks. In: Proceedings of the 6th European workshop—structural health monitoring 2012, EWSHM 2012 vol 2, pp 1583–1588
12. Panahi A (2011) All optical O2 sensors using innovative phase fluorimetry for monitoring of headspace in ullage for FAA mandated inerting fuel tanks of commercial airlines. In: Proceedings of SPIE—the international society for optical engineering 8026. https://doi.org/10.1117/12.886909
13. Panahi A (2009) Fiber optic oxygen sensor using fluorescence quenching for aircraft inerting fuel tank applications. In: Proceedings of SPIE—the international society for optical engineering 7314. https://doi.org/10.1117/12.821732
14. Saad S, Hassine L, Elfahem W (2014) Hydrogen FBG sensor using Pd/Ag film with application in propulsion system fuel tank model of aerospace vehicle. Photonic Sens 4(3):254–264. https://doi.org/10.1007/s13320-014-0164-5
15. Lakshmi MLSNS, Reddy C (2020) Measurement of fuel level in tank using IR sensors and reporting over IoT. Lect Notes Electr Eng 626:751–756. https://doi.org/10.1007/978-981-15-2256-7_69
16. Ray R (2021) An advanced review on visualisation of research trends in bacterial wilt. Plant Cell Biotechnol Mol Biol 22(15–16):124–132
17. Sharma A (2021) A bibliometric analysis and visualisation of research trends in cardiac pacemaker battery. Ann Rom Soc Cell Biol 25(1):3697–3704
18. Singh J (2021) A bibliometric analysis and visualisation of research trends in corrosion of implants. Ann Rom Soc Cell Biol 25(1):3512–3520
19. Kathuria D (2021) A bibliometric analysis and visualisation of research trends in leadless pacemaker. Ann Rom Soc Cell Biol 25(1):3649–3657
20. Singh H (2021) A bibliometric analysis and visualisation of research trends in implants and allergy. Ann Rom Soc Cell Biol 25(1):3640–3648
21. Mittal N (2020) Energy efficient strategy for prolonged lifetime in wireless sensor networks. J Green Eng. 10(11):11644–11664
22. Sharma A, Kakkar A (2020) A review on solar forecasting and power management approaches for energy-harvesting wireless sensor networks. Int J Commun Syst 33(8). https://doi.org/10.1002/dac.4366
23. Singh G, Virpal, Singh RC (2019) Highly sensitive gas sensor based on Er-doped SnO2 nanostructures and its temperature dependent selectivity towards hydrogen and ethanol. Sens Actuators B Chem vol 282, pp 373–383. https://doi.org/10.1016/j.snb.2018.11.086
24. Deshwal M, Arora A (2019) Annealing temperature optimization for highly sensitive ZnO based acetone gas sensor. Indian J Pure Appl Phys 57(6):422–426
25. Srivastava S (2020) Study of gas sensor detection for NOX gas: a review. In: Materials today: proceedings vol 37, no Part 2, pp 3709–3712. https://doi.org/10.1016/j.matpr.2020.10.161
26. Arkwright JW, Parkinson LA, Papageorgiou AW (2018) A distributed fluid level sensor suitable for monitoring fuel load on board a moving fuel tank. In: Proceedings of SPIE—the international society for optical engineering vol 10539. https://doi.org/10.1117/12.2288050

Application of Blockchain Technology for Defending the Pandemic: A Bibliometric Analysis

Rana Gill

Abstract The COVID-19 pandemic had disrupted the health system in the world, and modern technologies are essential for drug delivery, monitoring, sanitation, regulating access controls, and for many more initiatives. Blockchain technology plays a significant role in the development of vibrant health and preventive system against COVID-19. India and the USA lead in the research related to the application of blockchain technology. The leading research organizations and funding agencies are from the Middle East and China. The future researchers can concentrate on these leading institutions for sponsorships and collaborations. Future research can be on the innovative applications of blockchain technology in vaccine delivery, controlling, and education of people.

Keywords Blockchain · VOSviewer · Bibliometrics · Pandemic · Data · COVID-19

1 Introduction

The world is still in the clutches of the pandemic, COVID-19. Blockchain technology offers several supports to handle the crisis related to the pandemic [1]. Blockchain technology improves the quality of healthcare services in critical situations like in the case of pandemics [4]. One such support is the application of blockchain technology into the supply chain solutions of testing kits. The application of blockchain technology toward supply chain management of testing kits improves the visibility of the history of kits, avoids faulty kits, and enhances accuracy in the testing of samples [5]. Consortium blockchain technology can be used for contact tracing purposes [6–10], including the digital contact tracing apps based on blockchain technology [11] than allotting the work to a single person. In addition to contact tracing, the blockchain technology can be applied to public place access control and identifying safe places

R. Gill (✉)
Department of CSE-AIT, Chandigarh University, Mohali, Punjab, India
e-mail: rana.cse@cumail.com

J. R. Szymanski et al. (eds.), *Energy Systems, Drives and Automations*,
Lecture Notes in Electrical Engineering 1057,
https://doi.org/10.1007/978-981-99-3691-5_42

[6]. Blockchain technology can also be applied for multi-robot collaborations, for the works associated with spraying, disinfection, cleaning, treating, detecting high body temperature, mask absence, and delivering goods and medical supplies [12]. Similarly, the multi-drones can be used for monitoring and detecting, social distancing, sanitization, data analysis, delivering goods and medical supplies, and announcements [13]. The health information exchange is very critical in facing the pandemics like COVID-19, where healthcare information is to be transferred, even its sensitive data. Blockchain technology can be used for the decentralized and secured transfer of health information by protecting private and sensitive information [14].

Another application of blockchain technology in the post-pandemic world can be the smart contracts in the tourism industry [15]: application of blockchain technology for the agriculture food sector in post-pandemics [16]; for development of healthcare system to face the pandemic by blending the potential of blockchain technology and artificial intelligence [17, 18]; and for improvement of public distribution during pandemics [19].

There are several detailed reviews related to the application of blockchain technology during pandemics including systematic reviews. This paper addresses the research gap of a detailed bibliometric analysis on the application of blockchain technology for facing pandemics. This consolidates the literature on blockchain technology and pandemics in the Scopus database by using bibliometric analysis. The major parameters analyzed in this paper are the leading authors, funding organizations, research organizations, authors, and document sources. This structure of this paper was tailored with the sections of introduction, methodology of review, discussion, and conclusion.

2 Review Methodology

This paper adopted the review model from works [20–24]. Bibliometric analysis on the application of blockchain technology in the fight against pandemics was conducted in this paper. Boolean "Blockchain pandemic" is used from the database of Scopus for this review. The database was accessed on 31/08/2021.

3 Discussion

One hundred and ninety-eight documents from Scopus were bibliometrically analyzed in this paper. The various types of documents are shown in Fig. 1, and the details of the annual scientific production of this research segment are shown in Fig. 2. These documents were in English, Spanish, and Russian languages.

The leading countries of this research domain are shown in Fig. 3.

The top funding institutions of this research domain were shown in Fig. 5.

The details of the leading authors were shown in Fig. 5.

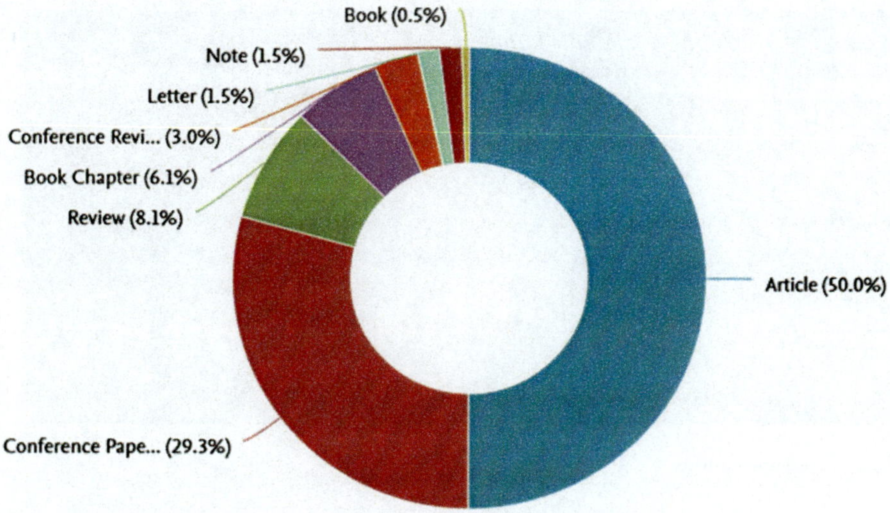

Fig. 1 Types of documents

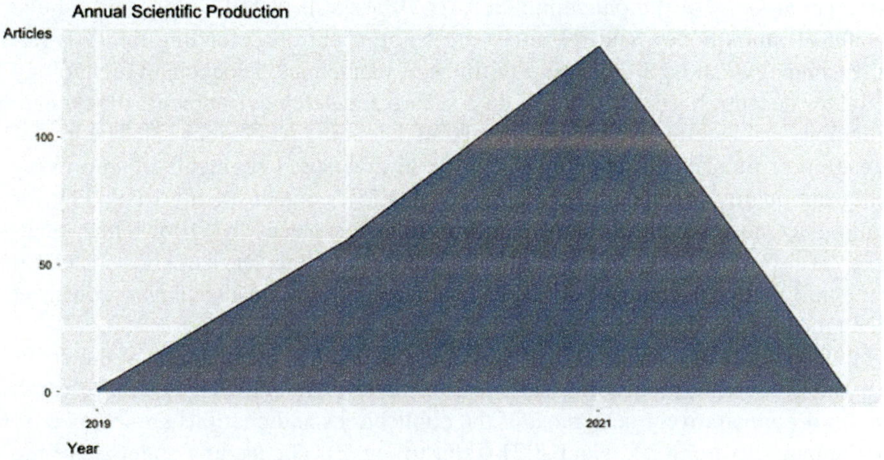

Fig. 2 Documents on an annual basis

The details of keywords were shown in Fig. 7.

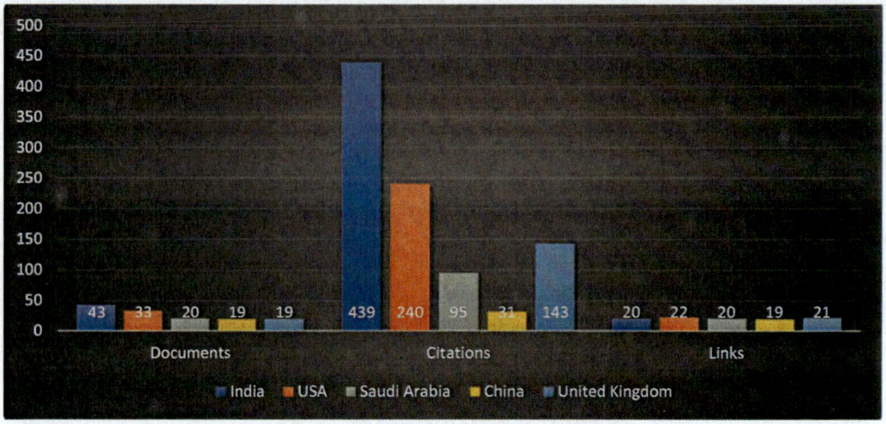

Fig. 3 Leading countries

4 Conclusions

The repercussions of the pandemic COVID-19 are still in full swing, and a huge amount of time, money, and resources are being spent on defending the COVID-19. Technology can be a great help in times of pandemics. Blockchain technology facilitates decentralized data sharing in a secure and safe environment. Blockchain technology can be effectively implemented in defending pandemics [2]. The implementation of blockchain technology can be in diversified segments of improving healthcare quality [4]; healthcare supply chains [5]; contact tracing [6–9, 11]; for public place access control; and identifying the safe places [6]. Blockchain technology is also applied for warning, educating, and sanitations services [12, 13] and for health information exchange [14]. Blockchain technology can also play a significant role in the post-pandemic eras [15–19].

In this bibliometric analysis of 198 Scopus documents, the majority of the documents are conference proceedings, articles, and reviews (refer to Fig. 1). This research domain is comparatively new, most of the publications and citations are from 2019, and the majority are in 2020 and 2021 (refer to Fig. 2). The leading countries of the research domain are shown in Figs. 3 and 4, with the research collaboration details. India leads the research on the application of blockchain technology for defending pandemics with 43 documents, 439 citations. However, the USA and the UK lead in collaborations. The USA follows India in the document publications and citations. The other leading countries of the research domain are Saudi Arabia and China. The leading research organizations are Khalifa University of Science and Technology, United Arab Emirates (4 publications), University of Petroleum and Energy Studies, India (5 documents), King Abdulaziz University, and the King Saud University of Saudi Arabia with 5 and 6 documents, respectively. The leading fund sponsors for the research in the application of blockchain technology for defending against the pandemic are the National Natural Science Foundation of China (8 funded documents) and Deanship of Scientific Research, King Saud University, Saudi Arabia (5 funded documents) (Figs. 5, 6, 7, 8, and 9).

The leading authors of this research domain are Muhammad G, Jayaram R, Salah K, and Kumar. All these authors are with four document publications and Muhammad G in citation and followed by Jayaraman R. The most influensive source of documents is IEEE Access (leads in document publications and citations), followed by IEEE Network. The other leading document sources are Studies in Computational Intelligence, Sustainability, and IEEE Internet of Things Journal.

Future researchers can concentrate on the innovative applications of blockchain technology for defending pandemics. The research can be on medicine delivery,

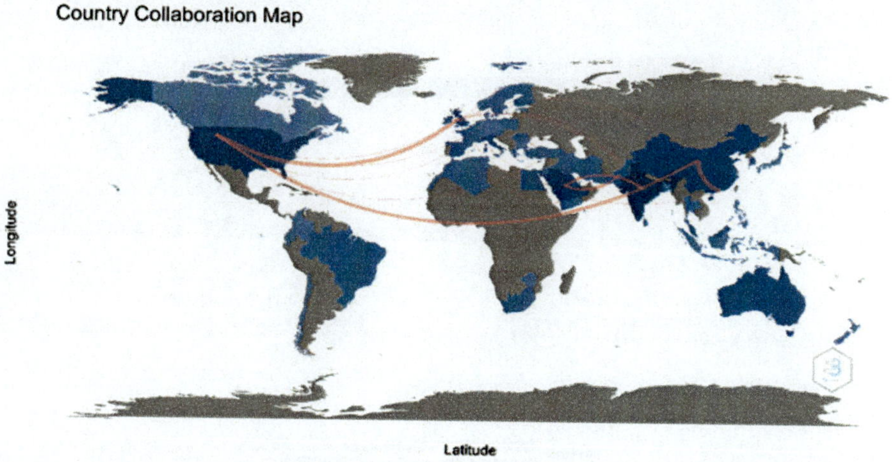

Fig. 4 Research networks of countries

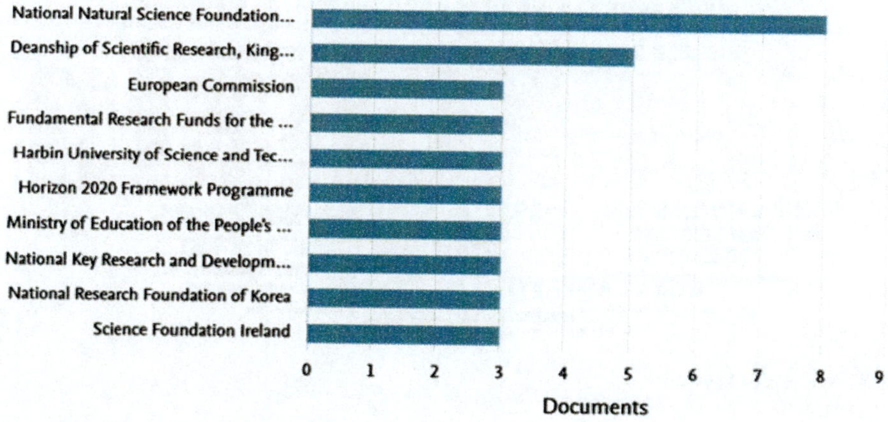

Fig. 5 Top funding agencies. *Source* www.scopus.com

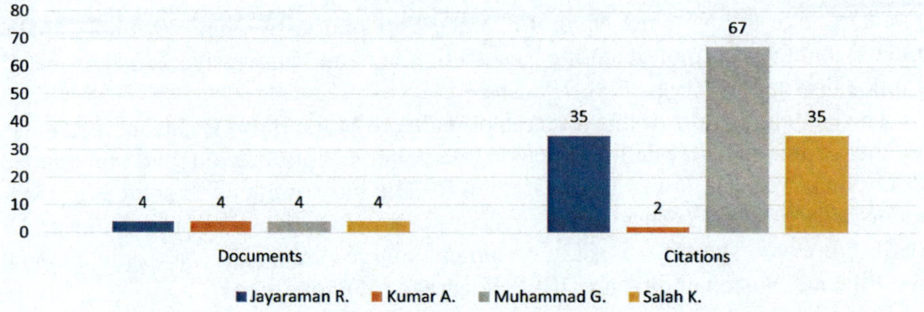

Fig. 6 Active authors

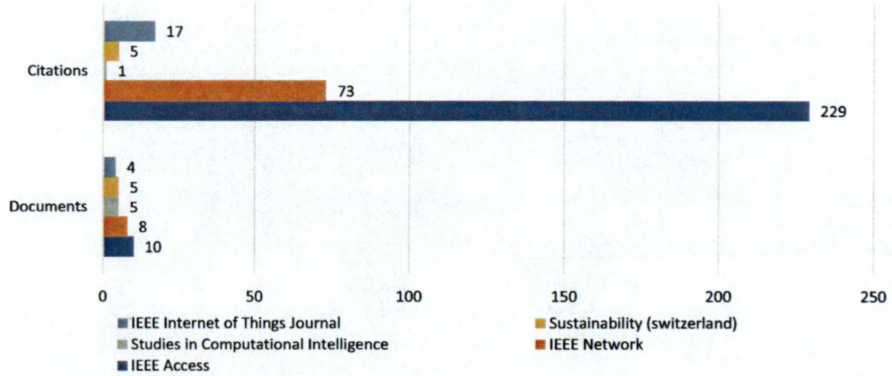

Fig. 7 Leading document sources

Fig. 8 Analysis of keywords

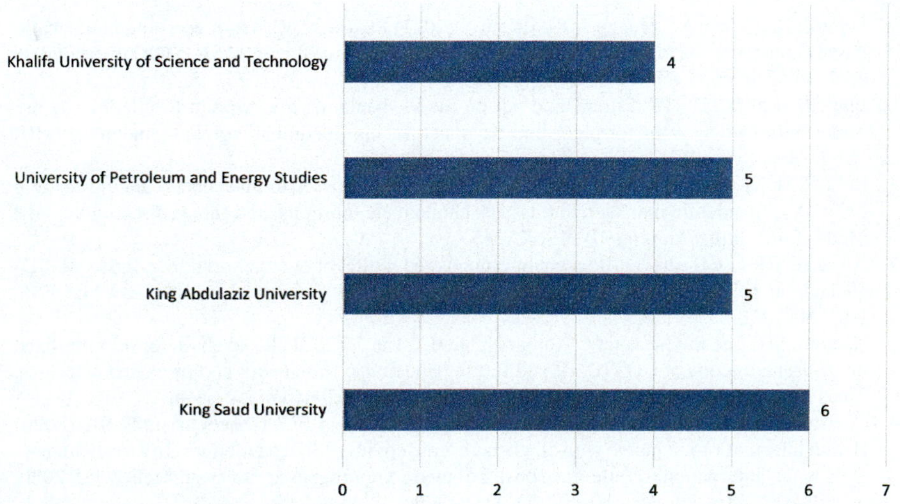

Fig. 9 Top organizations

education, vaccination, and monitoring. This paper can be useful for finding the leading fund sponsors, authors, and research institutions. These research leaders can be approached for funding and research collaborations. The influential journals can be targeted for future research publications and knowledge enhancement.

References

1. Chang MC, Park D (2020) How can blockchain help people in the event of pandemics such as the COVID-19? J Med Syst 44(5). https://doi.org/10.1007/s10916-020-01577-8
2. Chang MC, Park D (2020) How can blockchain help people in the event of pandemics such as the COVID-19? J Med Syst 44(5). https://doi.org/10.1007/s10916-020-01577-8
3. Khubrani MM, Alam S (2021) A detailed review of blockchain-based applications for protection against pandemic like COVID-19. Telkomnika Telecommun Comput Electron Control 19(4):1185–1196. https://doi.org/10.12928/TELKOMNIKA.v19i4.18465
4. Alanazi AMM (2020) How blockchain helps to enhance healthcare quality: potential impact on crisis situations like covid-19 pandemic. Int J Qual Res 14(4):1059–1080. https://doi.org/10.24874/IJQR14.04-05
5. Gupta M, Kumar V, Yadav V, Singh RK, Sadim M (2021) Proposed framework for dealing COVID-19 pandemic using blockchain technology. J Sci Ind Res 80(3):270–275
6. Alansari SA, Badr MM, Mahmoud M, Alasmary W, Alsolami F, Ali AM (2021) Efficient and privacy-preserving infection control system for covid-19-like pandemics using blockchain. IEEE Internet Things J. https://doi.org/10.1109/JIOT.2021.3092601
7. Aslam B, Javed AR, Chakraborty C, Nebhen J, Raqib S, Rizwan M (2021) Blockchain and ANFIS empowered IoMT application for privacy preserved contact tracing in COVID-19 pandemic. Pers Ubiquitous Comput. https://doi.org/10.1007/s00779-021-01596-3
8. Moraes TKL, Cernev AK, Diniz EH (2021) Contact tracing: achieving equilibrium between blockchain solutions and privacy amid the novel coronavirus pandemic. J Inf Technol Teach Cases. https://doi.org/10.1177/20438869211028869

9. Peng Z, Xu C, Wang J, Huang J, Xu J, Chu X (2021) PB-trace: privacy-preserving blockchain-based contact tracing to combat pandemics. In: Proceedings of the ACM SIGMOD international conference on management of data, pp 2389–2393 https://doi.org/10.1145/3448016.3459237
10. Platt M et al (2021) Test, trace, and put on the blockchain?: a viewpoint evaluating the use of decentralized systems for algorithmic contact tracing to combat a global pandemic. JMIR Public Heal Surveill 7(4). https://doi.org/10.2196/26460
11. Idrees SM, Nowostawski M, Jameel R (2021) Blockchain-based digital contact tracing apps for COVID-19 pandemic management: issues, challenges, solutions, and future directions. JMIR Med Inf 9(2) https://doi.org/10.2196/25245
12. Alsamhi SH, Lee B (2021) Blockchain-empowered multi-robot collaboration to fight COVID-19 and future pandemics. IEEE Access 9:44173–44197. https://doi.org/10.1109/ACCESS.2020.3032450
13. Alsamhi SH, Lee B, Guizani M, Kumar N, Qiao Y, Liu X (2021) Blockchain for decentralized multi-drone to combat COVID-19 and future pandemics: framework and proposed solutions. Trans Emerg Telecommun Technol. https://doi.org/10.1002/ett.4255
14. Christodoulou K, Christodoulou P, Zinonos Z, Carayannis EG, Chatzichristofis SA (2020) Health information exchange with blockchain amid covid-19-like pandemics. In Proceedings—16th annual international conference on distributed computing in sensor systems, DCOSS 2020, pp 412–417. https://doi.org/10.1109/DCOSS49796.2020.00071
15. Demirel E, Zeren SK, Hakan K (2021) Smart contracts in tourism industry: a model with blockchain integration for post pandemic economy. Curr Issues Tour. https://doi.org/10.1080/13683500.2021.1960280
16. Enescu FM, Ionescu VM (2020) Using Blockchain in the agri-food sector following SARS-CoV-2 pandemic. https://doi.org/10.1109/ECAI50035.2020.9223161
17. Jabarulla MY, Lee H-N (2021) A blockchain and artificial intelligence-based, patient-centric healthcare system for combating the covid-19 pandemic: opportunities and applications. Healthc 9(8). https://doi.org/10.3390/healthcare9081019
18. Islam A, Rahim T, Masuduzzaman M, Shin SY (2021) A blockchain-based artificial intelligence-empowered contagious pandemic situation supervision scheme using internet of drone things. IEEE Wirel Commun. https://doi.org/10.1109/MWC.001.2000429
19. Kumar A (2021) Improvement of public distribution system efficiency applying blockchain technology during pandemic outbreak (COVID-19). J Humanit Logist Supply Chain Manag 11(1):1–28. https://doi.org/10.1108/JHLSCM-06-2020-0050
20. Ray R (2021) An advanced review on visualisation of research trends in bacterial wilt. Plant Cell Biotechnol Mol Biol 22(15–16):124–132
21. Sharma A (2021) A bibliometric analysis and visualisation of research trends in cardiac pacemaker battery. Ann Rom Soc Cell Biol 25(1):3697–3704
22. Singh J (2021) A bibliometric analysis and visualisation of research trends in corrosion of implants. Ann Rom Soc Cell Biol 25(1):3512–3520
23. Kathuria D (2021) A bibliometric analysis and visualisation of research trends in leadless pacemaker. Ann Rom Soc Cell Biol 25(1):3649–3657
24. Singh H (2021) A bibliometric analysis and visualisation of research trends in implants and allergy. Ann Rom Soc Cell Biol 25(1):3640–3648

Multi-scale Multi-branch 3D-CNN for the Reduction of False Positives in the Detection of Pulmonary Nodules from the Chest CT Scans

V. N. Sukanya Doddavarapu, Giri Babu Kande, and NageshBabu Dasari

Abstract Early diagnosis of lung cancer is essential to increase the survival rate of the patients in computed tomography (CT) images. This can be achieved by reducing the false positive rate of lung nodules during detection. It is difficult to distinguish the true nodules from the several candidate nodules due to the heterogeneity of nodules and their similarity with the background. Also, the performance of the model is dependent on the size of the nodule. If the size of the input sample is too small, the contextual information of the nodule is missed, and if the size is too large, redundant data is added. To overcome these challenges, a multi-scale multi-branch 3D CNN is designed in this paper. The inputs are divided into three different scales with three different branches, and final probability is obtained by fusion the output of the three branches. To evaluate the proposed model two datasets: LIDC-IDRI dataset for internal validation and Japanese dataset for external validation are used. The proposed model gives a competition performance metric (CPM) of 94.6% (95% CI: 88.7–98.6%) for internal validation on 89 CT scans from the LIDC-IDRI dataset and 92.8% (95% CI: 82.1–89.6%) for external validation on 450 chest CT scans from the Japanese dataset.

Keywords Lung cancer · Computer-aided diagnosis (CAD) technology · Computed tomography (CT) · Lung nodule detection · Deep learning

V. N. S. Doddavarapu (✉)
Department of ECE, St. Ann's College of Engineering and Technology, Chirala, India
e-mail: sukanyadvn@gmail.com

G. B. Kande
ECE Department, VVIT, Nambur, Andrapradesh 522509, India

N. Dasari
Department of CSE, St. Ann's College of Engineering and Technology, Chirala, India

J. R. Szymanski et al. (eds.), *Energy Systems, Drives and Automations*,
Lecture Notes in Electrical Engineering 1057,
https://doi.org/10.1007/978-981-99-3691-5_43

1 Introduction

In the world, the deaths triggered by lung cancer are dominant compared to other cancers [1]. The abnormal cells developed in the lung cancer turn into tumor, if they are not predicted in the early stages. The signs of lung cancer are not apparent for early prediction resulting treatment tough and expensive. Nowadays, computed tomography (CT) screening is widely used for the analysis and diagnosis of lung cancer [2]. According to the National Lung Screening Trial Research team, the mortality of lung cancer is reduced when CT scanning is used for diagnosis. When measured on the chest, a pulmonary nodule, also known as rounded opacity, can have a diameter of up to 3 cm [3]. Pulmonary nodules are categorized based on internal texture as solid, part-solid, or non-solid as shown in Fig. 1, and location as juxta-pleural, juxta-vascular, or well-circumscribed is shown in Fig. 2.

The solid nodules fully cover the lung parenchyma inside them. Non-solid nodules are defined as localized nodular opacity with the hazy rise in attenuation that does not conceal the underlying parenchymal structure, which includes veins and airways. Part-solid nodules are a mixture of solid and non-solid nodules with considerable

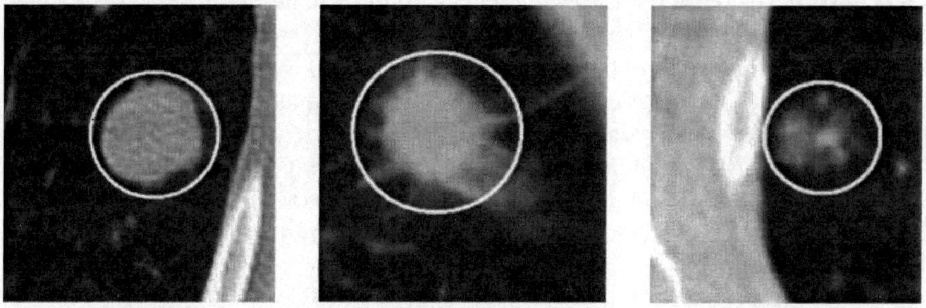

(a) Solid (b) Part-Solid (c) Non-Solid

Fig. 1 Classification of pulmonary nodules based on internal texture

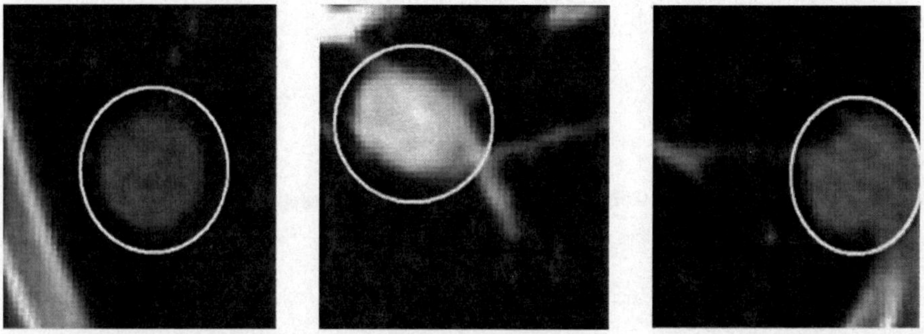

(a)Well-Circumscribed (b)Juxta-Vascular (c)Juxta-Pleural

Fig. 2 Classification of pulmonary nodules based on location fluctuations in intensity within them. A well-circumscribed lesion is not connected to any other lung structures. The juxta-vascular nodule has an uncertain connection with the surrounding arteries and vessels. A strong proportional relationship exists between the nodule and the pleural in the juxta-pleural nodule. To avoid missing of nodules in lung, the CT slice thickness should be less than diameter of the nodule. This will increase the total number of slices to 150 per patient.

Pulmonary nodules form a large number of images and convey the likelihood of malignancy based on the information contained in the nodules (morphology, shape and textural features, etc.). Their diagnosis will directly influence the treatment strategy that follows. The work is demanding, and numerous destabilizing variables, such as distraction and tiredness, as well as professional experience limits, may lead to misunderstanding of the given data. As a result, clinical computer-assistant technologies such as automated pulmonary nodule identification and diagnosis approaches are required to overcome these challenges and aid radiologists in making more accurate diagnoses. These approaches aid radiologists in terms of time, efficiency, and accuracy. As a result, in this paper, an efficient CADe system for the identification of pulmonary nodules of type solid, sub-solid, and large solid nodules, as well as false positive reduction, is designed using 3D convolution neural network (CNN).

2 Related Work

Earlier works are carried out for the classification of lung nodules based on machine learning algorithms. These techniques include steps such as lung segmentation [4, 5], nodule segmentation [6, 7], feature extraction, and classification using SVM classifier [5]. These techniques have many limitations such as SVM cannot handle multi-classification problems, and it cannot deal large amount of data [8]. These limitations can be overcome by using deep learning models. Since, these networks can extract features automatically and can train large amount of data. Recent works for lung nodule detection and classification using deep learning techniques are summarized in this section.

Suzuki et al. [9] designed 3D U-Net model for the classification of lung nodules using LIDC-IDRI dataset. Sreekumar et al. [10] designed a 3CD network, Doddavarapu et al. [11] designed a patch-based CNN, and Guo and Bai [12] designed a multi-scale pulmonary nodule detection system by fusion of cascade R-CNN and feature pyramid network (FPN). To reduce the size of the feature maps other than pooling, Ayachi et al. [13] proposed stride convolution, and Zhang et al. [14] proposed dilated convolution layers. The medical images in the datasets used for training are limited which cause over fitting. To solve these problems, Lu et al. [15] proposed data enhancement, Liu et al. [16] used transfer learning, Shi et al. [17] proposed unsupervised learning using sparse auto encoder, and Gupta et al. [18] proposed semi-supervised learning. Improved designs for lung nodule detection are designed using faster R-CNN [19–24]. U-Net finds application in the lung nodule detection [25–27], since it has symmetrical structure and skip connection. Improved version of U-Net for lung segmentation [28–31], and detection [32–34]. In addition, ResNet [35] and its extensions ResNeXt [36], SE-Net [37], SK-Net [38] are frequently used in lung nodule CADe systems [33, 34, 39] and CADx system [40–42].

Although the above discussed 2D or 3D CNN-based methods use spatial information of lung nodule, it has not been fully exploited and explored. Since the size of the sample plays avital role in determining the performance of the model. If the size of the training sample (i.e., input image patch) is too small, the contextual information is reduced. This finds difficulty in training with in sufficient data. On the other hand, if the training sample size is too large, redundant data including noise is added which will reduce the performance of the model. To overcome this disadvantage, a multi-scale multi-branch hybrid 3D CNN is designed in this paper.

3 Methods

The designed 3D CNN architecture is explained in this section. Section 3.1 gives the frame work of the proposed method, Sect. 3.2 gives selection of multi-branches for the architecture, and Sect. 3.3 gives training and implementation details.

3.1 Proposed Multi-scale Multi-branch 3D CNN Architecture

For our proposed 3D CNN, it consists of the Inception-ResNet model. To the IResNet, we extract features from the three different scales of 3D data from a given candidate nodule. On analyzing the result from the three branches by concatenating, the malignant or benign nodules are detected. The designed 3D CNN multi-branch architecture is as shown in Fig. 3a. It consists of three branches: Branch-S (small), Branch-M (medium), and Branch-L (large). The Branch-S consists of fifteen convolution layers followed by single Inception-ResNet (IRes) block. The Branch-M consists of twenty-eight convolution layers followed by two stacked Inception-ResNet (IRes) blocks such as IRes Block1 and IRes Block2. Branch-L and Branch-M are identical but differ in feature maps. The internal architecture of each Inception-ResNet (IRes) block is as shown in Fig. 3b. All the features obtained from the three branches are concatenated to get nodule probability.

3.2 Selection of Multiple Branches

In the traditional architectures ignoring the size of the nodule, networks are designed which gives false positives in nodule detection. To overcome this problem, a multi-branch 3D CNN architecture is designed in this paper. The selection of nodule size affects classification performance. If the nodule size is small, then the network is

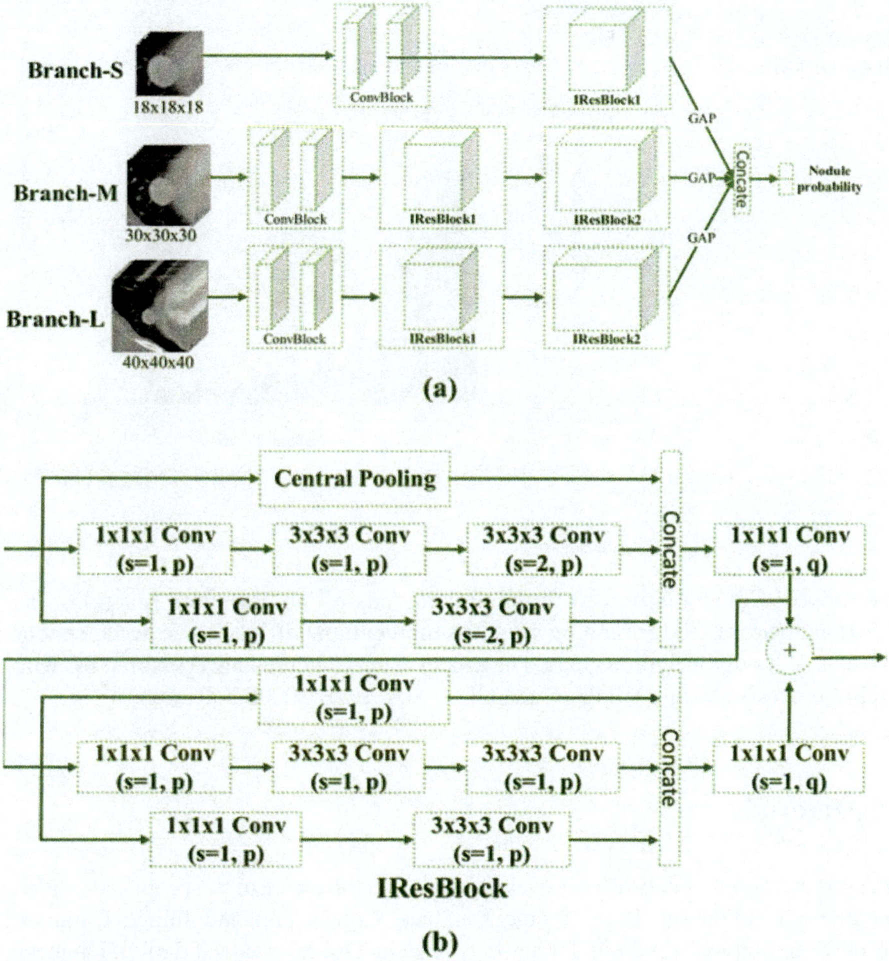

Fig. 3 Proposed architecture

not getting sufficient information for training. If the nodule size is too large, then the redundant data is added along with noise which affects the performance of the network. To overcome this challenge, a multi-branch network with multi-scale input is designed. Figure 4 gives size variation curve for the nodules available in LUNA16 database. This is available in the database separately in an excel file. We choose different input sizes on observation of several nodules with diameters, the peak nodule with a diameter of 9. So, the first input size is 18 × 18 × 18 for small nodules which contain sufficient information. Secondly, for the nodule diameter of < 15, the input size is 30 × 30 × 30 for the medium size nodules which contain rich information. At last, the input size of 40 × 40 × 40 for large nodules will also add some noise but is better handled for large nodules.

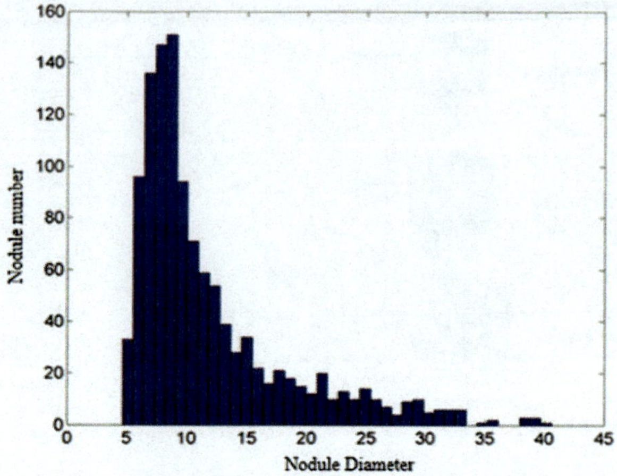

Fig. 4 Size distribution of lung nodule from LIDC-IDRI dataset

3.3 Training

The proposed model is trained by minimizing adam optimizer loss function. The hyperparameters used for training are set as momentum 0.9, batch size as 64, epochs as 500, and Xavier weights as bias. The model is implemented on a personal PC with intel i7 processor with NVIDIA Cuda GPU, MATLAB 2020b software.

4 Materials

We used two types of datasets for evaluating the performance of the proposed model. For internal validation, Lung Image Database Consortium and Image Database Resource Initiative (LIDC-IDRI) dataset is used, and for external validation, Japanese dataset is used. In this section, we discuss how the datasets are collected and how the radiologists annotated them as nodule or non-nodule.

4.1 LIDC-IDRI Dataset

To develop the nodule detection by the deep learning methods for automated detection of nodules from CT images, the LIDC-IDRI dataset is used. The Lung Imaging Database Consortium (LIDC) is a public dataset. It consists of 1018 CT scan images of the 1010 patients. The nodules in the dataset are annotated by four radiologists. The nodules in the dataset are examined independently by the radiologists. The nodules in the LIDC dataset are in the DICOM format. The annotation of the nodules by the radiologists is recorded in XML files. From the annotations of the radiologists of nodules in the dataset, the nodules are classified into three different categories as nodules $\geq$ 3 mm diameter, < 3 mm diameter, and non-nodules $\geq$ 3 mm diameter. According to the above classification, there are 1186 lung nodules. The LUNA 16 challenge which is the subset of LIDC-IDRI scans of thickness > 2–5 mm contains 888 CT scans.

4.2 JAPANESE Dataset

This dataset is used for external validation of the proposed model. It consists of 450 chest CT scans taken from Urban University Hospital in Japan. About 12–30 years of experienced radiologist annotated at least one nodule with diameter $\geq$ 5 mm.

5 Results and Discussion

The proposed multi-scale multi-branch 3D CNN model is evaluated described in the section. Figure 5 shows the FROC curve obtained when internal validation was performed using 89 CT scans taken from the LIDC-IDRI dataset. Figure 6 shows the FROC curve obtained when external validation was performed using 450 CT scans taken from the Japanese dataset.

The FROC curve yields a sensitivity of 95.9% with one false positive rate per scan using internal validation and a sensitivity of 92.8% with one false positive rate per scan using external validation. The proposed model gives a competition performance metric (CPM) of 94.6% (95% CI: 88.7–98.6%) for internal validation on 89 CT scans from the LIDC-IDRI dataset and 92.8% (95% CI: 82.1–89.6%) for external validation on 450 chest CT scans from the Japanese dataset. Figure 7 represents the true positives identified by the proposed model.

Fig. 5 FROC curve for internal validation

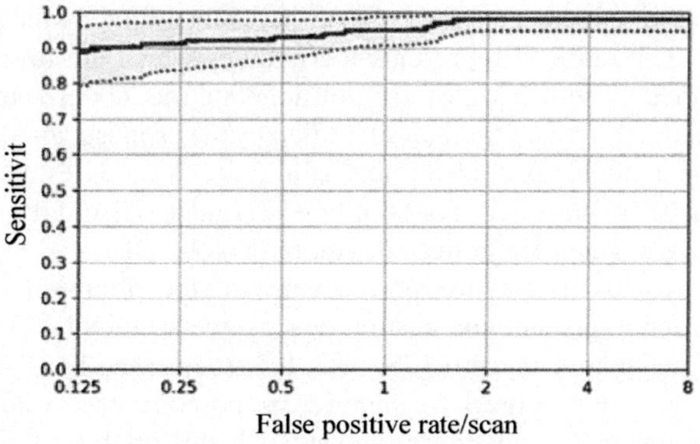

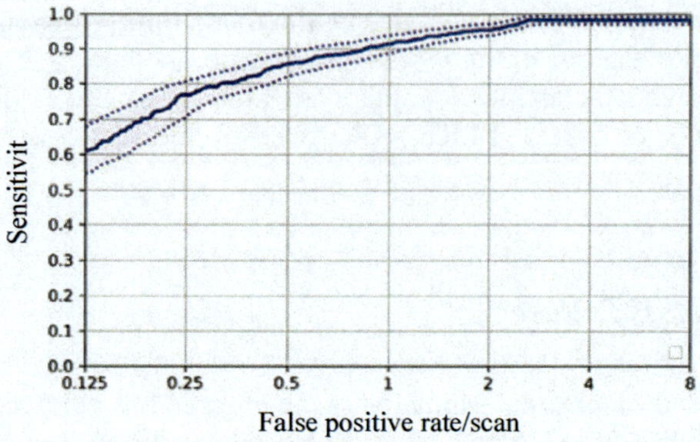

Fig. 6 FROC curve for external validation

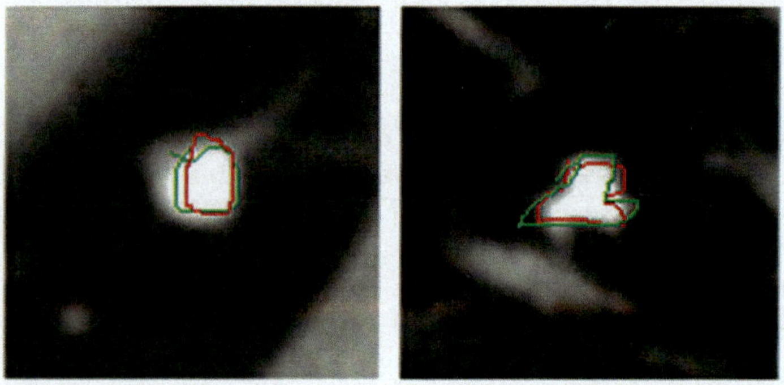

Fig. 7 True positives identified by the proposed model

Label A represents solid nodule, and B represents ground glass nodule. The red line indicates the region of interest (ROI) correctly identified by the proposed model. The green line indicates the region of interest (ROI) identified by the radiologist. Figure 8 represents the false positives identified by the proposed model.

The red ROI represents the nodules which are false identified by the proposed model. These nodules are not annotated as nodules by the radiologists. Figure 9 shows the false negatives identified by the proposed model. The green ROI represents the annotations made by the radiologists as nodules. No corresponding annotations made by proposed model in label A image. Also, false positives are identified in B image which are indicated with red color ROI.

To understand the performance of the proposed model, it is compared with different present and past models as given in Table 1, in terms of CPM with seven false positive rates (0.125, 0.25, 0.5, 1, 2, 4, and 8).

Where proposed-1 stands for proposed model evaluated on the internal dataset, proposed-2 stands for the model evaluated on the external dataset.

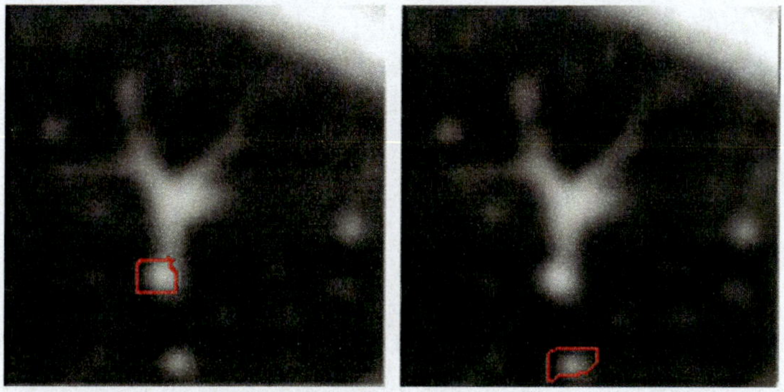

Fig. 8 False positives identified by the proposed model

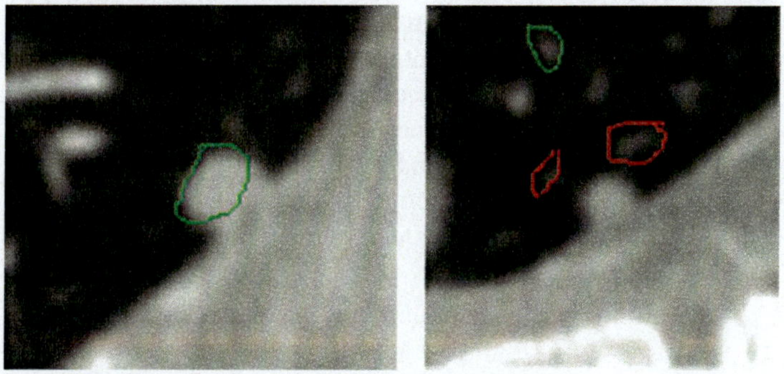

Fig. 9 False negatives identified by the proposed model

Table 1 Comparison with the present and past CNN architectures

Author	Data type	Number of false positives per scan							CPM
		0.125	0.25	0.5	1	2	4	8	
Lee et al. [43]	2D	0.262	0.359	0.472	0.586	0.674	0.713	0.766	0.548
Gong et al. [44]	2D	0.623	0.702	0.739	0.762	0.774	0.793	0.800	0.742
Huang et al. [45]	3D	0.817	0.851	0.869	0.883	0.891	0.907	0.914	0.876
Suzuki et al. [9]	3D	0.612	0.769	0.857	0.914	0.955	0.983	0.983	0.867
Proposed-1	**3D**	**0.926**	**0.947**	**0.956**	**0.961**	**0.986**	**0.988**	**0.995**	**0.959**
Proposed-2	**3D**	**0.823**	**0.869**	**0.925**	**0.948**	**0.968**	**0.986**	**0.988**	**0.928**

Bold values indicates the highlight proposed model

Lee et al. [43] used Gaussian weighted average image patches for the classification of lung nodules. Gong et al. [44] incorporated 3D tensor filtering with local feature image analysis for the automatic detection of pulmonary nodules from the CT scan images. Huang et al. [45] developed a 3D deep CNN and ensemble learning for pulmonary nodule prediction. Suzuki et al. [9] designed a 3D U-Net for the prediction of pulmonary nodules. The proposed multi-scale multi-branch 3D CNN model achieves better performance in terms of CPM compared to the existing techniques as given in the Table 1.

6 Conclusion

In this paper, a multi-scale multi-branch 3D CNN is designed for the false positive reduction of the lung nodules. The challenges in identifying the true nodules can be overcome by using the proposed model. The inputs are divided into three different scales with three different branches, and final probability is obtained by fusion the output of the three branches. This type of multi-scale division of the input reduces the redundancy in nodule information and also collects rich contextual information from the input sample. To evaluate the proposed model two datasets: LIDC-IDRI dataset for internal validation and Japanese dataset for external validation are used. The proposed model gives a competition performance metric (CPM) of 94.6% (95% CI: 88.7–98.6%) for internal validation on 89 CT scans from the LIDC-IDRI dataset and 92.8% (95% CI: 82.1–89.6%) for external validation on 450 chest CT scans from the Japanese dataset. In future, we will extend to more datasets.

References

1. Wild C, Weiderpass E, Stewart B (2020) World cancer report: cancer research for cancer prevention. International Agency for Research on Cancer, Lyon
2. Gao S, Li N, Wang S, Zhang F, Wei W, Li N, Bi N, Wang Z, He J (2020) Lung cancer in People's Republic of China. J Thorac Oncol 15:1567–1576
3. Zeng H, Chen W, Zheng R, Zhang S, Ji JS, Zou X, Xia C, Sun K, Yang Z, Li H (2018) Changing cancer survival in China during 2003–2015: a pooled analysis of 17 population-based cancer registries. Lancet Global Health 6:e555–e567
4. Wu J, Qian T (2019) A survey of pulmonary nodule detection, segmentation and classification in computed tomography with deep learning techniques. J Med Artif Intell 2:1–12
5. Pehrson LM, Nielsen MB, Ammitzbøl Lauridsen C (2019) Automatic pulmonary nodule detection applying deep learning or machine learning algorithms to the LIDC-IDRI database: a systematic review. Diagnostics 9:29
6. Niranjana G, Ponnavaikko M (2017) A review on image processing methods in detecting lung cancer using CT images. In: Proceedings of the 2017 international conference on technical advancements in computers and communications (ICTACC). IEEE, New York, pp 18–25
7. Kumar D, Wong A, Clausi DA (2015) Lung nodule classification using deep features in CT images. In: Proceedings of the 12th conference on computer and robot vision. IEEE, New York, pp 133–138
8. Chauhan VK, Dahiya K, Sharma A (2019) Problem formulations and solvers in linear SVM: a review. Artif Intell Rev 52:803–855
9. Suzuki K, Otsuka Y, Nomura Y, Kumamaru KK, Kuwatsuru R, Aoki S (2020) Development and validation of a modified three-dimensional U-Net deep-learning model for automated detection

of lung nodules on chest CT images from the lung image database consortium and Japanese datasets. Acad Radiol 29:S11–S17
10. Sreekumar A, Rajan Nair K, Sudheer S, Ganesh Nayar H, Nair JJ (2020) Malignant lung nodule detection using deep learning. In: Proceedings of the 2020 international conference on communication and signal processing (ICCSP). IEEE, New York, pp 0209–0212
11. Doddavarapu VNS, Kande GB, Rao BP (2021) Rotational invariant fractional derivative filters for lung tissue classification. IET Image Process 15(10):2202–2212
12. Guo N, Bai Z (2021) Multi-scale pulmonary nodule detection by fusion of cascade R-CNN and FPN. In: Proceedings of the 2021 international conference on computer communication and artificial intelligence (CCAI). IEEE, New York, pp 15–19
13. Ayachi R, Afif M, Said Y, Atri M (2018) Strided convolution instead of max pooling for memory efficiency of convolutional neural networks. International conference on the sciences of electronics, technologies of information and telecommunications. Springer, New York, pp 234–243
14. Zhang J, Xia Y, Zeng H, Zhang Y (2018) NODULe: combining constrained multi-scale LoG filters with densely dilated 3D deep convolutional neural network for pulmonary nodule detection. Neurocomputing 317:159–167
15. Lu X, Gu Y, Yang L, Zhang B, Zhao Y, Yu D, Zhao J, Gao L, Zhou T, Liu Y (2020) Multi-level 3D densenets for false-positive reduction in lung nodule detection based on chest computed tomography. Curr Med Imag 16:1004–1021
16. Liu M, Jiang X, Liu Y, Zhao F, Zhou H (2019) A semi-supervised convolutional transfer neural network for 3D pulmonary nodules detection. Neurocomputing 391:199–209
17. Shi Z, Hao H, Zhao M, Feng Y, He L, Wang Y, Suzuki K (2019) A deep CNN based transfer learning method for false positive reduction. Multimed Tool Appl 78:1017–1033
18. Gupta A, Saar T, Martens O, Le Moullec Y (2017) Unsupervised feature mapping via stacked sparse autoencoder for automated detection of large pulmonary nodules in CT images. Elektronika ir Elektrotechnika 23:59–63
19. Huang X, Sun W, Tseng TLB, Li C, Qian W (2019) Fast and fully-automated detection and segmentation of pulmonary nodules in thoracic CT scans using deep convolutional neural networks. Comput Med Imag Graph 74:25–36
20. Nasrullah N, Sang J, Alam MS, Xiang H (2019) Automated detection and classification for early stage lung cancer on CT images using deep learning. In: Pattern recognition and tracking XXX, international society for optics and photonics, p 109950S
21. Xie H, Yang D, Sun N, Chen Z, Zhang Y (2019) Automated pulmonary nodule detection in CT images using deep convolutional neural networks. Pattern Recogn 85:109–119
22. Tang H, Kim DR, Xie X (2018) Automated pulmonary nodule detection using 3D deep convolutional neural networks. In: Proceedings of the 2018 IEEE 15th international symposium on biomedical imaging (ISBI 2018). IEEE, New York, pp 523–526
23. Zhu W, Liu C, Fan W, Xie X (2018) Deeplung: deep 3D dual path nets for automated pulmonary nodule detection and classification. In: Proceedings of the 2018 IEEE winter conference on applications of computer vision (WACV). IEEE, New York, pp 673–681
24. Ding J, Li A, Hu Z, Wang L (2017) Accurate pulmonary nodule detection in computed tomography images using deep convolutional neural networks. International conference on medical image computing and computer-assisted intervention. Springer, New York, pp 559–567
25. Nemoto T, Futakami N, Yagi M, Kumabe A, Takeda A, Kunieda E, Shigematsu N (2020) Efficacy evaluation of 2D, 3D U-Net semantic segmentation and atlas-based segmentation of normal lungs excluding the trachea and main bronchi. J Radiat Res 61:257–264
26. Pang T, Guo S, Zhang X, Zhao L (2019) Automatic lung segmentation based on texture and deep features of HRCT images with interstitial lung disease. BioMed Res Int 2019:1–8
27. Zhao T, Gao D, Wang J, Tin Z (2018) Lung segmentation in CT images using a fully convolutional neural network with multi-instance and conditional adversary loss. IEEE 15th international symposium on biomedical imaging (ISBI 2018). IEEE, New York, pp 505–509
28. Keetha NV, Annavarapu CSR (2020) U-Det: a modified U-Net architecture with bidirectional feature network for lung nodule segmentation. arXiv preprint arXiv:2003.09293

29. Usman M, Lee BD, Byon SS, Kim SH, IlLee B (2019) Volumetric lung nodule segmentation using adaptive ROI with multi-view residual learning. arXiv preprint arXiv:1912.13335
30. Rocha J, Cunha A, Mendonça AM (2020) Conventional filtering versus U-Net based models for pulmonary nodule segmentation in CT images. J Med Syst 44:1–8
31. Wu B, Zhou Z, Wang J, Wang Y (2018) Joint learning for pulmonary nodule segmentation, attributes and malignancy prediction. In: Proceedings of the 2018 IEEE 15th international symposium on biomedical imaging (ISBI 2018). IEEE, New York, pp 1109–1113
32. Cao H, Liu H, Song E, Ma G, Xu X, Jin R, Liu T, Hung CC (2019) Two-stage convolutional neural network architecture for lung nodule detection. arXiv preprint arXiv:1905.03445
33. Gruetzemacher R, Gupta A, Paradice D (2018) 3D deep learning for detecting pulmonary nodules in CT scans. J Am Med Inf Assoc 25:1301–1310
34. Lan T, Li Y, Murugi JK, Ding Y, Qin Z (2018) RUN: residual U-Net for computer-aided detection of pulmonary nodules without candidate selection. arXiv preprint arXiv:1805.11856
35. He K, Zhang X, Ren S, Sun J (2016) Deep residual learning for image recognition. In: Proceedings of the IEEE conference on computer vision and pattern recognition, pp 770–778
36. Xie S, Girshick R, Dollar P, Tu Z, He K (2017) Aggregated residual transformations for deep neural networks. In: Proceedings of the IEEE conference on computer vision and pattern recognition, pp 1492–1500
37. Hu J, Shen L, Sun G (2018) Squeeze-and-excitation networks. In: Proceedings of the IEEE conference on computer vision and pattern recognition, pp 7132–7141
38. Li X, Wang W, Hu X, Yang J (2019) Selective kernel networks. In: Proceedings of the IEEE conference on computer vision and pattern recognition, pp 510–519
39. Gong L, Jiang S, Yang Z, Zhang G, Wang L (2019) Automated pulmonary nodule detection in CT images using 3D deep squeeze-and-excitation networks. Int J Comput Assist Radiol Surg 14:1969–1979
40. da Nobrega NVM, Peixoto SA, da Silva SPP, Rebouças Filho SP (2018) Lung nodule classification via deep transfer learning in CT lung images. IEEE 31st international symposium on computer-based medical systems (CBMS). IEEE, New York, pp 244–249
41. Xie Y, Xia Y, Zhang J, Feng DD, Fulham M, Cai W (2017) Transferable multi-model ensemble for benign-malignant lung nodule classification on chest CT. International conference on medical image computing and computer-assisted intervention. Springer, New York, pp 656–664
42. Nibali A, He Z, Wollersheim D (2017) Pulmonary nodule classification with deep residual networks. Int J Comput Assist Radiol Surg 12:1799–1808
43. Lee H, Lee H, Park M, Kim J (2017) Contextual convolutional neural networks for lung nodule classification using Gaussian-weighted average image patches. Proc SPIE 10134:23
44. Gong J, Liu J-YY, Wang L-JJ et al (2018) Automatic detection of pulmonary nodules in CT images by incorporating 3D tensor filtering with local image feature analysis. Phys Med 46:124–133
45. Huang W, Xue Y, Wu Y (2019) A CAD system for pulmonary nodule prediction based on deep three-dimensional convolutional neural networks and ensemble learning. PLoS ONE 14:e0219369

Heart Disorder Prediction Using 1D Convolutional Neural Network

Vedantham Ramachandran, Sudha Kishore Rekha, Barige Rajesh, Y. V. Narayana, and Parasa Sathish

Abstract Heart disorders are leading cause of deaths worldwide. If in case a risk of heart disease may include for the patient, the healthcare department people are mostly depending on the data of a patient. The data cannot be verified in-detail by the doctors all the time and predict accurately. However, it is concerning as the risky and time consuming as well. This paper aims at developing a deep learning-based heart disease prediction system that can be used to diagnose a patient's condition based on the medical record. Several parameters like fasting blood sugar, cholesterol, chest pain type, and resting blood pressure have been studied and used for the detection of heart disease. The parameters have been fed to a 1D convolutional neural network (CNN). The network is designed with convolution 2D layer, ReLU layer, softmax layer, fully-connected layer, and classification layer. The UCI heart dataset has been used for the experiments. The proposed system has produced an accuracy of 90% during training and 82% during testing.

Keywords Heart disorder · Blood pressure · Cholesterol · Convolutional neural network

---

V. Ramachandran (✉) · S. K. Rekha · Y. V. Narayana · P. Sathish
IT Department, Vasireddy Venkatadri Institute of Technology, Nambur, Guntur Dt., Andhra Pradesh, India
e-mail: vrc.bhatt@vvit.net

S. K. Rekha
e-mail: sudhakishore@vvit.net

Y. V. Narayana
e-mail: naarayanaa808@vvit.net

P. Sathish
e-mail: parasasathish@vvit.net

B. Rajesh
IT Department, PVPS Institute of Technology, Vijayawada, India
e-mail: barige.rajesh@pvpsit.net

J. R. Szymanski et al. (eds.), *Energy Systems, Drives and Automations*, Lecture Notes in Electrical Engineering 1057,
https://doi.org/10.1007/978-981-99-3691-5_44

1 Introduction

Any health problem that impacts the structure or functioning of the heart is a disease [1] of the heart. It is sometimes wrongly believed that there is only one type, when in fact, heart disease is numerous. The causes of these health problems are diverse. There are several kinds of heart diseases. They can be classified in particular according to how they affect the heart [2].

1. **Coronary artery disease and other vascular diseases**: These occur when the blood vessels harden (atherosclerosis). The coronary artery disease is characterized by narrowing or blockage of a coronary artery (of the heart). It is the most common form of heart disease and the major cause of heart attack and of angina (chest pain). A vascular disease tends to affect other vessels. It reduces blood flow and affects the functioning of the heart.
2. **Heart rhythm disturbances (arrhythmias)**: They make the heart beat too slowly, too quickly, or irregularly. Millions of people have arrhythmia which disrupts blood circulation. There are several types: some show no symptoms (asymptomatic) or warning signs; others are sudden and deadly.
3. **Structural heart disease**: These are caused by abnormalities in the structure of the heart, including valves, heart walls and muscles, or blood vessels near the heart. These heart diseases are sometimes existing at birth (congenital) or occur during life due to infection, wear, or other factors. People living with a heart defect, as well as their families, need lifelong support as they often have to receive ongoing medical care and undergo surgery.
4. **Heart failure**: This is a serious condition that occurs when the heart is weakened or damaged. The heart failure can be caused by two most common ways such as the heart attack and the high blood pressure. The disease is incurable, but a rapid diagnosis, the adoption of specific lifestyle habits, and the taking of adapted drugs help patients to lead a long normal and active life, without hospitalization.

The diagnosis of the diseases is done by a procedure known as angiogram. A coronary angiogram (also called a coronary angiogram) is a test that involves taking X-rays of the coronary arteries and vessels that helps to supply the blood for the heart. During this procedure, a special stain, an iodine product, is injected into the coronary arteries from a catheter (long and narrow tube) injected into a blood vessel; each of them then becomes visible on the X-ray. Angiography allows doctors to observe the blood flow in the heart separately and sometimes even to pinpoint possible problems with the coronary arteries.

Coronary angiography can be recommended for people with angina (chest pain) or those with symptoms of coronary artery disease [3]. It provides doctors with important information about the condition of the coronary arteries, which can be affected by atherosclerosis, regurgitation (blood pumped back through a damaged valve), or the accumulation of blood in a cavity caused by a malfunction of a heart valve.

Angiography is performed in the hospital or clinic. You will be asked to lie on a table and disinfect the area around where the catheter will be inserted (groin or arm). Through local anaesthesia that you will be given, your skin will be numb, preventing you from feeling any pain. Then, the catheter will be carefully guided through a vein or artery to near the heart [4]. Once the catheter is in place, it will release the special dye into the blood stream; this dye will facilitate the taking of clear and detailed X-rays of the coronary arteries. It should be noted that the injection of the dye could cause a brief feeling of heat, which should, however, dissipate fairly quickly. The duration of an angiography can vary between one and two hours; however, angiography is a very common procedure which, in general, is considered to be safe.

Some cardiologists and engineers invent solutions to better diagnose and predict heart problems by relying on artificial intelligence. As with radiology, only a handful of AI-based tools have so far passed the door of hospitals and received the green light from research, but many see these innovations in cardiology as having immense potential.

Deep learning is a new area of research in machine learning [5] which aims to bring machine learning closer to one of its initial objectives "intelligence artificial". It depends on manipulating large amounts of data by adding layers in the artificial neuron network to extract the properties of raw data through several processing layers composed of linear transformations and not linear. The large amount of healthcare data on the heart disease diagnosis in the world is produced by hospitals, and many medical laboratories, to be used in the field of scientific research, correct diagnosis and classification of diseases or to predict diseases.

For heart disease, the risk factors can be considered as alcohol intake, physical inactivity, obesity, high blood pressure, poor diet, cholesterol, family history, smoking, sex, and age. The hereditary risk factors are included diabetes and high blood pressure which leads to the heart disease [6]. It is very tough to evaluate the odds of getting heart disease based on risk factors. For predicting the output from existing information, machine learning methods can be used. In this paper, the application of one such type of machine learning technique like convolutional neural network (CNN) to forecast the heart disease risk from the risk factors is demonstrated.

2 Literature Survey

Sung and Yuan Lee have been utilized genetic algorithm (GA) and SVM to make the feature selection and to categorize the classification, respectively. The effectiveness of this method is validated with its results. In literature, it outclasses two other methods when SVM is implemented without GA by generating an accuracy of 96.38%. The accuracy is increased by 3.14% if GA is applied [7].

Davari et al. [8] have been proposed a method to diagnose the coronary artery disease (CAD) with automation. Using signals of heart rate variability (HRV), normal conditions are determined from an electrocardiogram (ECG). For reduction of dimensionality, the principal component analysis (PCA) is made utilized by the technique, and SVM is applied for the purpose of classification.

Boon et al. [9] have been presented a system to optimize the parameters and to make settings of feature extraction of HRV for multiple algorithms. To maximize the performance of prediction, the best subset of features is chosen by the algorithm, and SVM parameters are tuned at the same time.

Mustaqeem et al. [10] have been gathered sufficient information to create the own dataset. For analyzing the proposed algorithm's performance, two evaluation measures are utilized such as kappa and accuracy statistics. Purushottam et al. [15] have been ranked the features that are significant clinically. By using t value and classifiers such as SVM, K-nearest neighbour (KNN), and decision tree, these features are obtained in numerous ways. In prior to the SCD with an accuracy of 94.7, 89.4, 89.4, and 97.3%, the sudden cardiac death (SCD) is predicted by the technique for 4, 3, 2, and 1 min, respectively.

Acharya et al. [11] have been proposed nonlinear features to the classifiers such as SVM, decision tree (DT), and KNN. The combination of a nonlinear analysis and discrete wavelet transform (DWT) is utilized for predicting the accuracies of SCD of 92.11 (SVM), 93.42 (KNN), 98.68 (SVM), and 92.11% (KNN) before occurrence of SCD for 4, 3, 2, and 1 min, respectively.

In [12], a CNN-based technique has been proposed that involves 2 pooling layers and 2 convolution layers. Additionally, long short-term memory (LSTM) has been utilized in the sense that an RNN but with an output layer and hidden memory layer. Three gate units have been included in LSTM's memory block such as input, a self-recurrent connecting neuron, and forget and output.

In [13], a neural network-based technique has been proposed based on the electronic medical data of patients to predict the heart failure. To model and predict HF with LSTM, specifically word vectors and a one- hot encoding have been incorporated.

In [14], a novel technique based on CNN has been used that is targeted to make the classification of pathological and healthy people with the use of an auditory sensor for field programmable gate array (FPGA) which is exploited for decomposing audio to frequency bands in real-time.

Purushottam et al. [15] have been planned a framework to anticipate the patients' risk levels according to the provided metrics. This rules-based algorithm has been majorly influenced on assisting a non-specialized physician in such a way that taking the right decision in regard to the levels of risk. Yu and Lee [7] have been helped in identifying CAD at an early phase based on the technique of a fuzzy expert system and formulation of rules from doctors.

Long et al. [16] have been used fuzzy logic and presented a method known as the system of interval type-2 fuzzy logic system (IT2FLS). This method is used a procedure of blended learning which contains fuzzy c-mean clustering and tuned in parameters through GA and Firefly. The conducted experiments of IT2FLS are analyzed and are revealed that it outclasses other techniques of ML like ANN, SVM, and NB as this technique's computation cost is increased due to the involved high dimensionality.

In [17], the ranking of complete set of clinically essential features is done primarily to make easier the experimentation process. After that, these features are incorporated into various classifiers like SVM, DT, and K-nearest neighbour (KNN). The highest accuracy is achieved for the given dataset of SVM. In [11], these two classifiers like SVM and KNN have considered the input as non-linear features. The stated classifiers KNN and SVM are achieved the accuracies of 92.11 and 96.68%, respectively.

3 Proposed Heart Disease Prediction Using Deep Learning

In different classification tasks such as audio, image, and words, neural networks are utilized. For various purposes, different kinds of neural networks are made used. The recurrent neural networks are used more precisely to predict the sequence of words for an LSTM. In a similar manner, convolutional neural network has been utilized for classification of images. The establishment of basic building block for CNN is discussed in the blog.

Some concepts of neural network are revisited firstly before categorizing into the convolution neural network. Three types of layers are included in a regular neural network.

1. **Input Layers**: In the input layer, the input is given to the model. The total number of features in the data is equivalent to the number of neurons in this layer. In case of an image, consider the number of pixels.
2. **Hidden Layer**: From an input layer, the input is fed into the hidden layer. Based on the data size and type of model, different kinds of hidden layers are existed. Various number of neurons can be included in each hidden layer that are greater than the number of features in general. Based on the matrix multiplication of the previous layer output with that layer's learnable weights, the output from each layer is calculated. By adding the learnable biases that are followed by the function of activation, the network is made nonlinear.
3. **Output Layer**: As similar as sigmoid or softmax, the hidden layer's output is incorporated into a logistic function. However, the function of softmax or sigmoid is converted the each class output into the each class's probability score.

The data is incorporated into the model, and each layer's output is retrieved the step known as feedforward. Based on an error function, the error is determined. Some of the most widely used error functions are included square loss error, cross entropy, etc. With the calculation of derivatives, it's better to back propagate into the model. This phase is called as backpropagation which is utilized to reduce the loss basically (Fig. 1).

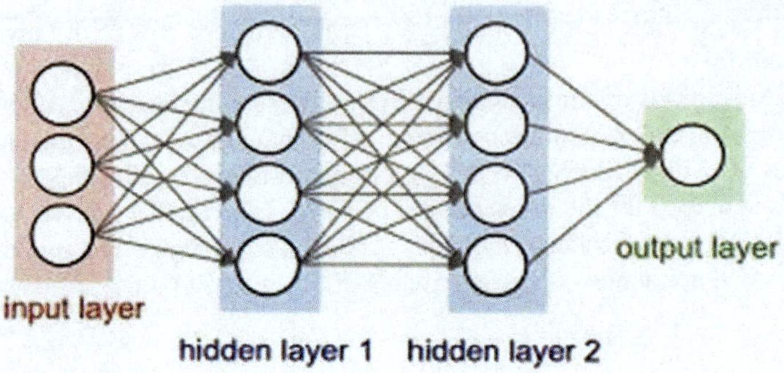

Fig. 1 Neural network structure

Convolution Neural Network

Convolution neural networks or CovNets are one of the type of neural networks that share their metrics. Here, an image is considered, and it is represented as a cuboid that includes the height (generally an image has blue, green, and red channels), width (image's dimension), and length. Let us assume this image with a small patch and implement a small neural network on it with k outputs with the vertical representation. Across the whole image, the neural network is operated which will lead to retrieve another image with different depth, height, and width. More channels but lesser height and width are considered rather than just B, G, and R channels. However, this operation is termed as convolution. When patch size is similar to that of the actual image, it will be concerned as a regular neural network. Fewer weights have included in the image due to this small patch (Fig. 2).

Fig. 2 Convolution process in CNN

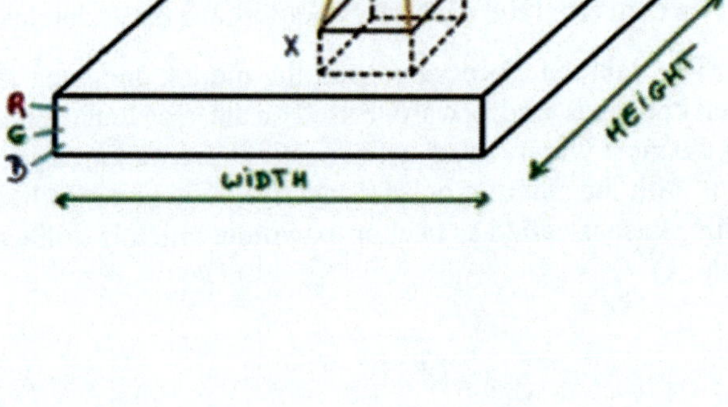

Now, it is time to discuss the mathematics part which is included in the complete process of convolution.

- A set of learnable filters are contained in the convolution layers for the patch in the above image. Small height and width and the same depth like the input volume have been involved in each filter (3 if the input layer is image input).
- Let us say, the filters' possible size can be $a \times a \times 3$ if the convolution is implemented on an image that has included the dimension of $34 \times 34 \times 3$. Here, "a" can be 3, 5, 7, etc., but it is small than the image dimension.
- For the whole input volume, each filter slide step by step during forward pass in which each step is called as stride (which can include the values like 2 or 3 or even 4 for images with higher dimensions). The dot product between the input volume's patch and filters' weights can be computed.

A 2D output will be retrieved for each filter and will stack them together as the filters slide. The output volume has included a depth which is equivalent to the number of filters as a result. All the filters will be learned by the network.

Layers Used to Build ConvNets

Based on differentiable function, the transformation of one volume to another has been made by each layer in ConvNets which has a sequence of layers. **Types of layers**: Let us consider an example of an image with the dimension of $32 \times 32 \times 3$ by running a ConvNets.

1. **Input Layer**: The raw input of image with depth 3, height 32, and width 32 is included in this layer.
2. **Convolution Layer**: Based on the determination of dot product between image path and all filters, the output volume is computed by this layer. The dimension for output volume $32 \times 32 \times 12$ will be received if the total number of 12 filters are used for this layer.
3. **Activation Function Layer**: To get the convolution layer's output, this layer will be implemented based on activation function in an element wise. Some of the different activation functions are included ReLU: Tanh, leaky ReLU, sigmoid: $1/(1 + e\wedge - x)$, and $\max(0, x)$, etc. The volume is remained in unchanged state so that the output volume will have included the dimension $32 \times 32 \times 12$.
4. **Pool Layer**: In the ConvNets, this layer is incorporated periodically, and the reduction of volume size is the essential function that helps to do the rapid computation based on preventing from over fitting and reducing memory. However, two different common types of pooling are involved **average pooling** and **max pooling**. The output volume with the dimension $16 \times 16 \times 12$ will be retrieved when using a max pool with stride 2 and 2×2 filters (Fig. 3).

Fully-Connected Layer: This layer is a type of regular neural network layer that can be used to compute the class scores by considering the input from the previous layer. Additionally, output is also determined for 1D array of size which is equal to the number of classes.

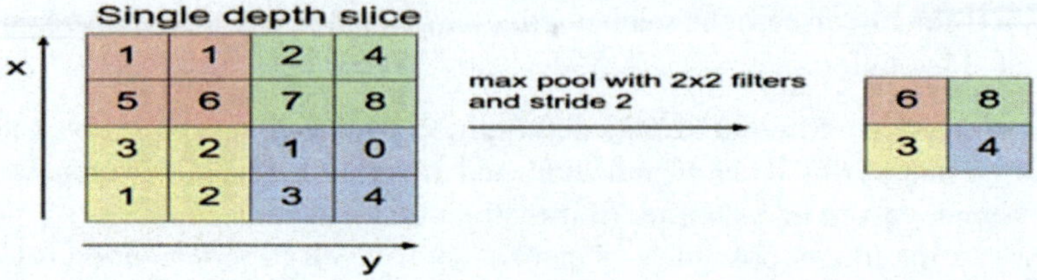

Fig. 3 Pooling process

The Convolution Operator

Convolutional neural networks are already called *convolution* in their name. This means that the so-called discrete convolution operator, which is common, is used in image and signal processing. Because this operator is the basis for CNNs, it is subsequently for the 1-dimensional and 2-dimensional case defined and illustrated.

1D convolution (discrete convolution) Let $A = (a1, ..., an)$ $A = (a_1, \dots, a_n)$ be an input vector. Continue to be F a so-called filter vector (also called kernel vector) with $F = (f_1, \dots, f_k)$ and $k < n$. Then, the 1-dimensional convolution operator $*$ is defined as follows:

$$1A * F := (a * f)_x = \sum_{i=1}^{k} a_{x+i-1} \cdot f_i \text{ with } x \in \{1, ..., n-k+1\}$$

Eq 1: 1D-Convolution operator To illustrate, consider the following example vectors for A and F:

$$A = \begin{matrix} 10 \\ 50 \\ 60 \\ 10 \\ 20 \\ [40] \\ [30] \end{matrix} \quad \text{and} \quad F = \begin{matrix} \frac{1}{s} \\ \frac{1}{|s|} \\ \frac{1}{|s|} \end{matrix}$$

Obviously $n = 7$ and $k = 3$. Then, the result vector after applying the convolution operators on A and F exactly $n - k + 1 = 7 - 3 + 1 = 5$ entries. According to the definition results:

$$A * F = \begin{array}{l} (a * f)_1 \\ (a * f)_2 \\ (a * f)_3 \\ I(a * f)_4 I \\ \left[(a * f)_5\right] \end{array} = \begin{array}{l} F \sum_{i=1}^{3} a_{1+i-1} \cdot f_i 1 \\ I \sum_{i=1}^{3} a_{2+i-1} \cdot f_i I \\ I \sum_{i=1}^{3} a_{3+i-1} \cdot f_i I \\ I \sum_{i=1}^{3} a_{4+i-1} \cdot f_i I \\ \left[\sum_{i=1}^{3} a_{5+i-1} \cdot f_i\right] \end{array}$$

$$= \begin{array}{l} F10 \cdot \frac{1}{3} + 50 \cdot \frac{1}{3} + 60 \cdot \frac{1}{3} 1 \\ I50 \cdot \frac{1}{3} + 60 \cdot \frac{1}{3} + 10 \cdot \frac{1}{3} I \\ I60 \cdot \frac{1}{3} + 10 \cdot \frac{1}{3} + 20 \cdot \frac{1}{3} I \\ I10 \cdot \frac{1}{3} + 20 \cdot \frac{1}{3} + 40 \cdot \frac{1}{3} I \\ \left[20 \cdot \frac{1}{3} + 40 \cdot \frac{1}{3} + 30 \cdot \frac{1}{3}\right] \end{array} = \begin{array}{l} 40 \\ 40 \\ 30 \\ 20 \\ [30] \end{array}$$

At first glance, the previous clear convolution operation. Obviously, the filter vector always overlaps a partial area the input. If, on the other hand, the input vector *A is taken*, for example, as a measurement curve of an electrical signals, the result of the convolution with this choice of *F could,* for example, be smoothing of the signal. Extreme outliers up (60) or down (10) result no longer included. A different choice of *F* might have "derived" filtered out. Applying the convolution operator to an input signal a particular filter is a common approach in signal processing.

In the 2-dimensional case, which is defined and explained below, this type opens up processing even more vivid.

4 Experimental Results

Two sections are involved in the experimental results such as training process and testing process. The dataset UDI heart disease repository has been downloaded from kaggle.com. The attributes used are listed below:

1. Age: age in years
2. Sex: (1 = male; 0 = female)
3. cp: chest pain type:
   - Value 1: typical angina
   - Value 2: atypical angina
   - Value 3: non-anginal pain
   - Value 4: asymptomatic
4. trestbps: resting blood pressure (in mm Hg on admission to the hospital)
5. chol: serum cholestoral
6. fbs: fasting blood sugar > 120 mg/dl) (1 = true; 0 = false)
7. restecg: restingelectrocardiographic results
8. thalach: maximum heart rate achieved
9. exang: exercise induced angina (1 = yes; 0 = no)
10. oldpeak: ST depression induced by exercise relative to rest

11. slope: the slope of the peak exercise ST segment
    - Value 1: upsloping
    - Value 2: flat
    - Value 3: downsloping
12. ca: number of major vessels (0–3) coloured by flourosopy
13. thal: 3 = normal; 6 = fixed defect; 7 = reversable defect
14. num: the predicted attribute

Training Process

Figure 4 is showed the 1D CNN network's training process. The network contains basically two graphs, which are accuracy and loss. If the validation accuracy will reach to final destination, the training process will be completed.

Training

Table 1 shows the confusion matrix of 1D CNNN of training process. The table describes, in which actual class contains two classes, and each class contains 400 samples. In predicted class, it shows predicted samples of actual class.

The overall accuracy of training model is **0.90**.

Table 2 describes the performance metrics of actual class. The table contains true negative, false negative, true positive, false positive, precision, sensitivity, and specificity of each class.

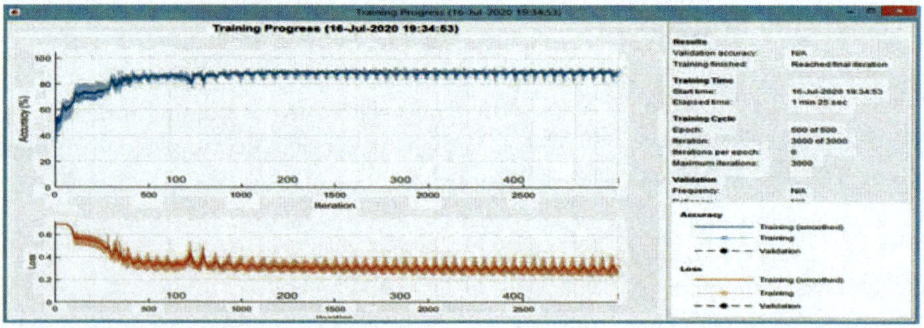

Fig. 4 Training process of 1D CNN network

Table 1 Confusion matrix-after testing

Actual class	Heart attack positive	Heart attack negative
Predicted class		
Heart attack positive	337	63
Heart attack negative	20	380

Table 2 Performance metrics of actual class in training process

Performance metrics	Actual class	
	Heart attack positive	Heart attack negative
True positive	337	380
False positive	63	20
False negative	20	63
True negative	380	337
Precision	0.84	0.95
Sensitivity	0.94	0.86
Specificity	0.86	0.94

Table 3 Confusion matrix-after testing

Actual class	Heart attack positive	Heart attack negative
Predicted class		
Heart attack positive	76	23
Heart attack negative	17	109

Testing

Table 3 shows the confusion matrix of 1D CNN in testing process. The table describes, in which actual class contains two classes, and each class contains 99 and 126 samples, respectively. In predicted class, it shows predicted samples of actual class.

The overall accuracy of testing model is **0.82**.

Table 4 describes the performance metrics of actual class. The table contains true positive, false positive, false negative, true negative, precision, sensitivity, and specificity of each class.

Table 4 Performance metrics of actual class in testing process

Performance metrics	Actual class	
	Heart attack positive	Heart attack negative
True positive	76	109
False positive	23	17
False negative	17	23
True negative	109	79
Precision	0.77	0.87
Sensitivity	0.82	0.83
Specificity	0.83	0.82

Table 5 Comparative results

S. No	Algorithm	Accuracy (%)
1	Feed forward neural network	70
2	Support vector machine	73
3	1D CNN	82

Table 5 illustrates the proposed method's results with the existing classification methods comparatively.

5 Conclusion

To predict the probabilities of having heart disease for a patient, the algorithm of 1D CNN-based heart disease prediction was implemented on the dataset in this paper. The accuracy of training obtained is 90%, and the accuracy of testing obtained is 82%. The precision, sensitivity, and specificity of testing are 0.77, 0.82, and 0.83 for positive heart attack and 0.87, 0.83, and 0.82 negative heart attack, respectively. The proposed method performed better when compared to the existing methods.

References

1. Desai SD, Giraddi S, Narayankar P, Pudakalakatti NR, Sulegaon S (2019) Back-propagation neural network versus logistic regression in heart disease classification. Advanced computing and communication technologies. Springer, Singapore, pp 133–144
2. Roca-Luque I, Rivas-Gándara N, Dos Subirà L, Francisco-Pascual J, Pijuan-Domenech A, Pérez-Rodon J, Teresa-Subirana M et al (2018) Mechanisms of intra-atrial re-entrant tachycardias in congenital heart disease: types and predictors. Am J Cardiol 122(4):672–682
3. Dube H, Madge S, Jagtap P, Potdar P, Bhandare N (2020) Review on heart disease classification. In: Proceedings of the 2020 5th international conference on communication and electronics systems (ICCES). IEEE, New York, pp 1169–1173
4. Ekız S, Erdoğmuş P (2017) Comparative study of heart disease classification. In: Proceedings of the 2017 electric electronics, computer science, biomedical engineering's meeting (EBBT). IEEE, New York, pp 1–4
5. Sowmiya C, Sumitra P (2017) Analytical study of heart disease diagnosis using classification techniques. In: Proceedings of the 2017 IEEE international conference on intelligent techniques in control, optimization and signal processing (INCOS). IEEE, New York, pp 1–5
6. Zhao T-T, Yuan Y-B, Wang Y-J, Gao J, He P (2017) Heart disease classification based on feature fusion. In: Proceedings of the 2017 international conference on machine learning and cybernetics (ICMLC), vol 1. IEEE, New York, pp 111–117
7. Yu S, Lee M (2012) Bispectral analysis and genetic algorithm for congestive heart failure recognition based on heart rate variability. Comput Biol Med 42(8):816–825
8. Davari DA et al (2017) Automated diagnosis of coronary artery disease (CAD) patients using optimized SVM. Comput Methods Prog Biomed 138:117–126
9. Boon K et al (2018) Paroxysmal atrial fibrillation prediction based on HRV analysis and non-dominated sorting genetic algorithm III. Comput Methods Prog Biomed 153:171–184

10. Mustaqeem A et al (2017) A statistical analysis based recommender model for heart disease patients. Int J Med Inform 108:134–145
11. Acharya U et al (2015) An integrated index for detection of sudden cardiac death using discrete wavelet transform and non-linear features. Knowl Based Syst 83:149–158
12. Tan J, Hagiwara Y, Pang W, Lim I, Oh S, Adam M, Tan R, Chen M, Acharya U (2018) Application of stacked convolutional and long short-term memory network for accurate identification of CAD ECG signals. Comput Biol Med 94:19–26
13. Jin B, Che C, Liu Z, Zhang S, Yin X, Wei X (2018) Predicting the risk of heart failure with EHR sequential data modeling. IEEE Access 6:9256–9261
14. Dominguez-Morales J, Jimenez-Fernandez A, Dominguez-Morales M, Jimenez-Moreno G (2018) Deep neural networks for the recognition and classification of heart murmurs using neuromorphic auditory sensors. IEEE Trans Biomed Circ Syst 12(1):24–34
15. Purushottam M et al (2016) Efficient heart disease prediction system. Proced Comput Sci 85:962–969
16. Long N et al (2015) A highly accurate firefly based algorithm for heart disease prediction. Exp Syst Appl 42(21):8221–8231
17. Fujita H et al (2016) Sudden cardiac death (SCD) prediction based on nonlinear heart rate variability features and SCD index. Appl Soft Comput 43:510–519

New Data Representation and Simulate Over Social Media Using Bayes Classifier of Machine Learning

Suresh Yallamati, Avinash Buradagunta, Shaik Mulla Almas, and Nagababu Pachhala

Abstract Nowadays, online social networks play a real job in the spread of information at an exceptionally vast scale, and it ends up basic to give intends to dissect this wonder. We address that the information dispersion process crosswise over networks normally reflects rich vicinity connections between clients. In the meantime, interpersonal organizations contain numerous networks regularizing correspondence pathways for information spread. We propose a probabilistic model to appraise the likelihood of a client's appropriation utilizing the guileless Bays classifier. Next, we manufacture a proposal strategy utilizing the probabilistic model. At long last, we show the viability of the proposed strategy with the information from Flicker and Movie focal point which are notable Web administrations. Viral advertising has connected to business-to-buyer exchanges and, however, has seen restricted selection in business-to-business exchanges. The writing audit uncovers an absence of new dissemination techniques, particularly in powerful and extensive scale networks. It likewise offers bits of knowledge into applying different digging strategies for viral showcasing. We present a methodology for online examination of information dissemination in Twitter. We recreate supposed information falls that demonstrate how information is being spread from client to client from the flood of messages and the social chart. The proposed model has been connected to the unfortunate impact minimization issue, where the time arrangement is thought about. The exploratory outcomes demonstrate our model can be used to limit the unfavorable effect of a specific impact by infusing different impacts. Besides, the proposed model additionally reveals insight into comprehension, examining and breaking down numerous impacts in informal communities.

Keywords Multiple influences · Agent-based modeling · Influence diffusion · Information diffusion · Viral marketing · Social media marketing · Social networks · Recommendation · Social networks

S. Yallamati (✉) · A. Buradagunta · S. M. Almas · N. Pachhala
Department of InformationTechnology, Vasireddy Venkatadri Institute of Technology, Guntur, India
e-mail: sureshyallamati@gmail.com

J. R. Szymanski et al. (eds.), *Energy Systems, Drives and Automations*,
Lecture Notes in Electrical Engineering 1057,
https://doi.org/10.1007/978-981-99-3691-5_45

1 Introduction

Recently, the number of clients utilizing informal organization administrations is becoming quickly one of the superficial qualities that can be extricated from socially bonded information which is the social impact [1]. Social impact happens when an individual is actuated by the activity of other people [2]. The enactment is any social activity that appeared other individuals, for example, composing and sending a record to companions. Social impact is additionally a key to clarifying information dissemination in interpersonal organizations. Since social impact between two clients depends on the social activity [3]. The dispersion of development over a network is one of the first explanations behind considering networks, and the spread of illness among a populace has been focused and concentrated for quite a long time [4]. Other pertinent inquiries incorporate "for what reason should we pick this specific online networking channel rather than others?" and "which clients have the ability to impact different clients." To spot new research progressions in information dissemination and dispersion for viral promoting, we played out a scan for different subjects identified with information dissemination for viral showcasing in business [5]. Our online strategy depends on a calculation and supporting system to deduce conceivable impact ways from the flood of messages and the fundamental social diagram. As far as anyone is concerned, no work exists on reproducing information falls and surmising impact ways on the Web while researching the effect of deficient and not cleaned datasets on such assessments. Such fragmented datasets get from API impediments absence of unequivocally noticeable client impact [6]. Our strategy can be utilized as a general model of deducing impact ways, limited to rewets, yet in addition to any sort of information that spreads over an interpersonal organization [7]. As such, every individual has a limited limit of considering and engrossing the effect of impacts, and the comparing consideration is constantly centered around specific impact messages. In the interim, the current information continues becoming dim of the general population consideration, particularly when other noteworthy impacts are infused into a similar setting. This component turns out to be increasingly unmistakable in time-delicate interpersonal organizations, for example, microblogging stages [8] (Fig. 1).

2 Related Work

Information diffusion and also information falls have been concentrated in the past in a disconnected route with generally little datasets. A rundown of models and strategies for information dissemination is depicted [9]. Two standard methodologies introduced there is the independent cascades (IC) and the linear threshold (LT) display. We proposed an agent-based multiple influences diffusion (AMID) model to dissect various impacts engendering in interpersonal organizations by thinking

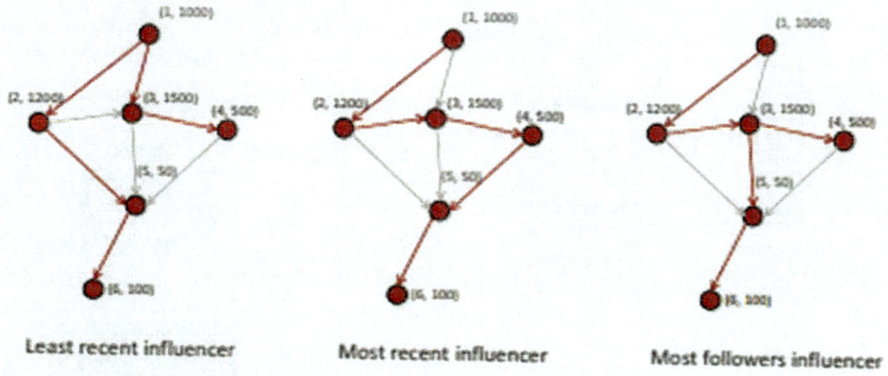

Fig. 1 Impact of influence models

about their connections. The impact dispersion process is displayed in a decentralized way by utilizing agent-based modeling (ABM) [10]. Informal organizations can be spoken to as a lot of hubs and edge shaping a network or chart, where the hubs are the members, and the edges are the sorts of associations. All the more explicitly in interpersonal organizations, individuals, or gatherings of individuals are the hubs, and the sorts of social connections they participate in are the edges [11]. Viral promoting is a method for publicizing items to a particular gathering of intrigued individuals utilizing on the Web interpersonal organization connections characterized it as focusing on a predetermined number of clients in the interpersonal organization by giving motivators and these focused on clients would then start the procedure of mindfulness spread by proliferating the information to their companions by means of their social connections [12]. To the best of our insight, the linear influence model created is the main genuine prescient model for the transient elements that has just been proposed. They considered the dispersion of hash labels [13]. Twitter proposed a model dependent on the suspicion that the impact of a hub relies upon what number of different hubs it affected in the past [14].

3 System Model

These works are characterized by the strategy utilized synchronous, nonconcurrent, impact amplification, and informal organization mining. There is an immense hole in creating viral showcasing strategies coordinated to business [15]. This article additionally addresses the confinements of some information dissemination models for viral advertising. We get great recreation results in instances of step-by-step evacuating impact edges, and we finished up and saw that there exist different impact ways, henceforth countless impacts. We ponder a significant number of the falls which are really influenced by various impact models, and how fall measurements and properties are influenced, in view of a solid physical structure. [16]. We influence

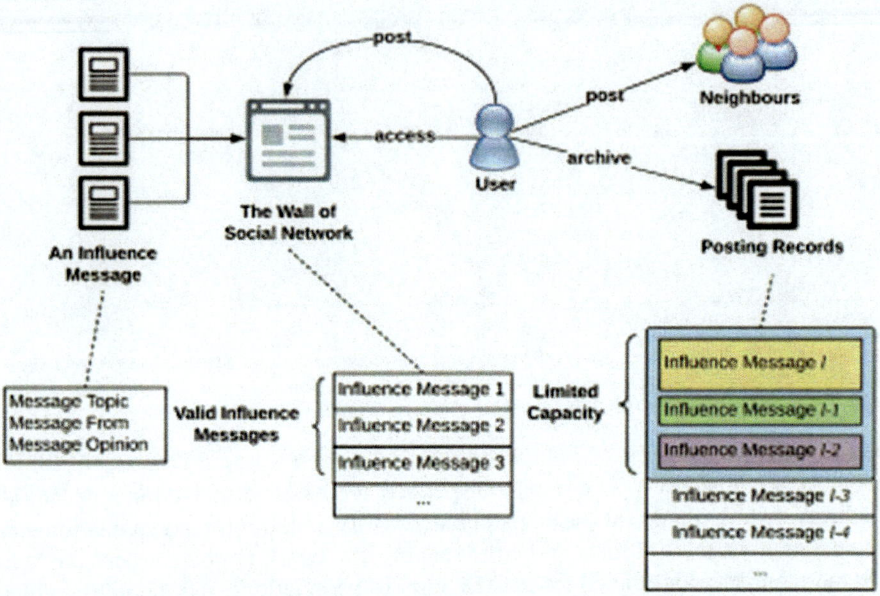

Fig. 2 Agent-based multiple influences diffusion model

an operator-based dispersion model to catch the transformative pattern of an interpersonal organization, just as the person's highlights and practices. Consequently, the numerous receptions of various developments by a specific client at various time steps can be empowered [17] (Fig. 2).

4 Proposed System

We propose to exploit a since quite a while ago followed conveyance, reverse Gaussian dispersion that is frequently utilized in survival information investigation to demonstrate the meager contamination timestamps with expansive vulnerability. We allude to the per user to model inference for more instincts [18]. Accordingly, let us recognize the presence and the level of the relationship in genuine information by connection examination. We acquaint the technique we propose with anticipate the dissemination marvel seen in online informal organizations. To begin, we formally characterize the time-based asynchronous independent cascades (T-BAsIC) model basic our methodology. At that point, we present the rundown of highlights registered for every individual from the network lastly we depict how they are utilized to gage the model parameters. We start by helping the definition to remember AsIC (for example, offbeat independent cascades model) as indicated, which is an augmentation of the IC display so the dissemination can unfurl inconsistent time. It displays the dispersion of information through a coordinated network $G = (V, E)$, where

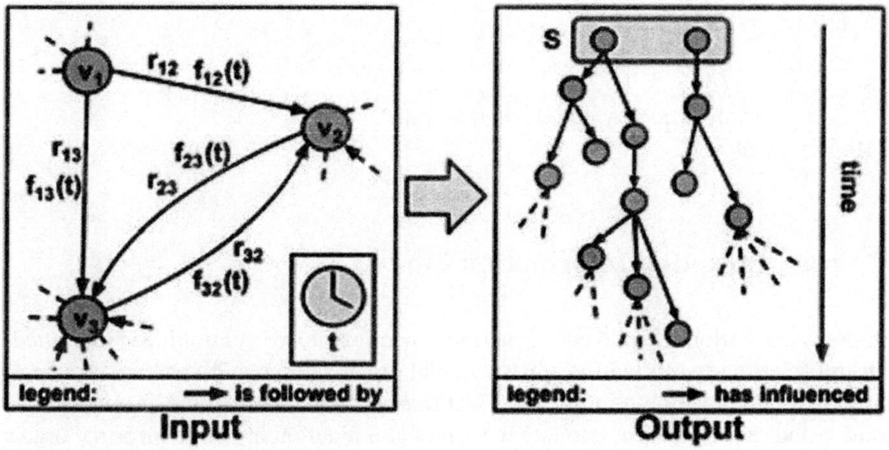

Fig. 3 Model predicts the cascade of diffusion

V is the arrangement of the considerable number of hubs, and $E(\subset V \times V)$ is the arrangement of the considerable number of connections (Fig. 3).

5 Recommendation Algorithm

Multiple activated neighbors. We have the likelihood that a client is initiated by one neighbor at any given moment. Give us a chance to consider the case that a hub v has different neighbors who are actuated to an archive. For the situation, one of the numerous neighbors can initiate v. To deal with this case, let Nv indicates the arrangement of the initiated neighbors of v. By and large, v will be initiated in the event that one of Nv enacts v. By accepting that each actuated neighbor of v autonomously enacts v from alternate neighbors, we can get the likelihood that v is initiated by somewhere around one of Nv. We consider it the actuation likelihood $\pi(v)$ of hub v and figure.

Algorithm: Ranking Procedure(G, d, t, H, C)

Input: G: an input graph, d: an input document, t: the current time, H: the history of users' activations before current time t, C: is a set of candidates

Output: L: A ordered list of candidates

1. Begin
2. $L = []$;
3. for $c \in C$ do
4. $\Pi(c) = 1$;
5. For $n \in$ activated neighbors(c, d, t) do

6. $\Pi(c) = \pi(c)(1 - \pi(n, c)$;
7. $\Pi(c) = 1 - \pi(c)$;
8. Insert $(c, \pi(c))$ into L;
9. Sort L for the second value of each topple;
10. Return L;

6 Viral Marketing Information Model

The previous sections displayed a general foundation of two central ideas identified with information dissemination: interpersonal organizations and viral showcasing. This article tends to the condition of Web-based social networking promoting in business and reveals insight into late work on dissemination in online informal organizations for viral advertising. These works are ordered by the strategy utilized: synchronous, nonconcurrent, impact boost, and informal organization mining.

6.1 Challenges in Information Diffusion for Viral Marketing

There are some general challenges identified with online informal organizations that influence the procedure of dispersion for viral showcasing. Loud and inadequate information are the most widely recognized difficulties in the online interpersonal organization structure recorded significant difficulties to dispersion in online informal communities, for example, gathering genuine examples, perceiving client standards of conduct, and huge scale recreation.

6.2 The VMID Model Structure

Addressing all the previously mentioned issues and difficulties requires an information systems model to keep up with the adequacy of viral advertising dissemination. Discoveries from the writing show the advantage of Web-based life showcasing in organizations. Viral promoting is a technique dependent on Web-based networking media advertising that will have better results contrasted and internet-based life showcasing in view of its one-of-a-kind method. The motivation behind the VMID demonstration is to address the issue of information dissemination for viral showcasing coordinated to mechanical B2B organizations. This model endeavors to fill the hole among organizations and Internet-based life advertising by embracing viral showcasing (Fig. 4).

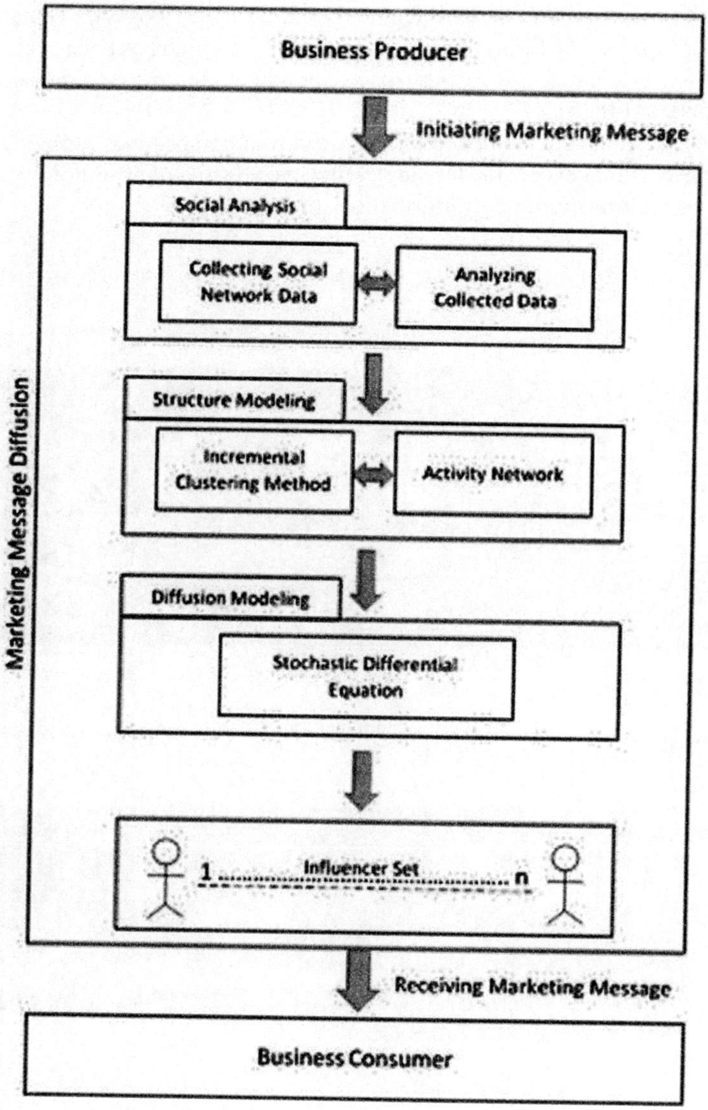

Fig. 4 VMID model

7 Experimental Results and Screenshots

We use the method proposed in [17] as a correlation strategy. The strategy predicts information dissemination with relational dispersion rate dependent on continuous-time Markov chain and can be connected specifically to our concern. The idea of dispersion was roused by common marvels, for example, sicknesses, waves, liquid, and water. For information dispersion models, this article proposes the selection of the factual idea "stochastic differential condition" (SDE), which has been systematically inspected for its adequacy in various spaces including buildings, connected science, and computers (Fig. 5).

We present on the course information spread four perspectives, utilizing the total model for everything except the last angle: first, we affirm our presumption that social connections are transporters of information (Fig. 6).

Fig. 5 Home page

Fig. 6 View information

Fig. 7 User and admin register form

We the nature of info information influences the reproduction impact ways. At that point, over remade falls we play out a fundamental examination and contrast it and past investigations on Twitter. As a fourth step, we explore how unique impact models portrayed in Sect. 3 apply an effect on remaking of falls (Fig. 7).

We have applied the AMID show in an all-inclusive adaptation of the impact amplification issue by thinking about how to stifle and limit the consistent effect of a specific impact message or feeling inside a settled time range. The goal of the analyzes is set to stifle a bothersome impact by using different techniques dependent on the AMID model. In the trials, three noteworthy kinds of impacts are included (Fig. 8).

The evolution of the network delays after some time steps. Next dependent on this state we endeavor different techniques to explore the heading of the networked advancement. Two situations are associated with the trials. (1) The informal organization is under the control of this association, having the benefits to control the topological structure of the interpersonal organization. (2) The association does not have any control to the social network.

Furthermore utilizing irrelevant impacts is not cost-proficient contrasted and others, yet it beats that of receiving the unfavorable feeling impact when the seed set size increments up to 25. We additionally measure and look at the spread of recently infused messages when any system has been embraced. It very well may be seen from Fig. 9 that the impact message topically applicable to different existing ones can without much of a stretch rule the interpersonal organization, and a low spending

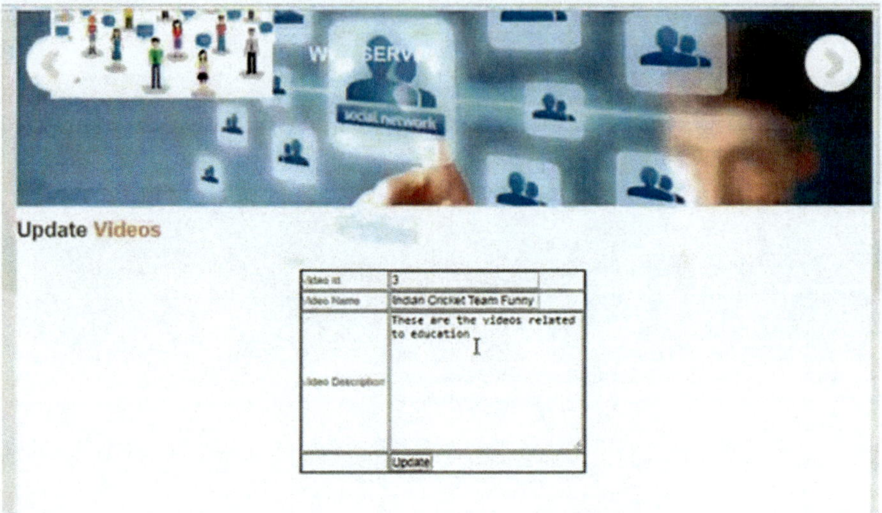

Fig. 8 Search results

plan of inexact ten seeds can nearly accomplish the greatest spread. Though, another message requires a much higher budget and appears not cost-efficient (Fig. 10).

Fig. 9 Information diffusion

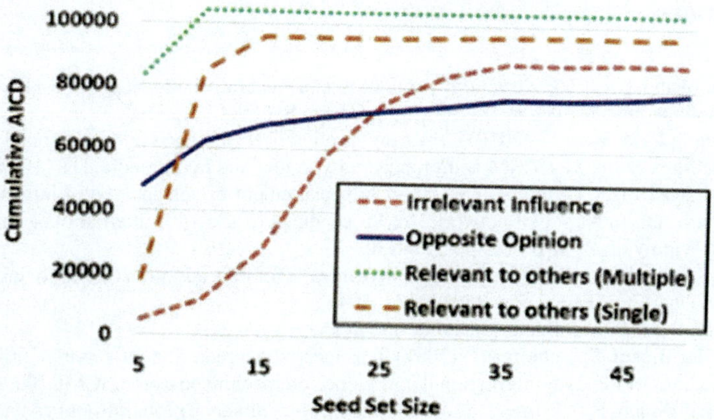

Fig. 10 Injected influence (cumulative AICD)

8 Conclusion and Future Work

We studied the problem of various impacts dispersion in interpersonal organizations and proposed an agent-based multiple influences diffusion (AMID) model to portray the issue by utilizing the ideas from multi-specialist systems. We demonstrated that such a remaking is possible, and social connections assume a transcendent job in information dispersion. Boisterous information has an effect, however, we comprehend which viewpoints are most basic and we take a shot at approaches to conquer these confinements. A primer examination of these falls demonstrates that they show essentially greater intricacy as past investigations showed, making ready for more extravagant investigations. In mechanical markets, it is a test to line up with a systematic strategy that encourages the association between accomplices. The VMID display requires more examinations and testing to guarantee its legitimacy in reality. Concerning the liberality of the model, another issue comprises in applying our methodology on other social information from different stages to examine both regular focuses and specificities of the information dispersion process as per the level. This model can be connected in different undertakings, for example, network discovery, information dissemination expectation, and perception. The test examines are directed on both engineered and genuine world datasets, showing its adequacy, strength, and versatility. Later on, we intend to research how to distinguish a fitting infusing impact message to limit the unwanted impact and investigate the circumstances when numerous new impacts can be infused into the social communication network.

References

1. Niu J, Wang L (2016) Structural properties and generative model of non-giant connected components in social networks. Sci China Inform Sci 59(12):123101
2. Au Yeung C-M, Iwata T (2010) Capturing implicit user influence in online social sharing. In: Proceedings of the 21st ACM conference on hypertext and hypermedia, HT '10, pp 245–254
3. Cha M, Mislove A, Gummadi KP (2009) A measurement-driven analysis of information propagation in the flickr social network. In: Proceedings of the 18th international conference on World Wide Web, WWW '09, pp 721–730
4. Chua F, Lauw H, Lim E-P (2013) Generative models for item adoptions using social correlation. Knowl Data Eng IEEE Trans 25(9):2036–2048
5. Cialdini RB (1993) Influence: science and practice
6. Dietz L, Bickel S, Scheffer T (2007) Unsupervised prediction of citation influences. In: Proceedings of the 24th international conference on machine learning, ICML '07, pp 233–240
7. Goyal A, Bonchi F, Lakshmanan LV (2010) Learning influence probabilities in social networks. In: Proceedings of the third ACM international conference on web search and data mining, WSDM '10, pp 241–250
8. Iwata T, Shah A, Ghahramani Z (2013) Discovering latent influence in online social activities via shared cascade poisson processes. In: Proceedings of the 19th ACM SIGKDD international conference on knowledge discovery and data mining, KDD '13, pp 266–274
9. Kraut RE, Rice RE, Cool C, Fish RS (1998) Varieties of social influence: the role of utility and norms in the success of a new communication medium. Org Sci 9:437–453
10. Kwak H, Lee C, Park H, Moon S (2010) What is twitter, a social network or a news media? In: Proceedings of the 19th international conference on World Wide Web, WWW '10, pp 591–600
11. Li D, Xu Z, Luo Y, Li S, Gupta A, Sycara K, Luo S, Hu L, Chen H (2013) Modeling information diffusion over social networks for temporal dynamic
12. Katz E, Lazarsfeld PF (1955) Personal influence: the part played by people in the flow of mass communications, vol 21. Free Press, New York
13. Kempe D (2003) Maximizing the spread of influence through a social network. In: KDD'03, pp 137–146
14. Kimura M, Saito K, Motoda H (2009) Blocking links to minimize contamination spread in a social network. ACM Trans Knowl Discov Data 3(9):1–9
15. Kwak H, Lee C, Park H, Moon S (2010) What is Twitter, a social network or a news media? In: WWW'10, pp 591–600
16. Mitchell TM (1997) Machine learning. McGraw Hill series in computer science
17. Newman MEJ (2003) The structure and function of complex networks. SIAM Rev 45:167–256
18. Romero DM, Meeder B, Kleinberg J (2011) Differences in the mechanics of information diffusion across topics: idioms, political hashtags, and complex contagion on twitter. In: WWW '11, pp 695–704
19. Hughes A, Palen L (2009) Twitter adoption and use in mass convergence and emergency events. Int J Emerg Manag 6(3):248–260

Regionalization of Precipitation in Andhra Pradesh and Telangana State by Using PCA

Gore Vikas Sudam, Ashutosh Chaturvedi, and K. V. Jayakumar

Abstract Precipitation is one of the most important climatic variables for sustaining civilization. It is characterized by extremely high variability in space and time. The magnitude and frequency of precipitation are of great significance in hydrologic and hydraulic design and has wide application in various areas. However, precipitation data is available only in those regions where the rain gauges are installed. The magnitude and frequency of precipitation in ungauged sites can be assessed by grouping areas with similar characteristics. The procedure of grouping of areas having similar behaviour is termed as "regionalization". Principal component analysis (PCA) which is based on consecutive sieving of stations was initiate for regionalization. PCA used for the reduction of the scale of the data set with minimum reducing of informations. Principal elements (PCs) that are orthogonal to every alternative are derived from inter-station correlation and/or variance matrix of precipitation within the study space. If the primary few leading PCs account for vital % of the full variance, their spatial patterns are analyzed to make undiversified precipitation regions. This approach may involve either plotting the unrotated and/or revolved computer loadings on the map of the study space or representing stations as points in two-dimensional house of leading PCs.

Keywords Precipitation · Regionalization · Principal component analysis · Normalization · Eigenvalue · ArcMap

---

G. V. Sudam (✉) · A. Chaturvedi
Department of Civil Engineering, Alamuri Ratnamala Institute of Engineering and Technology, Thane, Maharashtra 421601, India
e-mail: vikassgore807@gmail.com

K. V. Jayakumar
Water and Environmental Division, Department of Civil Engineering, National Institute of Technology Warangal, Hanamkonda, India
e-mail: kvj@nitw.ac.in

J. R. Szymanski et al. (eds.), *Energy Systems, Drives and Automations*,
Lecture Notes in Electrical Engineering 1057,
https://doi.org/10.1007/978-981-99-3691-5_46

1 Introduction

The present study is taken up the following objectives: to identify, analyze, and describe hydrologically homogenous regions using regionalization approaches, and the magnitude and frequency of precipitation in ungauged sites can be assessed by grouping areas with similar characteristics. The meaning of region is a set of sites depicting similar hydrological behaviour, and the meaning of regionalization is the process of identifying regions. Regional frequency analysis is the frequency analysis based on regional information.

The conventional apply in regionalization was to delineate regions as geographically contiguous areas supported physiographic and/or political/administrative boundaries. As such, regions might not have any definite relationship to causal/ explanatory variables influencing rain. Regions supported those factors need not be solid in rain. To delineate effective solid regions, a range of regionalization approaches are developed over the past six decades. They embody those supported correlation analysis (e.g. elementary linkage analysis and its variation) principal component analysis (PCA) additionally noted as eigenvector analysis or the connected, empirical orthogonal operate (EOF) analysis, divisor analysis, spectral analysis, cluster analysis, and PCA in association with cluster analysis, stratified approach, and region of influence.

Principal component analysis (PCA), approach of regionalization that is usually noted as eigenvector analysis or empirical orthogonal operate analysis uses the principal component analysis (PCA), PCs that area unit orthogonal to every different area unit derived from inter-station correlation and/or variance matrix of precipitation was enforced by Srinivas [1]. The author had used seasonally and monthly summer monsoon rain (SMR) of ninety stations. Six regions were known to be positively heterogeneous, another 3 were probably heterogeneous once homogeneity testing, and 10 afterwards coherent solid rain regions were known once applying the procedure to SMR. If the primary few leading PCs account for important PCs of the full variance, their spatial patterns area unit analyzed to make solid precipitation regions. This approach may involve either plotting the unrotated and/or turned computer loadings on the map of the study space, or representing stations as points in two-dimensional house of leading PCs. Within the context of regionalization of utmost rain, regional frequency (growth) curves of precipitation extremes area unit created for every of the described regions victimization pooled data from the region or by fitting regional regression relationship between PCs and parameters of the distribution. The developed curves or relationships area unit helpful to attain quantile estimates at sites that area unit ungauged or have inadequate knowledge.

Almazroui et al. [2] used the PCA for determination of the homogenous groups in Saudi Arabian climate were carried out by grouping similar long-term climatological features by using the two variables, precipitation, and air temperature. Five groups were identified: group A (Northern), B (Red Sea Coastal), C (Interior), D (Highland), and E (Southern). Each of the climatic groups exhibited a unique feature (with strong similarities among stations that belong to the same group), especially in terms of inter-annual fluctuations and the degree of seasonality. This simplified overview of climatic characteristics was beneficial for assisting in economic planning, especially for any semi-arid or arid areas that are naturally vulnerable to climatic variability and extremes. Darand and Daneshvar [3] used the S-mode PCA for seasonal correlation matrix with eight derived variables which had been mapped using GIS. Based on the eight eigenvalues, three factors were extracted between the component and varimax rotation to enhance interpretability of retained PCA scores. The authors considered seasonal precipitation data with eight derived variables and used mean seasonal precipitation amount and its percentage had been interpolated. Using correlation matrix, as opposed to the covariance matrix, the dry grids in the deserts and wet grids in the mountains directly compared.

The recommendations of Miller et al. [4] were to apply the rotated orthogonal based on varimax scores to achieve eigenvector-based regionalization regime of seasonal precipitation. Author considered three rotated components, based on screed plot and eigenvalues over 1.0, the percentage of variance of the data explained as 88.95%. Darand and Daneshvar [3] considered as important of the loading PCs and showed good correlations between the variables and the PCs which had the correlation coefficient > 0.7. Regression method used to calculate the standardized rotated PC scores. The PCs loading values got PC1 which was explained 47.2% of the total variance, PC2 was explained 23.77% of the total variance, and PC3 was explained 17.96% of the total variance. Munoz-Díaz and Rodrigo [5] said that the purpose of PCA was to identify the most important correlation structures between numbers of variables in order to obtain a description of the major part of the overall variance with few linear combinations based on the original variables. Richman [6] summarized the multiple of techniques which were used for rotations in a PCA in a climatological context. The goal of all these rotational strategies was to obtain a clear pattern of loading which was clearly marked by high loading for others. Typical rotational strategies were VARIMAX and QUARTIMAX. VARIMAX allowed maximization of the variances of each individual record for the various factors.

2 Study Area and Data

Meteorological Data: The Indian states like Telangana State (TS) and Andhra Pradesh (AP) were selected as the study area as shown in Fig. 1. Daily precipitation data sets were acquired for the period of 113 years ranging from 1900 to 2013 for the State of AP and TS. The data has been collected from Indian Meteorological Department (IMD) [7]. The precipitation data here was arranged in the grid patterns of $0.25° \times 0.25°$ grid cells. It could be implied that each grid cells will have an area of 625 km^2 (25 km $\times$ 25 km). The total number of grid cells taken in AP and TS is 433. The daily precipitation data for 433 rain stations, which is shown in Fig. 2, has been used for further analysis. If data is available which is not in required format and degree of resolution, then we have to interpolate the data in required format.

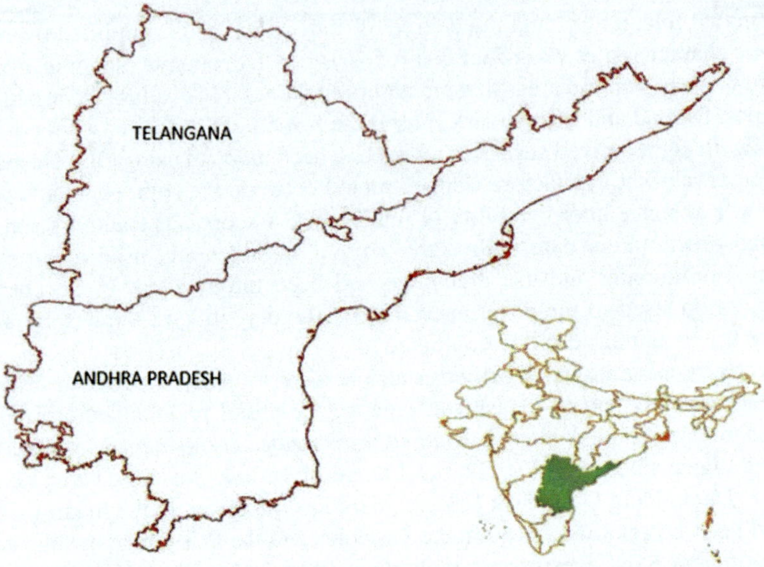

Fig. 1 Study area

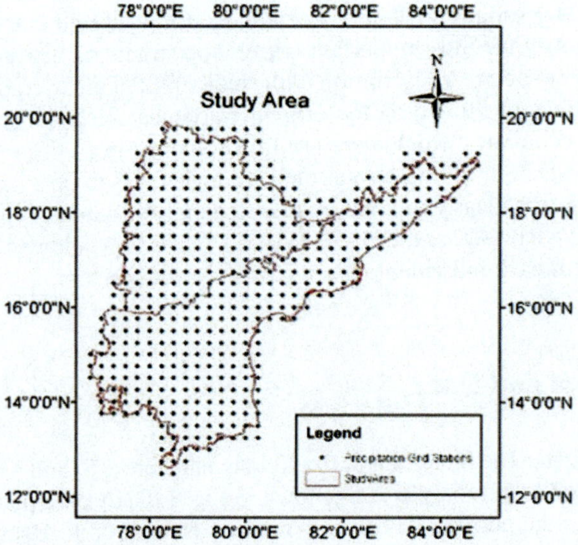

Fig. 2 Study area with 0.25° × 0.25° grid stations

3 Methodology

The methodology adopted in this research is based on the formulation of the research question, literature study, and concept built during the course period. To delineate effective homogeneous regions, a variety of regionalization technique have been developed over the six decades.

Principal component analysis (PCA) is a method that uses a corresponding degree conversion to change the position from a group of observations that may be related to a group of values of inconsistent variables called primary components. If there is a unit detection n variable p, then the number of distinct main components is min $(n - 1, p)$. This change is somehow explained by the fact that the main object has the greatest possible variation (i.e. calculating the highest number of possible variations of information), and everything in sequence has the best variation that can be made under the orthogonal limit in the previous parts. The following vectors (each is a combination of linear variations corresponding to the degreed containing the visibility) are the unrelated orthogonal base element. PCA is sensitive to the related growth of initial variability.

PCA was fictional in 1901 by Karl Pearson, as associate degree analogue of the axis theorem in mechanics; it absolutely was later severally developed and named by Harold Hotellingin within the PCA is often used as a tool to search for information analysis and to build prophetic models. PCA is caused by the eigenvalues decay of information variability (or correction) of the matrix or the single price decay of the information matrix, usually as a first step in the formation of information. The results of the PCA area unit are generally defined in terms of object scores, commonly referred to as output scores (variable values used are the same point of information), and loading (weight that all relevant initial variables should be increased to make feature points). If the element scores area of a unit is matched by unit variations, the loading should have a variety of information in them (and that is the size of the eigenvalues). If element scores do not appear to be rated (and therefore contain variations in data), then loading should be measured in units (standard), and this weight area is called eigenvectors; they are cosines for the orthogonal conversion of variables into main components (PCs) or back. PCA is a widely used method of mathematics in climate and climate. When one is faced with a very large database, one attempt to reduce the size of the database, while minimizing any loss of information, the purpose of these mathematical methods is to better understand and interpret the structure of the data. A standard database can be viewed as a variable measurement of p. For example, if 113-year daily rain data for 433 different recording stations is available, in the study area, there are $n = 41{,}273$ views (number of rainy days), of $p = 433$ variable (rain channel number); there are also other types of data, for example, climate change is measured at n stations simultaneously. Typically, the p variables are highly concentrated (especially if the p channels are locally close). One must therefore choose a variable ($m < p$) that displays all the information contained in the original matrix. This can be done by building new variables that are different from the original but built on them. The simplicity of the PCA approach lies in this restriction of specific functions of the original variable.

4 Results

In the study, area over 75% of the annual precipitation is received during the five rainy months of June–October, only thereby leading to large variations in rainfall on temporal scale. Monthly normal precipitation in the study area is presented in Table 1 shows the monthly normal precipitation data (in centimetre) varying with time of a few selected station.

The first step in building a PCA structure is to build a matrix coefficients matrix using a data set of 433 channels. Thereafter calculated eigenvalues and eigenvectors of the coefficient's matrix for merging. The number of PCs with functional value is determined using eigenvalues. Eigenvectors are made up of a combination of linear symbols in the data set. The coefficients for each original attribute provide an indication of agreement or inconsistency in the actual signals to the new size (PC). The familiarity of each attribute consists of the focus of the statement, subtracting the amount of data from each of its variables so that its approximate value (value) is zero and adjusting each variation to make it equal to 1. The normalized set of data gives the mean as zero and standard deviation as 1. Figure 3 illustrates the variation of normalized monthly normal precipitation data of a few selected stations.

Principal components (PCs) are new flexible elements formed as line combinations or first-generation compounds. This combination is made in such a way that the new variables (i.e. principal components) are incompatible, and most of the details within the original variable are highlighted or pressed into the original elements.

Table 1 Monthly normal precipitation data (in cm)

Month	Precipitation data (cm)	Percentage	Cumulative percentage
October	217	20	20
September	195	18	38
July	164	15	53
August	164	15	69
June	109	10	79
November	88	8	87
May	41	3	91
December	26	2	94
April	23	2	96
February	17	1	97
March	11	1	99
January	10	1	100

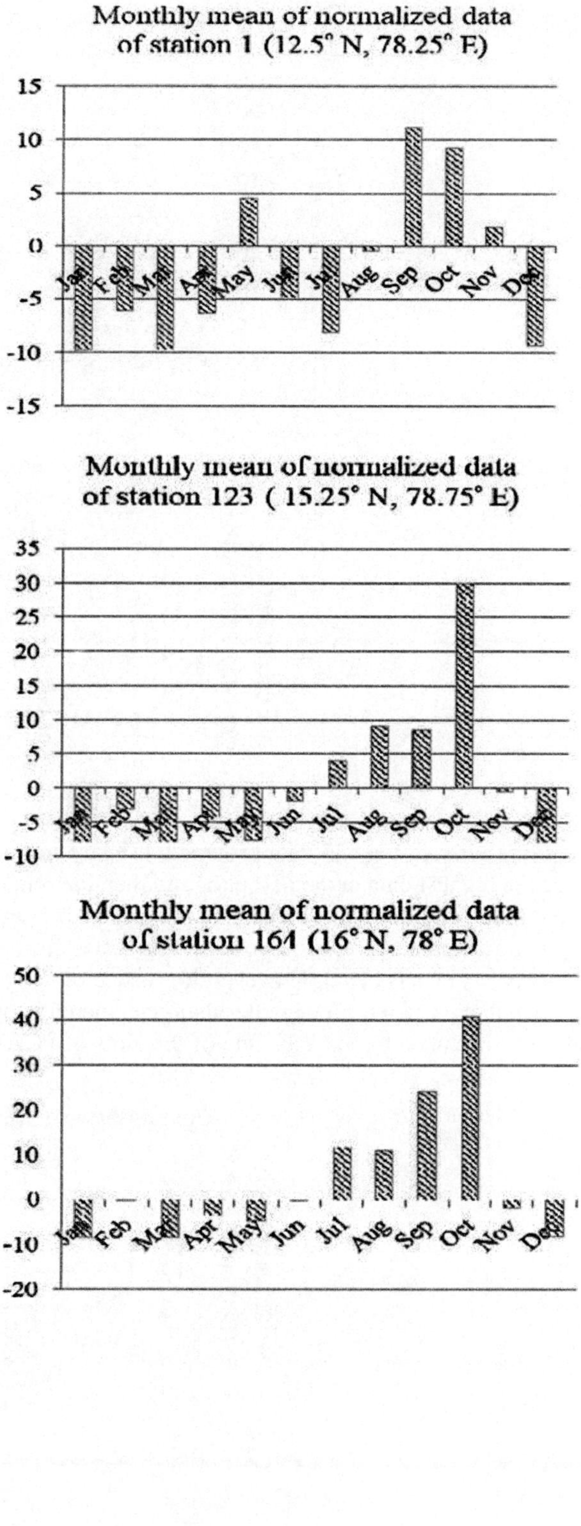

Fig. 3 Monthly mean of normalized data of few selected stations

Fig. 3 (continued)

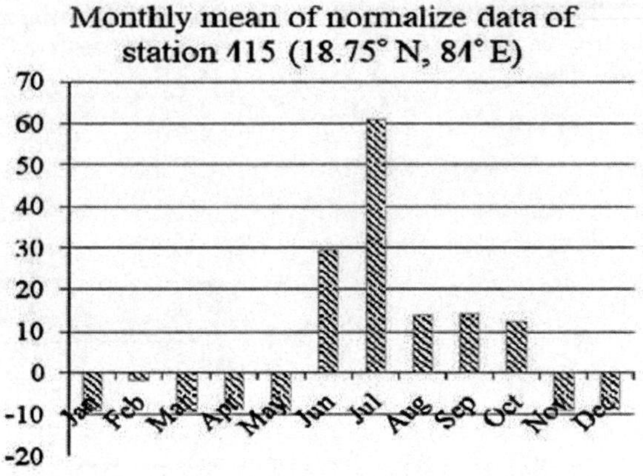

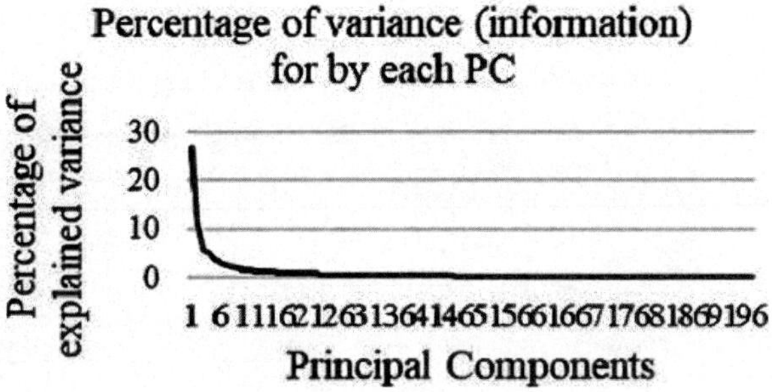

Thus, the view of the 433 data provided 433 key components, but the PCA tries to place the highest data in the first part, and then the remaining data in the second and so on, until something like screens in the scree structure of Fig. 4.

According to the scree plot shown in Fig. 4, it is not clear whether keeping fourteen or ninety-six PCs is better. However, if it is decided to choose to keep only fourteen PCs which have eigenvalue greater than one, then the total variability of the original data set is reduced to 66.67%. One of the aims of PCA is to reduce the dimensions

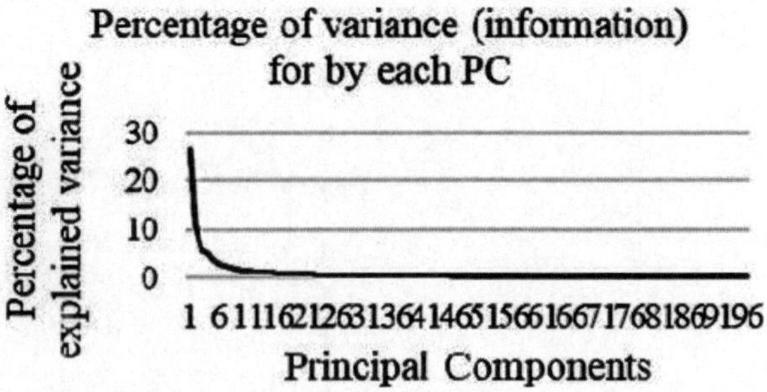

Fig. 4 Principal components

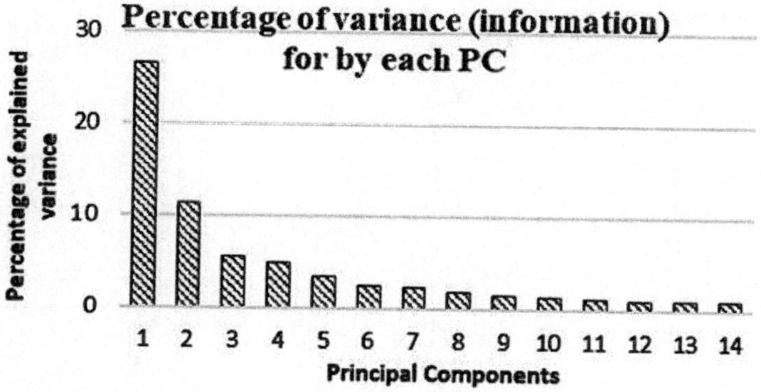

Fig. 5 First fourteen PCs

whiles keeping the maximum variability of the original data set, the number of PCs is too large if the ninety-six PCs are chosen, it is decided to take only first 14 PCs which have eigenvalues greater than one. Figure 5 shows the first 14 principal components, and the Table 2 shows the eigenvalues of them, ordered from the highest to the lowest variability of the data set. The first 14 PCs, which have the eigenvalue greater than one, which will covered the 66.67% of the total variance. For covering the 90% of the total variance, it gave the 97 eigenvalues or PCs. The first PC has the value of 114.95, which explains the 26.54% of the total variance. Similarly, the remaining 13 PCs, which have the values of 49.01, 24.11, 21.45, 15.46, 11.25, 10.47, 8.55, 7.17, 6.37, 5.41, 5.00, 4.80, and 4.58, explain the total variance of 11.32, 5.56, 4.95, 3.57, 2.59, 2.41, 1.97, 1.68, 1.47, 1.25, 1.15, 1.10, and 1.05%, respectively.

Figure 5 shows the first 14 PCs, and the Table 2 shows the values of the PC, explained variance and the cumulative variance of the highest 14 PCs.

The first principal component vector, i.e. the first column of the final data set, is correlated with the standardized original data set and using the correlation data the homogeneous region that has the strong correlation among them is identified.

The loading value of PC1 explains 26.54% of total variance and indicates positive correlation with all precipitation stations. The normalized PC score shows high positive values in the Papikondalu forest region in catchment area of Godavari River, East Godavari district as well as West Godavari district of mouth of the Godavari River, Krishna district in Krishna river catchment, Vishakhapatnam district of AP, Laknavaram forest area eastern part of the TS, and some south-west station of TS. The first factor strongly correlated with that region which has highest precipitation region of AP and TS. The Nellore district of southern part of the coastal AP and Chittoor and Anantapur district of southern part of the Rayalseema region of AP have a lowest PC score. Figure 6 illustrates the regions based on correlation ranges of PC1.

The loading value of PC2 explains 11.32% of the total variance. The standardized PC score shows high efficacious values in Penna river catchment area and Tirumala forest region of AP. The northern and the eastern part of the TS and the northern

Table 2 Eigenvalues, explained variance, and cumulative variance of PCs

Factor	Eigen values	Explained variance	Cumulative variance
1	114.95	26.54	26.54
2	49.01	11.32	37.86
3	24.11	5.56	43.4
4	21.45	4.95	48.39
5	15.46	3.57	51.96
6	11.25	2.59	54.56
7	10.47	2.41	56.98
8	8.55	1.97	58.95
9	7.17	1.65	60.61
10	6.37	1.47	62.08
11	5.41	1.25	63.33
12	5.00	1.15	64.49
13	4.80	1.10	65.60
14	4.58	1.05	66.66

part of the AP have negative correlation with second factor. Figure 7 illustrates the regions based on correlation ranges of PC2.

The loading value of PC3 explains 5.56% of the total variance. The standardized PC score shows high efficacious values in Hyderabad city area and Nallamala forest area of TS. Third factor negatively correlated with larger area of the northern part and the eastern part of the TS and the Rayalseema region, the Guntur, Prakasam, and Nellore district of the coastal AP. Figure 8 illustrates the regions based on correlation ranges of PC3.

The loading value PC4 explains 4.95% of the total variance. The standardized PC score shows high efficacious values in Tirumala forest region and coastal region near the Pulicat Lake south east part of the AP. The southern part and the central part of the TS and the Rayalseema region, the Guntur, Prakasam, Krishna, East Godavari, Vizianagaram, and Nellore district of the coastal AP have negative correlation with fourth factor. Figure 9 illustrates the regions based on correlation ranges of PC4.

Figure 10 shows the regions of PC5–PC14. The loading value of PC5 explains 3.57% the total variance. The standardized PC score shows zenith positive values in the Nallamalla forest region in the TS and Tirumala mountain ranges, and negative values in the Rayalaseema region, and Northern parts of the TS and AP. The PC6 loading value explains 2.59% of the total variance. The standardized PC score shows zenith positive values in the Palkonda forest ranges and has positive PC score Tirumala mountain ranges and Nallamala forest regions, the northern parts of the TS and AP.

The loading value PC7 explains 2.41% of total variance. The standardized PC score shows high positive values in the southern part Rayalaseema region, the Kondapalli forest region of the AP. It has positive values in the catchment area of the Musi

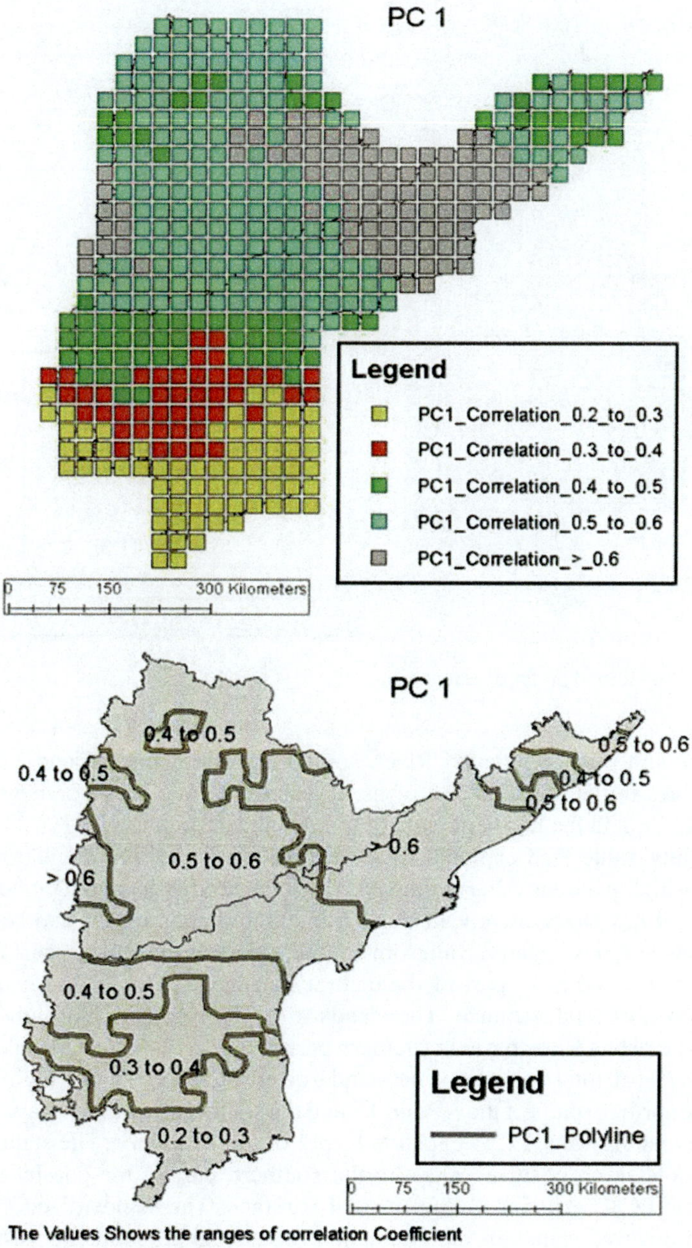

Fig. 6 Regions of first principal component

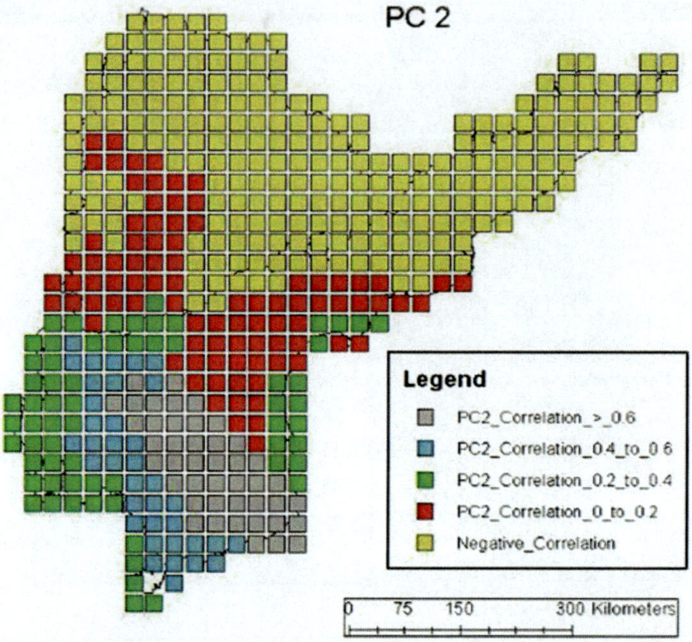

Fig. 7 Regions of second principal component

River, the mouth of the Godavari River, southern parts of the AP and TS. It has negative values in the northern part of the Rayalaseema region, catchment area of the Penna River, and the northern parts of the AP and TS.

The loading value PC8 explains 1.97% of total variance. The standardized PC score shows high positive values in the North Godavari district and Cumbun forest range in AP. It has the positive values in Tirumala mountain ranges and Nallamala forest regions and has negative values in southern part of the Rayalaseema region, north part of AP, and most part of the central Telangana. The loading value PC9 explains 1.65% of total variance. The standardized PC score shows high positive values in the Krishna River basin in southern part of the TS. The loading value PC10 explains 1.47% of total variance. The standardized PC score shows high positive values in the northern parts of the AP and TS and the southern part of the Rayalaseema region. The loading value PC11 explains 1.25% of total variance. The standardized PC score shows high positive values in the southern part of the coastal AP. The loading value PC12 explains 1.15% of total variance. The standardized PC score shows high positive values in the Nallamala forest region, southern parts of the Rayalaseema region, and the northern parts of the TS. The loading value PC13 explains 1.10% of total variance. The standardized PC score shows high positive values in the Koundinya forest region in southern part of the Rayalaseema. The loading value PC14 explains 1.05% of total variance. The standardized PC score shows high positive values in the Koundinya forest region in southern part of the

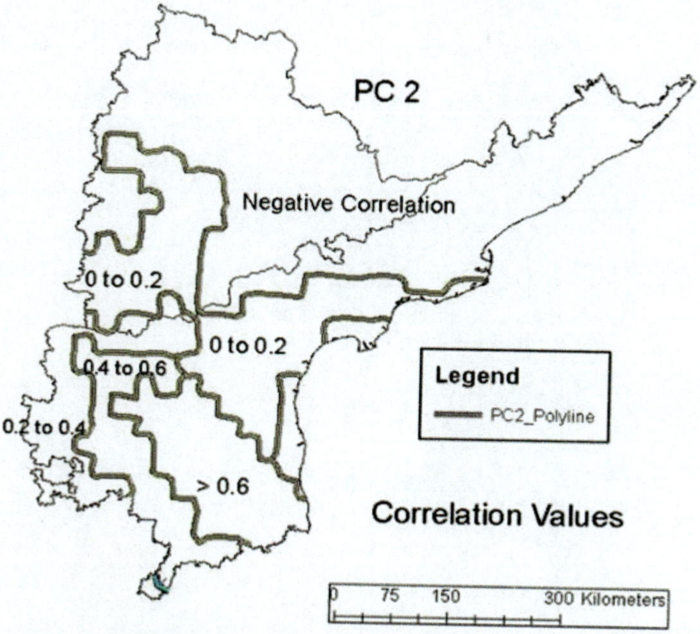

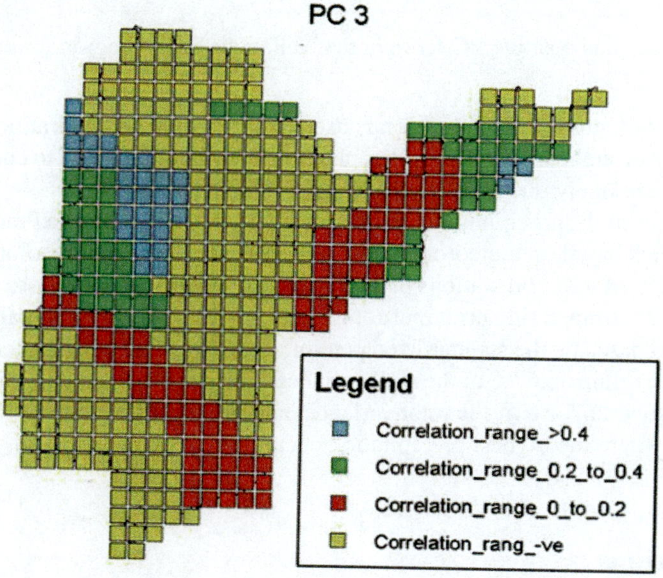

Fig. 8 Regions of third principal component

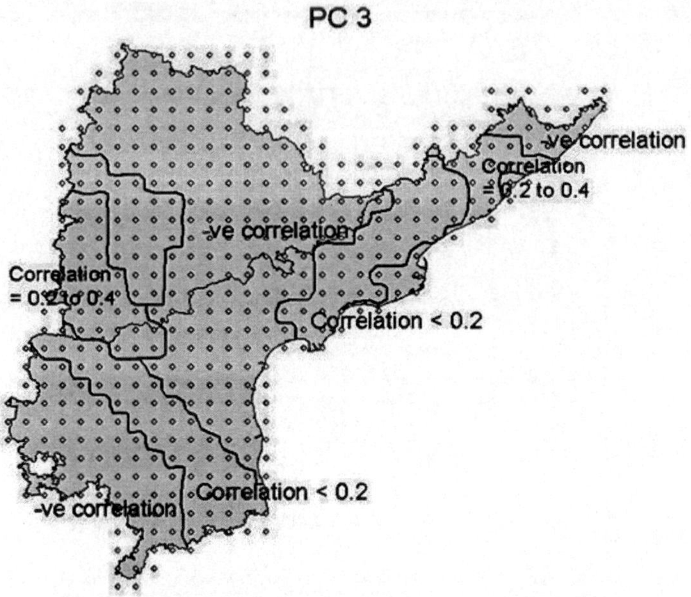

Fig. 8 (continued)

Rayalaseema and has positive PC score in the Nallamala Forest region, Laknavaram forest regions.

Summary: In this study, regionalization is used to hydrologic and hydraulic design, meteorological drought analysis as well as planning of agricultural area to cope water shortages that are likely due to low rains.

In the present study, precipitation regions are regionalized for Andhra Pradesh and Telangana States based on meteorological daily precipitation spatial resolution data of $0.25° \times 0.25°$ of 433 grid stations of 113 years during 1901 to 2013, provided by IMD by using ArcMap. Using principal component analysis, the 14 principal components extracted based on the standardized precipitation data which have eigenvalue > 1. The 14 PCs explain 66.67% of the total variance. The standardized PC score of the first PC explained 26.54% of the total variance, and highest PC score was obtained in the area of where the highest precipitation occurs in the study area.

5 Conclusions

The main objective of the study was to identify the homogeneous regions those that has similar characteristics of precipitation using regionalization approaches with the help of ArcMap. The states in study area are too largely extended to be considered as a similar characteristics (homogeneous) region.

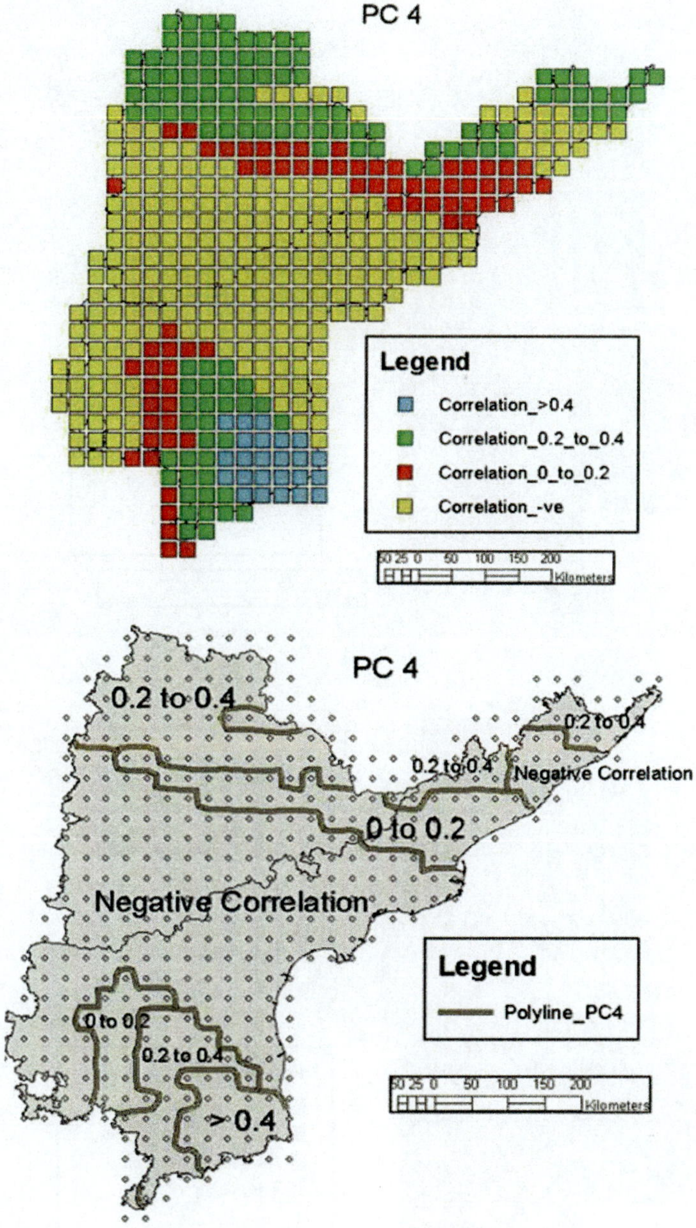

Fig. 9 Regions of fourth principal component

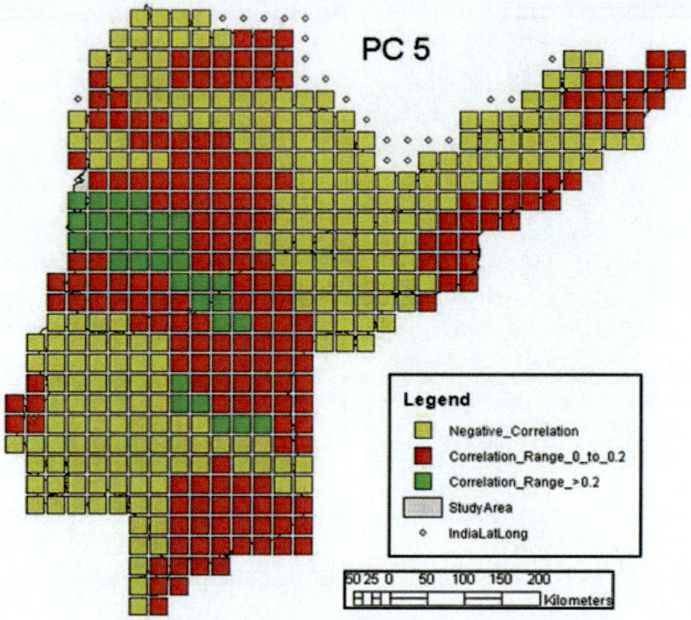

a

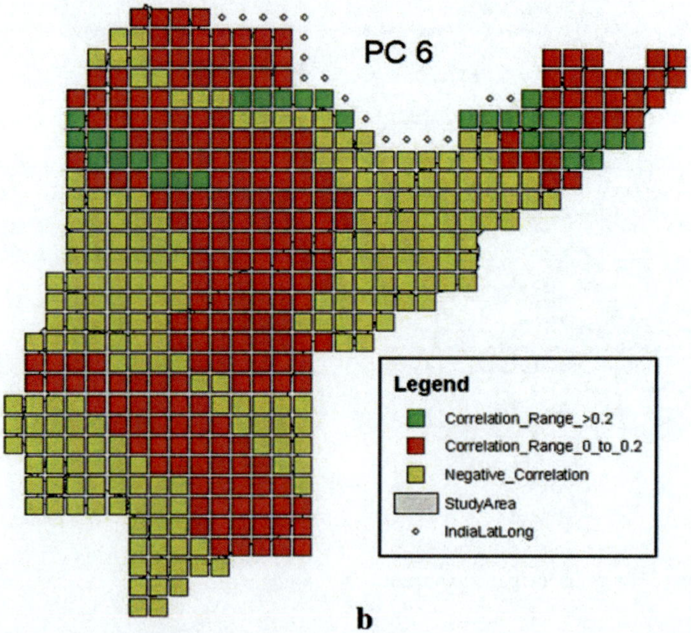

b

Fig. 10 Regions of 5th–14th PCs

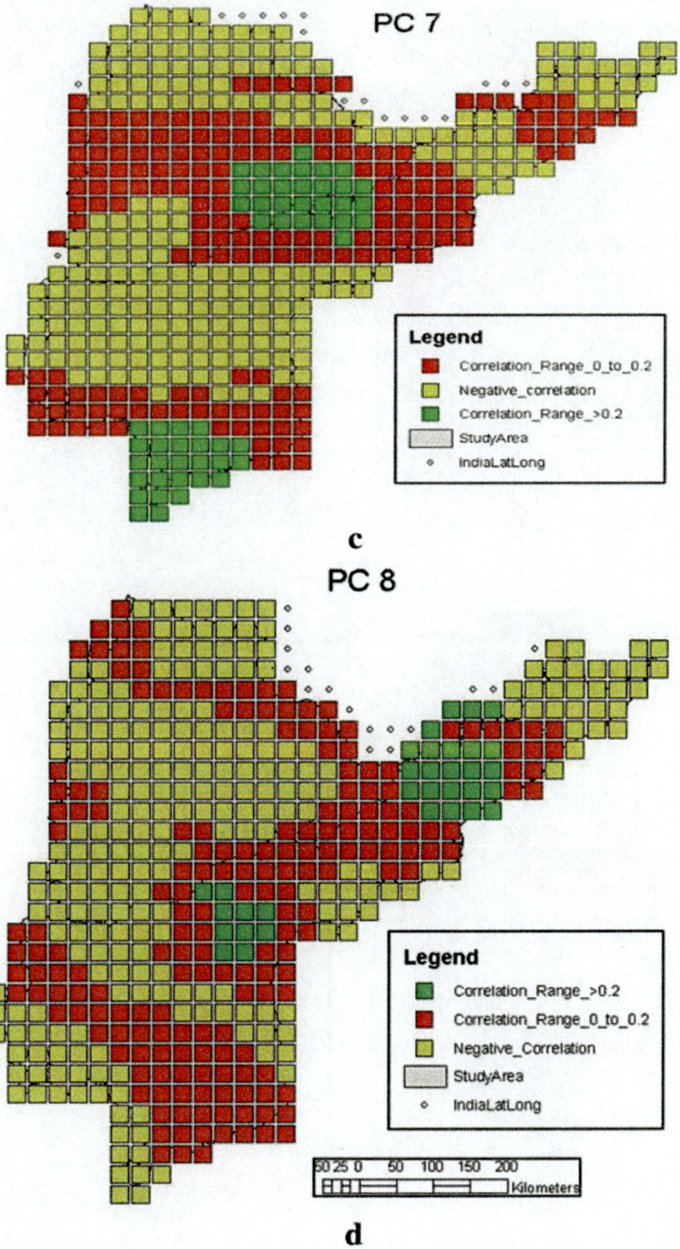

Fig. 10 (continued)

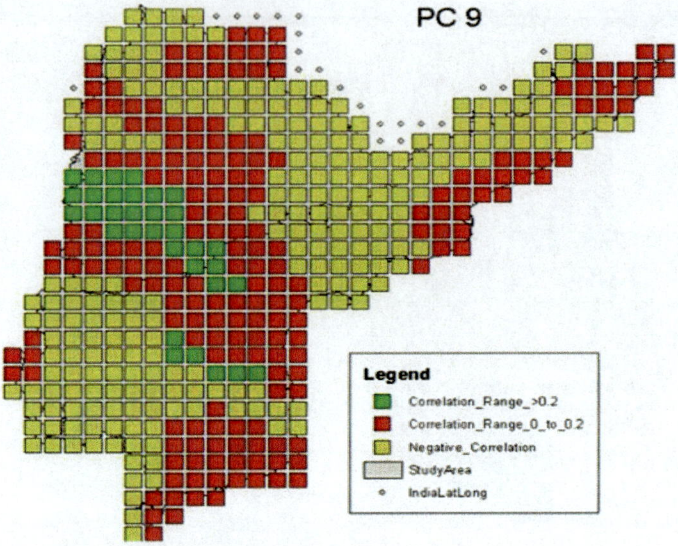

e

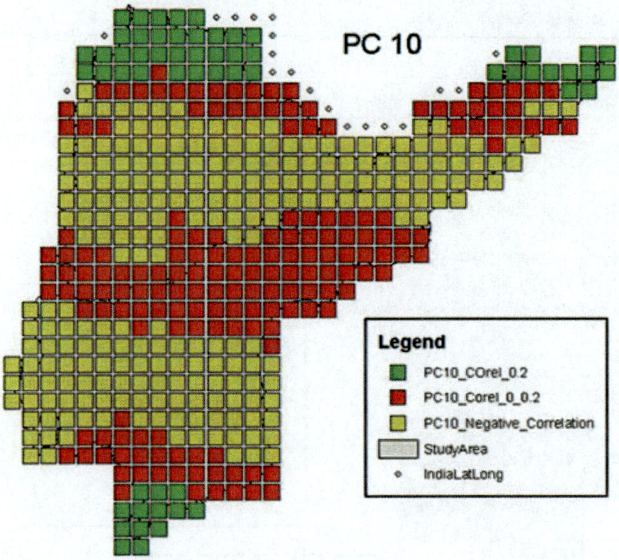

f

Fig. 10 (continued)

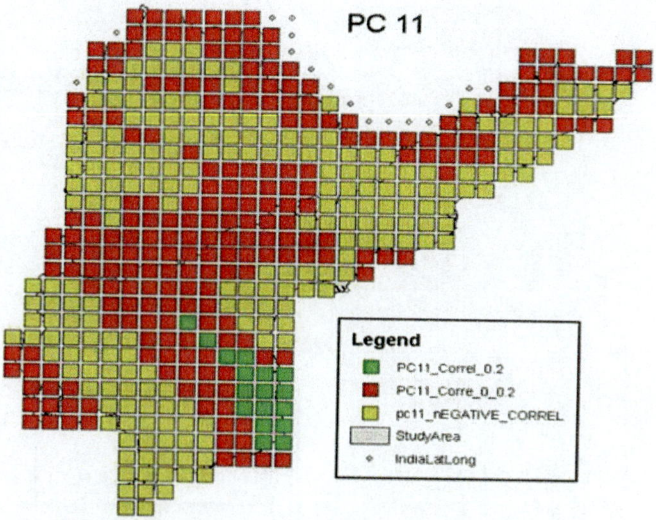

g

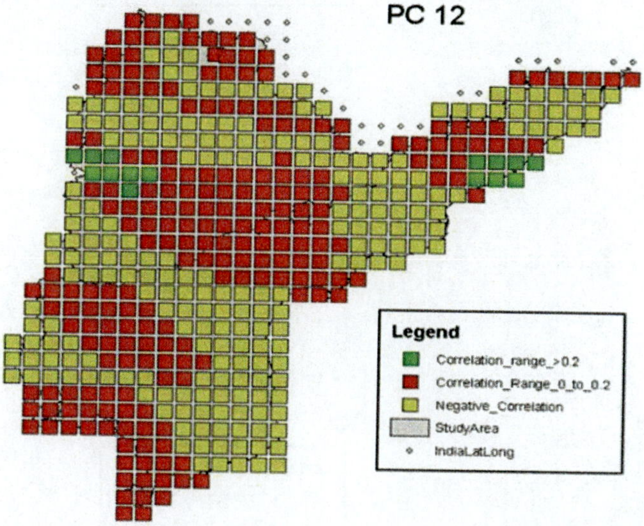

h

Fig. 10 (continued)

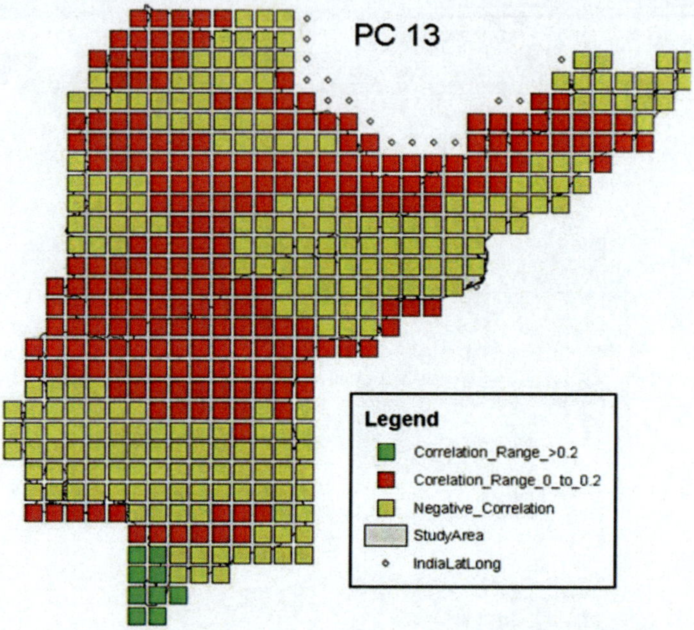

i

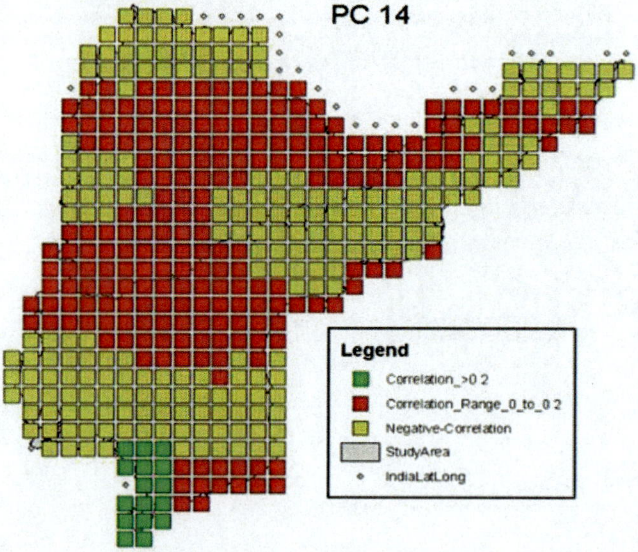

j

Fig. 10 (continued)

The 14 PCs have the eigenvalue > 1 which is explaining only 66.67% of the total variance of the daily precipitation data. For covering 90% of the total variance, it is taking 97 principal components. The main purpose of the PCA is to reduce the variables with minimum loss of information, so that with a few reduced variables, further analysis can be carried out. But here, the number of reduced variables still large. If only first 14 PCs which have eigenvalues more than 1 are considered, it covered only 66.67% of the information. So the PCA advise is not the best possible choice to descried homogeneous for daily precipitation data, because the first few leading PCs (derived from inter-station correlation and/or covariance matrix) do not cumulative account for a consequential per cent of the total variance shared by stations.

References

1. Srinivas VV (2013) Regionalization of precipitation in India: a review. J Indian Instit Sci 93(2):153–162
2. Almazroui M, Dambul R, Islam MN, Jones PD (2015) Principal components-based regionalization of the Saudi Arabian climate. Int J Climatol 35(9):2555–2573
3. Darand M, Daneshvar MRM (2014) Regionalization of precipitation regimes in Iran using principal component analysis and hierarchical clustering analysis. Environ Process 1(4):517–532
4. Miller SN, Semmens DJ, Goodrich DC, Hernandez M, Miller RC, Kepner WG, Guertin DP (2007) The automated geospatial watershed assessment tool. Environ Modell Softw 22(3):365–377
5. Munoz-Diaz D, Rodrigo FS (2004) Spatio-temporal patterns of seasonal rainfall in Spain (1912–2000) using cluster and principal component analysis: comparison. Ann Geophys 22(5):1435–1448
6. Richman MB (1986) Rotation of principal components. J Climatol 6(3):293–335
7. Indian Meteorology Department. http://www.imd.gov.in/

Speaker Recognition Using 3D Convolutional Neural Network and GMM

Manoj Rajput, Krishna Chauhan, Girdhar Gopal, Arun Kishor Johar, Ashutosh Tripathi, and Tarun Varma

Abstract Deep learning models are now regarded as state of the art in several fields of pattern recognition, including image identification and speech recognition. This paper implements a three-dimensional convolutional neural network-based model to identify a speaker with short audio signals from the available set of speakers enrolled during the model training. In order to improve the overall operation and performance of recognition, the proposed technique uses Log-Mel spectrograms and Mel-frequency cepstral coefficients (MFCCs) with their first and second differential features as input for the 3D CNN model. Furthermore, this work comprises both the design and implementation of the CNN model and employs the same prosody elements for a Gaussian mixture model (GMM) architecture. Before being retrieved for the final speaker recognition task, the audio features are pre-processed using a pre-emphasis filter for the spectral shape compensation. As compared to the relatively complex standard GMM designs, this is a significant finding in understanding how deep learning models can be applied to the problem of speaker recognition. The CNN model is trained and tested against a freely available and comprehensive VoxCeleb1 dataset.

Keywords 3D CNN · Gaussian mixture model · MFCC · Log-Mel spectrogram · Speaker recognition

M. Rajput · K. Chauhan · G. Gopal · A. K. Johar (✉) · T. Varma
Department of Electronics and Communication Engineering, Malaviya National Institute of Technology Jaipur, Jaipur, Rajasthan, India
e-mail: arun.e12006@cumail.in

A. K. Johar · A. Tripathi
Department of Electronics and Communication Engineering, Chandigarh University, Gharuan, Mohali, Punjab, India

J. R. Szymanski et al. (eds.), *Energy Systems, Drives and Automations*,
Lecture Notes in Electrical Engineering 1057,
https://doi.org/10.1007/978-981-99-3691-5_47

1 Introduction

The human speech signal carries a great deal of information on many different levels. At its most fundamental level, it conveys a message in words and provides information about the speaker's language and dialect, emotions, gender, and identity. These tasks can be further subdivided into two categories based on text dependency, text-dependent system where the recognition system has prior knowledge of the text being spoken and text-independent system in which the recognition system is not concerned about the text that is being recognized. Instead of attempting to recognize specific words said in a speech, the speaker identification system has to extract the identity of the person who is speaking. The task of determining who is speaking from a known collection of speakers is referred to as speaker recognition.

In this paper, only text-independent analysis has been done for speaker recognition for an audio length as short as 2–5 s, delivered by a set of speakers. It is a method of detecting the speaker's voice characteristics automatically. An anonymous speaker's voice is compared to the voice models of known speakers, called speaker recognition (SR). Speaker authentication that is not dependent on the text has been more versatile than speaker authentication dependent on text. Also, higher verification accuracy is easier to achieve in text-dependent mode.

2 Related Work

Speaker validation is often employed in forensic investigations and modifications, as well as in distant applications such as the telephone network, databases, websites, e-commerce, banking, and other private activities [1–3]. Speaker recognition is a significant area of study in speech processing, which is divided into two subfields: identification and authentication. The recognition method compares an unknown voice to a collection of *n* persons recorded in a speech database. Speaker verification checks the claimed speaker's identity by comparing it to voiceprints stored in the database and by confirming the claimed speaker's threshold value [4]. Numerous models are used to describe speakers in a speaker verification system.

2.1 GMM-UBM System

In a recent paper, Hossan et al. [5] suggested a novel method for extracting MFCC features using the discrete cosine transform (DCT-II) and found that the equal error rate gives 100% identification, whereas it is 90.36% in conventional MFCC, lower than the other feature extraction methods. This paper includes a detailed step-by-step explanation completed with basic algorithms and illustrations. Kumari and Jayanna [6] compare the equal error rate for two distinct modes of feature extraction and classification for a speaker verification system. They found that the feature extraction techniques using LPCC with GMM-UBM as model, with limited data constraint can be utilized for speaker verification. Omar and Pelecanos [7] address GMM and how it is used in speaker recognition, as well as the challenges that might arise. For example, to train a model using maximum likelihood, a lot of data is needed. Hasan and Hansen [8] presented UBM training for speaker verification systems and the GMM technique for speaker verification and recognition systems. This study's aim is to evaluate the system's performance using UBM data.

2.2 PLDA for I-Vector Model

Novoselov et al. [9] introduced two nonlinear PLDA algorithms for a verification system. EER and least normalized detection cost function were used to assess performance. The two strategies significantly enhance EER and min DCF. Kanagasundaram et al. [10] suggested a strategy for modifying the system's performance in response to a disparity between development and assessment data. To address this issue, a advanced unsupervised interdataset variability (IDV) correction technique was suggested. Kenny et al. [11] describe how to handle varied duration utterances. It was suggested in paper that the PLDA i-vector approach is used to verify speaker identity. PLDA, i-vectors generated from short sentences are less accurate than those generated from longer sentences, according to the research.

2.3 CNN-Based Systems

Salehghaffari et al. [12] designed a convolutional neural network architecture for voice verification. The i-vector and GMM-UBM are two previously discussed approaches in this study. They advise using Siamese neural networks rather than raw features to operate on a single typical speech feature such as Mel-frequency cepstral coefficients (MFCCs) in order to get a higher-level representation for speaker-associated properties. Muckenhirn et al. [13] demonstrated a method for speaker verification that extracts information about the speaker straight from the audio stream. Guo et al. [14] proposed a method which deals with the issue of speaker recognition system using short utterances and found that traditional i-vector and PLDA-based approaches are less successful with short utterances. Variani et al. [15] introduced a DNN-based methodology for speaker verification based on the text and found that it is much sound for additive-type noise that surpasses the i-vector system for operating points at low false rejection (FR). They reported that the combined system provides the enhancement by 14% for noiseless and 25% with noisy condition for EER.

3 Speaker Recognition Using DNN

The speaker recognition can be addressed using DNN. The three stages of training, enrollment, and verification are outlined as follows in this section.

Training: A background model for speaker representation gathered from the speakers' utterances must be constructed throughout the development period. The model generates the representation. The input data representation of a DNN may be formed using the extracted speech feature maps of utterances. In a perfect condition, the model loss such as Softmax drives the final representations as discriminative throughout the training. Several research projects have looked into this phase, with state-of-the-art methodologies like i-vectors and d-vectors. The fundamental aim is to employ a DNN architecture as a speaker feature extractor for speech categorization at the frame and utterance level.

Enrollment: The DNN network will create a unique model for each speaker during the enrollment phase. The utterances given by the targeted speaker will be used to build each speaker-specific model. In the development phase, each utterance or frame, as per the degree of representation, is sent to the supervised trained network. The final output (whatever layer previous to the softmax layer gives superior representation) will be aggregated to account for all the utterances or frames. The d-vector is the absolute representation of the speech as deduced by the DNN outputs. The speaker model can be constructed by averaging all the d-vectors of the targeted speaker utterances. The strategy in which the architecture creates the speaker model in one shot is presented by recording the same speaker utterances, rather than the averaging generally utilized in a *d*-vector system.

Verification: Each test utterance is put into the network, and its representation recovered at the model assessment step. The one-vs-all configuration is the most common verification approach. The different speaker model is compared with the test utterance features, and a judgment is based on the similarity score. The critical error indications in this arrangement are the false rejection and false acceptance rates. A predetermined threshold evaluates the rates of erroneous rejection/acceptance.

4 Methodology

4.1 Dataset

The dataset is part of the Voceleb1 collection, which is available for free and open source [16] under the Creative Commons Attribution version 4.0 International License. In its original form, the dataset was acquired by a fully automated procedure. The majority of the datasets for speaker identification are comprised of samples taken under extremely restrictive settings, like background noise. As they require manual annotation, they are typically limited in terms of size. The used dataset is collected from open-source media, like YouTube, and due to this, it is one of the largest available datasets for research in this domain. Originally, it contained over 100,000 utterances for 1251 celebrities, all of which were taken from actual real-life utterances. This is a gender-balanced dataset, with 690 males having 116 utterances per speaker. The audio files without label are removed from the experiments. The speakers come from a diverse spectrum of ethnic backgrounds, accents, professions, and ages. The audio files include background noise, laughter, overlapping speaking, and the acoustics of a room. The proposed model uses a subset of the dataset of 40 speakers, with the average length of audio files as 8 s long.

4.2 Pre-processing

After pre-processing is the most crucial part of any pattern recognition system. A pre-emphasis filter with coefficient value 0.95 is used on all audio utterances before feature extraction. This is used for the average spectral shape compensation in frequency domain analysis, which emphasizes higher frequencies. After applying the filter, the corresponding spectrum will be more flat and optimum for extracting the acoustic features. Another important step in audio analysis particularly voice detection is to remove the silence audio segments. The silence segments do not carry any information cues about speakers. Speech activity detection (SAD) is a technique for identifying the time intervals between speech and other sounds. Since the energy of the nonspeech interval is much lower, the easiest way is to apply an energy threshold over a short interval (up to 20 ms). If the energy of a particular segment is below the threshold, it is removed from audio utterance as a silence interval. Although an accurate detection is a challenging task under noise conditions, a simple energy-based SAD showed a good performance for DNN-based systems [17].

4.3 Feature Extraction

All speech samples have been down sampled to a sampling rate of 16 kHz. We conduct our study using log-Mel spectrograms and Mel-frequency cepstral coefficients (MFCCs). These audio characteristics are retrieved, together with their delta and delta-delta for log-Mel spectrograms and MFCCs, using the librosa library [18]. Frequencies are not perceived linearly by humans. Human ears are more sensitive to differences in lower frequencies than to variances in higher frequencies. For example, it is much easier to distinguish frequencies of between 500 and 1000 Hz; however, the frequencies of 10,000 and 10,500 Hz are difficult to distinguish as distinct, though the differences between frequency pair are same. Mel scale is defined as given below to mitigate this issue.

$$f_{\mathrm{mel}} = 2595 \log\left(1 + \frac{f_{\mathrm{in}}}{700}\right) \tag{1}$$

Table 1 Output parameters of the proposed CNN architecture

Layer	Input size	Output size	Kernel	Stride
1st Conv. layer	4 × 128 × 128	4 × 64 × 61	3 × 1 × 8	1 × 0 × 0
1st Max pool layer	4 × 64 × 61	4 × 32 × 30	1 × 2 × 2	1 × 2 × 2
2nd Conv. layer	4 × 32 × 30	4 × 14 × 13	3 × 5 × 5	1 × 2 × 2
2nd Max pool layer	4 × 14 × 13	4 × 7 × 6	1 × 2 × 2	1 × 2 × 2
3rd Conv. layer	4 × 7 × 6	4 × 7 × 6	3 × 3 × 3	1 × 1 × 1
3rd Max pool layer	4 × 7 × 6	4 × 3 × 3	1 × 2 × 2	1 × 2 × 2
4th Conv. layer	4 × 3 × 3	4 × 3 × 3	3 × 3 × 3	1 × 1 × 1
5th Conv. layer	4 × 3 × 3	4 × 3 × 3	3 × 3 × 3	1 × 1 × 1
Dense layer 1	4608	128		
Dense layer 2	128	8		

where f_{in} is the frequency on linear scale. Log-Mel spectrograms and MFCCs are widely utilized in deep learning frameworks for a variety of tasks, including emotion recognition [19, 20], audio classification [21], and automatic speech recognition [22]. For the generation of log-Mel spectrograms, we chose 128 Mel filter banks. The inclusion of 128 Mel filter banks enables us to apply the proposed 3D CNN model for optimum output feature map. We employ a Hanning window with a size of 2048 samples and a hop length of 512 as in [20]. MFCC features are also obtained for same window length and hop length. Log-Mel spectrogram, delta, and delta-delta features and MFCCs are fed, respectively, to the channels four channels of the CNN input layer. Thus, input to the CNN has a dimension of 1 × 4 × 128 × 128 (Table 1).

5 Proposed Architecture

Figure 1 shows the basic block diagram for an end-to-end speaker identification task. All audio features are extracted as described in section III and stacked sample wise in a 3D structure which is the initial input for the CNN model. The data formats are $D \times H \times W$, where D, H, and W represent depth of acoustic channels, height of modulation channels, and width of time sequence, respectively. The layer-wise CNN architecture is showed in Table 2.

Five 3-D convolution layers and two dense layers are used to deduce the acoustic representations from an utterance. The final feature map of the second dense layer is used to make the speaker model during the enrollment phase. The first convolution layer has a kernel size of 3 × 1 × 8, having one input channel and eight output channels, ensuring that the temporal features are preserved. The first convolution layer is followed by a Max pooling layer, which helps us decrease the computational complexity of the network. After that, four more convolution layers and two more

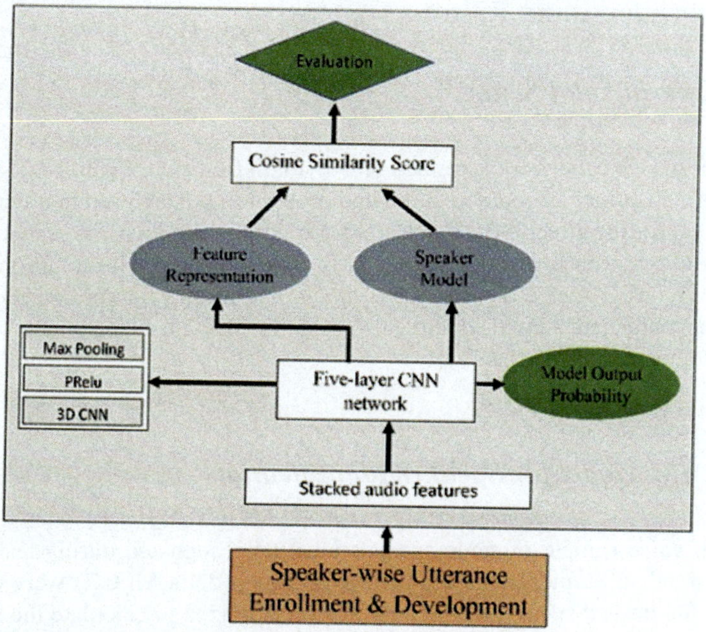

Fig. 1 Proposed architecture for end-to-end speaker recognition

Table 2 Development and test set statistics for speaker recognition in VoxCeleb1 dataset

Set	Speaker	Utterance
Development	1251	145,265
Test	1251	8251
Total	1251	153,516

Max pool layers are present, which help us extract valuable features from the given data. In each convolution layer, the number of output channels is double that of input channels. All the layers except the last are followed by PReLU activation. The output of the fifth convolution layer has 128 channels and a dimension of $4 \times 3 \times 3$, which is further flattened to 4608 features to make it compatible with the input of the dense layer.

6 Experiments and Results

6.1 Experimental Setup

The proposed CNN network is trained on a Pytorch framework with NVIDIA GPU. Batch normalization, L2 regularization and dropout of 0.2 are used in order to minimize the overfitting effects. The batch size for training the network is set to 64. In this section, the performance of devised CNN is also compared with traditional GMM architecture on VoxCeleb dataset. The training and testing for speaker recognition have been conducted using the identical speakers. Table 2 lists the utterance sets for the development and test sets.

6.2 GMM Model for Speaker Identification

The fresh voice sample of same speaker must be recognized during testing. The voice sample's 40-dimensional features (MFCCs + delta MFCC) were extracted first, and the trained speaker GMM models were utilized to calculate the scores of these features of all the models. The highest-scoring speaker model is projected to be the identified speaker. The Python operation includes the following steps.

1. 40-dimensional feature extraction.
2. Training speaker models.
3. Evaluating performance on test set.

7 Results

Results of the proposed 3D CNN and GMM model are given in Table 3 for speaker recognition task.

The recognition rates are 70.5 and 65.34% for 3D CNN and GMM, respectively. The performance shows that there is a 5% improvement over GMM using deep learning model. From the results, it can be concluded that CNN model is able to extract more relevant speech features.

Table 3 Results for speaker recognition on VoxCeleb

Method	Accuracy (%)
3D-CNN	70.05
GMM	65.34

8 Conclusion

The proposed model utilized a 3D CNN model and a GMM for speaker recognition task on a large-scale speaker identification dataset known as VoxCeleb 1 to perform the task successfully. For the purpose of evaluating model performance, a total of 1251 speakers and 153,516 utterances are used. For direct modeling of the speakers, a 3D-CNN architecture has been trained for feature extraction using training data from the speakers. The experimental results for the proposed 3D CNN architecture show that it outperforms the GMM model in terms of performance.

References

1. Ampbell JP (1997) Speaker recognition: a tutorial. Proceed IEEE 85(9):1437–1462
2. Bimbot F, Bonastre J-F, Fredouille C, Gravier G, Magrin-Chagnolleau I, Meignier S, Merlin T, Ortega-Garcıa J, Petrovska-Delacretaz D (2004) Reynolds DA (2004) A tutorial on text-independent speaker verification. EURASIP J Adv Signal Process 4:1–22
3. Petrovska-Delacretaz D, Cernocky J, Hennebert J, Chollet G (2000) Segmental approaches for automatic speaker verification. Dig Signal Process 10(1–3):198–212
4. Jayanna HS, Prasanna SRM (2009) Analysis, feature extraction, modeling and testing techniques for speaker recognition. IETE Techn Rev 26:181–190
5. Hossan MA, Memon S, Gregory MA (2010) A novel approach for MFCC feature extraction. In: Proceedings of the 2010 4th international conference on signal processing and communication systems. IEEE, New York, pp 1–5
6. Kumari TJ, Jayanna H (2014) Comparison of lpcc and mfcc features and gmm and gmm-ubm modeling for limited data speaker verification. In: Proceedings of the 2014 IEEE international conference on computational intelligence and computing research. IEEE, New York, pp 1–6
7. Omar MK, Pelecanos JW (2010) Training universal background models for speaker recognition. In: Odyssey, p 10
8. Hasan T, Hansen JH (2011) A study on universal background model training in speaker verification. IEEE Trans Audio Speech Lang Process 19(7):1890–1899
9. Novoselov S, Pekhovsky T, Oleg K, Mendelev V, Prudnikov A (2015) Non-linear plda for i-vector speaker verification, p 09
10. Kanagasundaram A, Dean D, Sridharan S (2015) Improving out-domain plda speaker verification using unsupervised inter-dataset variability compensation approach. In: Proceedings of the 2015 IEEE international conference on acoustics, speech and signal processing (ICASSP). IEEE, New York, pp 4654–4658
11. Kenny P, Stafylakis T, Ouellet P, Alam MJ, Dumouchel P (2013) Plda for speaker verification with utterances of arbitrary duration. In: Proceedings of the 2013 IEEE international conference on acoustics, speech and signal processing. IEEE, New York, pp 7649–7653
12. Salehghaffari H (2018) Speaker verification using convolutional neural networks
13. Muckenhirn H, Magimai-Doss M, Marcell S (2018) Towards directly modeling raw speech signal for speaker verification using cnns, vol 04, pp 4884–4888
14. Guo J, Nookala UA, Alwan A (2017) Cnn-based joint mapping of short and long utterance i-vectors for speaker verification using short utterances. In: INTERSPEECH, pp 3712–3716
15. Variani E, Lei X, McDermott E, Moreno IL, Gonzalez-Dominguez J (2014) Deep neural networks for small footprint text-dependent speaker verification. In: Proceedings of the 2014 IEEE international conference on acoustics, speech and signal processing (ICASSP). IEEE, New York, pp 4052–4056

16. Nagrani A, Chung JS, Zisserman A (2017) Voxceleb: a large-scale speaker identification dataset. arXiv preprint arXiv:1706.08612
17. Wang S, Rohdin J, Plchot O, Burget L, Yu K, Cernocky J (2020) Investigation of specaugment for deep speaker embedding learning. ICASSP 2020–2020 IEEE international conference on acoustics, speech and signal processing (ICASSP). IEEE, New York, pp 7139–7143
18. McFee B, Raffel C, Liang D, Ellis DP, McVicar M, Battenberg E, Nieto O (2015) Librosa: audio and music signal analysis in python. In: Proceedings of the 14th python in science conference, vol 8. Citeseer, pp 18–25
19. Jiang L, Tan P, Yang J, Liu X, Wang C (2021) Speech emotion recognition using emotion perception spectral feature. Concurr Comput Pract Exp 33(11):e5427
20. Chauhan K, Sharma KK, Varma T (2021) Speech emotion recognition using convolution neural networks. In: Proceedings of the 2021 international conference on artificial intelligence and smart systems (ICAIS). IEEE, New York, pp 1176–1181
21. Xu Y, Kong Q, Wang W, Plumbley MD (2018) Large-scale weakly supervised audio classification using gated convolutional neural network. In: Proceedings of the 2018 IEEE international conference on acoustics, speech and signal processing (ICASSP). IEEE, New York, pp 121–125
22. Park DS, Chan W, Zhang Y, Chiu C-C, Zoph B, Cubuk ED, Le QV (2019) Specaugment: a simple data augmentation method for automatic speech recognition. arXiv preprint arXiv:1904.08779

Fixed Points of Weak-Generalized Rational Type Contraction via Graph Structure

Kusuma Tummala, A. Sreerama Murthy, V. Ravindranath, P. Harikrishna, and N. V. V. S. Suryanarayana

Abstract In this paper, we prove the existence of fixed points of weak-generalized rational multivalued contraction mappings in metric spaces with a graph which extends some of known results.

Keywords Fixed point · Directed graph · Multivalued contraction · Metric space · AMS subject classification · 47H10 · 54H25

1 Introduction

Banach contraction is one of the most important result in fixed point theory; there are so many extensions and generalizations by extending contractive condition and the ambient space. Jachmski [9] extended fixed point theory results replacing the structure of orders by the structure of graphs on metric spaces. Fixed point theory and graph theory give the intersection of theories of fixed point results with single and multivalued mapping.

K. Tummala (✉)
Department of H&S, VNR Vignana Jyothi Institute of Engineering and Technology, Bachupally, Kukatpally, Hyderabad, Telangana State 500090, India
e-mail: kusumatummala9@gmail.com

A. S. Murthy
Kakinada, Andhra Pradesh 533003, India

V. Ravindranath
Department of Mathematics, JNTUK College of Engineering, JNTU, Kakinada, Andhra Pradesh 533003, India

P. Harikrishna · N. V. V. S. Suryanarayana
Department of Mathematics, N S Raju Institute of Technology, Visakhapatnam 531173, India

J. R. Szymanski et al. (eds.), *Energy Systems, Drives and Automations*,
Lecture Notes in Electrical Engineering 1057,
https://doi.org/10.1007/978-981-99-3691-5_48

Let (X, d) be a metric space, $CB(X)$ denotes the family of all nonempty, closed and bounded subsets of X, and $K(X)$ denotes the family of all nonempty compact subsets of X. The Pompeiu–Hausdorff metric $H : CB(X) \times CB(X) \to [0, \infty)$ is defined as $H(A, B) = \max\{\sup_{\alpha \in A} D(\alpha, B), \sup_{\beta \in B} D(\beta, B)\}$ where $A, B \in CB(X)$ and $D(\alpha, B) = \inf_{\beta \in B} d(\alpha, \beta)$. Nadlar [10] proved fixed point theorems in complete metric spaces using multivalued contraction mappings. In 2018, Jachymski [9] proved some fixed point results using graph structure in place of partial orders. In 2012, Bojar [5] proved some fixed point theorems in metric spaces endowed with a graph. For more results, we referred results on a metric space with graph structure [1, 3, 8, 12].

Section I This section is devoted to preliminary notions that are needed for the further study of this paper.

Definition 1 [9]: Let X be a non-empty set and Δ denote the diagonal of the Cartesian product $X \times X$. A directed graph or digraph G is characterized by a nonempty set $V(G)$ of its vertices and the set $E(G) \subset V(G) \subset V(G)$ of its directed edges. A digraph is reflexive if any vertex admits a loop. For a digraph $G = (V, E)$,

(i) If whenever $(a, b) \in E(G) \Rightarrow (b, a) \in E(G)$, then the digraph G is called an oriented graph
(ii) A digraph G is transitive whenever $[(a, b) \in E(G) \text{and} (b, c) \in E(G)] \Rightarrow (a, c) \in E(G)$, for any $a, b, c \in V(G)$
(iii) A path of G is a sequence $x_0, x_1, ...x_n \ldots$ with $(x_i, x_{i+1}) \in E(G)$ for each $i \in N$.
(iv) G is connected if there is a path between each two vertices, and it is weakly connected if $\overline{G}$ is connected, where $\overline{G}$ corresponds to the undirected graph obtained from G by ignoring the direction of edges.
(v) Let G^{-1} be the graph obtained of G by reversing the direction of edges. Thus, $E(G^{-1}) = \{(a, b)\} \in X \times X : (b, a) \in E(G)$.
(vi) We call (V', E') a subgraph of G if V′ $\subseteq$ V (G) and $E' \subseteq E(G)$ and for any edge $(a, b) \in E'$, $(a, b) \in V'$.

Definition 2 [1]: Let (X, d) be a metric space with endowed with a graph G such that $V(G) = X$ and let $T : X \to CB(X)$ be a multivalued mapping. T has the weakly graph-preserving (WGP) property, whenever for each $a \in X$ and $b \in T$ a with $(a, b) \in E(G)$ implies $(b, c) \in E(G)$ for all $c \in Tb$.

Definition 3 Let $F : (0, \infty) \to R$ be a function satisfying the following:

(F1) For all $\alpha, \beta \in (0, \infty)$ such that $\alpha < \beta$, $F(\alpha) < F(\beta)$.
(F2) For any positive real sequence $\{(an)\}$, $\lim_{n \to \infty} a_n = 0$ iff $\lim_{n \to \infty} F(a_n) = -\infty$
(F3) There exists $k \in (0, 1)$ such that $\lim_{n \to 0^+} F(\alpha) = 0$
(F4) $F(\inf A) = \inf F(A)$ for all $S \subset (0, \infty)$ with $\inf A > 0$.

Lemma [1]: Let (X, d) be a metric space and $T : X \to CB(X)$ be an upper semi-continuous mapping such that Tr is closed for all $r \in X$. If $r_n \to t_0$, and $tr \in Tr_n$, then $t_0 \in Tr_0$.

A function $\pi : R^+ \to R^+\big(\text{where } R^+ = [0, \infty]\big)$ is said to be a comparison function if it satisfies the following conditions (i) φ is monotonically increasing (ii) $\lim_{n\to\infty} \varphi_n(t) = 0$ for all $t \in R^+$. Meanwhile, if φ satisfies (i) and the following condition: (iii) $\sum_{0}^{\infty} \varphi_n(t)$ is convergent for each $t \geq 0$, then φ is named to be a (c)-comparison function.

Lemma [4]: let (X, d) be a metric space. Let $A, B \subset X$, and $q > 1$. Then, for every $a \in A$ there exists $b \in B$ such that $d(a, b) \leq H(A, B)$.

In 2021, [11] Ozlem Acar, Hassen Aydi, and Manuel De la Sen proved the existence of fixed point results for rational multivalued G- contraction and F-Khan-type multivalued contraction mappings on a metric space with a graph.

Theorem-1.1 Let G be a directed graph on a complete metric space (X, d) and f: X → CB(X) be a multivalued mapping. Suppose that f is upper semi-continuous and a WGP mapping. Suppose that: (i) f is a rational multivalued G-contraction of type I, that is, there is a strictly increasing (c)-comparison function φ such that $H(fa, Tfb) \leq \varphi(M(a, b))$ for all $(a, b) \in E(G)$ (1.1) where

$$M(a,b) = \max\left\{d(a,b), \frac{1}{2}[D(a, fa) + D(b, fb)], \frac{1}{2}[D(a, fb) + D(f, fa)], \frac{D(a, fa)D(b, fb)}{d(a,b)}, \frac{D(b, fb)[1 + D(a, fa)]}{1 + d(a,b)}\right\}$$

$Mf = \{a \in X : (a, u) \in E(G)$ for u ∈ Ta is nonempty, then f admits a fixed point.

In this section, we prove the existence of new fixed point results for multivalued mappings with rational type generalized weak contractions in metric spaces via a graph structure.

2 Main Results

Theorem 2.1 Let G be a directed graph on a complete metric space (X, d) and $T : X \to CB(X)$ be a multivalued mapping. Suppose that T is upper semi-continuous and a WGP mapping. Suppose that (i) T is a rational multivalued G-contraction of type I, that is, there is a strictly increasing (c)-comparison function φ such that

$$H(Ta, Tb) \leq \varphi(M(a, b)) + LN(a, b) \text{ for all } (a, b) \in E(G \quad (2.1)$$

Where $M(a, b) = \max\left\{d(a, b), \frac{1}{2}[D(a, Ta) + D(b, Tb)], \frac{1}{2}[D(b, Ta)\right.$
$\left. + D(a, Tb)], \frac{D(b, Ta)[1 + D(a, Ta)]}{1 + d(a, b)}, \frac{D(a, Ta)[1 + D(b, Tb)]}{1 + d(a, b)}\right.$
$\left., \frac{D(b, Ta)[1 + D(a, Tb)]}{1 + d(a, b)}\right\}$

$N(a, b) = \min\left\{D(a, Ta), D(a, Tb), D(b, Ta), \frac{D[b, Ta][1 + D(a, Tb)]}{1 + d(a, b)}\right\}$

$Mf = \{a \in X : (a, u) \in E(G) \text{ for } u \in Ta\}$

is nonempty, then T has a fixed point.

Proof Let $x_0 \in Mf$ then there exists $x_1 \in Tx_0$ such that $(x_0, x_1) \in E(G)$.

$$D(x_0, Tx_1) \leq H(Tx_0, Tx_1) \leq \varphi(M(x_0, x_1)) + LN(x_0, x_1) \forall (a, b) \in E(G)$$
$$\max\{d(x_0, x_1),$$
$$\frac{1}{2}[D(x_0, x_1) + D(x_1, Tx_1)], \frac{1}{2}[D(x_1, x_1)$$
$$+ D(x_0, Tx_1)], \frac{D(x_1, x_1)[1 + D(x_0, x_1)]}{1 + d(x_0, x_1)}, \frac{D(x_0 x_1)[1 + D(x_1, Tx_1)]}{1 + d(x_0, x_1)},$$
$$\left.\frac{D(x_1, x_1)[1 + D(x_0, Tx_1)]}{1 + d(x_0, x_1)}\right\} \leq \max\{d(x_0, x_1), D(x_1, Tx_1)\},$$

If $\max\{d(x_0, x_1), D(x_1, Tx_1)\} = D(x_1, Tx_1)$ then we get a contradiction and hence $\max\{d(x_0, x_1), D(x_1, Tx_1)\} = d(x_0, x_1)$.

$$N(x_0, x_1) =$$
$$\min\{D(x_0, Tx_0), D(x_0, Tx_1), D(x_1, Tx_0),$$
$$\left.\frac{D[x_1, Tx_0][1 + D(x_0, Tx_1)]}{1 + d(x_0, x_1)}\right\} = 0$$

from lemma 2, there exists $x_2 \in Tx_1$ such that $d(x_1, x_2) \leq \sqrt{q}(H(Tx_0, Tx_1))$

We have $H(Tx_0, Tx_1) \leq \varphi(d(x_0, x_1))$

Therefore $d(x_1, x_2) \leq \sqrt{q}(\varphi(d(x_0, x_1))) < q(\varphi(d(x_0, x_1)))$.

Since φ is strictly increasing, we have $0 < \varphi(d(x_1, x_2)) < \varphi(q(d(x_0, x_1)))$.

Take $q_1 = \frac{\varphi(q\varphi(d(x_0,x_1)))}{\varphi(d(x_1,x_2))}$, then $q_1 > 1$.
Since $(x_0, x_1) \in E(G)$, $x_1 \in Tx_0$ and $x_2 \in Tx_1$.
Using the WGP property, $(x_1, x_2) \in G(G)$, then

$$\text{We have } = D(x_2, Tx_2) \leq H(Tx_1, Tx_2) \leq \varphi(M(x_1, x_2))$$
$$L.N(x_1, x_2) \leq \varphi(\max\{d(x_1, x_2), D(x_2, Tx_2)\}$$
$$+L.0 \leq \varphi(d(x_1, x_2)) < \sqrt{q_1}\varphi(d(x_1, x_2))$$

Since φ is strictly increasing, we have $0 < \varphi(d(x_2, x_3)) < \varphi^2(q(d(x_0, x_1)))$.

Proceeding in this way, we can construct the sequence $\{x_n\}$ in X such that $x_{n+1} \in$ Tx_n, $(x_n,x_{n+1}) \in$ E(G) and $d(x_n, x_{n+1}) \leq \varphi^n(q\varphi(d(x_0, x_1)))$

Now we have to prove that $\{x_n\}$ is a Cauchy's sequence.

Let $m, n \in N$ with $m > n$.

Consider $d(x_n, x_m) \leq \sum_{k=n}^{m-1} d(x_k, x_{k+1}) \leq \sum_{k=n}^{m-1} \varphi^k(q\varphi(d(x_0, x_1)))$.

Since φ is a (C)-comparison function, the series on the right-hand side converges, and so we get $d(x_n, x_m)) \to 0$ as $n, m \to \infty$.

Hence $\{x_n\}$ is a Cauchy's sequence in (X, d), and which complete and hence $\{x_n\}$ is converges to some $z \in X$, i.e., $\lim_{n\to\infty} x_n = z$ Since T is upper semi-continuous, by Lemma 1, $z \in Tz$. Therefore, T has a fixed point in X.

Theorem 2.2. Let G be a directed graph on a complete metric space (X, d) and $T : X \to CB(X)$ be a multivalued mapping. Suppose that T is a WGP mapping. Suppose that:

(i) T is a rational multivalued G-contraction of type II, that is, there is a strictly increasing (c)-comparison function φ such that
$H(Ta, Tb) \leq \varphi((M(a, b)) + LN(a, b)$ for all $(a, b) \in E(G)$ (2.2)

$$\text{Where } M(a, b) = \max\left\{d(a, b), \frac{1}{2}[D(a, Ta) + D(b, Tb)], \frac{1}{2}[D(a, Tb) + D(b, Ta)], \frac{D(a, Ta)D(b, Tb)}{1 + H(Ta, Tb)}\right\}$$

$$N(a, b) = \min\left\{D(a, Ta), D(a, Tb), D(b, Ta), \frac{D(b, Ta)D(b, Tb)}{1 + H(Ta, Tb)}\right\}$$

(ii) $Mf = \{a \in X : (a, u) \in E(G) for u \in Ta\}$
(iii) The property P for any $\{a_n\}$ in X, if $a_n \to a$ and $(a_n a_{n+1}) \in E(G)$. Then there is a sequence $\{a_n\}$ with $(a_{nk}, a) \in E(G)$.
(iv) T is WGP mapping, then T has a fixed point.

Proof. Let $x_0 \in Mf$ then there exists $x_1 \in Tx_0$ such that $(x_0, x_1) \in E(G)$.

$$D(x_1, Tx_1) \leq H(Tx_0, Tx_1) \leq \varphi(M(x_0, x_1)) + LN(x_0, x_1),\ \forall (a, b) \in E(G)$$

$$M(x_0, x_1) = \max\left\{d(x_0, x_1), \frac{1}{2}[D(x_0, Tx_0) + D(x_1, Tx_1)],\right.$$

$$\left.\frac{1}{2}[D(x_0, Tx_1) + D(x_1, Tx_0)], \left\{\frac{D(x_0, Tx_0)[D(x_1, Tx_1)]}{1 + H(Tx_0, Tx_1)}\right\}\right.$$

$$\leq \max\{d(x_0, x_1), D(x_1, Tx_1)\}$$

If $\max\{d(x_0, x_1), D(x_1, Tx_1)\} = D(x_1, Tx_1)$ then we get a contradiction and hence $\max\{d(x_0, x_1), D(x_1, Tx_1)\}\} = d(x_0, x_1)$.

$$N(x_0, x_1) = \min\{D(x_0, Tx_0), D(x_0, Tx_1), D(x_1, Tx_0),$$

$$\left., \frac{D(x_1, Tx_0)D(x_1, Tx_1)}{1 + H(Tx_0, Tx_1)}\right\} = 0$$

$$\text{Hence },\ D(x_1, Tx_1) \leq H(Tx_0, Tx_1) \quad \leq \varphi(M(x_0, x_1))$$
$$+LN(x_0, x_1)D(x_1, Tx_1) \leq H(Tx_0, Tx_1) \leq \varphi(M(x_0, x_1))$$
$$+L.0D(x_1, Tx_1) \leq H(Tx_0, Tx_1) \leq \varphi(M(x_0, x_1))$$

$\leq \varphi(d(x_0, x_1)) \leq \sqrt{\sigma}\varphi(d(x_0, x_1))$ where $\sigma > 1$ is a constant, From lemma 2, there exists $x_2 \in Tx_1$ such that

$$d(x_1, x_2) \leq \sqrt{\sigma}(H(Tx_0, Tx_1)) < \sqrt{\sigma}\varphi(d(x_0, x_1))$$

Since φ is strictly increasing, we have $0 < \varphi(d(x_1, x_2)) < \varphi(\sigma(d(x_0, x_1)))$. Take $\sigma_1 = \frac{\varphi(\sigma\varphi(d(x_0, x_1)))}{\varphi(d(x_1, x_2))}$, and we have $\sigma_1 > 1$ and by using the WGP property we can write $(x_1, x_2) \in E(G).(x_1, x_2) \in E(G)$.

$$D(x_2, Tx_3) \leq (H(Tx_1, Tx_2)) \leq \varphi(d(x_1, x_2)) < \sqrt{\sigma}\varphi(d(x_1, x_2))$$

from lemma 2, there exists $x_3 \in Tx_2$ such that

$$d(x_2, x_3) \leq \sqrt{\sigma}(H(Tx_1, Tx_2)) < \sigma\varphi(d(x_1, x_2))$$

Since φ is strictly increasing, we have

$$0 < \varphi(d(x_1, x_2)) < \varphi^2(\sigma(d(x_0, x_1))).D(x_2, Tx_2)$$
$$\leq H(Tx_1, Tx_2) \leq \varphi(M(x_1, x_2)) + L.N(x_1, x_2)$$

$$M(x_1, x_2) = \max\left\{d(x_1, x_2), \frac{1}{2}[D(x_1, Tx1)+D(x_2, Tx_2)],\right.$$

$$\left.\frac{1}{2}[D(x_1, Tx_2) + D(x_2, Tx_1)], \frac{D(x_1, Tx_1)[D(x_2, Tx_2)]}{1 + H(Tx_1, Tx_1)}\right\}$$

$$\leq \max\left\{d(x_1, x_2), \frac{1}{2}[d(x_1, x_2) + D(x_2, Tx_2)], \frac{1}{2}D(x_1, Tx_2), d(x_1, x_2)\right\}$$

$$= max\{d(x_1, x_2), D(x_2, Tx_3),\} = d(x_1, x_2)$$

$$N(x_1, x_2) = \min\left\{D(x_1, Tx_1), D(x_1, Tx_2), D(x_2, Tx_1), \frac{D[x_2, Tx_1]D(x_2, Tx_2)}{1 + H(Tx_1, Tx_2)}\right\}$$

$$N(x_1, x_2) = \min\left\{d(x_1, x_2), D(x_1, Tx_2), 0, \frac{D[x_2, Tx_1]D(x_2, Tx_2)}{1 + H(Tx_1, Tx_2)}\right\} = 0$$

Take $\sigma_2 = \frac{\varphi^2(\sigma\varphi(d(x_0,x_1)))}{\varphi(d(x_2,x_3))}$, and we have $\sigma_2 > 1$

Proceeding in this way, we can construct the sequence $\{x_n\}$ in X such that $x_{n+1} \in$ T x_n, $(x_n,x_{n+1}) \in$ E(G) and $d(x_n,x_{n+1}) \leq \varphi^n(\sigma(d(x_0, x_1)))$.

Now we show that $\{x_n\}$ is a Cauchy's sequence. Let $m, n \in N$ with $m > n$.

Consider $d(x_n, x_m) \leq \sum_{k=n}^{m-1} d(x_k, x_{k+1}) \leq \sum_{k=n}^{m-1} \varphi^k(\sigma\varphi(d(x_0, x_1)))$.

Since φ is a (C)- comparison function, the series on the right-hand side converges, and so we get $d(x_n, x_m)) \to 0$ as $n, m \to \infty$. Hence $\{x_n\}$ is a Cauchy's sequence in (X, d), and which complete and hence $\{x_n\}$ is converges to some $t \in X$, i.e., $\lim_{n\to\infty} x_n = t$.

By the property-P, there is a subsequence $\{x_{nk}\}$ of $\{x_n\}$ such that $(x_{nk}, t) \in E(G)$ for each $k \in N$. Now we assume that $D(t, Tt) > 0$. Since $\lim_{n\to\infty} D(x_{nk}, x_{nk+1}) = 0$ $\lim_{n\to\infty} D(x_{nk}, 0) = 0$ there is $n_0 \in$ N, such that for $n_k > n_0$ and there is $n_1 \in N$ so that for $n_k > n$, $D(x_{nk}, t) < \frac{1}{3}D(t, Tt)$.

If we take $n_k > \max\{n_0, n_1\}$ then from the above two, we get

$$D(xx_{nk+1}, Tt) \leq H(Tx_{nk}, Tt) \leq \varphi(M(x_{nk}, t)) + LN(x_{nk}, t)$$

$$M(x_{nk}, t) = \max\left\{d(x_{nk}, t), \frac{1}{2}[D(x_{nk}, Tx_{nk}) + D(t, Tt)],\right.$$

$$\left.\frac{1}{2}[D(x_{nk}, Tt) + D(t, Tx_{nk})], \frac{D(x_{nk}, Tx_{nk})[D(t, Tt)]}{1 + H(Tx_{nk}, Tx_{nk})}\right]$$

$$\lim_{k\to\infty} M(x_{nk}, 0) = D(t, Tt)$$

$$N(x_{nk}, t) = \min\left\{D(x_{nk}, Tx_{nk}), D(x_{nk}, Tt), D(t, Tx_{nk}), \frac{D[x_{nk}, Tx_{nk}]D(t, Tt)]}{1 + H(Tx_{nk}, Tt)}\right\}$$

$\lim_{k \to \infty} N(x_{nk}, 0) = 0$ Taking k $\to \infty$ in the inequality

$$D(x_{nk+1}, Tt) \leq H(Tx_{nk}, Tt) \leq \varphi(M(x_{nk}, t)) + LN(x_{nk}, t)$$

We get $D(t, Tt) \leq \varphi(D(t, Tt)) + L.0D(t, Tt) < D(t, Tt)$,
which is a contradiction. Hence, $D(t, Tt) = 0$ So that T admits fixed point.

References

1. Acar O, Altun A (2016) Multi valued F-contractive mappings with a graph and some fixed point results. Pulb Math Debr 88:305–307
2. Ansari AH, Nantadilok J (2017) Best proximity points for proximal contractive type mappings with C-class functions in S-metric spaces. Fixed Point Theory Appl (12):17
3. Beg I, Butt AR, Rdojovic S (2010) The contraction principle for set valued mappings on a metric space with a graph. Comput Math Appl 60:1214–1219
4. Berinde M, Berinde V (2007) On genaral class of multi valued weakly Picard mappings. J Math Anal Appl 326:722–782
5. Bojor F (2012) Fixed point theorems for Reich type contractions on metric space with a graph. Nonlinear Anal 75:3895–3901
6. Babu GVR, Sailaja PD, Srichandana G (2020) Common fixed points via Ckclass functions in S-metric spaces. J Fixed Point Theory 1:22
7. Balraj D, Piramanantham V, Ansari AH (2018) Existence of best proximity points for proximal C-contractions satisfying rational expression on ordered metric spaces. Int J Math Appl 6(2-A):43–51
8. Chifu C, Petrusel G, Bota MF (2013) Fixed points and strict fixed points for multi valued contractions of Reich type on metric spaces endowed with a graph. Fixed point Theory Appl 2013.
9. Jachymski J (2018) The contraction principle for mappings on a complete metric space with graph. Proc Am Math Sco 136:1359–1373
10. Nadler SB (1969) Multi-valued contraction mappings. Pac J Math 30:475–488
11. Acar O, Aydi H, De la Sen M (2021) New fixed point results via a graph structures. E-Mathematics 9:1013
12. Qawaneh H, Noorani MS, Shatawani W, Aydi H, Alsamir H (2019) Fixed point results for multi valued contractions in b- metric spaces. Mathematics 7:132
13. Babu GVR, Sarma KKM, Krishna PH (2014) Fixed points of ψ—weak Geraghty contractions in partially ordered metric spaces. J Adv Res Pure Math 6(4):9–2. https://doi.org/10.5373/jarpm.1896.120413

A Survey on Challenges and Opportunities of Big Data Healthcare Analytics

Sk. Wasim Akram, M. Varalakshmi, Md. Salma Sultana, and M. S. N. D. Sowjanya

Abstract One of most technical and potential topics in research is healthcare big data analytics. Healthcare includes variety of data in terms of sensor data, clinical data, omics data, and so on. In healthcare, health records are maintained by physicians, health companies, patients, IT companies, and vendors, and it is very difficult for them to maintain this data manually. One of the biggest challenges faced by the researchers is how to handle this raw data with huge volume, variety, and velocity. A significant solution for this rapidly growing data is machine learning (ML). It contains advanced algorithms and various statistical measures for analyzing the datasets and accurately predicting the results for further processing. This paper intention is to analyze the potentials of big data techniques in healthcare. This analytic approach also helps researchers to focus on current opportunities and challenges in healthcare center. This paper discusses about role of big data analytics in healthcare and its future directions and challenges.

Keywords Big data · Predictive analytics · IoT · Healthcare system

1 Introduction

Artificial intelligence (AI) enables the machines to think and made more intelligent than human beings [1]. Machine learning is considered as a sub-branch in the field of AI study and allows the researcher to focus on numerous research works. In the view of a researcher, it is impossible to get the insights from the data without learning. Machine learning techniques like unsupervised, supervised, semi-supervised, and reinforcement also have deep learning and evolutionary learning techniques. These learnings are used to classify huge data very fastly. Most of the researchers use convolutional neural network, K-nearest Neighbor (KNN) machine learning algorithms [2] for classification of big data very fastly, and prediction of disease in a accurate way.

Sk. W. Akram (✉) · M. Varalakshmi · Md. S. Sultana · M. S. N. D. Sowjanya
Department of CSE, Vasireddy Venkatadri Institute of Technology, Guntur, A.P, India
e-mail: shaikwasimakram585@gmail.com

J. R. Szymanski et al. (eds.), *Energy Systems, Drives and Automations*,
Lecture Notes in Electrical Engineering 1057,
https://doi.org/10.1007/978-981-99-3691-5_49

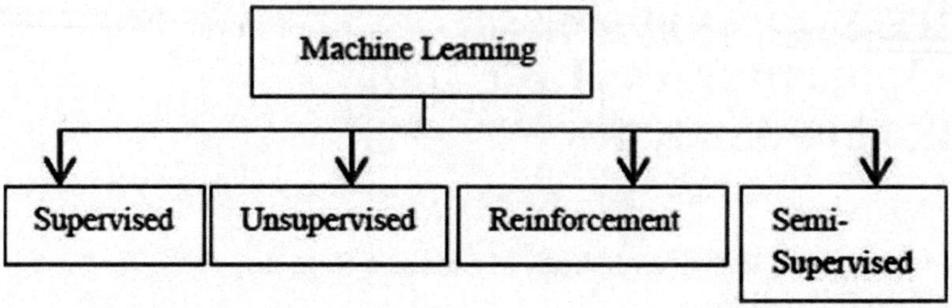

Fig. 1 Classification of machine learning techniques

Generally, medical data is tremendously increasing on daily basis so prediction of disease correctly in early stages is crucial task. At the same time, processing data in large volumes with variable formats is very difficult. With the help of data mining techniques, it is very easy to classify and analyze the large dataset in machine learning environment. The below diagram Fig. 1 depicts classification of machine learning techniques.

It is hard to accurately identify and diagnosis the patients by using clinical examination and evaluation process [3]. In the field of healthcare, it generates huge amount of information about evaluation of clinical decision support system, patient reports, medicine and cure, succeeding meet-ups, and so forth. It is very complicated to organize the data appropriately. Due to improper management of the information, it effects on the quality of the data association. More focus is to be maintained in measure of data needs and get the results in an efficient manner. To construct a better classifier, a machine learning classifier has to be chosen and segregates them into corresponding classifier based on their characteristics. Dataset is partitioned into two or more than two classes. These kinds of classifiers are employing for investigation of medical data and prediction of diseases. On a regular time interval, huge amount of data is collected, with the help of tools and techniques offered by analytics, it is possible to extract useful information from voluminous, complex, and huge velocity of data and finally translate it into information that helps to making decisions in healthcare.

There are four different phases of big data analytics that evolved in research and healthcare application. Particularly, it is used for exploration of a disease with the help of techniques and analytical tools, so these are able to perform tasks like data collection, transformation, modeling, and delivering the huge volumes of unstructured, semi-structured, and structured data created by current healthcare systems or health-related organizations. The components that are involved in healthcare big data analytics for knowledge discovery process are summarized as in below diagram (Fig. 2).

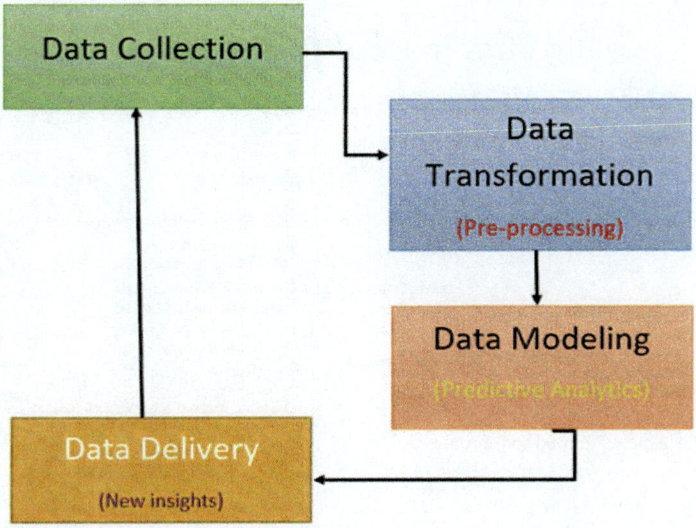

Fig. 2 Big data life cycle

2 Role of Big Data in Healthcare

Users are making interactions with online applications thereby generating huge amount of digital datasets in a continuous manner. Big data is ingested from multiple sources with different properties and managed by different managerial units in the organization that results in big data chain. While handling massive amount of big data [5], it possess challenges in terms of storage and processing. The concept of big data is growing exponentially and allows the researchers to focus on best techniques that extract value from it. In healthcare domain [6], data is categorized into different streams that include biometric data, clinical data, physiological data, social media data, and scientific research activities. Figure 3 depicts overview of applications, analytics, and techniques in big data healthcare analytics.

3 Tools Involved in Big Data Analytics

In real-time situations, using traditional tools and existing data processing techniques, it is immense difficult to process vast amount of healthcare data. There is a need of various analytical tools of big data for analyzing their functionalities. Today, various big analytical tools [7 and 8] are introduced for data analyzing, and this process aids in inspecting, cleaning, transforming, and modeling the data for extracting valuable information and helps in making profitable decisions and suggest optimal solutions. Table 1 contains some popular analytical tools to process big health data.

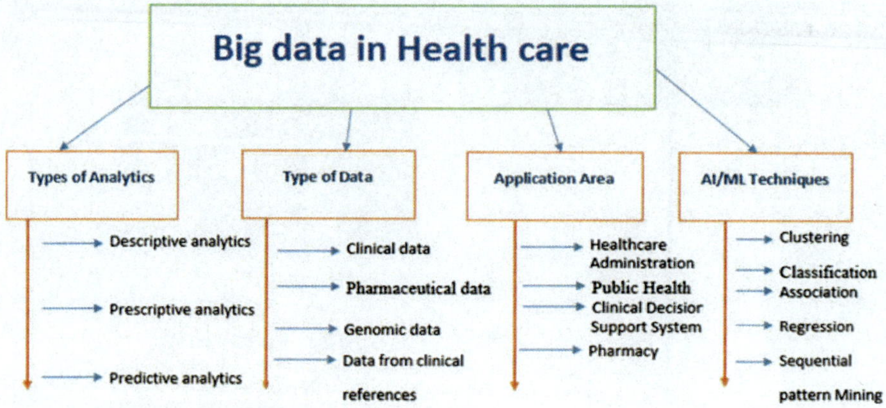

Fig. 3 Various aspects related to big data

Table 1 Tools of big data for analysis

S. no	Big data tool(s)	Purpose (or) functionality
1	Apache storm	It is real time distributed data processing framework which is designed primarily to process vast amount of data. It is similar to Hadoop framework
2	Tez	It is a framework that helps in developing real time batch and interactive big data applications
3	Sqoop	One of the popular tool developed to import data between RDBMS and Hadoop framework and export data between HDFS and RDBMS
4	Atlas	It is a data governance tool that helps to collect, analyze, manage, and exchange metadata with other tools within Hadoop stack as well as outside it
5	Accumulo	It is a data accessing tool that focuses on fast accessing the records of medical data from millions of rows and columns to an individual cell
6	Kafka	It is streaming tool runs on a cluster of commodity hardware, and it stores medical records in terms of streams. Each record of medical data consists of < key, value, timestamp >
7	Flume	It is an ingestion tool for collecting, aggregating, and transmitting huge volumes of medical data into Hadoop clusters for processing complex jobs
8	HCatlog	It is a database that stores meta data and table structure information, utilized by Hadoop framework to process both unstructured and structured medical data
9	Apache Mahout	It is machine learning tool for processing real-time health data
10	Apache Hadoop Yarn	This component is used for allocation of resources, monitoring the execution of MapReduce jobs and performs task scheduling on Hadoop clusters

(continued)

Table 1 (continued)

S. no	Big data tool(s)	Purpose (or) functionality
11	Avro	This component is designed to perform serialization and de-serialization of data in Hadoop system
12	Apache Solr	It is a non-relational (NOSQL) database designed to make distributed, fault tolerant, highly reliable, centralized, and self-respond to failure for Hadoop cluster
13	Apache Oozie	In order to manage the Hadoop tasks, there is a need of reliable workflow system. So Oozie is designed for this. It provides DAG's to implement the actions
14	Cassandra	It is file system in distributed fashion, designed primarily to process large volumes of health data which are located in many Hadoop servers
15	HBase	On the top of HDFS in Hadoop ecosystem, there exists a non-relational database called HBase that allows end users to provide random access and querying the data fastly
16	Zookeeper	This component is meant to provide synchronization and coordination between the nodes in a Hadoop cluster
17	Pig	It is a scripting language that allows the user to write a script for handling bulk amount of health data without knowing Java programming
18	Hive	It acts as a data warehouse that allows the user to summarize, query and analyzing huge datasets using a language called hive query language (HQL)
19	MapReduce	It is a programming aspect of Hadoop framework that divides the client job into small tasks and merge these intermediate results to produce the final result
20	Hadoop distributed file system (HDFS)	It is primary storage component for the applications of big data from where Hadoop framework is able to access the data for processing it

4 Predictive Analysis in Big Data

Predictive analytics becomes a dominant tool that helps the practitioners to find useful insights in healthcare data, thereby reducing cost while improving patient care. With the help of machine learning and AI [9, 10], predictive models are able to process vast amount of patient diverse data very quickly and predict a patient's disease response to offer certain treatments or prescribe some medicine and risk status of a particular disease [5] (Fig. 4).

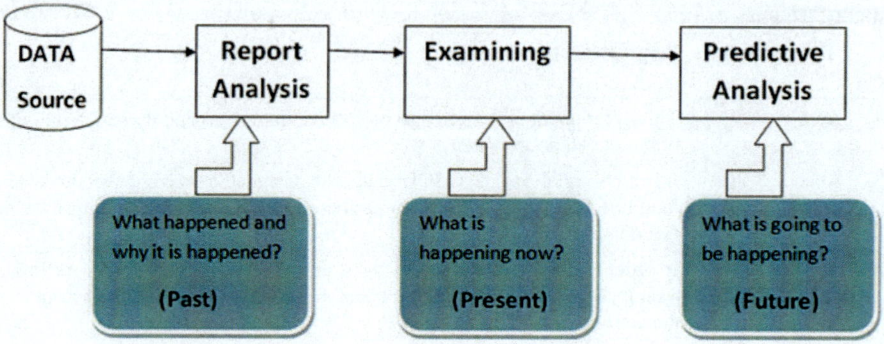

Fig. 4 Predictive analytics in big data

5 Key Values, Challenges, and Future Direction of Big Data:

Values	Challenges	Future direction
New approaches for the diagnosis and prognosis of personalized medicine	Lack of appropriate infrastructure for supporting BDA in health-related organization	Training and education of clinicians and public
Support human decision-making with automated algorithms	Data management, security, and privacy issues	Creation of new mechanism to accrue maximum value of data
Coordination of healthcare information	Limited awareness and support	Propose an approach that can be used in other healthcare application
Protecting privacy and innovation new business models, products, and services	Political barriers to adoption of IT and conflict of interests	More investment in infrastructure as well as National investments on health monitoring

5.1 Conclusion and Future Scope

At the present time, people are facing various diseases due to changes in environmental conditions and also because of their living habits. So it is very important to predict the diseases at earlier stages and increase the life span of patients. In some cases, it is complicated to the doctor to accurately predicting the disease on the basis of poor symptoms; the biggest challenge is to correctly predicting the disease in a right time. This paper focuses the role of machine learning techniques in healthcare data, challenges to the researchers while handling it. In future, ML-based disease prediction system is proposed, going to analyze various classifiers and finally predict exact disease with high accuracy rate.

References

1. Shaik Y, Parvathi VK, Biradar SR (2021) Role of IoT and big data analytics in healthcare for disease prediction. IEEE Xplore. https://doi.org/10.1109/ICCDW45521.2020.9318662
2. Chauhan R, Yafi E (2021) Big data analytics for Prediction modeling in Healthcare databases. IEEE Xplore https://doi.org/10.1109/IMCOM51814.2021.9377403
3. Wanga L, Alexanderb CA (2020) Big data analytics in medical engineering and healthcare: methods, advances and challenges. J Med Eng Technol. https://doi.org/10.1080/03091902.2020.1769758
4. Ravindran NJ, Gopalakrishnan P (2018) Predictive analysis for healthcare sector using big data technology. IEEE J. https://doi.org/10.1109/ICGCIoT.2018.8753090, ISBN:978-1-5386-5658-7
5. Onyemachi NC, Nonyelum OF (2020) Big data analytics in healthcare: a review. IEEE Xplore. https://doi.org/10.1109/ICECCO48375.2019.9043183
6. Galetsia P, Katsaliakia K, Kumarb S (2019) Values, challenges and future directions of big data analytics in healthcare: a systematic review. SCI J 241 https://doi.org/10.1016/j.socscimed.2019.112533
7. Ambigavathi M, Sridharan D (2018) Big data analytics in health care. IEEE Proceedings. https://doi.org/10.1109/ICoAC44903.2018.8939061
8. Sousa MJ, Pesqueira AM, Lemos C, Sousa M, Rocha Á (2019) Decision-Making based on big data analytics for people management in healthcare organizations. 9 https://doi.org/10.1007/s10916-019-1419-x. Springer
9. Islam MS, Hasan MM, Wang X, Germack HD, Alam MNE (2018) A systematic review on healthcare analytics: application and theoretical perspective of data mining. MPDI Health Care J. https://doi.org/10.3390/healthcare6020054
10. Sahoo PK, Mohapatra SK (2016) Analyzing healthcare big data with prediction for future health condition. IEEE Access 4 https://doi.org/10.1109/ACCESS.2016.2647619